Global Catastrophic Risks

Global Catastrophic Risks

Edited by

Nick Bostrom
Milan M. Ćirković

OXFORD
UNIVERSITY PRESS
Great Clarendon Street, Oxford OX2 6DP

Oxford University Press is a department of the University of Oxford.
It furthers the University's objective of excellence in research, scholarship,
and education by publishing worldwide in

Oxford New York

Auckland Cape Town Dar es Salaam Hong Kong Karachi
Kuala Lumpur Madrid Melbourne Mexico City Nairobi
New Delhi Shanghai Taipei Toronto

With offices in

Argentina Austria Brazil Chile Czech Republic France Greece
Guatemala Hungary Italy Japan Poland Portugal Singapore
South Korea Switzerland Thailand Turkey Ukraine Vietnam

Oxford is a registered trade mark of Oxford University Press
in the UK and in certain other countries

Published in the United States
by Oxford University Press Inc., New York

First published 2008

British Library Cataloguing in Publication Data
Data available

Library of Congress Cataloging in Publication Data
Data available

Typeset by Newgen Imaging Systems (P) Ltd., Chennai, India
Printed in Great Britain
on acid-free paper by
CPI Antony Rowe, Chippenham, Wiltshire

ISBN 978–0–19–960650–4 (Pbk)
ISBN 978–0–19–857050–9 (Hbk)

1 3 5 7 9 10 8 6 4 2

Acknowledgements

It is our pleasure to acknowledge the many people and institutions who have in one way or another contributed to the completion of this book. Our home institutions – the Future of Humanity Institute in the Oxford Martin School at Oxford University and the Astronomical Observatory of Belgrade – have offered environments conducive to our cross-disciplinary undertaking. Milan wishes to acknowledge the Oxford Colleges Hospitality Scheme and the Open Society Foundation of Belgrade for a pleasant time in Oxford back in 2004 during which this book project was conceived. Nick wishes to thank especially James Martin and Lou Salkind for their visionary support.

Physicist and polymath Cosma R. Shalizi gave an entire draft of the book a close, erudite and immensely helpful critical reading. We owe a great debt of gratitude to Alison Jones, Jessica Churchman and Dewi Jackson of Oxford University Press, who took so much interest in the project and helped shepherd it across a range of time scales. We are also appreciative of the scientific assistance by Peter Taylor and Rafaela Hillerbrand and for administrative support by Rachel Woodcock, Miriam Wood and Jo Armitage.

We thank John Leslie for stimulating our interest in extreme risk many years ago. We thank Mathew Gaverick, Julian Savulescu, Steve Rayner, Irena Diklić, Slobodan Popović, Tanja Berić, Ken D. Olum, Istvan Aranyosi, Max Tegmark, Vesna Milošević-Zdjelar, Toby Ord, Anders Sandberg, Bill Joy, Maja Bulatović, Alan Robertson, James Hughes, Robert J. Bradbury, Zoran Živković, Michael Vasser, Zoran Knežević, Ivana Dragićević, and Susan Rogers for pleasant and useful discussions of issues relevant to this book. Despairing of producing an exhaustive acknowledgement of even our most direct and immediate intellectual debts – which extend beyond science into the humanities and even music, literature, and art – we humbly apologize to all whom we have egregiously neglected.

Finally, let all the faults and shortcomings of this study be an impetus for others to do better. We thank in advance those who take up this challenge.

Foreword

In 1903, H.G. Wells gave a lecture at the Royal Institution in London, highlighting the risk of global disaster: 'It is impossible', proclaimed the young Wells, 'to show why certain things should not utterly destroy and end the human race and story; why night should not presently come down and make all our dreams and efforts vain. ... something from space, or pestilence, or some great disease of the atmosphere, some trailing cometary poison, some great emanation of vapour from the interior of the earth, or new animals to prey on us, or some drug or wrecking madness in the mind of man.' Wells' pessimism deepened in his later years; he lived long enough to learn about Hiroshima and Nagasaki and died in 1946.

In that year, some physicists at Chicago started a journal called the *Bulletin of Atomic Scientists*, aimed at promoting arms control. The 'logo' on the Bulletin's cover is a clock, the closeness of whose hands to midnight indicates the editor's judgement on how precarious the world situation is. Every few years the minute hand is shifted, either forwards or backwards.

Throughout the decades of the Cold War, the entire Western World was at great hazard. The superpowers could have stumbled towards Armageddon through muddle and miscalculation. We are not very rational in assessing relative risk. In some contexts, we are absurdly risk-averse. We fret about statistically tiny risks; carcinogens in food, a one-in-a-million chance of being killed in train crashes, and so forth. But most of us were 'in denial' about the far greater risk of death in a nuclear catastrophe.

In 1989, the Bulletin's clock was put back to seventeen minutes to midnight. There is now far less chance of tens of thousands of bombs devastating our civilization. But there is a growing risk of a few going off in a localized conflict. We are confronted by proliferation of nuclear weapons among more nations – and perhaps even the risk of their use by terrorist groups.

Moreover, the threat of global nuclear catastrophe could be merely in temporary abeyance. During the last century the Soviet Union rose and fell; there were two world wars. In the next hundred years, geopolitical realignments could be just as drastic, leading to a nuclear stand-off between new superpowers, which might be handled less adeptly (or less luckily) than the Cuba crisis, and the other tense moments of the Cold War era. The nuclear threat will always be with us – it is based on fundamental (and public) scientific ideas that date from the 1930s.

Despite the hazards, there are, today, some genuine grounds for being a techno-optimist. For most people in most nations, there has never been a better time to be alive. The innovations that will drive economic advance – information technology, biotechnology and nanotechnology – can boost the developing as well as the developed world. Twenty-first century technologies could offer lifestyles that are environmentally benign – involving lower demands on energy or resources than those demanded by what we consider a good life today. And we could readily raise the funds – were there the political will – to lift the world's two billion most-deprived people from their extreme poverty.

But, along with these hopes, twenty-first century technology will confront us with new global threats – stemming from bio-, cyber- and environmental-science, as well as from physics – that could be as grave as the bomb. The Bulletin's clock is now closer to midnight again. These threats may not trigger sudden worldwide catastrophe – the doomsday clock is not such a good metaphor – but they are, in aggregate, disquieting and challenging. The tensions between benign and damaging spin-offs from new technologies, and the threats posed by the Promethean power science, are disquietingly real. Wells' pessimism might even have deepened further were he writing today.

One type of threat comes from humanity's collective actions; we are eroding natural resources, changing the climate, ravaging the biosphere and driving many species to extinction.

Climate change looms as the twenty-first century's number-one environmental challenge. The most vulnerable people – for instance, in Africa or Bangladesh – are the least able to adapt. Because of the burning of fossil fuels, the CO_2 concentration in the atmosphere is already higher than it has ever been in the last half million years – and it is rising ever faster. The higher CO_2 rises, the greater the warming – and, more important still, the greater will be the chance of triggering something grave and irreversible: rising sea levels due to the melting of Greenland's icecap and so forth. The global warming induced by the fossil fuels we burn this century could lead to sea level rises that continue for a millennium or more.

The science of climate change is intricate. But it is simple compared to the economic and political challenge of responding to it. The market failure that leads to global warming poses a unique challenge for two reasons. First, unlike the consequences of more familiar kinds of pollution, the effect is diffuse: the CO_2 emissions from the UK have no more effect here than they do in Australia, and vice versa. That means that any credible framework for mitigation has to be broadly international. Second, the main downsides are not immediate but lie a century or more in the future: inter-generational justice comes into play; how do we rate the rights and interests of future generations compared to our own?

The solution requires coordinated action by all major nations. It also requires far-sightedness – altruism towards our descendants. History will judge us harshly if we discount too heavily what might happen when our grandchildren grow old. It is deeply worrying that there is no satisfactory fix yet on the horizon that will allow the world to break away from dependence on coal and oil – or else to capture the CO_2 that power stations emit. To quote Al Gore, 'We must not leap from denial to despair. We can do something and we must.'

The prognosis is indeed uncertain, but what should weigh most heavily and motivate policy-makers most strongly – is the 'worst case' end of the range of predictions: a 'runaway' process that would render much of the Earth uninhabitable.

Our global society confronts other 'threats without enemies', apart from (although linked with) climate change. High among them is the threat to biological diversity. There have been five great extinctions in the geological past. Humans are now causing a sixth. The extinction rate is one thousand times higher than normal and is increasing. We are destroying the book of life before we have read it. There are probably upwards of ten million species, most not even recorded – mainly insects, plants and bacteria.

Biodiversity is often proclaimed as a crucial component of human well-being. Manifestly it is: we are clearly harmed if fish stocks dwindle to extinction; there are plants in the rain forest whose gene pool might be useful to us. But for many of us these 'instrumental' – and anthropocentric – arguments are not the only compelling ones. Preserving the richness of our biosphere has value in its own right, over and above what it means to us humans.

But we face another novel set of vulnerabilities. These stem not from our collective impact but from the greater empowerment of individuals or small groups by twenty-first century technology.

The new techniques of synthetic biology could permit inexpensive synthesis of lethal biological weapons – on purpose, or even by mistake. Not even an organized network would be required: just a fanatic or a weirdo with the mindset of those who now design computer viruses – the mindset of an arsonist. Bio (and cyber) expertise will be accessible to millions. In our networked world, the impact of any runaway disaster could quickly become global.

Individuals will soon have far greater 'leverage' than present-day terrorists possess. Can our interconnected society be safeguarded against error or terror without having to sacrifice its diversity and individualism? This is a stark question, but I think it is a serious one.

We are kidding ourselves if we think that technical education leads to balanced rationality: it can be combined with fanaticism – not just the traditional fundamentalism that we are so mindful of today, but new age irrationalities too. There are disquieting portents – for instance, the Raelians (who claim to be cloning humans) and the Heavens Gate cult (who committed collective suicide in hopes that a space-ship would take them to a 'higher sphere'). Such cults claim to be 'scientific' but have a precarious foothold in reality. And there are extreme eco-freaks who believe that the world would be better off if it were rid of humans. Can the global village cope with its village idiots – especially when even one could be too many?

These concerns are not remotely futuristic – we will surely confront them within next ten to twenty years. But what of the later decades of this century? It is hard to predict because some technologies could develop with runaway speed. Moreover, human character and physique themselves will soon be malleable, to an extent that is qualitatively new in our history. New drugs (and perhaps even implants into our brains) could change human character; the cyberworld has potential that is both exhilarating and frightening.

We cannot confidently guess lifestyles, attitudes, social structures or population sizes a century hence. Indeed, it is not even clear how much longer our descendants would remain distinctively 'human'. Darwin himself noted that 'not one living species will transmit its unaltered likeness to a distant futurity'. Our own species will surely change and diversify faster than any predecessor – via human-induced modifications (whether intelligently controlled or unintended), not by natural selection alone. The post-human era may be only centuries away. And what about Artificial Intelligence? Superintelligent machine could be the last invention that humans need ever make. We should keep our minds open, or at least ajar, to concepts that seem on the fringe of science fiction.

These thoughts might seem irrelevant to practical policy – something for speculative academics to discuss in our spare moments. I used to think this. But humans are now, individually and collectively, so greatly empowered by rapidly changing technology that we can – by design or as unintended consequences – engender irreversible global changes. It is surely irresponsible not to ponder what this could mean; and it is real political progress that the challenges stemming from new technologies are higher on the international agenda and that planners seriously address what might happen more than a century hence.

We cannot reap the benefits of science without accepting some risks – that has always been the case. Every new technology is risky in its pioneering stages. But there is now an important difference from the past. Most of the risks encountered in developing 'old' technology were localized: when, in the early days of steam, a boiler exploded, it was horrible, but there was an 'upper bound' to just how horrible. In our ever more interconnected world, however, there are new risks whose consequences could be global. Even a tiny probability of global catastrophe is deeply disquieting.

We cannot eliminate all threats to our civilization (even to the survival of our entire species). But it is surely incumbent on us to think the unthinkable and study how to apply twenty-first century technology optimally, while minimizing the 'downsides'. If we apply to catastrophic risks the same prudent analysis that leads us to take everyday safety precautions, and sometimes to buy insurance – multiplying probability by consequences – we had surely conclude that some of the scenarios discussed in this book deserve more attention that they have received.

My background as a cosmologist, incidentally, offers an extra perspective – an extra motive for concern – with which I will briefly conclude.

The stupendous time spans of the evolutionary past are now part of common culture – except among some creationists and fundamentalists. But most educated people, even if they are fully aware that our emergence took billions of years, somehow think we humans are the culmination of the evolutionary tree. That is not so. Our Sun is less than half way through its life. It is slowly brightening, but Earth will remain habitable for another billion years. However, even in that cosmic time perspective – extending far into the future as well as into the past – the twenty-first century may be a defining moment. It is the first in our planet's history where one species – ours – has Earth's future in its hands and could jeopardise not only itself but also life's immense potential.

The decisions that we make, individually and collectively, will determine whether the outcomes of twenty-first century sciences are benign or devastating. We need to contend not only with threats to our environment but also with an entirely novel category of risks – with seemingly low probability, but with such colossal consequences that they merit far more attention than they have hitherto had. That is why we should welcome this fascinating and provocative book. The editors have brought together a distinguished set of authors with formidably wide-ranging expertise. The issues and arguments presented here should attract a wide readership – and deserve special attention from scientists, policy-makers and ethicists.

Martin J. Rees

Contents

·1·
Introduction

Nick Bostrom and Milan M. Ćirković

1.1 Why?

The term 'global catastrophic risk' lacks a sharp definition. We use it to refer, loosely, to a risk that might have the potential to inflict serious damage to human well-being on a global scale. On this definition, an immensely diverse collection of events could constitute global catastrophes: potential candidates range from volcanic eruptions to pandemic infections, nuclear accidents to worldwide tyrannies, out-of-control scientific experiments to climatic changes, and cosmic hazards to economic collapse. With this in mind, one might well ask, what use is a book on global catastrophic risk? The risks under consideration seem to have little in common, so does 'global catastrophic risk' even make sense as a topic? Or is the book that you hold in your hands as ill-conceived and unfocused a project as a volume on 'Gardening, Matrix Algebra, and the History of Byzantium'?

We are confident that a comprehensive treatment of global catastrophic risk will be at least somewhat more useful and coherent than the above-mentioned imaginary title. We also believe that studying this topic is highly important. Although the risks are of various kinds, they are tied together by many links and commonalities. For example, for many types of destructive events, much of the damage results from second-order impacts on social order; thus the risks of social disruption and collapse are not unrelated to the risks of events such as nuclear terrorism or pandemic disease. Or to take another example, apparently dissimilar events such as large asteroid impacts, volcanic super-eruptions, and nuclear war would all eject massive amounts of soot and aerosols into the atmosphere, with significant effects on global climate. The existence of such causal linkages is one reason why it is can be sensible to study multiple risks together.

Another commonality is that many methodological, conceptual, and cultural issues crop up across the range of global catastrophic risks. If our interest lies in such issues, it is often illuminating to study how they play out in different contexts. Conversely, some general insights – for example, into the biases of human risk cognition – can be applied to many different risks and used to improve our assessments across the board.

Beyond these theoretical commonalities, there are also pragmatic reasons for addressing global catastrophic risks as a single field. Attention is scarce. Mitigation is costly. To decide how to allocate effort and resources, we must make comparative judgements. If we treat risks singly, and never as part of an overall threat profile, we may become unduly fixated on the one or two dangers that happen to have captured the public or expert imagination of the day, while neglecting other risks that are more severe or more amenable to mitigation. Alternatively, we may fail to see that some precautionary policy, while effective in reducing the particular risk we are focusing on, would at the same time create new hazards and result in an increase in the overall level of risk. A broader view allows us to gain perspective and can thereby help us to set wiser priorities.

The immediate aim of this book is to offer an introduction to the range of global catastrophic risks facing humanity now or expected in the future, suitable for an educated interdisciplinary readership. There are several constituencies for the knowledge presented. Academics specializing in one of these risk areas will benefit from learning about the other risks. Professionals in insurance, finance, and business – although usually preoccupied with more limited and imminent challenges – will benefit from a wider view. Policy analysts, activists, and laypeople concerned with promoting responsible policies likewise stand to gain from learning about the state of the art in global risk studies. Finally, anyone who is worried or simply curious about what could go wrong in the modern world might find many of the following chapters intriguing. We hope that this volume will serve as a useful introduction to all of these audiences. Each of the chapters ends with some pointers to the literature for those who wish to delve deeper into a particular set of issues.

This volume also has a wider goal: to stimulate increased research, awareness, and informed public discussion about big risks and mitigation strategies. The existence of an interdisciplinary community of experts and laypeople knowledgeable about global catastrophic risks will, we believe, improve the odds that good solutions will be found and implemented to the great challenges of the twenty-first century.

1.2 Taxonomy and organization

Let us look more closely at what would, and would not, count as a global catastrophic risk. Recall that the damage must be serious, and the scale global. Given this, a catastrophe that caused 10,000 fatalities or 10 billion dollars worth of economic damage (e.g., a major earthquake) would not qualify as a global catastrophe. A catastrophe that caused 10 million fatalities or 10 trillion dollars worth of economic loss (e.g., an influenza pandemic) would count as a global catastrophe, even if some region of the world escaped unscathed. As for disasters falling between these points, the definition is vague. The stipulation of a precise cut-off does not appear needful at this stage.

Global catastrophes have occurred many times in history, even if we only count disasters causing more than 10 million deaths. A very partial list of examples might include the An Shi Rebellion (756–763), the Taiping Rebellion (1851–1864), and the famine of the Great Leap Forward in China, the Black Death in Europe, the Spanish flu pandemic, the two world wars, the Nazi genocides, the famines in British India, Stalinist totalitarianism, the decimation of the native American population through smallpox and other diseases following the arrival of European colonizers, probably the Mongol conquests, perhaps Belgian Congo – innumerable others could be added to the list depending on how various misfortunes and chronic conditions are individuated and classified.

We can roughly characterize the severity of a risk by three variables: its *scope* (how many people – and other morally relevant beings – would be affected), its *intensity* (how badly these would be affected), and its *probability* (how likely the disaster is to occur, according to our best judgement, given currently available evidence). Using the first two of these variables, we can construct a qualitative diagram of different types of risk (Fig. 1.1). (The probability dimension could be displayed along a z-axis were this diagram three-dimensional.)

The scope of a risk can be *personal* (affecting only one person), *local*, *global* (affecting a large part of the human population), or *trans-generational* (affecting

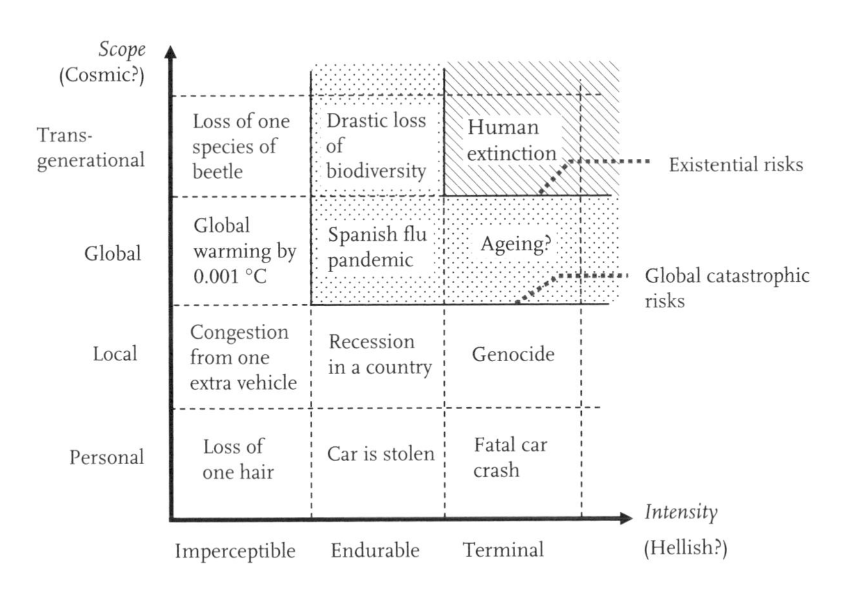

Fig. 1.1 Qualitative categories of risk. Global catastrophic risks are in the upper right part of the diagram. Existential risks form an especially severe subset of these.

not only the current world population but all generations that could come to exist in the future). The intensity of a risk can be classified as imperceptible (barely noticeable), endurable (causing significant harm but not destroying quality of life completely), or terminal (causing death or permanently and drastically reducing quality of life). In this taxonomy, global catastrophic risks occupy the four risks classes in the high-severity upper-right corner of the figure: a global catastrophic risk is of either global or trans-generational scope, and of either endurable or terminal intensity. In principle, as suggested in the figure, the axes can be extended to encompass conceptually possible risks that are even more extreme. In particular, trans-generational risks can contain a subclass of risks so destructive that their realization would not only affect or pre-empt future human generations, but would also destroy the potential of our future light cone of the universe to produce intelligent or self-aware beings (labelled 'Cosmic'). On the other hand, according to many theories of value, there can be states of being that are even worse than non-existence or death (e.g., permanent and extreme forms of slavery or mind control), so it could, in principle, be possible to extend the x-axis to the right as well (see Fig. 1.1 labelled 'Hellish').

A subset of global catastrophic risks is *existential risks*. An existential risk is one that threatens to cause the extinction of Earth-originating intelligent life or to reduce its quality of life (compared to what would otherwise have been possible) permanently and drastically.[1] Existential risks share a number of features that mark them out as deserving of special consideration. For example, since it is not possible to recover from existential risks, we cannot allow even one existential disaster to happen; there would be no opportunity to learn from experience. Our approach to managing such risks must be proactive. How much worse an existential catastrophe would be than a non-existential global catastrophe depends very sensitively on controversial issues in value theory, in particular how much weight to give to the lives of possible future persons.[2] Furthermore, assessing existential risks raises distinctive methodological problems having to do with observation selection effects and the need to avoid anthropic bias. One of the motives for producing this book is to stimulate more serious study of existential risks. Rather than limiting our focus to existential risk, however, we thought it better to lay a broader foundation of systematic thinking about big risks in general.

[1] (Bostrom, 2002, p. 381).

[2] For many aggregative consequentialist ethical theories, including but not limited to total utilitarianism, it can be shown that the injunction to *maximize expected value!* can be simplified – for all practical purposes – to the injunction to *minimize existential risk!* (Bostrom, 2003, p. 439). (Note, however, that aggregative consequentialism is threatened by the problem of infinitarian paralysis [Bostrom, 2007, p. 730].)

We asked our contributors to assess global catastrophic risks not only as they presently exist but also as they might develop over time. The temporal dimension is essential for a full understanding of the nature of the challenges we face. To think about how to tackle the risks from nuclear terrorism and nuclear war, for instance, we must consider not only the probability that something will go wrong within the next year, but also about how the risks will change in the future and the factors – such as the extent of proliferation of relevant technology and fissile materials – that will influence this. Climate change from greenhouse gas emissions poses no significant globally catastrophic risk now or in the immediate future (on the timescale of several decades); the concern is about what effects these accumulating emissions might have over the course of many decades or even centuries. It can also be important to anticipate hypothetical risks which will arise if and when certain possible technological developments take place. The chapters on nanotechnology and artificial intelligence are examples of such prospective risk analysis.

In some cases, it can be important to study scenarios which are almost certainly physically impossible. The hypothetical risk from particle collider experiments is a case in point. It is very likely that these experiments have no potential, whatever, for causing global disasters. The objective risk is probably zero, as believed by most experts. But just how confident can we be that there is no objective risk? If we are not certain that there is no objective risk, then there is a risk at least in a subjective sense. Such subjective risks can be worthy of serious consideration, and we include them in our definition of global catastrophic risks.

The distinction between objective and subjective (epistemic) risk is often hard to make out. The possibility of an asteroid colliding with Earth looks like a clear-cut example of objective risk. But suppose that in fact no sizeable asteroid is on collision course with our planet within a certain, sufficiently large interval of time. We might then say that there is no objective risk of an asteroid-caused catastrophe within that interval of time. Of course, we will not know that this is so until we have mapped out the trajectories of all potentially threatening asteroids and are able to calculate all perturbations, often chaotic, of those trajectories. In the meantime, we must recognize a risk from asteroids even though the risk might be purely subjective, merely reflecting our present state of ignorance. An empty cave can be similarly subjectively unsafe if you are unsure about whether a lion resides in it; and it can be rational for you to avoid the cave if you reasonably judge that the expected harm of entry outweighs the expected benefit.

In the case of the asteroid threat, we have access to plenty of data that can help us quantify the risk. We can estimate the probability of a catastrophic impact from statistics of past impacts (e.g., cratering data) and from observations sampling from the population of non-threatening asteroids. This particular risk, therefore, lends itself to rigorous scientific study, and the probability estimates we derive are fairly strongly constrained by hard evidence.[3]

For many other risks, we lack the data needed for rigorous statistical inference. We may also lack well-corroborated scientific models on which to base probability estimates. For example, there exists no rigorous scientific way of assigning a probability to the risk of a serious terrorist attack employing a biological warfare agent occurring within the next decade. Nor can we firmly establish that the risks of a global totalitarian regime arising before the end of the century are of a certain precise magnitude. It is inevitable that analyses of such risks will rely to a large extent on plausibility arguments, analogies, and subjective judgement.

Although more rigorous methods are to be preferred whenever they are available and applicable, it would be misplaced scientism to confine attention to those risks that are amenable to hard approaches.[4] Such a strategy would lead to many risks being ignored, including many of the largest risks confronting humanity. It would also create a false dichotomy between two types of risks – the 'scientific' ones and the 'speculative' ones – where, in reality, there is a continuum of analytic tractability.

We have, therefore, opted to cast our net widely. Although our topic selection shows some skew towards smaller risks that have been subject to more scientific study, we do have a range of chapters that tackle potentially large but more speculative risks. The page count allocated to a risk should not, of course, be interpreted as a measure of how seriously we believe the risk ought to be regarded. In some cases, we have seen it fit to have a chapter devoted to a risk that turns out to be quite small, because learning that a particular risk is small can be useful, and the procedures used to arrive at the conclusion might serve as a template for future risk research. It goes without saying that the exact composition of a volume like this is also influenced by many contingencies

[3] One can sometimes define something akin to objective physical probabilities ('chances') for deterministic systems, as is done, for example, in classical statistical mechanics, by assuming that the system is ergodic under a suitable course graining of its state space. But ergodicity is not necessary for there being strong scientific constraints on subjective probability assignments to uncertain events in deterministic systems. For example, if we have good statistics going back a long time showing that impacts occur on average once per thousand years, with no apparent trends or periodicity, then we have scientific reason – absent of more specific information – for assigning a probability of $\approx$0.1% to an impact occurring within the next year, whether we think the underlying system dynamic is indeterministic, or chaotic, or something else.

[4] Of course, when allocating research effort it is legitimate to take into account not just how important a problem is but also the likelihood that a solution can be found through research. The drunk who searches for his lost keys where the light is best is not necessarily irrational; and a scientist who succeeds in something relatively unimportant may achieve more good than one who fails in something important.

beyond the editors' control and that perforce it must leave out more than it includes.[5]

We have divided the book into four sections:

Part I: Background
Part II: Risks from Nature
Part III: Risks from Unintended Consequences
Part IV: Risks from Hostile Acts

This subdivision into three categories of risks is for convenience only, and the allocation of a risk to one of these categories is often fairly arbitrary. Take earthquakes which might seem to be paradigmatically a 'Risk from Nature'. Certainly, an earthquake is a natural event. It would happen even if we were not around. Earthquakes are governed by the forces of plate tectonics over which human beings currently have no control. Nevertheless, the risk posed by an earthquake is, to a very large extent, a matter of human construction. Where we erect our buildings and how we choose to construct them strongly influence what happens when an earthquake of a given magnitude occurs. If we all lived in tents, or in earthquake-proof buildings, or if we placed our cities far from fault lines and sea shores, earthquakes would do little damage. On closer inspection, we thus find that the earthquake risk is very much a joint venture between Nature and Man. Or take a paradigmatically anthropogenic hazard such as nuclear weapons. Again we soon discover that the risk is not as disconnected from uncontrollable forces of nature as might at first appear to be the case. If a nuclear bomb goes off, how much damage it causes will be significantly influenced by the weather. Wind, temperature, and precipitation will affect the fallout pattern and the likelihood that a fire storm will break out: factors that make a big difference to the number of fatalities generated by the blast. In addition, depending on how a risk is defined, it may also over time transition from one category to another. For instance, the risk of starvation might once have been primarily a Risk from Nature, when the main causal factors were draughts or fluctuations in local prey population; yet in the contemporary world, famines tend to be the consequences of market failures, wars, and social breakdowns, whence the risk is now at least as much one of Unintended Consequences or of Hostile Acts.

1.3 Part I: Background

The objective of this part of the book is to provide general context and methodological guidance for thinking systematically and critically about global catastrophic risks.

[5] For example, the risk of large-scale conventional war is only covered in passing, yet would surely deserve its own chapter in a more ideally balanced page allocation.

We begin at the end, as it were, with Chapter 2 by Fred Adams discussing the long-term fate of our planet, our galaxy, and the Universe in general. In about 3.5 billion years, the growing luminosity of the sun will essentially have sterilized the Earth's biosphere, but the end of *complex* life on Earth is scheduled to come sooner, maybe 0.9–1.5 billon years from now. This is the default fate for life on our planet. One may hope that if humanity and complex technological civilization survives, it will long before then have learned to colonize space.

If some cataclysmic event were to destroy *Homo sapiens* and other higher organisms on Earth tomorrow, there does appear to be a window of opportunity of approximately one billion years for another intelligent species to evolve and take over where we left off. For comparison, it took approximately 1.2 billion years from the rise of sexual reproduction and simple multicellular organisms for the biosphere to evolve into its current state, and only a few million years for our species to evolve from its anthropoid ancestors. Of course, there is no guarantee that a rerun of evolution would produce anything like a human or a self-aware successor species.

If intelligent life does spread into space by harnessing the powers of technology, its lifespan could become extremely long. Yet eventually, the universe will wind down. The last stars will stop shining 100 trillion years from now. Later, matter itself will disintegrate into its basic constituents. By 10^{100} years from now even the largest black holes would have evaporated. Our present understanding of what will happen at this time scale and beyond is quite limited. The current best guess – but it is really no more than that – is that it is not just technologically difficult but physically impossible for intelligent information processing to continue beyond some finite time into the future. If so, extinction is not a question of whether, but when.

After this peek into the extremely remote future, it is instructive to turn around and take a brief peek at the distant past. Some past cataclysmic events have left traces in the geological record. There have been about fifteen mass extinctions in the last 500 million years, and five of these eliminated more half of all species then inhabiting the Earth. Of particular note is the Permian – Triassic extinction event, which took place some 251.4 million years ago. This 'mother of all mass extinctions' eliminated more than 90% of all species and many entire phylogenetic families. It took upwards of 5 million years for biodiversity to recover.

Impacts from asteroids and comets, as well as massive volcano eruptions, have been implicated in many of the mass extinctions of the past. Other causes, such as variations in the intensity of solar illumination, may in some cases have exacerbated stresses. It appears that all mass extinctions have been mediated by atmospheric effects such as changes the atmosphere's composition or temperature. It is possible, however, that we owe our existence to mass extinctions. In particular, the comet that hit Earth 65 million years ago, which is believed to have been responsible for the demise of the dinosaurs, might have been a sine qua non for the subsequent rise of *Homo sapiens* by clearing an ecological niche that could be occupied by large mammals, including our ancestors.

At least 99.9% of all species that have ever walked, crawled, flown, swum, or otherwise abided on Earth are extinct. Not all of these were eliminated in cataclysmic mass extinction events. Many succumbed in less spectacular doomsdays such as from competition by other species for the same ecological niche. Chapter 3 reviews the mechanisms of evolutionary change. Not so long ago, our own species co-existed with at least one other hominid species, the Neanderthals. It is believed that the lineages of *H. sapiens* and *H. neanderthalensis* diverged about 800,000 years ago. The Neanderthals manufactured and used composite tools such as handaxes. They did not reach extinction in Europe until 33,000 to 24,000 years ago, quite likely as a direct result of competition with *Homo sapiens*. Recently, the remains of what might have been another hominoid species, *Homo floresiensis* – nicknamed 'the hobbit' for its short stature – were discovered on an Indonesian island. *H. floresiensis* is believed to have survived until as recently as 12,000 years ago, although uncertainty remains about the interpretation of the finds. An important lesson of this chapter is that extinction of intelligent species *has* already happened on Earth, suggesting that it would be naïve to think it may not happen again.

From a naturalistic perspective, there is thus nothing abnormal about global cataclysms including species extinctions, although the characteristic time scales are typically large by human standards. James Hughes in Chapter 4 makes clear, however, the idea of cataclysmic endings often causes a peculiar set of cognitive tendencies to come into play, what he calls 'the millennial, utopian, or apocalyptic psychocultural bundle, a characteristic dynamic of eschatological beliefs and behaviours'. The millennial impulse is pancultural. Hughes shows how it can be found in many guises and with many common tropes from Europe to India to China, across the last several thousand years. 'We may aspire to a purely rational, technocratic analysis', Hughes writes, 'calmly balancing the likelihoods of futures without disease, hunger, work or death, on the one hand, against the likelihoods of worlds destroyed by war, plagues or asteroids, but few will be immune to millennial biases, positive or negative, fatalist or messianic'. Although these eschatological tropes can serve legitimate social needs and help to mobilize needed action, they easily become dysfunctional and contribute to social disengagement. Hughes argues that we need historically informed and vigilant self-interrogation to help us keep our focus on constructive efforts to address real challenges.

Even for an honest, truth-seeking, and well-intentioned investigator it is difficult to think and act rationally in regard to global catastrophic risks and existential risks. These are topics on which it seems especially difficult to remain sensible. In Chapter 5, Eliezer Yudkowsky observes as follows:

> Substantially larger numbers, such as 500 million deaths, and *especially* qualitatively different scenarios such as the extinction of the entire human species, seem to trigger a *different mode of thinking* – enter into a 'separate magisterium'. People who would never dream of hurting a child hear of an existential risk, and say, 'Well, maybe the human species doesn't really deserve to survive'.

Fortunately, if we are ready to contend with our biases, we are not left entirely to our own devices. Over the last few decades, psychologists and economists have developed an extensive empirical literature on many of the common heuristics and biases that can be found in human cognition. Yudkowsky surveys this literature and applies its frequently disturbing findings to the domain of large-scale risks that is the subject matter of this book. His survey reviews the following effects: availability; hindsight bias; black swans; the conjunction fallacy; confirmation bias; anchoring, adjustment, and contamination; the affect heuristic; scope neglect; calibration and overconfidence; and bystander apathy. It behooves any sophisticated contributor in the area of global catastrophic risks and existential risks – whether scientist or policy advisor – to be familiar with each of these effects and we all ought to give some consideration to how they might be distorting our judgements.

Another kind of reasoning trap to be avoided is anthropic bias. Anthropic bias differs from the general cognitive biases reviewed by Yudkowsky; it is more theoretical in nature and it applies more narrowly to only certain specific kinds of inference. Anthropic bias arises when we overlook relevant observation selection effects. An observation selection effect occurs when our evidence has been 'filtered' by the precondition that a suitably positioned observer exists to have the evidence, in such a way that our observations are unrepresentatively sampled from the target domain. Failure to take observation effects into account correctly can result in serious errors in our probabilistic evaluation of some of the relevant hypotheses. Milan Ćirković, in Chapter 6, reviews some applications of observation selection theory that bear on global catastrophic risk and particularly existential risk. Some of these applications are fairly straightforward albeit not always obvious. For example, the tempting inference that certain classes of existential disaster must be highly improbable because they have never occurred in the history of our species or even in the history of life on Earth must be resisted. We are bound to find ourselves in one of those places and belonging to one of those intelligent species which have not yet been destroyed, whether planet or species-destroying disasters are common or rare: for the alternative possibility – that *our* planet has been destroyed or *our* species extinguished – is something that is unobservable for us, per definition. Other applications of anthropic reasoning – such as the Carter–Leslie Doomsday argument – are of disputed validity, especially in their generalized forms, but nevertheless worth knowing about. In some applications, such as the simulation argument, surprising constraints are revealed on what we can coherently assume about humanity's future and our place in the world.

There are professional communities that deal with risk assessment on a daily basis. The subsequent two chapters present perspectives from the systems engineering discipline and the insurance industry, respectively.

In Chapter 7, Yacov Haimes outlines some flexible strategies for organizing our thinking about risk variables in complex systems engineering projects. What knowledge is needed to make good risk management decisions? Answering this question, Haimes says, 'mandates seeking the "truth" about the unknowable complex nature of emergent systems; it requires intellectually bias-free modellers and thinkers who are empowered to experiment with a multitude of modelling and simulation approaches and to collaborate for appropriate solutions'. Haimes argues that organizing the analysis around the measure of the expected value of risk can be too constraining. Decision makers often prefer a more fine-grained decomposition of risk that allows them to consider separately the probability of outcomes in different severity ranges, using what Haimes calls 'the partitioned multi-objective risk method'.

Chapter 8, by Peter Taylor, explores the connections between the insurance industry and global catastrophic risk. Insurance companies help individuals and organizations mitigate the financial consequences of risk, essentially by allowing risks to be traded and shared. Peter Taylor argues that the extent to which global catastrophic risks can be privately insured is severely limited for reasons having to do with both their scope and their type.

Although insurance and reinsurance companies have paid relatively scant attention to global catastrophic risks, they have accumulated plenty of experience with smaller risks. Some of the concepts and methods used can be applied to risks at any scale. Taylor highlights the importance of the concept of *uncertainty*. A particular stochastic model of phenomena in some domain (such as earthquakes) may entail a definite probability distribution over possible outcomes. However, in addition to the chanciness described by the model, we must recognize two further sources of uncertainty. There is usually uncertainty in the values of the parameters that we feed into the model. On top of that, there is uncertainty about whether the model we use does, in fact, correctly describe the phenomena in the target domain. These higher-level uncertainties are often impossible to analyse in a statistically rigorous way. Analysts who strive for objectivity and who are expected to avoid making 'un-scientific' assumptions that they cannot justify face a temptation to ignore these subjective uncertainties. But such scientism can lead to disastrous misjudgements. Taylor argues that the distortion is often greatest at the tail end of exceedance probability curves, leading to an underestimation of the risk of extreme events.

Taylor also reports on two recent survey studies of perceived risk. One of these, conducted by Swiss Re in 2005, asked executives of multinationals about which risks to their businesses' financials were of greatest concern to them. Computer-related risk was rated as the highest priority risk, followed by foreign trade, corporate governance, operational/facility, and liability risk. Natural disasters came in seventh place, and terrorism in tenth place. It appears that, as far as financial threats to individual corporations are concerned, global catastrophic risks take the backseat to more direct and narrowly focused business hazards. A similar exercise, but with broader scope, is carried out annually by the World Economic Forum. Its 2007 Global Risk report classified risks by likelihood and severity based on opinions solicited from business leaders, economists, and academics. Risks were evaluated with a 10-year time frame. Two risks were given a severity rating of 'more than 1 trillion USD', namely, asset price collapse (10–20%) and retrenchment from globalization (1–5%). When severity was measured in number of deaths rather than economic losses, the top three risks were pandemics, developing world disease, and interstate and civil war. (Unfortunately, several of the risks in this survey were poorly defined, making it hard to interpret the reported opinions – one moral here being that, if one wishes to assign probabilities to risks or rank them according to severity or likelihood, an essential first step is to present clear definitions of the risks that are to be evaluated.[6])

The Background part of the book ends with a discussion by Richard Posner on some challenges for public policy in Chapter 9. Posner notes that governmental action to reduce global catastrophic risk is often impeded by the short decision horizons of politicians with their limited terms of office and the many competing demands on their attention. Furthermore, mitigation of global catastrophic risks is often costly and can create a free-rider problem. Smaller and poorer nations may drag their heels in the hope of taking a free ride on larger and richer countries. The more resourceful countries, in turn, may hold back because of reluctance to reward the free riders.

Posner also looks at several specific cases, including tsunamis, asteroid impacts, bioterrorism, accelerator experiments, and global warming, and considers some of the implications for public policy posed by these risks. Although rigorous cost–benefit analyses are not always possible, it is nevertheless important to attempt to quantify probabilities, potential harms, and the costs of different possible countermeasures, in order to determine priorities and optimal strategies for mitigation. Posner suggests that when

[6] For example, the risk 'Chronic disease in the developed world' is defined as 'Obesity, diabetes and cardiovascular diseases become widespread; healthcare costs increase; resistant bacterial infections rise, sparking class-action suits and avoidance of hospitals'. By most standards, obesity, diabetes, and cardiovascular disease are *already* widespread. And by *how much* would healthcare costs have to increase to satisfy the criterion? It may be impossible to judge whether this definition was met even after the fact and with the benefit of hindsight.

a precise probability of some risk cannot be determined, it can sometimes be informative to consider – as a rough heuristic – the 'implied probability' suggested by current expenditures on mitigation efforts compared to the magnitude of harms that would result if a disaster materialized. For example, if we spend one million dollars per year to mitigate a risk which would create 1 billion dollars of damage, we may estimate that current policies implicitly assume that the annual risk of the disaster is of the order of 1/1000. If this implied probability seems too small, it might be a sign that we are not spending enough on mitigation.[7] Posner maintains that the world is, indeed, under-investing in mitigation of several global catastrophic risks.

1.4 Part II: Risks from nature

Volcanic eruptions in recent historical times have had measurable effects on global climate, causing global cooling by a few tenths of one degree, the effect lasting perhaps a year. But as Michael Rampino explains in Chapter 10, these eruptions pale in comparison to the largest recorded eruptions. Approximately 75,000 years ago, a volcano erupted in Toba, Indonesia, spewing vast volumes of fine ash and aerosols into the atmosphere, with effects comparable to nuclear-winter scenarios. Land temperatures globally dropped by 5–15°C, and ocean-surface cooling of ≈2–6°C might have extended over several years. The persistence of significant soot in the atmosphere for one to three years might have led to a cooling of the climate lasting for decades (because of climate feedbacks such as increased snow cover and sea ice causing more of the sun's radiation to be reflected back into space). The human population appears to have gone through a bottleneck at this time, according to some estimates dropping as low as approximately five hundred reproducing females in a world population of approximately 4000 individuals. On the Toba catastrophe theory, the population decline was caused by the super-eruption, and the human species was teetering on the brink of extinction. This is perhaps the worst disaster that has ever befallen the human species, at least if severity is measured by how close to terminal was the outcome.

More than twenty super-eruption sites for the last two million years have been identified. This would suggest that, on average, a super-eruption occurs at least once every 50,000 years. However, there may well have been additional super-eruptions that have not yet been identified in the geological record.

[7] This heuristic is only meant to be a first stab at the problem. It is obviously not generally valid. For example, if one million dollars is sufficient to take all the possible precautions, there is no reason to spend more on the risk even if we think that its probability is much greater than 1/1000. A more careful analysis would consider the marginal returns on investment in risk reduction.

The global damage from super-volcanism would come chiefly from its climatic effects. The volcanic winter that would follow such an eruption would cause a drop in agricultural productivity which could lead to mass starvation and consequent social upheavals. Rampino's analysis of the impacts of super-volcanism is also relevant to the risks of nuclear war and asteroid or meteor impacts. Each of these would involve soot and aerosols being injected into the atmosphere, cooling the Earth's climate.

Although we have no way of preventing a super-eruption, there are precautions that we could take to mitigate its impacts. At present, a global stockpile equivalent to a two-month supply of grain exists. In a super-volcanic catastrophe, growing seasons might be curtailed for several years. A larger stockpile of grain and other foodstuffs, while expensive to maintain, would provide a buffer for a range of catastrophe scenarios involving temporary reductions in world agricultural productivity.

The hazard from comets and meteors is perhaps the best understood of all global catastrophic risks (which is not to deny that significant uncertainties remain). Chapter 11, by William Napier, explains some of the science behind the impact hazards: where comets and asteroids come from, how frequently impacts occur, and what the effects of an impact would be. To produce a civilization-disrupting event, an impactor would need a diameter of at least one or two kilometre. A ten kilometre impactor would, it appears, have a good chance of causing the extinction of the human species. But even sub-kilometre impactors could produce damage reaching the level of global catastrophe, depending on their composition, velocity, angle, and impact site.

Napier estimates that 'the per capita impact hazard is at the level associated with the hazards of air travel and the like'. However, funding for mitigation is meager compared to funding for air safety. The main effort currently underway to address the impact hazard is the *Spaceguard* project, which receives about four million dollars per annum from NASA besides in-kind and voluntary contributions from others. *Spaceguard* aims to find 90% of near-Earth asteroids larger than one kilometre by the end of 2008. Asteroids constitute the largest portion of the threat from near-Earth objects (and are easier to detect than comets) so when the project is completed, the subjective probability of a large impact will have been reduced considerably – unless, of course, it were discovered that some asteroid has a date with our planet in the near future, in which case the probability would soar.

Some preliminary study has been done of how a potential impactor could be deflected. Given sufficient advance warning, it appears that the space technology needed to divert an asteroid could be developed. The cost of producing an effective asteroid defence would be much greater than the cost of searching for potential impactors. However, if a civilization-destroying wrecking ball were found to be swinging towards the Earth, virtually any expense would be justified to avert it before it struck.

Asteroids and comets are not the only potential global catastrophic threats from space. Other cosmic hazards include global climatic change from fluctuations in solar activity, and very large fluxes from radiation and cosmic rays from supernova explosions or gamma ray bursts. These risks are examined in Chapter 12 by Arnon Dar. The findings on these risks are favourable: the risks appear to be very small. No particular response seems indicated at the present time beyond continuation of basic research.[8]

1.5 Part III: Risks from unintended consequences

We have already encountered climate change – in the form of sudden global cooling – as a destructive modality of super-eruptions and large impacts (as well as a possible consequence of large-scale nuclear war, to be discussed later). Yet it is the risk of gradual global warming brought about by greenhouse gas emissions that has most strongly captured the public imagination in recent years. Anthropogenic climate change has become the poster child of global threats. Global warming commandeers a disproportionate fraction of the attention given to global risks.

Carbon dioxide and other greenhouse gases are accumulating in the atmosphere, where they are expected to cause a warming of Earth's climate and a concomitant rise in seawater levels. The most recent report by the United Nations' Intergovernmental Panel on Climate Change (IPCC), which represents the most authoritative assessment of current scientific opinion, attempts to estimate the increase in global mean temperature that would be expected by the end of this century under the assumption that no efforts at mitigation are made. The final estimate is fraught with uncertainty because of uncertainty about what the default rate of emissions of greenhouse gases will be over the century, uncertainty about the climate sensitivity parameter, and uncertainty about other factors. The IPCC, therefore, expresses its assessment in terms of six different climate scenarios based on different models and different assumptions. The 'low' model predicts a mean global warming of +1.8°C (uncertainty range 1.1–2.9°C); the 'high' model predicts warming by +4.0°C (2.4–6.4°C). Estimated sea level rise predicted by the two most extreme scenarios of the six considered is 18–38 cm, and 26–59 cm, respectively.

Chapter 13, by David Frame and Myles Allen, summarizes some of the basic science behind climate modelling, with particular attention to the low-probability high-impact scenarios that are most relevant to the focus of this book. It is, arguably, this range of extreme scenarios that gives the greatest

[8] A comprehensive review of space hazards would also consider scenarios involving contact with intelligent extraterrestrial species or contamination from hypothetical extraterrestrial microorganisms; however, these risks are outside the scope of Chapter 12.

cause for concern. Although their likelihood seems very low, considerable uncertainty still pervades our understanding of various possible feedbacks that might be triggered by the expected climate forcing (recalling Peter Taylor's point, referred to earlier, about the importance of taking parameter and model uncertainty into account). David Frame and Myles Allen also discuss mitigation policy, highlighting the difficulties of setting appropriate mitigation goals given the uncertainties about what levels of cumulative emissions would constitute 'dangerous anthropogenic interference' in the climate system.

Edwin Kilbourne reviews some historically important pandemics in Chapter 14, including the distinctive characteristics of their associated pathogens, and discusses the factors that will determine the extent and consequences of future outbreaks.

Infectious disease has exacted an enormous toll of suffering and death on the human species throughout history and continues to do so today. Deaths from infectious disease currently account for approximately 25% of all deaths worldwide. This amounts to approximately 15 million deaths per year. About 75% of these deaths occur in Southeast Asia and sub-Saharan Africa. The top five causes of death due to infectious disease are upper respiratory infection (3.9 million deaths), HIV/AIDS (2.9 million), diarrhoeal disease (1.8 million), tuberculosis (1.7 million), and malaria (1.3 million).

Pandemic disease is indisputably one of the biggest global catastrophic risks facing the world today, but it is not always accorded its due recognition. For example, in most people's mental representation of the world, the influenza pandemic of 1918–1919 is almost completely overshadowed by the concomitant World War I. Yet although the WWI is estimated to have directly caused about 10 million military and 9 million civilian fatalities, the Spanish flu is believed to have killed at least 20–50 million people. The relatively low 'dread factor' associated with this pandemic might be partly due to the fact that only approximately 2–3% of those who got sick died from the disease. (The total death count is vast because a large percentage of the world population was infected.)

In addition to fighting the major infectious diseases currently plaguing the world, it is vital to remain alert to emerging new diseases with pandemic potential, such as SARS, bird flu, and drug-resistant tuberculosis. As the World Health Organization and its network of collaborating laboratories and local governments have demonstrated repeatedly, decisive early action can sometimes nip an emerging pandemic in the bud, possibly saving the lives of millions.

We have chosen to label pandemics a 'risk from unintended consequences' even though most infectious diseases (exempting the potential of genetically engineered bioweapons) in some sense arise from nature. Our rationale is that the evolution as well as the spread of pathogens is highly dependent on human civilization. The worldwide spread of germs became possible only after all the inhabited continents were connected by travel routes. By now, globalization in the form of travel and trade has reached such an extent that a highly contagious disease could spread to virtually all parts of the world within a matter of days or weeks. Kilbourne also draws attention to another aspect of globalization as a factor increasing pandemic risk: homogenization of peoples, practices, and cultures. The more the human population comes to resemble a single homogeneous niche, the greater the potential for a single pathogen to saturate it quickly. Kilbourne mentions the 'one rotten apple syndrome', resulting from the mass production of food and behavioural fads:

> If one contaminated item, apple, egg or, most recently, spinach leaf carries a billion bacteria – not an unreasonable estimate – and it enters a pool of cake mix constituents then packaged and sent to millions of customers nationwide, a bewildering epidemic may ensue.

Conversely, cultural as well as genetic diversity reduces the likelihood that any single pattern will be adopted universally before it is discovered to be dangerous – whether the pattern be virus RNA, a dangerous new chemical or material, or a stifling ideology.

By contrast to pandemics, artificial intelligence (AI) is not an ongoing or imminent global catastrophic risk. Nor is it as uncontroversially a serious cause for concern. However, from a long-term perspective, the development of general artificial intelligence exceeding that of the human brain can be seen as one of the main challenges to the future of humanity (arguably, even as *the* main challenge). At the same time, the successful deployment of friendly superintelligence could obviate many of the other risks facing humanity. The title of Chapter 15, 'Artificial Intelligence as a positive and negative factor in global risk', reflects this ambivalent potential.

As Eliezer Yudkowsky notes, the prospect of superintelligent machines is a difficult topic to analyse and discuss. Appropriately, therefore, he devotes a substantial part of his chapter to clearing common misconceptions and barriers to understanding. Having done so, he proceeds to give an argument for giving serious consideration to the possibility that radical superintelligence could erupt very suddenly – a scenario that is sometimes referred to as the 'Singularity hypothesis'. Claims about the steepness of the transition must be distinguished from claims about the timing of its onset. One could believe, for example, that it will be a long time before computers are able to match the general reasoning abilities of an average human being, but that once that happens, it will only take a short time for computers to attain radically superhuman levels.

Yudkowsky proposes that we conceive of a superintelligence as an enormously powerful optimization process: 'a system which hits small targets in large search spaces to produce coherent real-world effects'. The superintelligence will be able to manipulate the world (including human beings) in such a way as to achieve its goals, whatever those goals might be.

To avert disaster, it would be necessary to ensure that the superintelligence is endowed with a 'Friendly' goal system: that is, one that aligns the system's goals with genuine human values.

Given this set-up, Yudkowsky identifies two different ways in which we could fail to build Friendliness into our AI: philosophical failure and technical failure. The warning against philosophical failure is basically that we should be careful what we wish for because we might get it. We might designate a target for the AI which at first sight seems like a nice outcome but which in fact is radically misguided or morally worthless. The warning against technical failure is that we might fail to get what we wish for, because of faulty implementation of the goal system or unintended consequences of the way the target representation was specified. Yudkowsky regards both of these possible failure modes as very serious existential risks and concludes that it is imperative that we figure out how to build Friendliness into a superintelligence before we figure out how to build a superintelligence.

Chapter 16 discusses the possibility that the experiments that physicists carry out in particle accelerators might pose an existential risk. Concerns about such risks prompted the director of the Brookhaven Relativistic Heavy Ion Collider to commission an official report in 2000. Concerns have since resurfaced with the construction of more powerful accelerators such as CERN's Large Hadron Collider. Following the Brookhaven report, Frank Wilczek distinguishes three catastrophe scenarios:

1. Formation of tiny black holes that could start accreting surrounding matter, eventually swallowing up the entire planet.
2. Formation of negatively charged stable strangelets which could catalyse the conversion of all the ordinary matter on our planet into strange matter.
3. Initiation of a phase transition of the vacuum state, which would propagate outward in all directions at near light speed and destroy not only our planet but the entire accessible part of the universe.

Wilczek argues that these scenarios are exceedingly unlikely on various theoretical grounds. In addition, there is a more general argument that these scenarios are extremely improbable which depends less on arcane theory. Cosmic rays often have energies far greater than those that will be attained in any of the planned accelerators. Such rays have been bombarding the Earth's atmosphere (and the moon and other astronomical objects) for billions of years without a single catastrophic effect having been observed. Assuming that collisions in particle accelerators do not differ in any unknown relevant respect from those that occur in the wild, we can be very confident in the safety of our accelerators.

By everyone's reckoning, it is highly improbable that particle accelerator experiments will cause an existential disaster. The question is *how* improbable? And what would constitute an 'acceptable' probability of an existential disaster?

In assessing the probability, we must consider not only how unlikely the outcome seems given our best current models but also the possibility that our best models and calculations might be flawed in some as-yet unrealized way. In doing so we must guard against overconfidence bias (compare Chapter 5 on biases). Unless we ourselves are technically expert, we must also take into account the possibility that the experts on whose judgements we rely might be consciously or unconsciously biased.[9] For example, the physicists who possess the expertise needed to assess the risks from particle physics experiments are part of a professional community that has a direct stake in the experiments going forward. A layperson might worry that the incentives faced by the experts could lead them to err on the side of downplaying the risks.[10] Alternatively, some experts might be tempted by the media attention they could get by playing up the risks. The issue of how much and in which circumstances to trust risk estimates by experts is an important one, and it arises quite generally with regard to many of the risks covered in this book.

Chapter 17 (by Robin Hanson) from Part III on Risks from unintended consequences focuses on social collapse as a devastation multiplier of other catastrophes. Hanson writes as follows:

> The main reason to be careful when you walk up a flight of stairs is not that you might slip and have to retrace one step, but rather that the first slip might cause a second slip, and so on until you fall dozens of steps and break your neck. Similarly we are concerned about the sorts of catastrophes explored in this book not only because of their terrible direct effects, but also because they may induce an even more damaging collapse of our economic and social systems.

This argument does not apply to some of the risks discussed so far, such as those from particle accelerators or the risks from superintelligence as envisaged by Yudkowsky. In those cases, we may be either completely safe or altogether doomed, with little probability of intermediary outcomes. But for many other types of risk – such as windstorms, tornados, earthquakes, floods, forest fires, terrorist attacks, plagues, and wars – a wide range of outcomes are possible, and the potential for social disruption or even social collapse constitutes a major part of the overall hazard. Hanson notes that many of these risks appear to follow a power law distribution. Depending on the characteristic exponent of such a power law distribution, most of the damage expected from a given

[9] Even if we ourselves are expert, we must still be alert to unconscious biases that may influence our judgment (e.g., anthropic biases, see Chapter 6).

[10] If experts anticipate that the public will not quite trust their reassurances, they might be led to try to sound even more reassuring than they would have if they had believed that the public would accept their claims at face value. The public, in turn, might respond by discounting the experts' verdicts even more, leading the experts to be even more wary of fuelling alarmist overreactions. In the end, experts might be reluctant to acknowledge any risk at all for fear of a triggering a hysterical public overreaction. Effective risk communication is a tricky business, and the trust that it requires can be hard to gain and easy to lose.

type of risk may consist either of frequent small disturbances or of rare large catastrophes. Car accidents, for example, have a large exponent, reflecting the fact that most traffic deaths occur in numerous small accidents involving one or two vehicles. Wars and plagues, by contrast, appear to have small exponents, meaning that most of the expected damage occurs in very rare but very large conflicts and pandemics.

After giving a thumbnail sketch of economic growth theory, Hanson considers an extreme opposite of economic growth: sudden reduction in productivity brought about by escalating destruction of social capital and coordination. For example, 'a judge who would not normally consider taking a bribe may do so when his life is at stake, allowing others to expect to get away with theft more easily, which leads still others to avoid making investments that might be stolen, and so on. Also, people may be reluctant to trust bank accounts or even paper money, preventing those institutions from functioning.' The productivity of the world economy depends both on scale and on many different forms of capital which must be delicately coordinated. We should be concerned that a relatively small disturbance (or combination of disturbances) to some vulnerable part of this system could cause a far-reaching unraveling of the institutions and expectations upon which the global economy depends.

Hanson also offers a suggestion for how we might convert some existential risks into non-existential risks. He proposes that we consider the construction of one or more continuously inhabited refuges – located, perhaps, in a deep mineshaft, and well-stocked with supplies – which could preserve a small but sufficient group of people to repopulate a post-apocalyptic world. It would obviously be preferable to prevent altogether catastrophes of a severity that would make humanity's survival dependent on such modern-day 'Noah's arks'; nevertheless, it might be worth exploring whether some variation of this proposal might be a cost-effective way of somewhat decreasing the probability of human extinction from a range of potential causes.[11]

1.6 Part IV: Risks from hostile acts

The spectre of nuclear Armageddon, which so haunted the public imagination during the Cold War era, has apparently entered semi-retirement. The number of nuclear weapons in the world has been reduced to half, from a Cold War high of 65,000 in 1986 to approximately 26,000 in 2007, with approximately

[11] Somewhat analogously, we could prevent much permanent loss of biodiversity by moving more aggressively to preserve genetic material from endangered species in biobanks. The Norwegian government has recently opened a seed bank on a remote island in the arctic archipelago of Svalbard. The vault, which is dug into a mountain and protected by steel-reinforced concrete walls one metre thick, will preserve germplasm of important agricultural and wild plants.

96% of these weapons held by the United States and Russia. Relationships between these two nations are not as bad as they once were. New scares such as environmental problems and terrorism compete effectively for media attention. Changing winds in horror-fashion aside, however, and as Chapter 18 makes it clear, nuclear war remains a very serious threat.

There are several possibilities. One is that relations between the United States and Russia might again worsen to the point where a crisis could trigger a nuclear war. Future arms races could lead to arsenals even larger than those of the past. The world's supply of plutonium has been increasing steadily to about 2000 tons – about ten times as much as remains tied up in warheads – and more could be produced. Some studies suggest that in an all-out war involving most of the weapons in the current US and Russian arsenals, 35–77% of the US population (105–230 million people) and 20–40% of the Russian population (28–56 million people) would be killed. Delayed and indirect effects – such as economic collapse and a possible nuclear winter – could make the final death toll far greater.

Another possibility is that nuclear war might erupt between nuclear powers other than the old Cold War rivals, a risk that is growing as more nations join the nuclear club, especially nations that are embroiled in volatile regional conflicts, such as India and Pakistan, North Korea, and Israel, perhaps to be joined by Iran or others. One concern is that the more nations get the bomb, the harder it might be to prevent further proliferation. The technology and know-how would become more widely disseminated, lowering the technical barriers, and nations that initially chose to forego nuclear weapons might feel compelled to rethink their decision and to follow suit if they see their neighbours start down the nuclear path.

A third possibility is that global nuclear war could be started by mistake. According to Joseph Cirincione, this almost happened in January 1995:

> Russian military officials mistook a Norwegian weather rocket for a US submarine-launched ballistic missile. Boris Yelstin became the first Russian president to ever have the 'nuclear suitcase' open in front of him. He had just a few minutes to decide if he should push the button that would launch a barrage of nuclear missiles. Thankfully, he concluded that his radars were in error. The suitcase was closed.

Several other incidents have been reported in which the world, allegedly, was teetering on the brink of nuclear holocaust. At one point during the Cuban missile crisis, for example, President Kennedy reportedly estimated the probability of a nuclear war between the United States and the USSR to be 'somewhere between one out of three and even'.

To reduce the risks, Cirincione argues, we must work to resolve regional conflicts, support and strengthen the Nuclear Non-proliferation Treaty – one of the most successful security pacts in history – and move towards the abolition of nuclear weapons.

William Potter and Gary Ackerman offer a detailed look at the risks of nuclear terrorism in Chapter 19. Such terrorism could take various forms:

- Dispersal of radioactive material by conventional explosives ('dirty bomb')
- Sabotage of nuclear facilities
- Acquisition of fissile material leading to the fabrication and detonation of a crude nuclear bomb ('improvised nuclear device')
- Acquisition and detonation of an intact nuclear weapon
- The use of some means to trick a nuclear state into launching a nuclear strike.

Potter and Ackerman focus on 'high consequence' nuclear terrorism, which they construe as those involving the last three alternatives from the above list. The authors analyse the demand and supply side of nuclear terrorism, the consequences of a nuclear terrorist attack, the future shape of the threat, and conclude with policy recommendations.

To date, no non-state actor is believed to have gained possession of a fission weapon:

> There is no credible evidence that either al Qaeda or Aum Shinrikyo were able to exploit their high motivations, substantial financial resources, demonstrated organizational skills, far-flung network of followers, and relative security in a friendly or tolerant host country to move very far down the path toward acquiring a nuclear weapons capability. As best one can tell from the limited information available in public sources, among the obstacles that proved most difficult for them to overcome was access to the fissile material needed . . .

Despite this track record, however, many experts remain concerned. Graham Allison, author of one of the most widely cited works on the subject, offers a standing bet of 51 to 49 odds that 'barring radical new anti-proliferation steps' there will be a terrorist nuclear strike within the next ten years. Other experts seem to place the odds much lower, but have apparently not taken up Allison's offer.

There is wide recognition of the importance of prevention nuclear terrorism, and in particular of the need to prevent fissile material from falling into the wrong hands. In 2002, the G-8 Global Partnership set a target of 20 billion dollars to be committed over a ten-year period for the purpose of preventing terrorists from acquiring weapons and materials of mass destruction. What Potter and Ackerman consider most lacking, however, is the sustained high-level leadership needed to transform rhetoric into effective implementation.

In Chapter 20, Christopher Chyba and Ali Nouri review issues related to biotechnology and biosecurity. While in some ways paralleling nuclear risks – biological as well as nuclear technology can be used to build weapons of mass destruction – there are also important divergences. One difference is that biological weapons can be developed in small, easily concealed facilities and require no unusual raw materials for their manufacture. Another difference is that an infectious biological agent can spread far beyond the site of its original release, potentially across the entire world.

Biosecurity threats fall into several categories, including naturally occurring diseases, illicit state biological weapons programmes, non-state actors and bio-hackers, and laboratory accidents or other inadvertent release of disease agents. It is worth bearing in mind that the number of people who have died in recent years from threats in the first of these categories (naturally occurring diseases) is six or seven orders of magnitudes larger than the number of fatalities from the other three categories combined. Yet biotechnology does contain brewing threats which look set to expand dramatically over the coming years as capabilities advance and proliferate. Consider the following sample of recent developments:

- A group of Australian researchers, looking for ways of controlling the country's rabbit population, added the gene for interleukin-4 to a mousepox virus, hoping thereby to render the animals sterile. Unexpectedly, the virus inhibited the host's immune system and all the animals died, including individuals who had previously been vaccinated. Follow-up work by another group produced a version of the virus that was 100% lethal in vaccinated mice despite the antiviral medication given to the animals.
- The polio virus has been synthesized from readily purchased chemical supplies. When this was first done, it required a protracted cutting-edge research project. Since then, the time needed to synthesize a virus genome comparable in size to the polio virus has been reduced to weeks. The virus that caused the Spanish flu pandemic, which was previously extinct, has also been resynthesized and now exists in laboratories in the United States and in Canada.
- The technology to alter the properties of viruses and other microorganisms is advancing at a rapid pace. The recently developed method of RNA interference provides researchers with a ready means of turning off selected genes in humans and other organisms. 'Synthetic biology' is being established as new field, whose goal is to enable the creation of small biological devices and ultimately new types of microbes.

Reading this list, while bearing in mind that the complete genomes from hundreds of bacteria, fungi, viruses – including Ebola, Marburg, smallpox, and the 1918 Spanish influenza virus – have been sequenced and deposited in a public online database, it is not difficult to concoct in one's imagination frightening possibilities. The technological barriers to the production of super bugs are being steadily lowered even as the biotechnological know-how and equipment diffuse ever more widely.

The dual-use nature of the necessary equipment and expertise, and the fact that facilities could be small and easily concealed, pose difficult challenges for would-be regulators. For any regulatory regime to work, it would also have to strike a difficult balance between prevention of abuses and enablement of research needed to develop treatments and diagnostics (or to obtain other medical or economic benefits). Chyba and Nouri discuss several strategies for promoting biosecurity, including automated review of gene sequences submitted for DNA-synthesizing at centralized facilities. It is likely that biosecurity will grow in importance and that a multipronged approach will be needed to address the dangers from designer pathogens.

Chris Phoenix and Mike Treder (Chapter 21) discuss nanotechnology as a source of global catastrophic risks. They distinguish between 'nanoscale technologies', of which many exist today and many more are in development, and 'molecular manufacturing', which remains a hypothetical future technology (often associated with the person who first envisaged it in detail, K. Eric Drexler). Nanoscale technologies, they argue, appear to pose no new global catastrophic risks, although such technologies could in some cases either augment or help mitigate some of the other risks considered in this volume. Phoenix and Treder consequently devote the bulk of their chapter to considering the capabilities and threats from molecular manufacturing. As with superintelligence, the *present* risk is virtually zero since the technology in question does not yet exist; yet the future risk could be extremely severe.

Molecular nanotechnology would greatly expand control over the structure of matter. Molecular machine systems would enable fast and inexpensive manufacture of microscopic and macroscopic objects built to atomic precision. Such production systems would contain millions of microscopic assembly tools. Working in parallel, these would build objects by adding molecules to a workpiece through positionally controlled chemical reactions. The range of structures that could be built with such technology greatly exceeds that accessible to the biological molecular assemblers (such as ribosome) that exist in nature. Among the things that a nanofactory could build: another nanofactory. A sample of potential applications:

- microscopic nanobots for medical use
- vastly faster computers
- very light and strong diamondoid materials
- new processes for removing pollutants from the environment
- desktop manufacturing plants which can automatically produce a wide range of atomically precise structures from downloadable blueprints
- inexpensive solar collectors
- greatly improved space technology

- mass-produced sensors of many kinds
- weapons, both inexpensively mass-produced and improved conventional weapons, and new kinds of weapons that cannot be built without molecular nanotechnology.

A technology this powerful and versatile could be used for an indefinite number of purposes, both benign and malign.

Phoenix and Treder review a number of global catastrophic risks that could arise with such an advanced manufacturing technology, including war, social and economic disruption, destructive forms of global governance, radical intelligence enhancement, environmental degradation, and 'ecophagy' (small nanobots replicating uncontrollably in the natural environment, consuming or destroying the Earth's biosphere). In conclusion, they offer the following rather alarming assessment:

> In the absence of some type of preventive or protective force, the power of molecular manufacturing products could allow a large number of actors of varying types – including individuals, groups, corporations, and nations – to obtain sufficient capability to destroy all unprotected humans. The likelihood of at least one powerful actor being insane is not small. The likelihood that devastating weapons will be built and released accidentally (possibly through overly sensitive automated systems) is also considerable. Finally, the likelihood of a conflict between two [powers capable of unleashing a mutually assured destruction scenario] escalating until one feels compelled to exercise a doomsday option is also non-zero. This indicates that unless adequate defences can be prepared against weapons intended to be ultimately destructive – a point that urgently needs research – the number of actors trying to possess such weapons must be minimized.

The last chapter of the book, authored by Bryan Caplan, addresses totalitarianism as a global catastrophic risk. The totalitarian governments of Nazi Germany, Soviet Russia, and Maoist China were responsible for tens of millions of deaths in the last century. Compared to a risk like that of asteroid impacts, totalitarianism as a global risk is harder to study in an unbiased manner, and a cross-ideological consensus about how this risk is best to be mitigated is likely to be more elusive. Yet the risks from oppressive forms of government, including totalitarian regimes, must not be ignored. Oppression has been one of the major recurring banes of human development throughout history, it largely remains so today, and it is one to which the humanity remains vulnerable.

As Caplan notes, in addition to being a misfortune in itself, totalitarianism can also amplify other risks. People in totalitarian regimes are often afraid to publish bad news, and the leadership of such regimes is often insulated from criticism and dissenting views. This can make such regimes more likely to overlook looming dangers and to commit serious policy errors (even as evaluated from the standpoint of the self-interest of the rulers). However, as

Caplan notes further, for some types of risk, totalitarian regimes might actually possess an advantage compared to more open and diverse societies. For goals that can be achieved by brute force and massive mobilization of resources, totalitarian methods have often proven effective.

Caplan analyses two factors which he claims have historically limited the durability of totalitarian regimes. The first of these is the problem of succession. A strong leader might maintain a tight grip on power for as long as he lives, but the party faction he represents often stumbles when it comes to appointing a successor that will preserve the status quo, allowing a closet reformer – a sheep in wolf's clothing – to gain the leadership position after a tyrant's death. The other factor is the existence of non-totalitarian countries elsewhere in the world. These provide a vivid illustration to the people living under totalitarianism that things could be much better than they are, fuelling dissatisfaction and unrest. To counter this, leaders might curtail contacts with the external world, creating a 'hermit kingdom' such as Communist Albania or present-day North Korea. However, some information is bound to leak in. Furthermore, if the isolation is too complete, over a period of time, the country is likely to fall far behind economically and militarily, making itself vulnerable to invasion or externally imposed regime change.

It is possible that the vulnerability presented by these two Achilles heels of totalitarianism could be reduced by future developments. Technological advances could help solve the problem of succession. Brain scans might one day be used to screen out closet sceptics within the party. Other forms of novel surveillance technologies could also make it easier to control population. New psychiatric drugs might be developed that could increase docility without noticeably reducing productivity. Life-extension medicine might prolong the lifespan of the leader so that the problem of succession comes up less frequently. As for the existence of non-totalitarian outsiders, Caplan worries about the possible emergence of a world government. Such a government, even if it started out democratic, might at some point degenerate into totalitarianism; and a worldwide totalitarian regime could then have great staying power given its lack of external competitors and alien exemplars of the benefits of political freedom.

To have a productive discussion about matters such as these, it is important to recognize the distinction between two very different stances: 'here a valid consideration in favour of some position *X*' versus '*X* is all-things-considered the position to be adopted'. For instance, as Caplan notes:

> If people lived forever, stable totalitarianism would be a little more likely to emerge, but it would be madness to force everyone to die of old age in order to avert a small risk of being murdered by the secret police in a thousand years.

Likewise, it is possible to favour the strengthening of certain new forms global governance while also recognizing as a legitimate concern the danger of global totalitarianism to which Caplan draws our attention.

1.7 Conclusions and future directions

The most likely global catastrophic risks all seem to arise from human activities, especially industrial civilization and advanced technologies. This is not necessarily an indictment of industry or technology, for these factors deserve much of the credit for creating the values that are now at risk – including most of the people living on the planet today, there being perhaps 30 times more of us than could have been sustained with primitive agricultural methods, and hundreds of times more than could have lived as hunter–gatherers. Moreover, although new global catastrophic risks have been created, many smaller-scale risks have been drastically reduced in many parts of the world, thanks to modern technological society. Local and personal disasters – such as starvation, thirst, predation, disease, and small-scale violence – have historically claimed many more lives than have global cataclysms. The reduction of the aggregate of these smaller-scale hazards may outweigh an increase in global catastrophic risks. To the (incomplete) extent that true risk levels are reflected in actuarial statistics, the world is a safer place than it has ever been: world life expectancy is now sixty-four years, up from fifty in the early twentieth century, thirty-three in Medieval Britain, and an estimated eighteen years during the Bronze Age. Global catastrophic risks are, by definition, the largest in terms of *scope* but not necessarily in terms of their expected severity (probability × harm). Furthermore, technology and complex social organizations offer many important tools for managing the remaining risks. Nevertheless, it is important to recognize that the biggest global catastrophic risks we face today are not purely external; they are, instead, tightly wound up with the direct and indirect, the foreseen and unforeseen, consequences of our own actions.

One major current global catastrophic risk is infectious pandemic disease. As noted earlier, infectious disease causes approximately 15 million deaths per year, of which 75% occur in Southeast Asia and sub-Saharan Africa. These dismal statistics pose a challenge to the classification of pandemic disease as a global catastrophic risk. One could argue that infectious disease is not so much a *risk* as an *ongoing global catastrophe*. Even on a more fine-grained individuation of the hazard, based on specific infectious agents, at least some of the currently occurring pandemics (such as HIV/AIDS, which causes nearly 3 million deaths annually) would presumably qualify as global catastrophes. By similar reckoning, one could argue that cardiovascular disease (responsible for approximately 30% of world mortality, or 18 million deaths per year) and cancer (8 million deaths) are also ongoing global catastrophes. It would be perverse if the study of possible catastrophes that *could* occur were to drain attention away from actual catastrophes that *are* occurring.

It is also appropriate, at this juncture, to reflect for a moment on the biggest cause of death and disability of all, namely ageing, which accounts for perhaps two-thirds of the 57 million deaths that occur each year, along with an enormous loss of health and human capital.[12] If ageing were not certain but merely probable, it would immediately shoot to the top of any list of global catastrophic risks. Yet the fact that ageing is not just a possible cause of future death, but a certain cause of present death, should not trick us into trivializing the matter. To the extent that we have a realistic prospect of mitigating the problem – for example, by disseminating information about healthier lifestyles or by investing more heavily in biogerontological research – we may be able to save a much larger expected numbers of lives (or quality-adjusted life-years) by making partial progress on this problem than by completely eliminating some of the global catastrophic risk discussed in this volume.

Other global catastrophic risks which are either already substantial or expected to become substantial within a decade or so include the risks from nuclear war, biotechnology (misused for terrorism or perhaps war), social/economic disruption or collapse scenarios, and maybe nuclear terrorism. Over a somewhat longer time frame, the risks from molecular manufacturing, artificial intelligence, and totalitarianism may rise in prominence, and each of these latter ones is also potentially existential.

That a particular risk is larger than another does not imply that more resources ought to be devoted to its mitigation. Some risks we might not be able to do anything about. For other risks, the available means of mitigation might be too expensive or too dangerous. Even a small risk can deserve to be tackled as a priority if the solution is sufficiently cheap and easy to implement – one example being the anthropogenic depletion of the ozone layer, a problem now well on its way to being solved. Nevertheless, as a rule of thumb it makes sense to devote most of our attention to the risks that are largest and/or most urgent. A wise person will not spend time installing a burglar alarm when the house is on fire.

Going forward, we need continuing studies of individual risks, particularly of potentially big but still relatively poorly understood risks, such as those from biotechnology, molecular manufacturing, artificial intelligence, and systemic risks (of which totalitarianism is but one instance). We also need studies to identify and evaluate possible mitigation strategies. For some risks and ongoing disasters, cost-effective countermeasures are already known; in these cases, what is needed is leadership to ensure implementation of the appropriate programmes. In addition, there is a need for studies to clarify methodological problems arising in the study of global catastrophic risks.

[12] In mortality statistics, deaths are usually classified according to their more proximate causes (cancer, suicide, etc.). But we can estimate how many deaths are due to ageing by comparing the age-specific mortality in different age groups. The reason why an average 80-year-old is more likely to die within the next year than an average 20-year-old is that senescence has made the former more susceptible to a wide range of specific risk factors. The surplus mortality in older cohorts can therefore be attributed to the negative effects of ageing.

The fruitfulness of further work on global catastrophic risk will, we believe, be enhanced if it gives consideration to the following suggestions:

- In the study of individual risks, focus more on producing actionable information such as early-warning signs, metrics for measuring progress towards risk reduction, and quantitative models for risk assessment.
- Develop and implement better methodologies and institutions for information aggregation and probabilistic forecasting, such as prediction markets.
- Put more effort into developing and evaluating possible mitigation strategies, both because of the direct utility of such research and because a concern with the policy instruments with which a risk can be influenced is likely to enrich our theoretical understanding of the nature of the risk.
- Devote special attention to existential risks and the unique methodological problems they pose.
- Build a stronger interdisciplinary and international risk community, including not only experts from many parts of academia but also professionals and policymakers responsible for implementing risk reduction strategies, in order to break out of disciplinary silos and to reduce the gap between theory and practice.
- Foster a critical discourse aimed at addressing questions of prioritization in a more reflective and analytical manner than is currently done; and consider global catastrophic risks and their mitigation within a broader context of challenges and opportunities for safeguarding and improving the human condition.

Our hopes for this book will have been realized if it adds a brick to the foundation of a way of thinking that enables humanity to approach the global problems of the present era with greater maturity, responsibility, and effectiveness.

PART I
Background

·2·

Long-term astrophysical processes

Fred C. Adams

2.1 Introduction: physical eschatology

As we take a longer-term view of our future, a host of astrophysical processes are waiting to unfold as the Earth, the Sun, the Galaxy, and the Universe grow increasingly older. The basic astronomical parameters that describe our universe have now been measured with compelling precision. Recent observations of the cosmic microwave background radiation show that the spatial geometry of our universe is flat (Spergel et al., 2003). Independent measurements of the red-shift versus distance relation using Type Ia supernovae indicate that the universe is accelerating and apparently contains a substantial component of dark vacuum energy (Garnavich et al., 1998; Perlmutter et al., 1999; Riess et al., 1998).[1] This newly consolidated cosmological model represents an important milestone in our understanding of the cosmos. With the cosmological parameters relatively well known, the future evolution of our universe can now be predicted with some degree of confidence (Adams and Laughlin, 1997). Our best astronomical data imply that our universe will expand forever or at least live long enough for a diverse collection of astronomical events to play themselves out.

Other chapters in this book have discussed some sources of cosmic intervention that can affect life on our planet, including asteroid and comet impacts (Chapter 11, this volume) and nearby supernova explosions with their accompanying gamma-rays (Chapter 12, this volume). In the longer-term future, the chances of these types of catastrophic events will increase. In addition, taking an even longer-term view, we find that even more fantastic events could happen in our cosmological future. This chapter outlines some of the astrophysical events that can affect life, on our planet and perhaps

[1] 'Dark energy' is a common term unifying different models for the ubiquitous form of energy permeating the entire universe (about 70% of the total energy budget of the physical universe) and causing accelerated expansion of space time. The most famous of these models is Einstein's *cosmological constant*, but there are others, going under the names of *quintessence*, *phantom energy*, and so on. They are all characterized by negative pressure, in sharp contrast to all other forms of energy we see around us.

elsewhere, over extremely long time scales, including those that vastly exceed the current age of the universe.

These projections are based on our current understanding of astronomy and the laws of physics, which offer a firm and developing framework for understanding the future of the physical universe (this topic is sometimes called *Physical Eschatology* – see the review of Ćirković, 2003). Notice that as we delve deeper into the future, the uncertainties of our projections must necessarily grow. Notice also that this discussion is based on the assumption that the laws of physics are both known and unchanging; as new physics is discovered, or if the physical constants are found to be time dependent, this projection into the future must be revised accordingly.

2.2 Fate of the Earth

One issue of immediate importance is the fate of Earth's biosphere and, on even longer time scales, the fate of the planet itself. As the Sun grows older, it burns hydrogen into helium. Compared to hydrogen, helium has a smaller partial pressure for a given temperature, so the central stellar core must grow hotter as the Sun evolves. As a result, the Sun, like all stars, is destined to grow brighter as it ages. When the Sun becomes too bright, it will drive a runaway greenhouse effect through the Earth's atmosphere (Kasting et al., 1988). This effect is roughly analogous to that of global warming driven by greenhouse gases (see Chapter 13, this volume), a peril that our planet faces in the near future; however, this later-term greenhouse effect will be much more severe. Current estimates indicate that our biosphere will be essentially sterilized in about 3.5 billion years, so this future time marks the end of life on Earth. The end of *complex* life may come sooner, in 0.9–1.5 billion years owing to the runaway greenhouse effect (e.g., Caldeira and Kasting, 1992).

The biosphere represents a relatively small surface layer and the planet itself lives comfortably through this time of destruction. Somewhat later in the Sun's evolution, when its age reaches 11–12 billion years, it eventually depletes its store of hydrogen in the core region and must readjust its structure (Rybicki and Denis, 2001; Sackmann et al., 1993). As it does so, the outer surface of the star becomes somewhat cooler, its colour becomes a brilliant red, and its radius increases. The red giant Sun eventually grows large enough to engulf the radius of the orbit of Mercury, and that innermost planet is swallowed with barely a trace left. The Sun grows further, overtakes the orbit of Venus, and then accretes the second planet as well. As the red giant Sun expands, it loses mass so that surviving planets are held less tightly in their orbits. Earth is able to slip out to an orbit of larger radius and seemingly escape destruction. However, the mass loss from the Sun provides a fluid that the Earth must plough through as it makes its yearly orbit. Current calculations indicate that the frictional forces acting on Earth through its interaction with the solar outflow cause the planet to experience enough orbital decay that it is dragged back into the Sun. Earth is thus evaporated, with its legacy being a small addition to the heavy element supply of the solar photosphere. This point in future history, approximately 7 billion years from now, marks the end of our planet.

Given that the biosphere has at most only 3.5 billion years left on its schedule, and Earth itself has only 7 billion years, it is interesting to ask what types of 'planet-saving' events can take place on comparable time scales. Although the odds are not good, the Earth has some chance of being 'saved' by being scattered out of the solar system by a passing star system (most of which are binary stars). These types of scattering interactions pose an interesting problem in solar system dynamics, one that can be addressed with numerical scattering experiments. A large number of such experiments must be run because the systems are chaotic, and hence display sensitive dependence on their initial conditions, and because the available parameter space is large. Nonetheless, after approximately a half million scattering calculations, an answer can be found: the odds of Earth being ejected from the solar system before it is accreted by the red giant Sun is a few parts in 10^5 (Laughlin and Adams, 2000).

Although sending the Earth into exile would save the planet from eventual evaporation, the biosphere would still be destroyed. The oceans would freeze within a few million years and the only pockets of liquid water left would be those deep underground. The Earth contains an internal energy source – the power produced by the radioactive decay of unstable nuclei. This power is about 10,000 times smaller than the power that Earth intercepts from the present-day Sun, so it has little effect on the current operation of the surface biosphere. If Earth were scattered out of the solar system, then this internal power source would be the only one remaining. This power is sufficient to keep the interior of the planet hot enough for water to exist in liquid form, but only at depths 14 km below the surface. This finding, in turn, has implications for present-day astronomy: the most common liquid water environments may be those deep within frozen planets, that is, those that have frozen water on their surfaces and harbour oceans of liquid water below. Such planets may be more common than those that have water on their surface, like Earth, because they can be found in a much wider range of orbits about their central stars (Laughlin and Adams, 2000).

In addition to saving the Earth by scattering it out of the solar system, passing binaries can also capture the Earth and thereby allow it to orbit about a new star. Since most stars are smaller in mass than our Sun, they live longer and suffer less extreme red giant phases. (In fact, the smallest stars with less than one-fourth of the mass of the Sun will never become red giants – Laughlin et al., 1997.) As a result, a captured Earth would stand a better chance of long-term survival. The odds for this type of planet-saving event taking place while the biosphere remains intact are exceedingly slim – only about one in three million (Laughlin and Adams, 2000), roughly the odds of winning a big state lottery.

For completeness, we note that in addition to the purely natural processes discussed here, human or other intentional intervention could potentially change the course of Earth's orbit given enough time and other resources. As a concrete example, one could steer an asteroid into the proper orbit so that gravitational scattering effectively transfers energy into the Earth's orbit, thereby allowing it to move outward as the Sun grows brighter (Korycansky et al., 2001). In this scenario, the orbit of the asteroid is chosen to encounter both Jupiter and Saturn, and thereby regain the energy and angular momentum that it transfers to Earth. Many other scenarios are possible, but the rest of this chapter will focus on physical phenomena not including intentional actions.

2.3 Isolation of the local group

Because the expansion rate of the universe is starting to accelerate (Garnavich et al., 1998; Perlmutter et al., 1999; Riess et al., 1998), the formation of galaxies, clusters, and larger cosmic structures is essentially complete. The universe is currently approaching a state of exponential expansion and growing cosmological fluctuations will freeze out on all scales. Existing structures will grow isolated. Numerical simulations illustrate this trend (Fig. 2.1) and show how the universe will break up into a collection of 'island universes', each containing one bound cluster or group of galaxies (Busha et al., 2003; Nagamine and Loeb, 2003). In other words, the largest gravitationally bound structures that we see in the universe today are likely to be the largest structures that ever form. Not only must each group of galaxies (eventually) evolve in physical isolation, but the relentless cosmic expansion will stretch existing galaxy clusters out of each others' view. In the future, one will not even be able to see the light from galaxies living in other clusters. In the case of the Milky Way, only the Local Group of Galaxies will be visible. Current observations and recent numerical studies clearly indicate that the nearest large cluster – Virgo – does not have enough mass for the Local Group to remain bound to it in the future (Busha et al., 2003; Nagamine and Loeb, 2003). This local group consists of the Milky Way, Andromeda, and a couple of dozen dwarf galaxies (irregulars and spheroidals). The rest of the universe will be cloaked behind a cosmological horizon and hence will be inaccessible to future observation.

2.4 Collision with Andromeda

Within their clusters, galaxies often pass near each other and distort each other's structure with their strong gravitational fields. Sometimes these

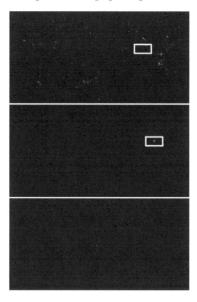

Fig. 2.1 Numerical simulation of structure formation in an accelerating universe with dark vacuum energy. The top panel shows a portion of the universe at the present time (cosmic age 14 Gyr). The boxed region in the upper panel expands to become the picture in the central panel at cosmic age 54 Gyr. The box in the central panel then expands to become the picture shown in the bottom panel at cosmic age 92 Gyr. At this future epoch, the galaxy shown in the centre of the bottom panel has grown effectively isolated. (Simulations reprinted with permission from Busha, M.T., Adams, F.C., Evrard, A.E., and Wechsler, R.H. (2003). Future evolution of cosmic structure in an accelerating universe. *Astrophys. J.*, 596, 713.)

interactions lead to galactic collisions and merging. A rather important example of such a collision is coming up: the nearby Andromeda galaxy is headed straight for our Milky Way. Although this date with our sister galaxy will not take place for another 6 billion years or more, our fate is sealed – the two galaxies are a bound pair and will eventually merge into one (Peebles, 1994).

When viewed from the outside, galactic collisions are dramatic and result in the destruction of the well-defined spiral structure that characterizes the original galaxies. When viewed from within the galaxy, however, galactic collisions are considerably less spectacular. The spaces between stars are so vast that few, if any, stellar collisions take place. One result is the gradual brightening of the night sky, by roughly a factor of two. On the other hand, galactic collisions are frequently associated with powerful bursts of star formation. Large clouds of molecular gas within the galaxies merge during such collisions and produce new stars at prodigious rates. The multiple supernovae resulting from the deaths of the most massive stars can have catastrophic consequences and represent a significant risk to any nearby biosphere (see Chapter 12, this volume), provided that life continues to thrive in thin spherical layers on terrestrial planets.

2.5 The end of stellar evolution

With its current age of 14 billion years, the universe now lives in the midst of a Stelliferous Era, an epoch when stars are actively forming, living, and dying. Most of the energy generated in our universe today arises from nuclear fusion that takes place in the cores of ordinary stars. As the future unfolds, the most common stars in the universe – the low-mass stars known as red dwarfs – play an increasingly important role. Although red dwarf stars have less than half the mass of the Sun, they are so numerous that their combined mass easily dominates the stellar mass budget of the galaxy. These red dwarfs are parsimonious when it comes to fusing their hydrogen into helium. By hoarding their energy resources, they will still be shining trillions of years from now, long after their larger brethren have exhausted their fuel and evolved into white dwarfs or exploded as supernovae. It has been known for a long time that smaller stars live much longer than more massive ones owing to their much smaller luminosities. However, recent calculations show that red dwarfs live even longer than expected. In these small stars, convection currents cycle essentially all of the hydrogen fuel in the star through the stellar core, where it can be used as nuclear fuel. In contrast, our Sun has access to only about 10% of its hydrogen and will burn only 10% of its nuclear fuel while on the main sequence. A small star with 10% of the mass of the Sun thus has nearly the same fuel reserves and will shine for tens of trillions of years (Laughlin et al., 1997). Like all stars, red dwarfs get brighter as they age. Owing to their large population, the brightening of red dwarfs nearly compensates for the loss of larger stars, and the galaxy can maintain a nearly constant luminosity for approximately one trillion years (Adams et al., 2004).

Even small stars cannot live forever, and this bright stellar era comes to a close when the galaxies run out of hydrogen gas, star formation ceases, and the longest-lived red dwarfs slowly fade into oblivion. As mentioned earlier, the smallest stars will shine for trillions of years, so the era of stars would come to an end at a cosmic age of several trillion years if new stars were not being manufactured. In large spiral galaxies like the Milky Way, new stars are being made from hydrogen gas, which represents the basic raw material for the process. Galaxies will continue to make new stars as long as the gas supply holds out. If our Galaxy were to continue forming stars at its current rate, it would run out of gas in 'only' 10–20 billion years (Kennicutt et al., 1994), much shorter than the lifetime of the smallest stars. Through conservation practices – the star formation rate decreases as the gas supply grows smaller – galaxies can sustain normal star formation for almost the lifetime of the longest-lived stars (Adams and Laughlin, 1997; Kennicutt et al., 1994). Thus, both stellar evolution and star formation will come to an end at approximately the same time in our cosmic future. The universe will be about 100 trillion (10^{14}) years old when the stars finally stop shining. Although our Sun will have long since burned out, this time marks an important turning point for any surviving biospheres – the power available is markedly reduced after the stars turn off.

2.6 The era of degenerate remnants

After the stars burn out and star formation shuts down, a significant fraction of the ordinary mass will be bound within the degenerate remnants that remain after stellar evolution has run its course. For completeness, however, one should keep in mind that the majority of the baryonic matter will remain in the form of hot gas between galaxies in large clusters (Nagamine and Loeb, 2004). At this future time, the inventory of degenerate objects includes brown dwarfs, white dwarfs, and neutron stars. In this context, degeneracy refers to the state of the high-density material locked up in the stellar remnants. At such enormous densities, the quantum mechanical exclusion principle determines the pressure forces that hold up the stars. For example, when most stars die, their cores shrink to roughly the radial size of Earth. With this size, the density of stellar material is about one million times greater than that of the Sun, and the pressure produced by degenerate electrons holds up the star against further collapse. Such objects are white dwarfs and they will contain most of the mass in stellar bodies at this epoch. Some additional mass is contained in brown dwarfs, which are essentially failed stars that never fuse hydrogen, again owing to the effects of degeneracy pressure. The largest stars, those that begin with masses more than eight times that of the Sun, explode at the end of their lives as supernovae. After the explosion, the stellar cores are compressed to densities about one quadrillion times that of the Sun. The resulting stellar body is a neutron star, which is held up by the degeneracy pressure of its constituent neutrons (at such enormous densities, typically *a few* $\times\ 10^{15}$ g/cm^3, electrons and protons combine to form neutrons, which make the star much like a gigantic atomic nucleus). Since only three or four out of every thousand stars are massive enough to produce a supernova explosion, neutron stars will be rare objects.

During this Degenerate Era, the universe will look markedly different from the way it appears now. No visible radiation from ordinary stars will light up the skies, warm the planets, or endow the galaxies with the faint glow they have today. The cosmos will be darker, colder, and more desolate. Against this stark backdrop, events of astronomical interest will slowly take place. As dead stars trace through their orbits, close encounters lead to scattering events, which force the galaxy to gradually readjust its structure. Some stellar remnants are ejected beyond the reaches of the galaxy, whereas others fall in toward the centre. Over the next 10^{20} years, these interactions will enforce the dynamical destruction of the entire galaxy (e.g., Binney and Tremaine, 1987; Dyson, 1979).

In the meantime, brown dwarfs will collide and merge to create new low-mass stars. Stellar collisions are rare because the galaxy is relentlessly empty. During this future epoch, however, the universe will be old enough so that some collisions will occur, and the merger products will often be massive enough to sustain hydrogen fusion. The resulting low-mass stars will then burn for trillions of years. At any given time, a galaxy the size of our Milky Way will harbour a few stars formed through this unconventional channel (compare this stellar population with the approximately 100 billion stars in the Galaxy today).

Along with the brown dwarfs, white dwarfs will also collide at roughly the same rate. Most of the time, such collisions will result in somewhat larger white dwarfs. More rarely, white dwarf collisions produce a merger product with a mass greater than the Chandrasekhar limit. These objects will result in a supernova explosion, which will provide spectacular pyrotechnics against the dark background of the future galaxy.

White dwarfs will contain much of the ordinary baryonic matter in this future era. In addition, these white dwarfs will slowly accumulate weakly interacting dark matter particles that orbit the galaxy in an enormous diffuse halo. Once trapped within the interior of a white dwarf, the particles annihilate each other and provide an important source of energy for the cosmos. Dark matter annihilation will replace conventional nuclear burning in stars as the dominant energy source. The power produced by this process is much lower than that produced by nuclear burning in conventional stars. White dwarfs fuelled by dark matter annihilation produce power ratings measured in quadrillions of Watts, roughly comparable to the total solar power intercepted by Earth (approximately 10^{17} Watts). Eventually, however, white dwarfs will be ejected from the galaxy, the supply of the dark matter will get depleted, and this method of energy generation must come to an end.

Although the proton lifetime remains uncertain, elementary physical considerations suggest that protons will not live forever. Current experiments show that the proton lifetime is longer than about 10^{33} years (Super-Kamiokande Collaboration, 1999), and theoretical arguments (Adams et al., 1998; Ellis et al., 1983; Hawking et al., 1979; Page, 1980; Zeldovich, 1976) suggest that the proton lifetime should be less than about 10^{45} years. Although this allowed range of time scales is rather large, the mass-energy stored within white dwarfs and other degenerate remnants will eventually evaporate when their constituent protons and neutrons decay. As protons decay inside a white dwarf, the star generates power at a rate that depends on the proton lifetime. For a value near the centre of the (large) range of allowed time scales (specifically 10^{37} years), proton decay within a white dwarf generates approximately 400 Watts of power – enough to run a few light bulbs. An entire galaxy of these stars will appear dimmer than our present-day Sun. The process of proton decay converts the mass energy of the particles into radiation, so the white dwarfs evaporate away. As the proton decay process grinds to completion, perhaps 10^{40} years from now, all of the degenerate stellar remnants disappear from the universe. This milestone marks a definitive end to life as we know it, as no carbon-based life can survive the cosmic catastrophe induced by proton decay. Nonetheless, the universe continues to exist, and astrophysical processes continue beyond this end of known biology.

2.7 The era of black holes

After the protons decay, the universe grows even darker and more rarefied. At this late time, roughly when the universe is older than 10^{45} years, the only stellar-like objects remaining are black holes. They are unaffected by proton decay and slide unscathed through the end of the previous era. These objects are often defined to be regions of space-time with such strong gravitational fields that even light cannot escape from their surfaces. But at this late epoch, black holes will be the brightest objects in the sky. Thus, even black holes cannot last forever. They shine ever so faintly by emitting a nearly thermal spectrum of photons, gravitons, and other particles (Hawking, 1974). Through this quantum mechanical process – known as Hawking radiation – black holes convert their mass into radiation and evaporate at a glacial pace (Fig. 2.2). In the far future, black holes will provide the universe with its primary source of power.

Although their energy production via Hawking radiation will not become important for a long time, the production of black holes, and hence the black hole inventory of the future, is set by present-day (and past) astrophysical processes. Every large galaxy can produce millions of stellar black holes, which result from the death of the most massive stars. Once formed, these black holes will endure for up to 10^{70} years. In addition, almost every galaxy harbours a super-massive black hole anchoring its centre; these monsters were produced during the process of galaxy formation, when the universe was only a billion years old, or perhaps even younger. They gain additional mass with time and provide the present-day universe with accretion power. As these large black holes evaporate through the Hawking process, they can last up to 10^{100} years. But even the largest black holes must ultimately evaporate. This Black Hole Era will be over when the largest black holes have made their explosive exits from our universe.

2.8 The Dark Era and beyond

When the cosmic age exceeds 10^{100} years, the black holes will be gone and the cosmos will be filled with the leftover waste products from previous

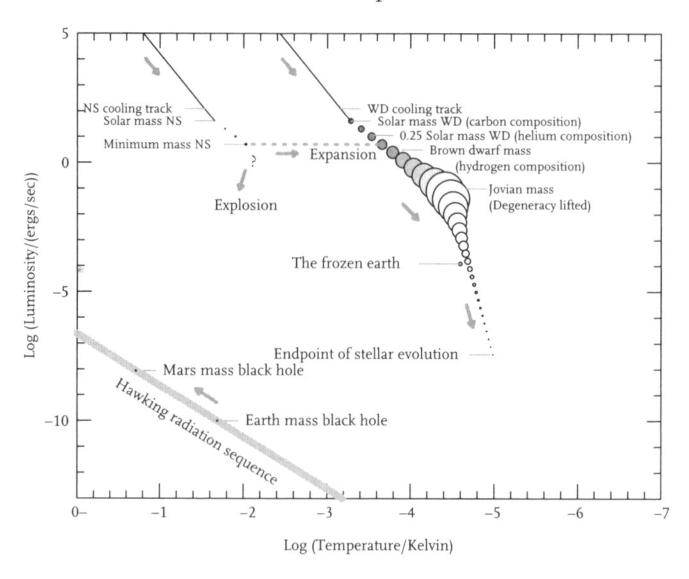

Fig. 2.2 This plot shows the long-term evolution of cold degenerate stars in the H-R diagram. After completing the early stages of stellar evolution, white dwarfs and neutron stars cool to an equilibrium temperature determined by proton decay. This figure assumes that proton decay is driven by gravity (microscopic black holes) on a time scale of 10^{45} years. The white dwarf models are plotted at successive twofold decrements in mass. The mean stellar density (in $\log[\rho/g]$) is indicated by the grey scale shading, and the sizes of the circles are proportional to stellar radius. The relative size of the Earth and its position on the diagram are shown for comparison. The evaporation of a neutron star, starting with one solar mass, is illustrated by the parallel sequence, which shows the apparent radial sizes greatly magnified for clarity. The Hawking radiation sequence for black holes is also plotted. The arrows indicate the direction of time evolution. (Reprinted with permission from Adams, F.C., Laughlin, G., Mbonye, M., and Perry, M.J. (1998). Gravitational demise of cold degenerate stars. *Phys. Rev. D*, 58, 083003.)

eras: neutrinos, electrons, positrons, dark matter particles, and photons of incredible wavelength. In this cold and distant Dark Era, physical activity in the universe slows down, almost (but not quite) to a standstill. The available energy is limited and the expanses of time are staggering, but the universe doggedly continues to operate. Chance encounters between electrons and positrons can forge positronium atoms, which are exceedingly rare in an accelerating universe. In addition, such atoms are unstable and eventually decay. Other low-level annihilation events also take place, for example, between any surviving dark matter particles. In the poverty of this distant epoch, the generation of energy and entropy becomes increasingly difficult.

At this point in the far future, predictions of the physical universe begin to lose focus. If we adopt a greater tolerance for speculation, however, a number of possible events can be considered. One of the most significant potential events is that the vacuum state of the universe could experience a phase transition to a lower energy state. Our present-day universe is observed to be accelerating, and one possible implication of this behaviour is that empty space has a non-zero energy associated with it. In other words, empty space is not really empty, but rather contains a positive value of vacuum energy. If empty space is allowed to have a non-zero energy (allowed by current theories of particle physics), then it remains possible for empty space to have two (or more) different accessible energy levels. In this latter case, the universe could make a transition from its current (high energy) vacuum state to a lower-energy state sometime in the future (the possibility of inducing such a phase transition is discussed in Chapter 16). As the universe grows increasingly older, the probability of a spontaneous transition grows as well. Unfortunately, our current understanding of the vacuum state of the universe is insufficient to make a clear predictions on this issue – the time scale for the transition remains enormously uncertain. Nonetheless, such a phase transition remains an intriguing possibility. If the universe were to experience a vacuum phase transition, it remains possible (but is not guaranteed) that specific aspects of the laws of physics (e.g., the masses of the particles and/or the strengths of the forces) could change, thereby giving the universe a chance for a fresh start.

2.9 Life and information processing

The discussion in this chapter has focused on physical processes that can take place in the far future. But what about life? How far into the future can living organisms survive? Although this question is of fundamental importance and holds enormous interest, our current understanding of biology is not sufficiently well developed to provide a clear answer. To further complicate matters, protons must eventually decay, as outlined above, so that carbon-based life will come to a definitive end. Nonetheless, some basic principles can be discussed if we are willing to take a generalized view of life, where we consider life to be essentially a matter of information processing. This point of view has been pioneered by Freeman Dyson (1979), who argued that the rate of metabolism or information processing in a generalized life form should be proportional to its operating temperature.

If our universe is accelerating, as current observations indicate, then the amount of matter and hence energy accessible to a given universe will be finite. If the operating temperature of life remains constant, then this finite free energy would eventually be used up and life would come to an end. The only chance for continued survival is to make the operating temperature of life decrease. More specifically, the temperature must decrease fast enough to allow for an infinite amount of information processing with a finite amount of free energy.

According to the Dyson scaling hypothesis, as the temperature decreases, the rate of information processing decreases, and the quality of life decreases accordingly. Various strategies to deal with this problem have been discussed, including the issue of digital versus analogous life, maintaining long-term survival by long dormant periods (hibernation), and the question of classical versus quantum mechanical information processing (e.g., Dyson, 1979; Krauss and Starkman, 2000). Although a definitive conclusion has not been reached, the prospects are rather bleak for the continued (infinite) survival of life. The largest hurdle seems to be continued cosmic acceleration, which acts to limit the supply of free energy. If the current acceleration comes to an end, so that the future universe expands more slowly, then life will have a better chance for long-term survival.

2.10 Conclusion

As framed by a well-known poem by Robert Frost, the world could end either in fire or in ice. In the astronomical context considered here, Earth has only a small chance of escaping the fiery wrath of the red giant Sun by becoming dislodged from its orbit and thrown out into the icy desolation of deep space. Our particular world is thus likely to end its life in fire. Given that humanity has a few billion years to anticipate this eventuality, one can hope that migration into space could occur, provided that the existential disasters outlined in other chapters of this book can be avoided. One alternative is for a passing star to wander near the inner portion of our solar system. In this unlikely event, the disruptive gravitational effects of the close encounter could force Earth to abandon its orbit and be exiled from the solar system. In this case, our world would avoid a scalding demise, but would face a frozen future.

A similar fate lies in store for the Sun, the Galaxy, and the Universe. At the end of its life as an ordinary star, the Sun is scheduled to become a white dwarf. This stellar remnant will grow increasingly cold and its nuclei will atrophy to lower atomic numbers as the constituent protons decay. In the long run, the Sun will end up as a small block of hydrogen ice. As it faces its demise, our Galaxy will gradually evaporate, scattering its stellar bodies far and wide. The effective temperature of a stellar system is given by the energies of its stellar orbits. In the long term, these energies will fade to zero and the galaxy will end its life in a cold state. For the universe as a whole, the future is equally bleak, but far more drawn out. The currently available astronomical data indicate that the universe will expand forever, or at least for long enough that the timeline outlined above can play itself out. As a result, the cosmos, considered as a whole, is likely to grow ever colder and face an icy death.

In the beginning, starting roughly fourteen billion years ago, the early universe consisted of elementary particles and radiation – essentially because the background was too hot for larger structures to exist. Here we find that the universe of the far future will also consist of elementary particles and radiation – in this case because the cosmos will be too cold for larger entities to remain intact. From this grand perspective, the galaxies, stars, and planets that populate the universe today are but transient phenomena, destined to fade into the shifting sands of time. Stellar remnants, including the seemingly resilient black holes, are also scheduled to decay. Even particles as fundamental as protons will not last forever. Ashes to ashes, dust to dust, particles to particles – such is the ultimate fate of our universe.

Suggestions for further reading

Adams, F.C. and Laughlin, G. (1997). A dying universe: the long term fate and evolution of astrophysical objects. *Rev. Mod. Phys.*, **69**, pp. 337–372. This review outlines the physics of the long-term future of the universe and its constituent astrophysical objects (advanced level).

Adams, F.C. and Laughlin, G. (1999). *Five Ages of the Universe: Inside the Physics of Eternity* (New York: The Free Press). Provides a popular level account of the future history of the universe.

Ćirković, M.M. (2003). Resource letter Pes-1: physical eschatology. *Am. J. Phys.*, **71**, pp. 122–133. This paper provides a comprehensive overview of the scientific literature concerning the future of the universe (as of 2003). The treatment is broad and also includes books, popular treatments, and philosophical accounts (advanced level).

Dyson, F.J. (1979). Time without end: physics and biology in an open universe. *Rev. Mod. Phys.*, **51**, pp. 447–460. This review represents one of the first comprehensive treatments of the future of the universe and includes discussion of the future of both communication and biology (advanced level).

Islam, J.N. (1983). *The Ultimate Fate of the Universe* (Cambridge: Cambridge University Press). This book provides one of the first popular level accounts of the future of the universe and raises for the first time many subsequently discussed questions.

Rees, M. (1997). *Before the Beginning: Our Universe and Others* (Reading, MA: Addision-Wesley). This book provides a popular level treatment of the birth of the universe and hence the starting point for discussions of our cosmic future.

References

Adams, F.C. and Laughlin, G. (1997). A dying universe: the long term fate and evolution of astrophysical objects. *Rev. Mod. Phys.*, **69**, 337–372.

Adams, F.C., Laughlin, G., and Graves, G.J.M. (2004). Red dwarfs and the end of the main sequence. *Rev. Mexican Astron. Astrophys.*, **22**, 46–49.

Adams, F.C., Laughlin, G., Mbonye, M., and Perry, M.J. (1998). Gravitational demise of cold degenerate stars. *Phys. Rev. D*, **58**, 083003 (7 pages).

Binney, J. and Tremaine, S. (1987). *Galactic Dynamics* (Princeton, NJ: Princeton University Press).

Busha, M.T., Adams, F.C., Evrard, A.E., and Wechsler, R.H. (2003). Future evolution of cosmic structure in an accelerating universe. *Astrophys. J.*, **596**, 713–724.

Caldeira, K. and Kasting, J.F. (1992). The life span of the biosphere revisited. *Nature*, **360**, 721–723.

Ćirković, M.M. (2003). Resource letter PEs-1: physical eschatology. *Am. J. Phys.*, **71**, 122–133.

Dyson, F.J. (1979). Time without end: physics and biology in an open universe. *Rev. Mod. Phys.*, **51**, 447–460.

Ellis, J., Hagelin, J.S., Nanopoulos, D.V., and Tamvakis, K. (1983). Observable gravitationally induced baryon decay. *Phys. Lett.*, **B**124, 484–490.

Garnavich, P.M., Jha, S., Challis, P., Clocchiatti, A., Diercks, A., Filippenko, A.V., Gilliland, R.L., Hogan, C.J., Kirshner, R.P., Leibundgut, B., Phillips, M.M., Reiss, D., Riess, A.G., Schmidt, B.P., Schommer, R.A., Smith, R.C., Spyromilio, J., Stubbs, C., Suntzeff, N.B., Tonry, J., and Carroll, S.M. (1998). Supernova limits on the cosmic equation of state. *Astrophys. J.*, **509**, 74–79.

Hawking, S.W. (1974). Black hole explosions? *Nature*, **248**, 30–31.

Hawking, S.W., Page, D.N., and Pope, C.N. (1979). The propagation of particles in space-time foam. *Phys. Lett.*, **B**86, 175–178.

Kasting, J.F. (1988). Runaway and moist greenhouse atmospheres and the evolution of Earth and Venus. *Icarus*, **74**, 472–494.

Kennicutt, R.C., Tamblyn, P., and Congdon, C.W. (1994). Past and future star formation in disk galaxies. *Astrophys. J.*, **435**, 22–36.

Korycansky, D.G., Laughlin, G., and Adams, F.C. (2001). Astronomical Engineering: a strategy for modifying planetary orbits. *Astrophys. Space Sci.*, **275**, 349–366.

Krauss, L.M. and Starkman, G.D. (2000). Life, the Universe, and Nothing: life and death in an ever-expanding universe. *Astrophys. J.*, **531**, 22–30.

Laughlin, G. and Adams, F.C. (2000). The frozen Earth: binary scattering events and the fate of the Solar System. *Icarus*, **145**, 614–627.

Laughlin, G., Bodenheiemr, P., and Adams, F.C. (1997). The end of the main sequence. *Astrophys. J.*, **482**, 420.

Nagamine, K. and Loeb, A. (2003). Future evolution of nearby large-scale structures in a universe dominated by a cosmological constant. *New Astron.*, **8**, 439–448.

Nagamine, K. and Loeb, A. (2004). Future evolution of the intergalactic medium in a universe dominated by a cosmological constant. *New Astron.*, **9**, 573–583.

Page, D. N. (1980). Particle transmutations in quantum gravity. *Phys. Lett.*, **B**95, 244–246.

Peebles, P.J.E. (1994). Orbits of the nearby galaxies. *Astrophys. J.*, **429**, 43–65.

Perlmutter, S., Aldering, G., Goldhaber, G., Knop, R.A., Nugent, P., Castro, P.G., Deustua, S., Fabbro, S., Goobar, A., Groom, D.E., Hook, I.M., Kim, A.G., Kim, M.Y., Lee, J.C., Nunes, N.J., Pain, R., Pennypacker, C.R., Quimby, R., Lidman, C., Ellis, R.S., Irwin, M., McMahon, R.G., Ruiz-Lapuente, P., Walton, N., Schaefer, B., Boyle, B.J., Filippenko, A.V., Matheson, T., Fruchter, A.S., Panagia, N., Newberg, H.J.M., and Couch, W.J. (1999). Measurements of Ω and Λ from 42 high-redshift supernovae. *Astrophys. J.*, **517**, 565–586.

Riess, A.G., Filippenko, A.V., Challis, P., Clocchiatti, A., Diercks, A., Garnavich, P.M., Gilliland, R.L., Hogan, C.J., Jha, S., Kirshner, R.P., Leibundgut, B., Phillips, M.M.,

Reiss, D., Schmidt, B.P., Schommer, R.A., Smith, R.C., Spyromilio, J., Stubbs, C., Suntzeff, N.B., and Tonry, J. (1998). Observational evidence from supernovae for an accelerating universe and a cosmological constant. *Astron. J.*, **116**, 1009–1038.

Rybicki, K.R. and Denis, C. (2001). On the final destiny of the Earth and the Solar System. *Icarus*, **151**, 130–137.

Sackmann, I.J., Boothroyd, A.I., and Kramer, K.E. (1993). Our Sun III: present and future. *Astrophys. J.*, **418**, 457–468.

Spergel, D.N., Verde, L., Peiris, H.V., Komatsu, E., Nolta, M.R., Bennett, C.L., Halpern, M., Hinshaw, G., Jarosik, N., Kogut, A., Limon, M., Meyer, S.S., Page, L., Tucker, G.S., Weiland, J.L., Wollack, E., and Wright, E.L. (2003). First-year Wilkinson microwave anisotropy probe (WMAP) Observations: determination of cosmological parameters. *Astrophys. J. Suppl.*, **148**, 175–194.

Super-Kamiokande Collaboration. (1999). Search for proton decay through $p \to \nu K^+$ in a large water Cherenkov detector. *Phys. Rev. Lett.*, **83**, 1529–1533.

Zeldovich, Ya.B. (1976). A new type of radioactive decay: gravitational annihilation of baryons. *Phys. Lett.*, **A**59, 254.

·3·

Evolution theory and the future of humanity

Christopher Wills

3.1 Introduction

No field of science has cast more light on both the past and the future of our species than evolutionary biology. Recently, the pace of new discoveries about how we have evolved has increased (Culotta and Pennisi, 2005).

It is now clear that we are less unique than we used to think. Genetic and palaeontological evidence is now accumulating that hominids with a high level of intelligence, tool-making ability, and probably communication skills have evolved independently more than once. They evolved in Africa (our own ancestors), in Europe (the ancestors of the Neanderthals) and in Southeast Asia (the remarkable 'hobbits', who may be miniaturized and highly acculturated *Homo erectus*).

It is also becoming clear that the genes that contribute to the characteristics of our species can be found and that the histories of these genes can be understood. Comparisons of entire genomes have shown that genes involved in brain function have evolved more quickly in hominids than in more distantly related primates.

The genetic differences among human groups can now be investigated. Characters that we tend to think of as extremely important markers enabling us to distinguish among different human groups now turn out to be understandable at the genetic level, and their genetic history can be traced. Recently a single allelic difference between Europeans and Africans has been found (Lamason et al., 2005). This functional allelic difference accounts for about a third of the differences in skin pigmentation in these groups. Skin colour differences, in spite of the great importance they have assumed in human societies, are the result of natural selection acting on a small number of genes that are likely to have no effects beyond their influence on skin colour itself.

How do these and other recent findings from fields ranging from palaeontology to molecular biology fit into present-day evolution theory, and what light do they cast on how our species is likely to evolve in the future?

I will introduce this question by examining briefly how evolutionary change takes place. I will then turn to the role of environmental changes that have resulted in evolutionary changes in the past and extrapolate from those past changes to the changes that we can expect in the short-term and long-term future. These changes will be placed in the context of what we currently know about the evolution of our species. I will group these changes into physical changes and changes that stem from alterations of our own intellectual abilities. I will show that the latter have played and will continue to play a large role in our evolution and in the evolution of other animal and plant species with which we interact. Finally, I will turn to a specific examination of the probable course of future evolution of our species and of the other species on which we depend.

3.2 The causes of evolutionary change

Evolutionary changes in populations, of humans and all other organisms, depend on five factors.

The first and perhaps the most essential is mutation. Evolution depends on the fact that genetic material does not replicate precisely, and that errors are inevitably introduced as genes are passed from one generation to the next. In the absence of mutation, evolutionary change would slow and eventually stop.

The effects of mutations are not necessarily correlated with the sizes of the mutational changes themselves. Single changes in the base sequence of DNA can have no effect or profound effects on the phenotype – the allelic differences that affect skin colour, as discussed in Section 3.1, can be traced to a single alteration in a base from G to A, changing one amino acid in the protein from an alanine to a threonine. At the other end of the spectrum, entire doublings of chromosome number, which take place commonly in plants and less often in animals, can disturb development dramatically – human babies who have twice the normal number of chromosomes die soon after birth. But such doubling can sometimes have little effect on the organism.

A fascinating source of mutation-like changes has recently been discovered. Viruses and other pieces of DNA can transfer genes from one animal, plant, or bacterial species to another, a process known as horizontal gene transfer. Such transfers appear to have played little part in our own recent history, but they have been involved in the acquisition of important, new capabilities in the past: the origin of our adaptive immune system is one remarkable example (Agrawal et al., 1998).

The most important mechanism that decides which of these mutational changes are preserved and which are lost is natural selection. We normally think of natural selection as taking place when the environment changes. But environmental change is not essential to evolution. Darwin realized that natural selection is taking place all the time. In each generation, even if the environment is unchanged, the fittest organisms are the most likely to survive and produce offspring. New mutations will continue to arise, a few of which will enable their carriers to take greater advantage of their environment even if it is not changing.

It is now realized that natural selection often acts to preserve genetic variation in populations. This type of selection, called balancing selection, results from a balance of selective pressures acting on genetic variation. It comes in many forms (Garrigan and Hedrick, 2003). Heterozygote advantage preserves the harmful sickle cell allele in human populations because people who are heterozygous for the allele are better able to resist the effects of malaria. A more prevalent type of balancing selection is frequency-dependent selection, in which a mutant allele may be beneficial when it is rare but loses that benefit as it rises in frequency. Such selection has the capability of maintaining many alleles at a genetic locus in a population. It also has the intriguing property that as alleles move to their internal equilibrium frequencies, the cost of maintaining the polymorphism goes down. This evolutionary "freebie" means that many frequency-dependent polymorphisms can be maintained in a population simultaneously.

Three other factors play important but usually subordinate roles in evolutionary change: genetic recombination, the chance effects caused by genetic drift, and gene flow between populations.

Arguments have been made that these evolutionary processes are having little effect on our species at the present time (Jones, 1991). If so, this is simply because our species is experiencing a rare halcyon period in its history. During the evolutionary eye blink of the last 10,000 years, since the invention of agriculture and the rise of technology, our population has expanded dramatically. The result has been that large numbers of individuals who would otherwise have died have been able to survive and reproduce. I have argued elsewhere (Wills, 1998) and will explore later in this chapter the thesis that even this halcyon period may be largely an illusion. Powerful psychological pressures and new environmental factors (Spira and Multigner, 1998) are currently playing a major role in determining who among us reproduces.

It seems likely that this halcyon period (if it really qualifies as one) will soon come to an end. This book examines many possible scenarios for such resurgence in the strength of natural selection, and in this chapter I will examine how these scenarios might affect our future evolution.

3.3 Environmental changes and evolutionary changes

As I pointed out earlier, evolutionary change can continue even in the absence of environmental change, but its pace is likely to be slow because it primarily 'fine-tunes' the adaptation of organisms that are already well adapted to their environment. When environmental changes occur, they can spur the pace of evolutionary change and can also provide advantages to new adaptations that would not have been selected for in an unchanging environment. What will be the evolutionary effects of environmental change that we can expect in the future? To gauge the likelihood of such effects, we must begin by examining the evolutionary consequences of changes that took place in the past. Let us look first at *completed evolutionary changes*, in which the evolutionary consequences of an environmental change have been fully realized, and then examine *ongoing evolutionary changes* in which the evolutionary changes that result from environmental change have only begun to take place.

3.3.1 Extreme evolutionary changes

Throughout the history of life environmental changes, and the evolutionary changes that result, have sometimes been so extreme as to cause massive extinctions. Nonetheless, given enough time, our planet's biosphere can recover and regain its former diversity. Consider the disaster that hit the Earth 65 million years ago. A brief description of what happened can hardly begin to convey its severity.

One day approximately 65 ± 1 million years ago, without warning, a ten kilometre wide asteroid plunged into the atmosphere above the Yucatan peninsula at a steep angle from the southeast. It traversed the distance from upper stratosphere to the shallow sea in 20 seconds, heating the air ahead of it to a blue-hot plasma. The asteroid hit the ocean and penetrated through the sea bottom, first into the Earth's crust and then into the molten mantle beneath the crust. As it did so, it heated up and exploded. The energy released, equivalent to 100 million megatons of TNT, was at least a million times as great as the largest hydrogen bomb that we humans have ever exploded. The atmospheric shock wave moved at several times the speed of sound across North America, incinerating all the forests in its path. Crust and mantle material erupted towards the sky, cooling and forming an immense choking cloud as it spread outwards. Shock waves raced through the crust, triggering force-10 earthquakes around the planet, and a 300m high tsunami spread further destruction across a wide swath of all the Earth's coastal regions. Volcanoes erupted along the planet's great fault lines, adding their own noxious gases and dust to the witches' brew that was accumulating in the atmosphere.

Most of the animals and plants that lived in southern North America and the northern part of South America were killed by the direct effects of the impact. As the great cloud of dust blanketed the Earth over the next six months, blocking out the sun, many more animals and plants succumbed. There was no safe place on land or sea. Carbon dioxide released from the bolide impact caused a spike in temperatures worldwide (Beerling et al., 2002). All the dinosaurs perished, along with all the large flying and ocean-going reptiles and all the abundant nautilus-like ammonites that had swum in the oceans; of the mammals and birds, only a few survived.

When the dust eventually began to clear the landscape was ghastly and moon-like, with only a few timid ferns poking out of cracks in the seared rocks and soil, and a few tiny mammals and tattered birds surviving on the last of their stores of seeds. It took the better part of a million years for the planet to recover its former verdant exuberance. And it took another 4 million years before new species of mammals filled all the ecological niches that had been vacated by the ruling reptiles.

Asteroid impacts as large as the one that drove the dinosaurs to extinction are rare and have probably happened no more than two or three times during the last half billion years. But at least seven smaller impacts, each sufficiently severe to result in a wave of extinction, have occurred during that period. Each was followed by a recovery period, ranging up to a few million years (though many of the recovery times may have been less than that (Alroy et al., 2001)). During these recovery periods further extinctions took place and new clades of animals and plants appeared.

Asteroids are not the only source of environmental devastation. Massive volcanic eruptions that took place some 251 million years ago were the probable cause of the most massive wave of extinctions our planet has ever seen, the Permian–Triassic extinction (e.g., Benton, 2003). As befits the violence of that extinction event, the resulting alterations in the biosphere were profound. The event set in motion a wave of evolutionary change leading to mammal-like therapsid reptiles (although therapsids with some mammalian characteristics had already appeared before the extinction event). It also gave the ancestors of the dinosaurs an opportunity to expand into vacant ecological niches, though the earliest dinosaurs of which we have records did not appear in the fossil record until 20 million years after the extinction event (Flynn et al., 1999).

Completed catastrophic events are characterized by both mass extinctions and sufficient time for recovery to take place. In general, the more severe the extinction event, more are the differences found to separate the pre-event world from the recovered world. The characteristics of the recovered world are largely shaped by the types of organisms that survived the catastrophe, but complex and unexpected subsequent interactions can take place. Therapsids with some mammalian characteristics survived the Permian–Triassic extinction, and the descendants of these surviving therapsids dominated the world for much of the Triassic period. Nonetheless, halfway through the Triassic, the therapsids began to lose ground. Dinosaurs began to dominate the niches for large land animals, with the result that the mammalian lineages that survived this conflict were primarily small herbivores and insectivores. However, some mammals were able to occupy specialized niches and grow quite large (Ji et al., 2006), and others were sufficiently big and fierce that they were able to prey on small dinosaurs (Hu et al., 2005). The later Cretaceous–Tertiary extinction provided the mammals with the opportunity to take over from the dinosaurs once more.

3.3.2 Ongoing evolutionary changes

A massive extinction event such as the Cretaceous–Tertiary event has a low likelihood of occurrence during any given short time period, and that probability has fluctuated only moderately over the course of the history of Earth, at least before the advent of humanity. But even though such major events are unlikely in the near future, less dramatic environmental changes, many driven by our own activities, are taking place at the present time. Glaciers have advanced and retreated at least eleven times during the last 2.4 million years. The diversity of vertebrates before the onset of this series of ice ages was far greater than the diversity of vertebrates of the present time (Barnosky et al., 2004; Zink and Slowinski, 1995). More recently, human hunting resulted in a wave of large mammal and bird extinctions in the late Pleistocene (Surovell et al., 2005).

There has been a relatively mild decrease in the number of species of mammal, compared with their great abundance in the early Ploicene. This decrease has resulted from Pliocene and Pleistocene climate change and from the human Pleistocene overkill. Now, the decrease is likely to become substantially steeper. It is clear that the relatively mild decrease in the number of species resulting from Pliocene and Pleistocene climate change and from the human Pleistocene overkill is likely to become substantially steeper. Some of the often discussed worst-case scenarios for the future range from the onset of an ice age of such severity that the planet freezes from poles to equator, to a series of nuclear wars or volcanic eruptions that irreversibly poison the atmosphere and oceans. If such terrifying scenarios do not transpire, however, we have a good chance of coming to terms with our environment and slowing the rate of extinction.

As we have seen, alterations in the environment can open up opportunities for evolutionary change as well as close them off through extinction. Even small environmental changes can sometimes have dramatic evolutionary consequences over short spans of time. Three species of diploid flowering plant (*Tragopogon*, Asteraceae) were introduced into western Washington State, North America, from Europe about 100 years ago. Two tetraploid species, arising from different combinations of these diploids, arose without human intervention soon afterwards and have thrived in this area (though such tetraploids have not appeared in their native Europe). DNA studies have shown that a variety of genetic modifications have occurred in these two tetraploids over a few decades (Cook et al., 1998). This report and many similar such stories of rapid recent evolution in both animals and plants indicate that evolutionary changes can take place within the span of a human lifetime.

New analyses of the fossil record suggest that recovery to former diversity levels from even severe environmental disasters may be more rapid than had previously been thought (Alroy et al., 2001). Present-day ecosystem diversity may also be regained rapidly after minor disturbances. We have recently shown that the diversity levels of tropical forest ecosystems are resilient, and that while these forests may not recover easily from severe environmental disasters there is an advantage to diversity that can lead to rapid recovery after limited damage (Wills et al., 2006).

Nothing illustrates the potential for rapid evolutionary response to environmental change in our own species more vividly than the discovery in 2004 of a previously unknown group of hominids, the 'hobbits'. These tiny people, one metre tall, lived on the island of Flores and probably on other islands of what is now Indonesia, as recently as 12,000 years ago (Brown et al., 2004). They have been given the formal name *Homo floresiensis*, but I suspect that it is the name hobbit that will stick. Sophisticated tools found near their remains provide strong evidence that these people, who had brains no larger than those of chimpanzees, were nonetheless expert tool users and hunters. Stone points and blades, including small blades that showed signs of being hafted, were found in the same stratum as the skeletal remains (Brown et al., 2004). Using these tools the hobbits might have been able to kill (and perhaps even help to drive extinct!) the pygmy mastodons with which they shared the islands.

It is probable that the hobbits had physically larger ancestors and that the hobbits themselves were selected for reduced stature when their ancestors reached islands such as Flores, where food was limited. It was this relatively minor change in the physical environment, one that nonetheless had a substantial effect on survival, that selected for the hobbits' reduction in stature. At the same time, new hunting opportunities and additional selective pressures must have driven their ability to fashion sophisticated weapons.

The ancestor of the hobbits may have been *Homo erectus*, a hominid lineage that has remained distinct from ours for approximately 2 million years. But, puzzlingly, features of the hobbits' skeletons indicate that they had retained a mix of different morphologies, some dating back to a period 3 million years ago – long before the evolution of the morphologically different genus *Homo*. The history of the hobbits is likely to be longer and more complex than we currently imagine (Dennell and Roebroeks, 2005).

Determining whether the hobbits are descendants of *Homo erectus*, or of earlier lineages such as *Homo habilis*, requires DNA evidence. No DNA has yet been isolated from the hobbit bones that have been found so far, because the bones are water-soaked and poorly preserved. In the absence of such evidence it is not possible to do more than speculate how long it took the hobbits to evolve from larger ancestors. But, when better-preserved hobbit remains are discovered and DNA sequences are obtained from them, much light will be cast on the details of this case of rapid and continuing evolvability of some of our closest relatives.

The evolution of the hobbits was strongly influenced by the colonization of islands by their ancestors. Such colonizations are common sources of evolutionary changes in both animals and plants. Is it possible that similar colonization events in the future could bring about a similar diversification of human types?

The answer to this question depends on the extent and effect of gene flow among the members of our species. At the moment the differences among human groups are being reduced because of gene flow that has been made possible by rapid and easy travel. Thus it is extremely unlikely that different human groups will diverge genetically because they will not be isolated. But widespread gene flow may not continue in the future. Consider one possible scenario described in the next paragraph.

If global warming results in a planet with a warm pole-to-pole climate, a pattern that was typical of the Miocene, there will be a rapid rise in sea level by eighty metres as all the world's glaciers melt (Williams and Ferrigno, 1999). Depending on the rapidity of the melt, the sea level rise will be accompanied by repeated tsunamis as pieces of the Antarctic ice cap that are currently resting on land slide into the sea. There will also be massive disturbance of the ocean's circulation pattern, probably including the diversion or loss of the Gulf Stream. Such changes could easily reduce the world's arable land substantially – for example, all of California's Central Valley and much of the southeastern United States would be under water. If the changes occur swiftly, the accompanying social upheavals would be substantial, possibly leading to warfare over the remaining resources. Almost certainly there would be substantial loss of human life, made far worse if atomic war breaks out. If the changes occur sufficiently slowly it is possible that alterations in our behaviour and the introduction of new agricultural technologies may soften their impact.

What will be the evolutionary consequences to our species of such changes? Because of the wide current dispersal and portability of human technology, the ability to travel and communicate over long distances is unlikely to be lost completely as a result of such disasters unless the environmental disruption is extreme. But if a rapid decrease in population size were accompanied by societal breakdown and the loss of technology, the result could be geographic fragmentation of our species. There would then be a resumption of the genetic divergence among different human groups that had been taking place before the Age of Exploration (one extreme example of which is the evolution of the hobbits). Only under extremely adverse conditions, however, would the fragmentation persist long enough for distinctly different combinations of genes to become fixed in the different isolated groups. In all but the most extreme scenarios, technology and communication would become re-established over a span of a few generations, and gene flow between human groups would resume.

3.3.3 Changes in the cultural environment

Both large and small changes in the physical environment can bring about evolutionary change. But even in the absence of such changes, cultural selective pressures that have acted on our species have had a large effect on our evolution and will continue to do so. To understand these pressures, we must put them into the context of hominid history. As we do so, we will see that cultural change has been a strong driving force of human evolution, and has also affected many other species with which we are associated.

About 6 million years ago, in Africa, our evolutionary lineage separated from the lineage that led to chimpanzees and bonobos. The recent discovery in Kenya of chimpanzee teeth that are half a million years old (McBrearty and Jablonski, 2005) shows that the chimpanzee lineage has remained distinct for most of that time from our own lineage, though the process of actual separation of the gene pools may have been a complicated one (Patterson et al., 2006). There is much evidence that our remote ancestors were morphologically closer to chimpanzees and bonobos than to modern humankind ourselves. The early hominid *Ardipithecus ramidus*, living in East Africa 4.4 million years ago, had skeletal features and a brain size resembling those of chimpanzees (White et al., 1994). Its skeleton differed from those of chimpanzees in only two crucial respects: a slightly more anterior position of the foramen magnum, which is the opening at the bottom of the skull through which the spinal cord passes, and molars with flat crowns like those of modern humans rather than the highly cusped molars of chimpanzees. If we could resurrect an *A. ramidus* it would probably look very much like a chimpanzee to us – though there is no doubt that *A. ramidus* and present-day chimpanzees would not recognize each other as members of the same species.

Evolutionary changes in the hominid line include a gradual movement in the direction of upright posture. The changes required a number of coordinated alterations in all parts of the skeleton but in the skull and the pelvis in particular. Perhaps the most striking feature of this movement towards upright posture is how gradual it has been. We can trace the change in posture through the gradual anterior movement of the skull's foramen magnum that can be seen to have taken place from the oldest hominid fossils down to the most recent.

A second morphological change is of great interest. Hominid brains have undergone a substantial increase in size, with the result that modern human brains have more than three times the volume of a chimpanzee brain. Most of these increases have taken place during the last 2.5 million years of our history. The increases took place not only in our own immediate lineage but also in at least one other extinct lineage that branched off about a million years ago – the lineage of Europe and the Middle East that included the pre-Neanderthals and Neanderthals. It is worth emphasizing, however, that this overall evolutionary 'trend' may have counterexamples, in particular the apparent substantial reduction in both brain and body size in the *H. floresiensis* lineage. The remarkable abilities of the hobbits make it clear that brain size is not the only determiner of hominid success.

Our ability to manipulate objects and thus alter our environment has also undergone changes during the same period. A number of changes in the structure of hominid hands, such as the increase in brain size beginning at least 2.5 million years ago, have made them more flexible and sensitive. And our ability to communicate, too, has had a long evolutionary history, reflected in both physical and behavioural characteristics. The Neanderthals had a voice box indistinguishable from our own, suggesting that they were capable of speech (Arensburg et al., 1989). The ability of human children to learn a complex language quickly (and their enthusiasm for doing so) has only limited counterparts in other primates. Although some chimpanzees, bonobos and gorillas have shown remarkable understanding of human spoken language, their ability to produce language and to teach language to others is severely limited (Tagliatela et al., 2003).

Many of the changes in the hominid lineage have taken place since the beginning of the current series of glaciations 2.5 million years ago. There has been an accelerating increase in the number of animal and plant extinctions worldwide during this period. These include extinctions in our own lineage, such as those of *H. habilis*, *H. erectus* and *H. ergaster*, and more recently the Neanderthals and *H. floresiensis*. These extinctions have been accompanied by a rapid rate of evolution in our own lineage.

What are the cultural pressures that have contributed to this rapid evolution? I have argued elsewhere (Wills, 1993) that a feedback loop involving our brains, our bodies, our genes, and our rapidly changing cultural environment has been an important contributor to morphological and behavioural changes. Feedback loops are common in evolution and have led to many extreme results of sexual selection in animals and of interactions with pollinators among flowering plants. A 'runaway brain' feedback can explain why rapid changes have taken place in the hominid lineage.

The entire human and chimpanzee genomes are now available for comparison, opening up an astounding new world of possibilities for scientific investigation (Chimpanzee Sequencing and Analysis Consortium, 2005). Overall comparisons of the sequences show that some 10 million genetic changes separate us from chimpanzees. We have hardly begun to understand which of these changes have played the most essential role in our evolution. Even at such early stages in these genome-wide investigations, however, we can measure the relative rate of change of different classes of genes as they have diverged in the two lineages leading to humans and chimpanzees. It is now possible to examine the evolution of genes that are involved in brain function in the hominid lineage and to compare these changes with the evolution of the equivalent (homologous) genes in other primates.

The first such intergenomic comparisons have now been made between genes that are known to be involved in brain growth and metabolism and genes that affect development and metabolic processes in other tissues of the body. Two types of information have emerged, both of which demonstrate the rapid evolution of the hominid lineage.

First, the genes that are expressed in brain tissue have undergone more regulatory change in the human lineage than they have in other primate lineages. Gene regulation determines whether and when a particular gene is expressed in a particular tissue. Such regulation, which can involve many different interactions between regulatory proteins and stretches of DNA, has a strong influence on how we develop from embryo to adult. As we begin to understand some of these regulatory mechanisms, it is becoming clear that they have played a key role in many evolutionary changes, including major changes in morphology and behaviour. We can examine the rate of evolution of these regulatory changes by comparing the ways in which members of this class of genes are expressed in the brains of ourselves and of our close relatives (Enard et al., 2002). One pattern that often emerges is that a given gene may be expressed at the same level (say high or low) in both chimpanzees and rhesus monkeys, but at a different level (say intermediate) in humans. Numerous genes show similar patterns, indicating that their regulation has undergone significantly more alterations in our lineage than in those of other primates. Unlike the genes involved in brain function, regulatory changes have not occurred preferentially in the hominid lineage in genes expressed in the blood and liver.

Second, genes that are implicated in brain function have undergone more meaningful changes in the human lineage than in other lineages. Genes that code for proteins undergo two types of changes: *non-synonymous* changes that alter the proteins that the genes code for, possibly changing their function, and *synonymous* changes that change the genes but have no effect on the proteins. When genes that are involved in brain function are compared among different mammalian lineages, significantly more potentially functional changes have occurred in the hominid lineage than in the other lineages (Clark et al., 2003). This finding shows clearly that in the hominid lineage strong natural selection has changed genes involved in brain function more rapidly than the changes that have taken place in other lineages.

Specific changes in genes that are involved in brain function can now be followed in detail. Evidence is accumulating that six *microcephalin* genes are involved in the proliferation of neuroblasts during early brain development. One of these genes, *MCPH*1, has been found to carry a specific haplotype at high frequency throughout human populations (Evans et al., 2005), and it has reached highest frequency in Asia (Fig. 3.1). The haplotype has undergone some further mutations and recombinations since it first arose about 37,000 years ago, and these show strong linkage disequilibrium. Other alleles at this

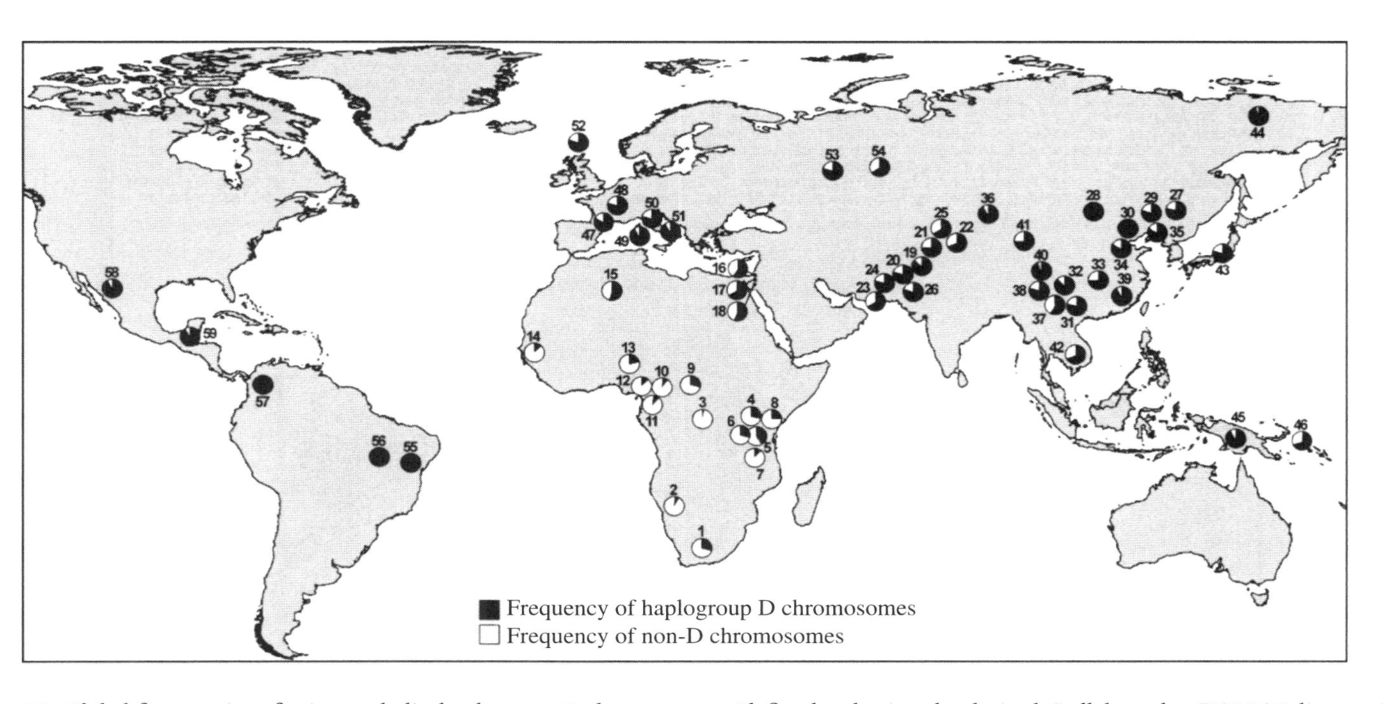

Fig. 3.1 Global frequencies of microcephalin haplogroup D chromosomes (defined as having the derived C allele at the G37995C diagnostic SNP) in a panel of 1184 individuals.

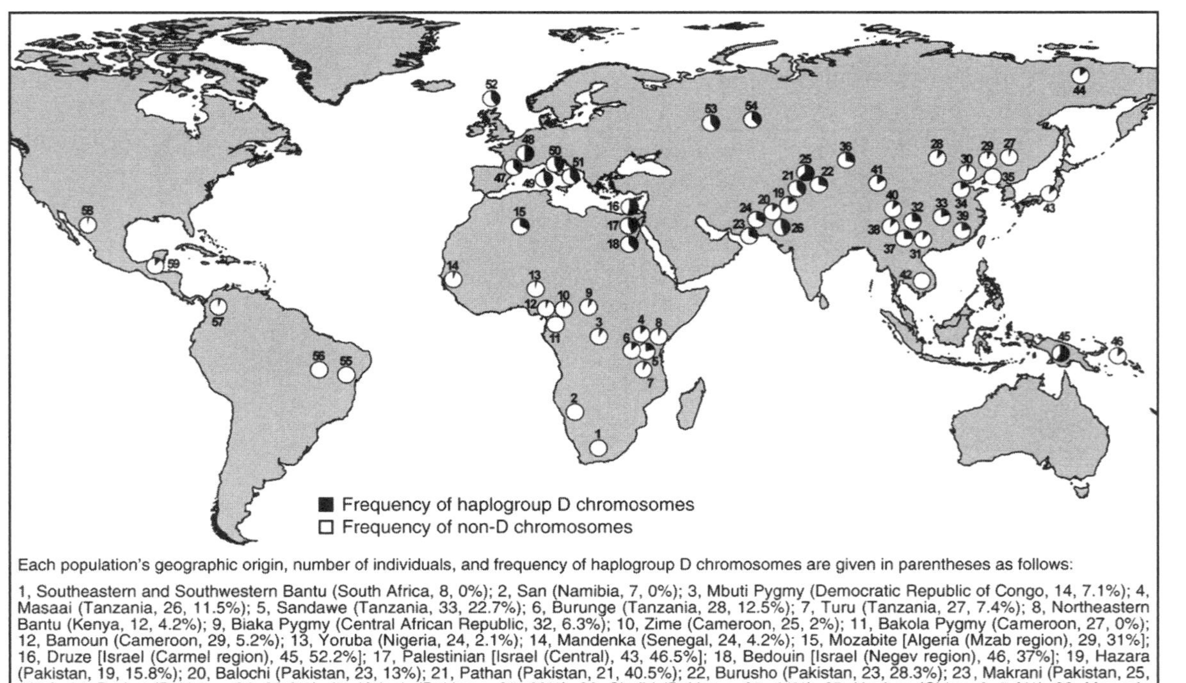

Frequency of haplogroup D chromosomes
Frequency of non-D chromosomes
Each population's geographic origin, number of individuals, and frequency of haplogroup D chromosomes are given in parentheses as follows:
1, Southeastern and Southwestern Bantu (South Africa, 8, 0%); 2, San (Namibia, 7, 0%); 3, Mbuti Pygmy (Democratic Republic of Congo, 14, 7.1%); 4, Masaai (Tanzania, 26, 11.5%); 5, Sandawe (Tanzania, 33, 22.7%); 6, Burunge (Tanzania, 28, 12.5%); 7, Turu (Tanzania, 27, 7.4%); 8, Northeastern Bantu (Kenya, 12, 4.2%); 9, Biaka Pygmy (Central African Republic, 32, 6.3%); 10, Zime (Cameroon, 25, 2%); 11, Bakola Pygmy (Cameroon, 27, 0%); 12, Bamoun (Cameroon, 29, 5.2%); 13, Yoruba (Nigeria, 24, 2.1%); 14, Mandenka (Senegal, 24, 4.2%); 15, Mozabite [Algeria (Mzab region), 29, 31%]; 16, Druze [Israel (Carmel region), 45, 52.2%]; 17, Palestinian [Israel (Central), 43, 46.5%]; 18, Bedouin [Israel (Negev region), 46, 37%]; 19, Hazara (Pakistan, 19, 15.8%); 20, Balochi (Pakistan, 23, 13%); 21, Pathan (Pakistan, 21, 40.5%); 22, Burusho (Pakistan, 23, 28.3%); 23, Makrani (Pakistan, 25, 32%); 24, Brahui (Pakistan, 24, 33.3%); 25, Kalash (Pakistan, 25, 60%); 26, Sindhi (Pakistan, 25, 44%); 27, Hezhen (China, 9, 5.6%); 28, Mongola (China, 9, 11.1%); 29, Daur (China, 8, 6.3%); 30, Orogen (China, 10, 5%); 31, Miaozu (China, 10, 10%); 32, Yizu (China, 10, 25 %); 33, Tujia (China, 10, 20%); 34, Han (China, 42, 17.9%); 35, Xibo (China, 9, 0%); 36, Uygur (China, 10, 30%); 37, Dai (China, 10, 25%); 38, Lahu (China, 10, 10%); 39, She (China, 7, 21.4%); 40, Naxi (China, 9, 11.1%); 41, Tu (China, 9, 16.7%); 42, Cambodian (Cambodia, 10, 0%); 43, Japanese (Japan, 28, 10.7%); 44, Yakut [Russia (Siberia region), 24, 12.5%]; 45, Papuan (New Guinea, 16, 59.4%); 46, NAN Melanesian (Bougainville, 18, 11.1%); 47, French Basque (France, 15, 40%); 48, French (France, 29, 50%); 49, Sardinian (Italy, 27, 46.3%); 50, North Italian [Italy (Bergamo region), 12, 45.8%]; 51, Tuscan (Italy, 8, 37.5%); 52, Orcadian (Orkney Islands, 16, 40.6%); 53, Russian (Russia, 25, 38%); 54, Adygei [Russia (Caucasus region), 15, 40%]; 55, Karitiana (Brazil, 24, 0%); 56, Surui (Brazil, 21, 0%); 57, Colombian (Colombia, 13, 3.8%); 58, Pima (Mexico, 25, 2%); 59, Maya (Mexico, 24, 12.5%).

Fig. 3.2 Worldwide frequencies of ASPM haplogroup D chromosomes (defined as having the derived G allele at the A44871G diagnostic polymorphism), based on a panel of 1186 individuals.

locus do not show such a pattern of disequilibrium. Because disequilibrium breaks down with time, it is clear that this recent haplotype has spread as a result of strong natural selection.

Another gene also associated with microcephaly, *abnormal spindle-like microcephaly-associated* (*ASPM*), shows even more recent evidence of extremely strong selection (Mekel-Bobrov et al., 2005). An allele found chiefly in Europe and the Middle East, but at much lower frequencies in Asia (Fig. 3.2), appears to have arisen as recently as 5800 years ago. The spread of this allele has been so rapid that it must confer a selective advantage of several percent on its carriers.

Although both these alleles carry non-synonymous base changes, the allelic differences that are being selected may not be these changes but may be in linked regulatory regions. Further, a direct effect of these alleles on brain function has not been demonstrated. Nonetheless, the geographic patterns seen in these alleles indicate that natural selection is continuing to act powerfully on our species.

Hawks and coworkers (Hawks et al. 2007) have recently shown that the pattern of strong recent selection detectable by linkage disequilibrium extends to more than 2,000 regions of the human genome. They calculate that over the last 40,000 years our species has evolved at a rate 100 times as fast as our previous evolution. Such a rapid rate may require that many of the newly selected genes are maintained by frequency-dependent selection, to reduce the selective burden on our population.

3.4 Ongoing human evolution

All these pieces of evidence from our past history show that humans have retained the ability to evolve rapidly. But are we continuing to evolve at the present time? Without a doubt! Although the evolutionary pressures that are acting on us are often different from those that acted on our ancestors a million years ago or even 10,000 years ago, we are still exposed to many of the same pressures. Selection for resistance to infectious disease continues. We have conquered, at least temporarily, many human diseases, but many others – such as tuberculosis, AIDS, malaria, influenza, and many diarrhoeal diseases – continue to be killers on a massive scale (see Chapter 14, this volume). Selection for resistance to these diseases is a continuing process because the disease organisms themselves are also evolving.

3.4.1 Behavioural evolution

Although the selective effects of psychological pressures are difficult to measure, they must also play a role. Michael Marmot and his colleagues have shown that an individual's position in a work hierarchy has an impact on his or her health (Singh-Manoux et al., 2005). The members at the top level of the hierarchy live healthier lives than those at the bottom level – or even those who occupy positions slightly below the top level!

Reproduction also applies psychological pressures. The introduction of effective means of birth control has provided personal choice in reproduction for more people than at any time in the past. We can only surmise how selection for genes that influence reproductive decision-making will affect our future evolution. There is some evidence from twin studies, however, that heritability for reproductive traits, especially for age at first reproduction, is substantial (Kirk et al., 2001). Thus, rapid changes in population growth rates such as those being experienced in Europe and the former Soviet Union are likely to have evolutionary consequences.

Concerns over dysgenic pressures on the human population resulting from the build-up of harmful genes (Muller, 1950) have abated. It is now realized that even in the absence of selection, harmful mutant alleles will accumulate only slowly in our gene pool and that many human characteristics have a complex genetic component so that it is impossible to predict the effects of selection on them. Alleles that are clearly harmful – sickle cell, Tay-Sachs, muscular dystrophy and others – will soon be amenable to replacement by functional alleles through precise gene surgery performed on germ line cells. Such surgery, even though it will be of enormous benefit to individuals, is unlikely to have much of an effect on our enormous gene pool unless it becomes extremely inexpensive.

One intriguing direction for current and future human evolution that has received little or no attention is selection for intellectual diversity. The measurement of human intellectual capabilities, and their heritability, is at a primitive stage. The heritability of IQ has received a great deal of attention, but a recent meta-analysis estimates broad heritability of IQ to be 0.5 and narrow heritability (the component of heritability that measures selectable phenotypic variation) to be as low as 0.34 (Devlin et al., 1997). But IQ is only one aspect of human intelligence, and other aspects of intelligence need investigation. Daniel Goleman has proposed that social intelligence, the ability to interact with others, is at least as important as IQ (Goleman, 1995), and Howard Gardner has explored multiple intelligences ranging from artistic and musical through political to mechanical (Gardner, 1993). All of us have a different mix of such intelligences. To the extent that genes contribute to them, these genes are likely to be polymorphic – that is, to have a number of alleles, each at appreciable frequency, in the human population.

This hypothesis of polymorphic genes involved in behaviour leads to two predictions. First, loci influencing brain function in humans should have more alleles than the same loci in chimpanzees. Note, however, that most of the functional polymorphic differences are likely to be found, not in structural genes, but in the regulatory regions that influence how these genes are expressed. Because of the difficulty of determining which genetic differences in the polymorphisms are responsible for the phenotypic effects, it may be some time before this prediction can be tested.

The second prediction is that some type of balancing selection, probably with a frequency-dependent component, is likely to be maintaining these alleles in the human population.

When an allele has an advantage if it is rare but loses that advantage if it is common, it will tend to be maintained at the frequency at which there is neither an advantage nor a disadvantage. If we suppose that alleles influencing many behaviours or skills in the human population provide an advantage when they are rare, but lose that advantage when they are common, there will be a tendency for the population to accumulate these alleles at such intermediate frequencies. And, as noted earlier, if these genes are maintained by frequency dependence, the cost to the population of maintaining this diversity can be low.

Numerous examples of frequency-dependent balancing selection have been found in populations. One that influences behaviours has been found in *Drosophila melanogaster*. Natural populations of this fly are polymorphic for two alleles at a locus (*for*, standing for forager) that codes for a protein kinase. A recessive allele at this locus, *sitter*, causes larvae to sit in the same spot while feeding. The dominant allele, *rover*, causes its carriers to move about while feeding. Neither allele can take over (reach fixation) in the population. *Rover* has an advantage when food is scarce, because *rover* larvae can find more food and grow more quickly than *sitter* larvae. *Sitter* has an advantage when food is plentiful. If they are surrounded by abundance, sitter larvae that do not waste time and effort moving about can mature more quickly than rovers (Sokolowski et al., 1997).

It will be fascinating to see whether behaviour-influencing polymorphisms such as those at the *for* locus are common in human populations. If so, one type of evolutionary change may be the addition of new alleles at these loci as our culture and technology become more complex and opportunities for new types of behaviours arise. It is striking that the common *MCPH*1 allele has not reached fixation in any human population, even though it has been under positive selection since before modern humans spread to Europe. It may be that this allele is advantageous when it is rare but loses that advantage when it becomes common. There appear to be no behavioral effects associated with this allele (Mekel-Bobrov et al., 2007), but detailed studies may reveal small differences. Natural selection can work on small phenotypic differences as well as large, and much of our recent evolution may have resulted from selection for genes with small effects.

3.4.2 The future of genetic engineering

There has been much speculation about the effects of genetic engineering on the future of our species, including the possibility that a 'genetic elite' may emerge that would benefit from such engineering to the exclusion of other human groups (e.g., Silver, 1998). Two strong counterarguments to this viewpoint can be made.

First, the number of genes that can potentially be modified in our species is immense. Assuming 50,000 genes per diploid human genome and 6 billion individuals, the number of genes in our gene pool is 3×10^{14}. The task of changing even a tiny fraction of these genes would be enormous, especially since each such change could lead to dangerous and unexpected side effects. It is far more likely that our growing understanding of gene function will enable us to design specific drugs and other compounds that can produce desirable changes in our phenotypes and that these changes will be sufficiently easy and inexpensive that they will not be confined to a genetic elite (Wills, 1998). Even such milder phenotypic manipulations are fraught with danger, however, as we have seen from the negative effects that steroid and growth hormone treatments have had on athletes.

Second, the likelihood that a 'genetic elite' will become established seems remote. The modest narrow (selectable) heritability of IQ mentioned earlier shows the difficulty of establishing a genetic elite through selection. Heritabilities that are even lower are likely to be the rule for other physical or behavioural characteristics that we currently look upon as desirable.

Attempts to establish groups of clones of people with supposedly desirable characters would also have unexpected and unpredictable effects, in this case because of the environment. Clones of Bill Gates or Mother Teresa, growing up at a different time and in a different place, would turn into people who reflected the influences of their unique upbringing, just as the originals of such hypothetical clones did. And, luckily, environmental effects can work to defang evil dysgenic schemes as well as utopian eugenic ones. 'The Boys from Brazil' notwithstanding, it seems likely that if clones of Adolf Hitler were to be adopted into well-adjusted families in healthy societies they would grow up to be nice, well-adjusted young men.

3.4.3 The evolution of other species, including those on which we depend

Discussions of human evolution have tended to ignore the fact that we have greatly influenced the evolution of other species of animals and plants. These species have in turn influenced our own evolution. The abundant cereals that made the agricultural revolution possible were produced by unsung generations of primitive agriculturalists who carried out a process of long-continued artificial selection. Some results of such extremely effective selection are seen in Indian corn, which is an almost unrecognizable descendent of the wild grass teosinte, and domesticated wheat, which is an allohexaploid with genetic contributions from three different wild grasses. The current immense human population depends absolutely on these plants and also on other plants and animals that are the products of thousands of generations of artificial selection.

One consequence of climate change such as global warming is that the agriculture of the future will have to undergo rapid adaptations (see also Chapter 13, this volume). Southern corn leaf blight, a fungus that severely damaged corn production in the Southeast United States during the 1970s, was controlled by the introduction of resistant strains, but only after severe losses. If the climate warms, similar outbreaks of blight and other diseases that are prevalent in tropical and subtropical regions will become a growing threat to the world's vast agricultural areas.

Our ability to construct new strains and varieties of animals and plants that are resistant to disease, drought, and other probable effects of climate change depends on the establishment and maintenance of stocks of wild ancestral species. Such stocks are difficult to maintain for long periods because governments and granting agencies tend to lose interest and because societal upheavals can sometimes destroy them. Some of the stocks of wild species related to domestic crops that were collected by Russian geneticist Nikolai Vavilov in the early part of the twentieth century have been lost, taking with them an unknown number of genes of great potential importance. It may be possible to avoid such losses in the future through the construction of multiple seed banks and gene banks to safeguard samples of the planet's genetic diversity. The Norwegian government recently opened a bunker on Svalbard, an Arctic island, designed to hold around two million seeds, representing all known varieties of the world's crops. Unfortunately, however, there are no plans to replicate this stock centre elsewhere.

Technology may aid us in adjusting to environmental change, provided that our technological capabilities remain intact during future periods of rapid environmental change. To cite one such example, a transgenic tomato strain capable of storing excess salt in its leaves while leaving its fruit relatively salt-free has been produced by overexpression of an *Arabidopsis* transporter gene. The salt-resistant plants can grow at levels of salt fifty times higher than those found in normal soil (Zhang and Blumwald, 2001). The ability to produce crop plants capable of growing under extreme environmental conditions may enable us to go on feeding our population even as we are confronted with shrinking areas of arable land.

3.5 Future evolutionary directions

Even a large global catastrophe such as a ten kilometre asteroidal/cometary impact would not spell doom for our species if we could manage to spread to other solar systems by the time the impactor arrives. We can, however, postulate a number of scenarios, short of extinction, that will test our ability to survive as a species. I will not discuss here scenarios involving intelligent machines or more radical forms of technology-enabled human transformation.

3.5.1 Drastic and rapid climate change without changes in human behaviour

If climatic change either due to slow (e.g., anthropogenic) or due to sudden (e.g., supervolcanic) causative agents is severe and the survivors are few in number, there may be no time to adapt. The fate of the early medieval Norse colonists in Greenland, who died out when the climate changed because they could not shift from eating meat to eating fish (Berglund, 1986) stands as a vivid example. Jared Diamond (2005) has argued in a recent book that climate change has been an important factor in several cases of societal collapse in human history.

3.5.2 Drastic but slower environmental change accompanied by changes in human behaviour

If the environmental change occurs over generations rather than years or decades, there may be time for us to alter our behaviours deliberately. These behavioural changes will not be evolutionary changes, at least at first, although as we will see the ability to make such changes depends on our evolutionary and cultural history. Rather they will consist of changes in *memes* (Dawkins, 1976), the non-genetic learned or imitated behaviours that form an essential part of human societies and technology. The change that will have the largest immediate effect will be population control, through voluntary or coerced means or both. Such changes are already having an effect. The one-child policy enforced by the Chinese government, imperfect though it is in practice, has accelerated that country's demographic transition and helped it to gain a 20-fold increase in per capita income over the last quarter century.

Such demographic transitions are taking place with increasing rapidity in most parts of the world, even in the absence of government coercion. Predictions of a human population of 12 billion by 2050 were made by the United Nations in 1960. These frightening projections envisioned that our population would continue to increase rapidly even after 2050. These estimates have now been replaced by less extreme predictions that average nine billion by 2050, and some of these revised projections actually predict a slow decline in world population after mid-century. Demographic transitions in sub-Saharan Africa and South Asia will lag behind the rest of the planet, but there is no reason to suppose that these regions will not catch up during this century as education, particularly the education of women, continues to spread. These demographic transitions are unlikely to be reversed in the future, as long as education continues to spread. Recently the chief of a small village on the remote island of Rinca in Indonesia complained to me that all his six children wanted to go to medical school, and he could not imagine how he could send them all.

Accompanying the demographic transitions will be technological changes in how the planet is fed. If rising ocean levels cause the loss of immense amounts of arable land (including major parts of entire countries, such as low-lying Bangladesh), hordes of refugees will have to be fed and housed. Technology and alterations in our eating habits hold out some hope. Soy protein is similar in amino acid content to animal proteins, and its production has increased 400% in the past thirty years. This crop is already beginning to change our eating habits. And new agricultural infrastructure, such as intense hydroponic agriculture carried out under immense translucent geodesic domes with equipment for recycling water, will rapidly become adopted when the alternative is starvation.

Little will be done to confront these problems without a series of catastrophic events that make it clear even to the most reactionary societies and governments that drastic change is needed. These catastrophes are already beginning to throw into sharp relief current social behaviours that are inadequate for future challenges, such as a disproportionate use of the world's limited resources by particular countries and restrictions on the free flow of information by dictatorial governments. Societal models based on national self-interest or on the preservation of power by a few will prove inadequate in the face of rapid and dramatic environmental changes. I will predict that – in spite of the widespread resistance to the idea – a global governmental organization with super-national powers, equivalent on a global scale to the European Union, will inevitably emerge. As we confront repeated catastrophes, we will see played out in the economic realm a strong selection against individual and societal behaviours that cannot be tolerated in a world of scarcity.

Discomfiting as such predictions may be to some, the accelerating rate of environmental change will make them inevitable. If we are to retain a substantial human population as the planet alters, our behaviours must alter as well. Otherwise, our societal upheavals may result in long-term ecological damage that can be reversed only after tens or hundreds of thousands of years.

Scream and argue and fight as we may about how to behave in the future, our past evolutionary history has provided most of us with the ability to change how we behave. This is our remarkable strength as a species.

3.5.3 Colonization of new environments by our species

A third future scenario, that of the colonization of other planets, is rapidly moving from the realm of science fiction to a real possibility. More than 200 extrasolar planets have been discovered in the last decade. These are mostly Jupiter-sized or larger, but it is safe to predict that within years or decades Earth-sized extrasolar planets, some of them showing evidence of life, will be found. The smallest extrasolar planet yet found is a recently discovered mere

5-Earth masses companion to Gliese 581 which lies in the habitable zone and is likely to possess surface water (Beaulieu et al., 2006). In view of the selection effects applicable to the surveys thus far, the discoveries of even smaller planets are inevitable.

This prediction is at variance with the argument presented by Ward and Brownlee (2000) that planets harbouring complex life are likely to be extremely rare in our galaxy. However, that argument was based on a biased interpretation of the available data and the most restrictive view on how planetary systems form (Kasting, 2001).

No more exciting moment of scientific discovery can be imagined than when we first obtain an image or spectrum of an Earth-sized planet with an oxygen-rich atmosphere circling a nearby star (soon to become possible with the advent of Darwin, Kepler, Gaia and several other terrestrial planet-seeking missions in the next several years). It is likely that the challenge of visiting and perhaps colonizing these planets is one that we as a species will be unable to resist.

The colonization of other planets will result in an explosive Darwinian adaptive radiation, involving both our species and the animals and plants that accompany us. Just as the mammals radiated into new ecological niches after the extinction of the dinosaurs, and the finches and land tortoises that Darwin encountered on the Galápagos Islands radiated adaptively as they spread to different islands, we will adapt in different ways to new planets that we explore and colonize.

The new planetary environments will be different indeed. How will we be able to colonize new planets peopled with indigenous life forms that have a different biochemistry and mechanism of inheritance from us? Could we, and the animals and plants that we bring with us, coexist with these indigenous and highly adapted life forms? Could we do so without damaging the ecology of the planets we will be occupying? And how will competition with these life forms change us? Will we be able to direct and accelerate these changes to ourselves by deliberately modifying the genes of these small populations of colonists?

If we are able to adapt to these new environments, then 10,000 or 100,000 years from now our species will be spread over so wide a region that no single environment-caused disaster would be able to wipe us all out. But our continued existence would still be fraught with danger. Will we, collectively, still recognize each other as human? What new and needless prejudices will divide us, and what new misunderstandings will lead to pointless conflict?

Suggestions for further reading

Kareiva, P., Watts, S., McDonald, R., and Boucher, T. (2007). Domesticated nature: shaping landscapes and ecosystems for human welfare. *Science*, **316**, 1866–1869.

The likelihood that we will permanently alter the world's ecosystems for our own benefit is explored in this paper.

Myers, N. and Knoll, A.H. (2001). The biotic crisis and the future of evolution. *Proc. Natl. Acad. Sci. (USA)*, **98**, 5389–5392. Discomfiting predictions about the course of future evolution can be found in this paper.

Palumbi, S.R. (2001). Humans as the world's greatest evolutionary force. *Science*, **293**, 1786–1790. The influence of humans on the evolution of other organisms is examined.

Unfortunately, none of these authors deals with the consequences of changes in evolutionary pressures on our own species.

References

Agrawal, A., Eastman, Q.M., and Schatz, D.G. (1998). Implications of transposition mediated by V(D) J-recombination proteins RAG1 and RAG2 for origins of antigen-specific immunity. *Nature*, **394**, 744–751.

Alroy, J., Marshall, C.R., Bambach, R.K., Bezusko, K., Foote, M., Fursich, F.T., Hansen, T.A., Holland, S.M., Ivany, L.C., Jablonski, D., Jacobs, D.K., Jones, D.C., Kosnik, M.A., Lidgard, S., Low, S., Miller, A.I., Novack-Gottshall, P.M., Olszewski, T.D., Patzkowsky, M.E., Raup, D.M., Roy, K., Sepkoski, J.J., Jr, Ommers, M.G., Wagner, P.J., and Webber, A. (2001). Effects of sampling standardization on estimates of Phanerozoic marine diversification. *Proc. Natl. Acad. Sci. (USA)*, **98**, 6261–6266.

Arensburg, B., Tillier, A.M., Vandermeersch, B., Duday, H., Schepartz, L.A., and Rak, Y. (1989). A Middle Paleolithic hyoid bone. *Nature*, **338**, 758–760.

Barnosky, A.D., Bell, C.J., Emslie, S.D., Goodwin, H.T., Mead, J.I., Repenning, C.A., Scott, E., and Shabel, A.B. (2004). Exceptional record of mid-Pleistocene vertebrates helps differentiate climatic from anthropogenic ecosystem perturbations. *Proc. Natl. Acad. Sci. (USA)*, **101**, 9297–9302.

Beaulieu, J.-P., Bennett, D.P., Fouque, P., Williams, A., Dominik, M., Jørgensen, U.G., Kubas, D., Cassan, A., Coutures, C., Greenhill, J., Hill, K., Menzies, J., Sackett, P.D., Albrow, M., Brillant, S., Caldwell, J.A.R., Calitz, J.J., Cook, K.H., Corrales, E., Desort, M., Dieters, S., Dominis, D., Donatowicz, J., Hoffman, M., Kane, S., Marquette, J.-B., Martin, R., Meintjes, P., Pollard, K., Sahu, K., Vinter, C., Wambsganss, J., Woller, K., Horne, K., Steele, I., Bramich, D.M., Burgdorf, M., Snodgrass, C., Bode, M., Udalski, A., Szymaski, M.K., Kubiak, M., Wickowski, T., Pietrzyski, G., Soszyski, I., Szewczyk, O., Wyrzykowski, L., Paczyski, B., Abe, F., Bond, I.A., Britton, T.R., Gilmore, A.C., Hearnshaw, J.B., Itow, Y., Kamiya, K., Kilmartin, P.M., Korpela, A.V., Masuda, K., Matsubara, Y., Motomura, M., Muraki, Y., Nakamura, S., Okada, C., Ohnishi, K., Rattenbury, N.J., Sako, T., Sato, S., Sasaki, M., Sekiguchi, T., Sullivan, D.J., Tristram, P.J., Yock, P.C.M., and Yoshioka, T. (2006). Discovery of a cool planet of 5.5 Earth masses through gravitational microlensing. *Nature*, **439**, 437.

Beerling, D.J., Lomax, B.H., Royer, D.L., Upchurch, G.R., Jr, and Kump, L.R. (2002). An atmospheric pCO_2 reconstruction across the Cretaceous–Tertiary boundary from leaf megafossils. *Proc. Natl. Acad. Sci. (USA)*, **99**, 7836–7840.

Benton, M.J., and Twitchett, R.J. (2003) How to kill (almost) all life: the end-Permian extinction event, *Trends Ecol. Evol.* **18**, 358–365.

Berglund, J. (1986). The decline of the Norse settlements in Greenland. *Arc. Anthropol.*, **23**, 109–136.

Brown, P., Sutikna, T., Morwood, M.J., Soejono, R.P., Jatmiko, Saptomo, E.W., and Due, R.A. (2004). A new small-bodied hominin from the Late Pleistocene of Flores, Indonesia. *Nature*, **431**, 1055–1061.

Clark, A.G., Glanowski, S., Nielsen, R., Thomas, P.D., Kejariwal, A., Todd, M.A., Tanenbaum, D.M., Civello, D., Lu, F., Murphy, B., Ferriera, S., Wang, G., Zheng, X., White, T.J., Sninsky, J.J., Adams, M.D., and Cargill, M. (2003). Inferring nonneutral evolution from human–chimp–mouse orthologous gene trios. *Science*, **302**, 1960–1963.

Chimpanzee Sequencing and Analysis Consortium, T.C.S. (2005). Initial sequence of the chimpanzee genome and comparison with the human genome. *Nature*, **437**, 69–87.

Cook, L.M., Soltis, P.M., Brunsfeld, S.J., and Soltis, D.E. (1998). Multiple independent formations of *Tragopogon* tetraploids (Asteraceae): evidence from RAPD markers. *Mol. Ecol.*, **7**, 1293–1302.

Culotta, E. and Pennisi, E. (2005). Evolution in action. *Science*, **310**, 1878–1879.

Dawkins, R. (1976). *The Selfish Gene* (New York, NY: Oxford University Press).

Dennell, R. and Roebroeks, W. (2005). An Asian perspective on early human dispersal from Africa. *Nature*, **438**, 1099–1104.

Devlin, B., Daniels, M., and Roeder, K. (1997). The heritability of IQ. *Nature*, **388**, 468–471.

Diamond, Jared (2005). Collapse: How Societies Choose to Fail or Succeed. Viking, New York.

Enard, W., Khaitovich, P., Klose, J., Zöllner, S., Heissig, F., Giavalisco, P., Nieselt-Struwe, K., Muchmore, E., Varki, A., Ravid, R., Doxiadis, G.M., Bontrop, R.E., and Pääbo, S. (2002). Intra- and interspecific variation in primate gene expression patterns. *Science*, **296**, 340–343.

Evans, P.D., Gilbert, S.L., Mekel-Bobrov, N., Vallender, E.J., Anderson, J.R., Vaez-Azizi, L.M., Tishkoff, S.A., Hudson, R.R., and Lahn, B.T. (2005). Microcephalin, a gene regulating brain size, continues to evolve adaptively in humans. *Science*, **309**, 1717–1720.

Flynn, J.J., Parrish, J.M., Rakotosamimanana, B., Simpson, W.F., Whatley, R.L., and Wyss, A.R. (1999). A Triassic fauna from Madagascar, including early dinosaurs. *Science*, **286**, 763–765.

Gardner, H. (1993). *Multiple Intelligences: The Theory in Practice* (New York: Basic Books).

Garrigan, D. and Hedrick, P.W. (2003). Perspective: detecting adaptive molecular polymorphism: lessons from the MHC. *Evolution*, **57**, 1707–1722.

Goleman, D. (1995). *Emotional Intelligence* (New York: Bantam Books).

Hawks, J., Wang, E.T., Cochran, G.M., Harpending, H.C. and Moyzis, R.K. (2007) Recent acceleration of human adaptive evolution. Proceedings of the National Academy of Sciences (US), **104**, 20753–20758.

Hu, Y., Meng, J., Wang, Y., and Li, C. (2005). Large Mesozoic mammals fed on young dinosaurs. *Nature*, **433**, 149–152.

Ji, Q., Luo, Z.-X., Yuan, C.-X., and Tabrum, A.R. (2006). A swimming mammaliaform from the middle Jurassic and ecomorphological diversification of early mammals. *Science*, **311**, 1123–1127.

Jones, J.S. (1991). Is evolution over? If we can be sure about anything, it's that humanity won't become superhuman. *New York Times*, p. E17.

Kasting, J.F. (2001). Peter Ward and Donald Brownlee's 'Rare Earth'. *Persp. Biol. Med.*, **44**, 117–131.

Kirk, K.M., Blomberg, S.P., Duffy, D.L., Heath, A.C., Owens, I.P.F., and Martin, N.G. (2001). Natural selection and quantitative genetics of life-history traits in Western women: a twin study. *Evolution*, **55**, 423–435.

Lamason, R.L., Mohideen, M.-A.P.K., Mest, J.R., Wong, A.C., Norton, H.L., Aros, M.C., Jurynec, M.J., Mao, X., Humphreville, V.R., Humbert, J.E., Sinha, S., Moore, J.L., Jagadeeswaran, P., Zhao, W., Ning, G., Makalowska, I., McKeigue, P.M., O'Donnell, D., Kittles, R., Parra, J., Mangini, N.J., Grunwald, D.J., Shriver, M.D., Canfield, V.A., and Cheng, K.C. (2005). SLC24A5, a putative cation exchanger, affects pigmentation in zebrafish and humans. *Science*, **310**, 1782–1786.

McBrearty, S. and Jablonski, N.G. (2005). First fossil chimpanzee. *Nature*, **437**, 105–108.

Mekel-Bobrov, N., Posthuma, D., Gilbert, S.L., Lind, P., Gosso, M.F., Luciano, M., Harris, S.E., Bates, T.C., Polderman, T.J.C., Whalley, L.J., Fox, H., Starr, J.M., Evans, P.D., Montgomery, G.W., Fernandes, C., Heutink, P., Martin, N.G., Boomsma, D.I., Deary, I.J., Wright, M.J., de Geus, E.J.C. and Lahn, B.T. (2007) The ongoing adaptive evolution of *ASPM* and *Microcephalin* is not explained by increased intelligence. Human Molecular Genetics 16, 600–608.

Mekel-Bobrov, N., Gilbert, S.L., Evans, P.D., Vallender, E.J., Anderson, J.R., Hudson, R.R., Tishkoff, S.A., and Lahn, B.T. (2005). Ongoing adaptive evolution of ASPM, a brain size determinant in *Homo sapiens*. *Science*, **309**, 1720–1722.

Muller, H.J. (1950). Our load of mutations. *Am. J. Human Genet.*, **2**, 111–176.

Patterson, N., Richter, D.J., Gnerre, S., Lander, E.S., and Reich, D. (2006). Genetic evidence for complex speciation of humans and chimpanzees. *Nature*, **441**, 1103–1108.

Silver, L. (1998). *Remaking Eden* (New York: Harper).

Singh-Manoux, A., Marmot, M.G., and Adler, N.E. (2005). Does subjective social status predict health and change in health status better than objective status? *Psychosomatic Medicine*, **67**, 855–861.

Sokolowski, M.B., Pereira, H.S., and Hughes, K. (1997). Evolution of foraging behavior in *Drosophila* by density-dependent selection. *Proc. Natl. Acad. Sci. (USA)*, **94**, 7373–7377.

Spira, A. and Multigner, L. (1998). Environmental factors and male infertility. *Human Reprod.*, **13**, 2041–2042.

Surovell, T., Waguespack, N., and Brantingham, P.J. (2005). Global archaeological evidence for proboscidean overkill. *Proc. Natl. Acad. Sci. (USA)*, **102**, 6231–6236.

Tagliatela, J.P., Savage-Rumbaugh, S., and Baker, L.A. (2003). Vocal production by a language-competent *Pan paniscus*. *Int. J. Primatol.*, **24**, 1–17.

Ward, P. and Brownlee, D. (2000). *Rare Earth: Why Complex Life Is Uncommon in the Universe* (New York: Copernicus Books).

White, T.D., Suwa, G., and Asfaw, B. (1994). *Australopithecus ramidus*, a new species of early hominid from Aramis, Ethiopia. *Nature*, **371**, 306–312.

Williams, R.S. and Ferrigno, J. (1999). Estimated present-day area and volume of glaciers and maximum sea level rise potential. pp. 1–10. Satellite Image Atlas of Glaciers of the World. Washington DC: US Geological Survey.

Wills, C. (1993). *The Runaway Brain: The Evolution of Human Uniqueness* (New York: Basic Books).

Wills, C. (1998). *Children of Prometheus: The Accelerating Pace of Human Evolution* (Reading, MA: Perseus Books [formerly Addison-Wesley]).

Wills, C., Harms, K.E., Condit, R., King, D., Thompson, J., He, F., Muller-Landau, H.C., Ashton, P., Losos, E., Comita, L., Hubbell, S., LaFrankie, J., Bunyavejchewin, S., Dattaraja, H.S., Davies, S., Esufali, S., Foster, R., Gunatilleke, N., Gunatilleke, S., Hall, P., Itoh, A., John, R., Kiratiprayoon, S., de Lao, S.L., Massa, M., Nath, C., Noor, M.N.S., Kassim, A.R., Sukumar, R., Suresh, H.S., Sun, I.-F., Tan, S., Yamakura, T., and Zimmerman, J. (2006). Nonrandom processes maintain diversity in tropical forests. *Science*, **311**, 527–531.

Zhang, H.-X. and Blumwald, E. (2001). Transgenic salt-tolerant tomato plants accumulate salt in foliage but not in fruit. *Nat. Biotechnol.*, **19**, 765–768.

Zink, R.M. and Slowinski, J.B. (1995). Evidence from molecular systematics for decreased avian diversification in the Pleistocene epoch. *Proc. Natl. Acad. Sci. (USA)*, **92**, 5832–5835.

·4·

Millennial tendencies in responses to apocalyptic threats

James J. Hughes

4.1 Introduction

Aaron Wildavsky proposed in 1987 that cultural orientations such as egalitarianism and individualism frame public perceptions of technological risks, and since then a body of empirical research has grown to affirm the risk-framing effects of personality and culture (Dake, 1991; Gastil et al., 2005; Kahan, 2008). Most of these studies, however, have focused on relatively mundane risks, such as handguns, nuclear power, genetically modified food, and cellphone radiation. In the contemplation of truly catastrophic risks – risks to the future of the species from technology or natural threats – a different and deeper set of cognitive biases come into play, the millennial, utopian, or apocalyptic psychocultural bundle, a characteristic dynamic of eschatological beliefs and behaviours. This essay is an attempt to outline the characteristic forms millennialism has taken, and how it biases assessment of catastrophic risks and the courses of action necessary to address them.

Millennialism is the expectation that the world as it is will be destroyed and replaced with a perfect world, that a redeemer will come to cast down the evil and raise up the righteous (Barkun, 1974; Cohn, 1970). Millennialism is closely tied to other historical phenomena: utopianism, apocalypticism, messianism, and millenarian violence. Western historians of millenialism have focused the most attention on the emergence of Christianity out of the messianic expectations of subjugated Jewry and subsequent Christian movements based on exegesis of the Book of Revelations expecting the imminent return of Christ. But the millennial impulse is pancultural, found in many guises and with many common tropes from Europe to India to China, across the last several thousand years. When Chinese peasants followed religio-political revolutionaries claiming the mantle of the Coming Buddha, and when Mohammed birthed Islam preaching that the Last Judgement was imminent, they exhibited many similar features to medieval French peasants leaving their fields to follow would-be John the Baptists. Nor is the millennial impulse restricted to religious movements and beliefs in magical or supernatural agency. Revolutionary socialism and fascism embodied the same impulses and promises, although purporting to be based on science, das Volk, and the secular state instead of prophecy, the body of believers, and the Kingdom of Heaven (Rhodes, 1980; Rowley, 1983).

In this essay I will review some of the various ways in which the millennial impulse has manifested. Then I will parse contemporary secular expectations about catastrophic risks and utopian possibility for signs of these characteristic millennial dynamics. Finally, I will suggest that by avoiding the undertow of the psychocultural dysfunctions and cognitive biases that often accompany millennialism, we may be able to better anticipate the real benefits and threats that we face in this era of accelerating change and take appropriate prophylactic action to ensure a promising future for the human race.

4.2 Types of millennialism

Western scholars have pointed to three theological positions among Christian millennialists that appear to have some general applicability to millennial typology, based on the role of human agency in ending Tribulations and bringing the Millennium.

4.2.1 Premillennialism

Premillennialism is the most familiar form of millennial thought in the United States and Europe today, characterized by the belief that everything will get awful before the millennium makes them better (Whalen, 2000). Christian premillennialists, among them many early Christians, have based their eschatological expectations on the Book of Revelations. They believed that the Antichrist will preside over a period of Tribulations, followed by God's rescue of the righteous, the Rapture. Eventually Christ returns, defeats evil, judges all resurrected souls, and establishes a reign of the Kingdom of Heaven on Earth. Saved people will spend eternity in this new kingdom, and the unsaved will spend an eternity in damnation.

This doctrine was reintroduced among Protestants in the 1830s in the United States as 'dispensationalism' (Boyer, 1994; Crutchfield, 1992). Some dispensationalists became 'Millerites', influenced by the exegetical efforts of a nineteenth century lay scholar, William Miller, to interpret contemporary events as a fulfilment of a prophetic timeline in the Book of Revelations. Dispensationalism and Millerism inspired some successful sects that have flourished to this day, such as the Seventh-day Adventists and the Jehovah's Witnesses, despite their failed prophecies of specific dates for the Second Coming.

Premillennialism gained general acceptance among Christian evangelicals only in the twentieth century. Hal Lindsey's *The Late Great Planet Earth* (1970) popularized the thinking of modern Millerites who saw the European Union, the re-creation of the state of Israel and other modern trends as fulfilment of the millennial timeline. Christian Right politicians such as Pat Robertson, a former Republican candidate for President, are premillennialists, interpreting daily events in the Middle East for the millions in their television and radio audience through the lens of Biblical exegesis. Premillennialism is also the basis of the extremely popular *Left Behind* novels by Jerry Jenkins and Tim LaHaye, and their film adaptations.

Premillennialists are generally fatalists who do not believe human beings can influence the timing or outcome of the Tribulations, Rapture, and Second Coming (Wojcik, 1997). The best the believer can do is to save as many souls as possible before the end. A very similar doctrine can be found among certain Mahayana Buddhists who held that, after the passing of each Buddha, the world gradually falls into the age of *mappo* or the degeneracy of the Buddhist teachings. In the degenerate age enlightenment is nearly impossible, and the best that we can hope for is the intercession of previously enlightened, and now divine, beings to bring us to a Pure Land (Blum, 2002). The Japanese monk Nichiren Daishonin (1222–1282) founded one of the most successful schools of Japanese Buddhism using the idea of *mappo* as the rationale for his new dispensation. For Nichiren Buddhists, Japan will play a millennial role in the future as the basis for the conversion of the entire world to the Buddhist path (Stone, 1985).

In a secular context, Marxist futurism has often been appropriated into a form of premillennial expectation. According to classical Marxist eschatology the impersonal workings of capitalism and technological innovation will immiserate all the people of the world, wipe out all pre-existing identities and institutions, and then unite the world into revolutionary working-class movement. The role of self-conscious revolutionaries is only to explain this process to workers so that they understand their role in the unfolding of the inevitable historical telos and hastening the advent of the millennial worker's paradise.

4.2.2 Amillennialism

Amillennialists believe that the millennial event has already occurred, or is occurring, in the form of some movement or institution, even though there are still bad things happening in the world (Hoekema, 2007; Riddlebarger, 2003). For Christian amillennialists the Millennium is actually the ongoing establishment of righteousness on Earth through the agency of the Church, struggling to turn back Satan and the Tribulations. Augustinian amillennialism was made the official doctrine of the early Christian Church, and premillennialism was declared heresy. The subsequent millenarian rebellions against church and civil authority, inspired by the Book of Revelations, reinforced the Catholic Church's insistence that the Millennium would not be an abrupt, revolutionary event, but the gradual creation of the Kingdom of Heaven in each believer's heart in a church-ruled world. In abstract, when the 'Church Age' ends Christ will return to judge humanity and take the saved to heaven for eternity, but attempts to predict the timeline are proscribed.

Another, more radical and recent example of amillennialism was found in the Oneida Community (1848–1881) in upstate New York, founded by John Humphrey Noyes (Klaw, 1993). Noyes believed that the Second Coming had occurred in 70 AD, and that all believers should begin to live as if in the Kingdom of Heaven, including forbidding monogamy and private property. Orthodox Communist or Maoists outside of the Soviet Union or China can be seen as secular amillennialists, believing Stalin's and Mao's regimes were paradises in which eventually all humanity would be able to share.

4.2.3 Post-millennialism

Post-millennialists believe that specific human accomplishments are necessary to bring the Millennium (Bock, 1999). Post-millennialist Christians argue that if Christians establish the Kingdom of Heaven on Earth through Christian rule this will hasten or be synonymous with the second coming. This doctrine has inspired both progressive 'Social Gospel' theologies, such as slavery abolitionism and anti-alcohol temperance, as well as theocratic movements such as the contemporary far Right 'Christian Reconstruction' and Dominionism.

The Buddhist Pali canon scriptures that describe the coming Buddha, Maitreya, are an example of post-millennialism (Hughes, 2007). The scripture foretells that humanity, repulsed by the horrors of an apocalyptic war, will build a utopian civilization thickly populated with billions of happy, healthy people who live for thousands of years in harmony with one another and with nature. The average age at marriage will be 500 years. The climate will always be good, neither too hot nor too cold. Wishing trees in the public squares will provide anything you need. The righteous king dissolves the government and turns over the property of the state to Maitreya. These millennial beliefs inspired a series of Buddhist uprisings in China (Naquin, 1976), and helped bring the Buddhist socialist movement in Burma to power from 1948 to 1962 (Malalgoda, 1970).

This worldview also corresponds to revolutionary Marxist–Leninism. Although the march of history is more or less assured, the working class may wander for centuries through the desert until the revolutionary vanguard can lead them to the Promised Land. Once socialism is established, it will gradually evolve into communism, in which the suppressed and distorted true human nature will become non-acquisitive and pro-social. Technology will provide such abundance that conflict over things will be unnecessary, and the state will wither. But revolutionary human agency is necessary to fulfil history.

4.3 Messianism and millenarianism

Messianism and millenarianism are defined here as forms of millennialism in which human agency, magical or revolutionary, is central to achieving the Millennium. Messianic movements focus on a particular leader or movement, while millenarian movements, such as many of the peasant uprisings of fifteenth and sixteenth Europe, believe revolutionary violence to smash the evil old order will help usher in the Millennium (Rhodes, 1980; Smith, 1999; Mason, 2002). Al Qaeda is, for instance, rooted in Islamic messianic eschatology; the Jihad to establish the global Caliphate is critical in the timeline for the coming of the Mahdi, or messiah. Osama bin Laden is only the latest of a long line of Arab leaders claiming, or being ascribed, the mantle of Mahdism (Cook, 2005; Furnish, 2005). In the Hindu and Buddhist messianic tradition there is the belief in the periodic emergence of Maha Purushas, 'great men', who arrive in times of need to provide either righteous rule or saintliness. Millenarian uprisings in China were often led by men claiming to be Maitreya, the next Buddha. Secular messiahs and revolutionary leaders are similarly often depicted in popular mythology as possessing extraordinary wisdom and abilities from an early age, validating their unique eschatological role. George Washington could never tell a lie and showed superhuman endurance at Valley Forge, whereas Chairman Mao was a modern Moses, leading his people on a Long March to Zion and guiding the masses with the wisdom in his Little Red Book.

4.4 Positive or negative teleologies: utopianism and apocalypticism

Utopianism and apocalypticism are defined here as the millennial impulse with, respectively, an optimistic and pessimistic eschatological expectation. By utopianism I mean the belief that historical trends are inevitably leading to a wonderful millennial outcome (Manuel and Manuel, 1979), including the Enlightenment narrative of inevitable human progress (Nash, 2000; Tuveson, 1949). By apocalypticism I do not mean simply the belief that something very bad may happen, since very bad events are simply a prelude to very good events for most millennialists, but that the bad event will be cataclysmic, or even the end of history.

In that sense, utopianism is the default setting of most millennial movements, even if the Tribulations are expected to be severe and indeterminately long. The promise of something better, at least for the righteous, is far more motivating than a guaranteed bad end. Even the most depressing religious eschatology, the Norse *Ragnarok* – at which humans and gods are defeated, and Earth and the heavens are destroyed – holds out a millennial promise that a new earth and Sun will emerge, and the few surviving gods and humans with live in peace and prosperity (Crossley-Holland, 1981).

Millennial expectations of better times have not only been a comfort to people with hard, sad lives, an 'opium for the masses', but also, because of their mobilizing capacity, an essential catalyst of social change and political reform (Hobsbawm, 1959; Jacoby, 2005; Lanternari, 1965). From Moses' mobilization of enslaved Jewry with a promise of a land of milk and honey, to medieval millenarian peasant revolts, to the Sioux Ghost Dance, to the integrationist millennialism of the African-American Civil Rights Movement, millenarian leaders have arisen out of repressive conditions to preach that they could lead their people to a new Zion. Sometimes the millennial movements are disastrously unsuccessful when they rely on supernatural methods for achieving their ends, as with the Ghost Dance (Mooney, 1991). Sometimes utopian and millennial currents contribute to social reform even in their defeat, as they did from the medieval peasant revolts through the rise of revolutionary socialism (Jacoby, 2005). Although movements for utopian social change were most successful when they focused on temporal, rather than millennial, goals through human, rather than supernatural, agency, expectations of utopian outcomes helped motivate participants to take risks on collective action against large odds.

Although there have been few truly apocalyptic movements or faiths, those which foretell an absolute, unpleasant and unredeemed end of history, there have been points in history with widespread apocalyptic expectation. The stories of the Biblical flood and the destruction of Sodom and Gomorrah alerted Christians to the idea that God was quite willing to destroy almost all of humanity for our persistent sinfulness, well before the clock starts on the Tribulation–Millennium timeline. Although most mythic beliefs include apocalyptic periods in the past and future, as with *Ragnarok* or the Hindu–Buddhist view of a cyclical destruction – recreation of the universe, most myths make apocalypse a transient stage in human history.

It remained for more secular times for the idea of a truly cataclysmic end of history, with no redeeming Millennium, to become a truly popular current of thought (Heard, 1999; Wagar, 1982; Wojcik, 1997, 1999). Since the advent of the Nuclear Age, one apocalyptic threat after another, natural and man-made, has been added to the menu of ways that human history could end, from environmental destruction and weapons of mass destruction, to plague and asteroid strikes (Halpern, 2001; Leslie, 1998; Rees, 2004). In a sense, long-term apocalypticism is also now the dominant scientific worldview, insofar as most scientists see no possibility for intelligent life to continue after the Heat Death of the Universe (2002, Ellis; see also the Chapter 2 in this volume).

4.5 Contemporary techno-millennialism

4.5.1 The singularity and techno-millennialism

Joel Garreau's (2006) recent book on the psychoculture of accelerating change, *Radical Evolution: The Promise and Peril of Enhancing Our Minds, Our Bodies – and What It Means to Be Human*, is structured in three parts: Heaven, Hell and Prevail. In the Heaven scenario he focuses on the predictions of inventor Ray Kurzweil, summarized in his 2005 book, *The Singularity Is Near*. The idea of a techno-millennial 'Singularity' was coined in a 1993 paper by mathematician and science fiction author Vernor Vinge. In physics 'singularities' are the centres of black holes, within which we cannot predict how physical laws will work. In the same way, Vinge said, greater-than-human machine intelligence, multiplying exponentially, would make everything about our world unpredictable. Most Singularitarians, like Vinge and Kurzweil, have focused on the emergence of superhuman machine intelligence. But the even more fundamental concept is exponential technological progress, with the multiplier quickly leading to a point of radical social crisis. Vinge projected that self-willed artificial intelligence would emerge within the next 30 years, by 2023, with either apocalyptic or millennial consequences. Kurzweil predicts the Singularity for 2045.

The most famous accelerating trend is 'Moore's Law', articulated by Intel co-founder Gordon Moore in 1965, which is the observation that the number of transistors that can be fit on a computer chip has doubled about every 18 months since their invention. Kurzweil goes to great lengths to document that these trends of accelerating change also occur in genetics, mechanical miniaturization, and telecommunications, and not just in transistors. Kurzweil projects that the 'law of accelerating returns' from technological change is 'so rapid and profound it represents a rupture in the fabric of human history'. For instance, Kurzweil predicts that we will soon be able to distribute trillions of nanorobots in our brains, and thereby extend our minds, and eventually upload our minds into machines. Since lucky humans will at that point merge with superintelligence or become superintelligent, some refer to the Singularity as the 'Techno-rapture', pointing out the similarity of the narrative to the Christian Rapture; those foresighted enough to be early adopters of life extension and cybernetics will live long enough to be uploaded and 'vastened' (given vastly superior mental abilities) after the Singularity. The rest of humanity may however be 'left behind'.

This secular 'left behind' narrative is very explicit in the Singularitarian writings of computer scientist Hans Moravec (1990, 2000). For Moravec the human race will be superseded by our robot children, among whom some of us may be able to expand to the stars. In his *Robot: Mere Machine to Transcendent Mind*, Moravec (2000, pp. 142–162) says

'Our artificial progeny will grow away from and beyond us, both in physical distance and structure, and similarity of thought and motive. In time their activities may become incompatible with the old Earth's continued existence . . . An entity that fails to keep up with its neighbors is likely to be eaten, its space, materials, energy, and useful thoughts reorganized to serve another's goals. Such a fate may be routine for humans who dally too long on slow Earth before going Ex.

Here we have Tribulations and damnation for the late adopters, in addition to the millennial utopian outcome for the elect.

Although Kurzweil acknowledges apocalyptic potentials – such as humanity being destroyed by superintelligent machines – inherent in these technologies, he is nonetheless uniformly utopian and enthusiastic. Hence Garreau's labelling Kurzweil's the 'Heaven' scenario. While Kurzweil (2005) acknowledges his similarity to millennialists by, for instance, including a tongue-in-cheek picture in *The Singularity Is Near* of himself holding a sign with that slogan, referencing the classic cartoon image of the EndTimes street prophet, most Singularitarians angrily reject such comparisons insisting their expectations are based solely on rational, scientific extrapolation.

Other Singularitarians, however, embrace parallels with religious millennialism. John Smart, founder and director of the California-based Acceleration Studies Foundation, often notes the similarity between his own 'Global Brain' scenario and the eschatological writings of the Jesuit palaeontologist Teilhard de Chardin (1955). In the Global Brain scenario, all human beings are linked to one another and to machine intelligence in the emerging global telecommunications web, leading to the emergence of collective intelligence. This emergent collectivist form of Singularitarianism was proposed also by Peter Russell (1983) in *The Global Brain*, and Gregory Stock (1993) in *Metaman*. Smart (2007) argues that the scenario of an emergent global human-computer meta-mind is similar to Chardin's eschatological idea of humanity being linked in a global 'noosphere', or info-sphere, leading to a post-millennial 'Omega Point' of union with God. Computer scientist Juergen Schmidhuber (2006) also has adopted Chardin's 'Omega' to refer to the Singularity.

For most Singularitarians, as for most millennialists, the process of technological innovation is depicted as autonomous of human agency, and wars, technology bans, energy crises or simple incompetence are dismissed as unlikely to slow or stop the trajectory. Kurzweil (2006) insists, for instance, that the accelerating trends he documents have marched unhindered through wars, plagues and depressions. Other historians of technology (Lanier, 2000; Seidensticker, 2006; Wilson, 2007) argue that Kurzweil ignores techno-trends which did stall, due to design challenges and failures, and to human factors that slowed the diffusion of new technologies, factors which might also slow or avert greater-than-human machine intelligence. Noting that most predictions of electronic transcendence fall within the predictor's expected lifespan, technology writer Kevin Kelly (2007) suggests that people who make such predictions have a cognitive bias towards optimism.

The point of this essay is not to parse the accuracy or empirical evidence for exponential change or catastrophic risks, but to examine how the millennialism that accompanies their consideration biases assessment of their risks and benefits, and the best courses of action to reduce the former and ensure the latter. There is of course an important difference between fear of a civilization-ending nuclear war, grounded in all-too-real possibility, and fear of the end of history from a prophesied supernatural event. I do not mean to suggest that all discussion of utopian and catastrophic possibilities are merely millennialist fantasies, but rather that recognizing millennialist dynamics permits more accurate risk/benefit assessments and more effective prophylactic action.

4.6 Techno-apocalypticism

In *Radical Evolution* Joel Garreau's Hell scenario is centred on the Luddite apocalypticism of the techno-millennial apostate, Bill Joy, former chief scientist and co-founder of Sun Microsystems. In the late 1990s Joy began to believe that genetics, robotics, and nanotechnology posed novel apocalyptic risks to human life. These technologies, he argued, posed a different kind of threat because they could self-replicate; guns do not breed and shoot people on their own, but a rogue bioweapon could. His essay 'Why the Future Doesn't Need Us,' published in April 2000 in *Wired* magazine, called for a global, voluntary 'relinquishment' of these technologies.

Greens and others of an apocalyptic frame of mind were quick to seize on Joy's essay as an argument for the enacting of bans on technological innovation, invoking the 'precautionary principle', the idea that a potentially dangerous technology should be fully studied for its potential impacts before being deployed. The lobby group ETC argued in its 2003 report 'The Big Down' that nanotechnology could lead to a global environmental and social catastrophe, and should be placed under government moratorium. Anxieties about the apocalyptic risks of converging bio-, nano- and information technologies have fed a growing Luddite strain in Western culture (Bailey 2001a, 2001b), linking Green and anarchist advocates for neo-pastoralism (Jones, 2006; Mander, 1992; Sale, 2001; Zerzan, 2002) to humanist critics of techno-culture (Ellul, 1967; Postman, 1993; Roszak, 1986) and apocalyptic survivalists to Christian millennialists. The neo-Luddite activist Jeremy Rifkin has, for instance, built coalitions between secular and religious opponents of reproductive and agricultural biotechnologies, arguing that the encroachment into the natural order will have apocalyptic consequences. Organizations such as the Chicago-based Institute on Biotechnology & The Human Future, which brings together bio- and nano-critics from the Christian Right and the secular Left, represent the institutionalization of this new Luddite apocalypticism advocating global bans on 'genocidal' lines of research (Annas, Andrews, and Isasi, 2002).

Joy has, however, been reluctant to endorse Luddite technology bans. Joy and Kurzweil are entrepreneurs and distrust regulatory solutions. Joy and Kurzweil also share assumptions about the likelihood and timing of emerging technologies, differing only in their views on the likelihood of millennial or apocalyptic outcomes. But they underlined their *similarity* of worldview by issuing a startling joint statement in 2005 condemning the publication of the genome of the 1918 influenza virus, which they viewed as a cookbook for a potential bioterror weapon (Kurzweil and Joy, 2005). Disturbing their friends in science and biotech, leery of government mandates for secrecy, they called for 'international agreements by scientific organizations to limit such publications' and 'a new Manhattan Project to develop specific defences against new biological viral threats'.

In the 1990s anxieties grew about the potential for terrorists to use recombinant bioengineering to create new bioweapons, especially as bioweapon research in the former Soviet Union came to light. In response to these threats the Clinton administration and US Congress started major bioterrorism preparedness initiatives in the 1990s, despite warnings from public health advocates such as Laurie Garrett (1994, 2000) that monies would be far better spent on global public health initiatives to prevent, detect, and combat emerging infectious diseases. After 9/11 the Bush administration, motivated in part by the millennial expectations of both the religious Right and secular neo-conservatives, focused even more attention on the prevention of relatively low probability/low lethality bioterrorism than on the higher probability/lethality prospects of emerging infectious diseases such as pandemic flu. Arguably apocalyptic fears around bioterrorism, combined with the influence of the neo-conservatives and biotech lobbies, distorted public health priorities. Perhaps conversely we have not yet had *sufficient* apocalyptic anxiety about emerging plagues to force governments to take a comprehensive, proactive approach to public health. (Fortunately efforts at infectious disease monitoring, gene sequencing and vaccine production are advancing nonetheless; a year after Kurzweil and Joy's letter a team at the US National Institutes of Health had used the flu genome to develop a vaccine for the strain [NIH, 2006].)

An example of a more successful channelling of techno-apocalyptic energies into effective prophylaxis was the Millennium Bug or Y2K phenomenon. In the late 1990s a number of writers began to warn that a feature of legacy software systems from the 1960s and 1970s, which coded years with two digits instead of four, would lead to widespread technology failure in the first seconds of 2000. The chips controlling power plants, air traffic, and the sluice gates in sewer systems would suddenly think the year was 1900 and freeze. Hundreds of thousands of software engineers around the world were trained to analyse 40-year-old software languages and rewrite them. Hundreds of billions of dollars were spent worldwide on improving information systems, disaster preparedness, and on global investment in new hardware and software, since it was often cheaper simply to replace than to repair legacy systems (Feder, 1999; Mussington, 2002). Combined with the imagined significance of the turn of the Millennium, Christian millennialists saw the crisis as a portent of the EndTimes (Schaefer, 2004), and secular apocalyptics bought emergency generators, guns and food in anticipation of a prolonged social collapse (CNN, 1998; Kellner, 1999; Tapia, 2003). Some anti-technology Y2K apocalyptics argued for widespread technological relinquishment – getting off the grid and returning to a nineteenth century lifestyle.

The date 1 January 2000 was as unremarkable as all predicted millennial dates have been, but in this case, many analysts believe potential catastrophes were averted due to the proactive action from governments, corporations, and individual consumers (Special Committee on the Year 2000 Technology Problem, 2000), motivated in part by millennial anxieties. Although the necessity and economic effects of pre-Y2K investments in information technology modernization remain controversial, some subsequent economic and productivity gains were probably accrued (Kliesen, 2003). Althoughthe size and cost of the Y2K preparations may not have been optimal, the case is still one of proactive policy and technological innovation driven in part by millennial/apocalyptic anxiety. Similar dynamics can be observed around the apocalyptic concerns over 'peak oil', 'climate change', and the effects of environmental toxins, which have helped spur action on conservation, alternative energy sources, and the testing and regulation of novel industrial chemicals (Kunstler, 2006).

4.7 Symptoms of dysfunctional millennialism in assessing future scenarios

Some critics denigrate utopian, millennial, and apocalyptic impulses, both religious and secular, seeing them as irrational at best, and potentially murderous and totalitarian at worst. They certainly can manifest in the dangerous and irrational ways as I have catalogued in this essay. But they are also an unavoidable accompaniment to public consideration of catastrophic risks and techno-utopian possibilities. We may aspire to a purely rational, technocratic analysis, calmly balancing the likelihoods of futures without disease, hunger, work or death, on the one hand, against the likelihoods of worlds destroyed by war, plagues or asteroids, but few will be immune to millennial biases, positive or negative, fatalist or messianic. Some of these effects can be positive. These mythopoetic interpretations of the historical moment provide hope and meaning to the alienated and lost. Millennialist energies can overcome social inertia and inspire necessary prophylaxis and force recalcitrant institutions to necessary action and reform. In assessing the prospects for catastrophic risks, and potentially revolutionary social and technological progress, can we embrace millennialism and harness its power without giving in to magical thinking, sectarianism, and overly optimistic or pessimistic cognitive biases?

I believe so: understanding the history and manifestations of the millennial impulse, and scrutinizing even our most purportedly scientific and rational ideas for their signs, should provide some correction for their downsides. Based on the discussion so far, I would identify four dysfunctional manifestations of millennialism to watch for. The first two are the manic and depressive errors of millennialism, tendencies to utopian optimism and apocalyptic pessimism. The other two dysfunctions have to do with the role of human agency, a tendency towards fatalist passivity on the one hand, believing that human action can have no effect on the inevitable millennial or apocalyptic outcomes, and the messianic tendency on the other hand, the conviction that specific individuals, groups, or projects have a unique historical role to play in securing the Millennium.

Of course, one may acknowledge these four types of millennialist biases without agreeing whether a particular assessment or strategy reflects them. A realistic assessment may in fact give us reasons for great optimism or great pessimism. Apocalyptic anxiety during the 1962 Cuban missile confrontation between the United States and the Soviet Union was entirely warranted, whereas historical optimism about a New World Order was understandable during the 1989–1991 collapse of the Cold War. Sometimes specific individuals (Gandhis, Einsteins, Hitlers, etc.) do have a unique role to play in history, and sometimes (extinction from gamma-ray bursts from colliding neutron stars or black holes) humanity is completely powerless in the face of external events. The best those who ponder catastrophic risks can do is practise a form of historically informed cognitive therapy, interrogating our responses to see if we are ignoring counterfactuals and alternative analyses that might undermine our manic, depressive, fatalist, or messianic reactions.

One symptom of dysfunctional millennialism is often dismissal of the possibility that political engagement and state action could affect the outcome of future events. Although there may be some trends or cataclysms that are beyond all human actions, all four millennialist biases – utopian, apocalyptic, fatalist, and messianic – underestimate the potential and importance of collective action to bring about the Millennium or prevent apocalypse. Even messianists are only interested in public approbation of their own messianic mission, not winning popular support for a policy. So it is always incumbent on us to ask how engaging with the political process, inspiring collective action, and changing state policy could steer the course of history. The flip side of undervaluing political engagement as too uncertain, slow, or ineffectual is a readiness to embrace authoritarian leadership and millenarian violence in order to achieve quick, decisive, and far-sighted action.

Millennialists also tend to reduce the complex socio-moral universe into those who believe in the eschatological worldview and those who do not, which also contributes to political withdrawal, authoritarianism and violence. For millennialists society collapses into friends and enemies of the Singularity, the Risen Christ, or the Mahdi, and their enemies may be condemning themselves or all of humanity to eternal suffering. Given the stakes on the table – the future of humanity – enemies of the *Ordo Novum* must be swept aside. Apostates and the peddlers of mistaken versions of the salvific faith are even more dangerous than outright enemies, since they can fatally weaken and mislead the righteous in their battle against Evil. So the tendency to demonize those who deviate can unnecessarily alienate potential allies and lead to tragic violence. The Jones Town suicides, the Oklahoma City bombing, Al Qaeda, and Aum Shinrikyo are contemporary examples of a millennial logic in which murder is required to fight evil and heresies, and wake complacent populations to imminent millennial threats or promises (Hall, 2000; Mason, 2002; Whitsel, 1998). Whenever contemporary millenarians identify particular scientists, politicians, firms, or agencies as playing a special role in their eschatologies, as specific engineers did for the Unabomber, we can expect similar violence in the future. A more systemic and politically engaged analysis, on the other hand, would focus on regulatory approaches addressed at entire fields of technological endeavours rather than specific actors, and on the potential for any scientist, firm, or agency to contribute to both positive and negative outcomes.

4.8 Conclusions

The millennial impulse is ancient and universal in human culture and is found in many contemporary, purportedly secular and scientific, expectations about the future. Millennialist responses are inevitable in the consideration of potential catastrophic risks and are not altogether unwelcome. Secular techno-millennials and techno-apocalyptics can play critical roles in pushing reluctant institutions towards positive social change or to enact prophylactic policies just as religious millennialists have in the past. But the power of millennialism comes with large risks and potential cognitive errors which require vigilant self-interrogation to avoid.

Suggestions for further reading

Baumgartner, F.J. (1999). *Longing for the End: A History of Millennialism in Western Civilization* (London: St. Martin's Press). A history of apocalyptic expectations from Zoroastrianism to Waco.

Cohn, N. (1999). *The Pursuit of the Millennium: Revolutionary Millenarians and Mystical Anarchists of the Middle Ages* (New York: Oxford University Press, 1961, 1970, 1999). The classic text on medieval millennialism. Devotes much attention to Communism and Nazism.

Heard, A. (1999). *Apocalypse Pretty Soon: Travels in End-time America* (New York: W. W. Norton & Company). A travelogue chronicling a dozen contemporary millennial groups, religious and secular, from UFO cults and evangelical premillennialists to transhumanists and immortalists.

Leslie, J. (1998). *The End of the World: The Science and Ethics of Human Extinction.* London, New York. Routledge. A catalogue of real apocalyptic threats facing humanity.

Manuel, F.E. and Fritzie, P.M. (1979). *Utopian Thought in the Western World* (Cambridge: Harvard University Press). The classic study of utopian thought and thinkers, from Bacon to Marx.

Noble, D. (1998). *The Religion of Technology: The Divinity of Man and the Spirit of Invention* (New York: Alfred A. Knopf). A somewhat overwrought attempt to unveil the millennial and religious roots of the space programme, artificial intelligence, and genetic engineering. Argues that there is a continuity of medieval pro-technology theologies with contemporary techno-millennialism.

Olson, T. (1982). *Millennialism, Utopianism, and Progress* (Toronto: University of Toronto Press). Argues millennialism and utopianism were separate traditions that jointly shaped the modern secular idea of social progress, and post-millennialist Social Gospel religious movements.

References

Annas, G.J., Andrews, L., and Isasi, R. (2002). Protecting the endangered human: toward an international treaty prohibiting cloning and inheritable alterations. *Am. J. Law Med.*, **28**, 151–178.

Bailey, R. (2001a). Rebels against the future: witnessing the birth of the global anti-technology movement. *Reason,* 28 February. http://www.reason.com/news/show/34773.html

Bailey, R. (2001b). Rage against the machines: witnessing the birth of the neo-Luddite movement. *Reason,* July. http://www.reason.com/news/show/28102.html

Barkun, M. (1974). *Disaster and the Millennium* (New Haven, IN: Yale University Press).

Blum, M.L. (2002). *The Origins and Development of Pure Land Buddhism* (New York: Oxford University Press).

Bock, D.L. (1999). *Three Views on the Millennium and Beyond. Premillennialism, Postmillennialism, Amillennialism.* Zondervan. Grand Rapids, Michigca (Mi).

Boyer, P. (1994). *When Time Shall Be No More: Prophecy Belief in Modern American Culture*. Belknap, Cambridge MA.
Camp, G.S. (1997). *Selling Fear: Conspiracy Theories and End-time Paranoia*. Baker. Grand Rapids MI.
CNN. (1998). Survivalists try to prevent millennium-bug bite. October 10. http://www.cnn.com/US/9810/10/y2k.survivalists/
Cohn, N. (1970). *The Pursuit of the Millennium: Revolutionary Millenarians and Mystical Anarchists of the Middle Ages* (New York: Oxford University Press).
Cook, D. (2005). *Contemporary Muslim Apocalyptic Literature* (Syracuse, NY: Syracuse University Press).
Crossley-Holland, K. (1981). *The Norse Myths* (Tandem Library).
Crutchfield, L. (1992). *Origins of Dispensationalism: The Darby Factor* (University Press of America).
Dake, K. (1991). Orienting dispositions in the perception of risk: an analysis of contemporary worldviews and cultural biases. *J. Cross-cultural Psychol.*, **22**, 61–82.
Ellis, G.F.R. (2002). *The Far-future Universe: Eschatology from a Cosmic Perspective* (Templeton Foundation).
Ellul, J. (1967). *The Technological Society* (Vintage).
ETC. (2003). The Big Down. ETC. http://www.etcgroup.org/documents/TheBigDown.pdf
Feder, B.J. (1999). On the year 2000 front, humans are the big wild cards. New York Times, 28 December.
Furnish, T. (2005). *Holiest Wars: Islamic Mahdis, Their Jihads, and Osama bin Laden* (Westport: Praeger).
Garreau, J. (2006). *Radical Evolution: The Promise and Peril of Enhancing Our Minds, Our Bodies-and What It Means to Be Human* (New York: Broadway).
Garrett, L. (1994). *The Coming Plague: Newly Emerging Diseases in a World Out of Balance* (New York: Farrar, Straus & Cudahy).
Garrett, L. (2000). *Betrayal of Trust: The Collapse of Global Public Health* (Hyperion).
Gastil, J., Braman, D., Kahan, D.M., and Slovic, P. (2005). The 'Wildavsky Heuristic': The Cultural Orientation of Mass Political Opinion. Yale Law School, Public Law Working Paper No. 107. http://ssrn.com/abstract=834264
Hall, J.R. (2000). *Apocalypse Observed: Religious Movements and Violence in North America, Europe, and Japan* (London: Routledge).
Halpern, P. (2001). *Countdown to Apocalypse: A Scientific Exploration of the End of the World* (New York, NY: Basic Books).
Heard, A. (1999). *Apocalypse Pretty Soon: Travels in End-time America* (New York: W. W. Norton & Company).
Hobsbawm, E.J. (1959). *Primitive Rebels: Studies in Archaic Forms of Social Movement in the 19th and 20th Centuries* (New York: W. W. Norton & Company).
Hoekema, A. (2007). Amillenialism. http://www.the-highway.com/amila_Hoekema.html
Hughes, J.J. (2007). The compatibility of religious and transhumanist views of metaphysics, suffering, virtue and transcendence in an enhanced future. *The Global Spiral*, **8**(2). http://metanexus.net/magazine/tabid/68/id/9930/Default.aspx

Jacoby, R. (2005). *Picture Imperfect: Utopian Thought for an Anti-Utopian Age* (New York: Columbia University Press).
Jones, S. (2006). *Against Technology: From the Luddites to Neo-Luddism* (NY: Routledge).
Joy, B. (2000). Why the future doesn't need us. *Wired*, April. http://www.wired.com/wired/archive/8.04/joy.html
Kahan, D.M. (2008). Two conceptions of emotion in risk regulation. *University of Pennsylvania Law Review*, **156**. http://ssrn.com/abstract=962520
Kaplan, J. (1997). *Radical Religion in America: Millenarian Movements from the Far Right to the Children of Noah* (Syracuse, NY: Syracuse University Press).
Katz, D.S. and Popkin, R.H. (1999). *Messianic Revolution: Radical Religious Politics to the End of the Second Millennium* (New York: Hill and Wang).
Kellner, M.A. (1999). *Y2K: Apocalypse or Opportunity?* (Wheaton Illinois: Harold Shaw Publications).
Kelly, K. (2007). The Maes-Garreau Point. *The Technium*, March 14. http://www.kk.org/thetechnium/archives/2007/03/the_maesgarreau.php
King, M.L. (1950). The Christian Pertinence of Eschatological Hope. http://www.stanford.edu/group/King/publications/papers/vol1/500215-The_Christian_Pertinence_of_Eschatological_Hope.htm
Klaw, S. (1993). *Without Sin: The Life and Death of the Oneida Community* (New York: Allen Lane, Penguin Press).
Kliesen, K.L. (2003). Was Y2K behind the business investment boom and bust? Review, Federal Reserve Bank of St. Louis, January: 31-42.
Kunstler, J.H. (2006). *The Long Emergency: Surviving the End of Oil, Climate Change, and Other Converging Catastrophes of the Twenty-First Century* (New York: Grove Press).
Kurzweil, R. (2005). *The Singularity is Near* (New York: Viking).
Kurzweil, R. (2006). Questions and answers on the singularity. *Non-Prophet*, January 8. http://nonprophet.typepad.com/nonprophet/2006/01/guest_blogger_r.html
Kurzweil, R. and Joy, B. (2005). Recipe for Destruction. *New York Times*, 17 October. http://www.nytimes.com/2005/10/17/opinion/17kurzweiljoy.html
Lanier, J. (2000). One half of a manifesto. *Edge*, **74**. http://www.edge.org/documents/archive/edge74.html
Lanternari, V. (1965). *The Religions of the Oppressed: A Study of Modern Messianic Cults* (New York: Alfred A. Knopf).
Leslie, J. (1998). *The End of the World: The Science and Ethics of Human Extinction* (London, New York: Routledge).
Lindsey, H. (1970). *The Late Great Planet Earth* (Zondervan).
Malalgoda, K. (1970). Millennialism in Relation to Buddhism. *Comp. Studies Soc. History*, **12**(4), 424–441.
Mander, J. (1992). *In the Absence of the Sacred: The Failure of Technology and the Survival of the Indian Nations* (San Francisco, California: Sierra Club Books).
Manuel, F.E. and Manuel, F.P. (1979). *Utopian Thought in the Western World* (Cambridge, MA: Belknap Press).
Mason, C. (2002). *Killing for Life: The Apocalyptic Narrative of Pro-life Politics* (Cornell: Cornell University Press).

Mooney, J. (1991). *The Ghost-Dance Religion and the Sioux Outbreak of* 1890 (University of Nebraska Press).
Moravec, H. (1990). *Mind Children: The Future of Robot and Human Intelligence* (Harvard: Harvard University Press).
Moravec, H. (2000). *Robot: Mere Machine to Transcendent Mind* (OX: Oxford University Press).
Mussington, D. (2002). *Concepts for Enhancing Critical Infrastructure Protection: Relating Y2K to Cip Research and Development* (Washington, DC: Rand Corporation).
Naquin, S. (1976). *Millenarian Rebellion in China: The Eight Trigrams Uprising of* 1813 (New Haven, CT: Yale University Press).
Nash, D. (2000). The failed and postponed millennium: secular millennialism since the enlightenment. *J. Religious History*, **24**(1), 70–86.
National Institute of Allergy and Infectious Diseases. (2006). Experimental Vaccine Protects Mice Against Deadly 1918 Flu Virus. *NIH News*. http://www.nih.gov/news/pr/oct2006/niaid-17.htm
Postman, N. (1993). *Technopoly: The Surrender of Culture to Technology* (NY: Vintage).
Rhodes, J.M. (1980). *The Hitler Movement: A Modern Millenarian Revolution* (Stanford, CA: Hoover Institution Press, Stanford University).
Rees, M. (2004). *Our Final Hour* (New York, NY: Basic Books).
Riddlebarger, K. (2003). *A Case for Amillennialism: Understanding the End Times* (Grand Rapids, MI: Baker Books).
Roszak, T. (1986). *The Cult of Information: A Neo-Luddite Treatise on High-tech, Artificial Intelligence, and the True Art of Thinking* (Berkeley: University of California Press).
Rowley, D. (1983). Redeemer Empire: Russian Millenarianism. *Am. Historical Rev.*, **104**(5). http://www.historycooperative.org/journals/ahr/104.5/ah001582.html
Russell, P. (1983). *The Global Brain: Speculation on the Evolutionary Leap to Planetary Consciousness* (Los Angels: Tarcher).
Sale, K. (2001). *Rebels Against the Future: The Luddites and Their War on the Industrial Revolution: Lessons for the Computer Age* (New York, NY: Basic Books).
Schaefer, N.A. (2004). Y2K as an endtime sign: apocalypticism in America at the fin-de-millennium. *J. Popular Cult.*, **38**(1), 82–105.
Schmidhuber, J. (2006). New millennium AI and the convergence of history. In Duch, W. and Mandziuk, J. (eds.), *Challenges to Computational Intelligence* (Berlin: Springer). http://arxiv.org/abs/cs.AI/0606081
Seidensticker, B. (2006). *Future Hype: The Myths of Technology Change* (San Francisco, CA: Berrett-Koehler Publishers).
Seidensticker, B. (2005). Brief History of Intellectual Discussion Of Accelerating Change. http://www.accelerationwatch.com/history_brief.html
Smart, J. (2007). Why 'Design' (A Universe Tuned for Life and Intelligence) Does Not Require a Designer, and Teleology (a Theory of Destiny) is Not a Theology – Understanding the Paradigm of Evolutionary Development. http://www.accelerationwatch.com/
Smith, C. (1999). 'Do Apocalyptic World Views Cause Violence?' The Religious Movements Homepage Project at the University of Virginia. http://religiousmovements.lib.virginia.edu/nrms/millennium/violence.html

Special Committee on the Year 2000 Technology Problem. (2000). *Y2K Aftermath – Crisis Averted: Final Committee Report. February* 2000. U.S. Senate.

Stock, G. (1993). *Metaman: The Merging of Humans and Machines into a Global Superorganism* (New York: Simon & Schuster).

Stone, J. (1985). Seeking enlightenment in the last age: Mappo Thought in Kamakura Buddhism. *Eastern Buddhist*, **18**(1), 28–56.

Tapia, A.H. (2003). Technomillennialism: a subcultural response to the technological threat of Y2K. *Sci., Technol. Human Values*, **28**(4), 483–512.

Teilhard de Chardin, P. 1955. *The Phenomenon of Man.* (New York: Harper & Row).

Thompson, D. (1997). *The End of Time: Faith and Fear in the Shadow of the Millennium* (Hanover, NH: University Press of New England).

Tuveson, E.L. (1949). *Millennium and Utopia: A Study in the Background of the Idea of Progress* (Berkeley, CA: University of California Press).

Vinge, V. (1993). The Coming Technological Singularity: How to Survive in the Post-Human Era. Presented at the VISION-21 Symposium sponsored by NASA Lewis Research Center and the Ohio Aerospace Institute, 30–31 March 1993. http://www-rohan.sdsu.edu/faculty/vinge/misc/singularity.html

Wagar, W.W. (1982). *Terminal Visions: The Literature of Last Things* (Bloomington: Indiana University Press).

Whalen, R.K. (2000). Premillennialism. In Landes, R.A. (ed.), *The Encyclopedia of Millennialism and Millennial Movements*, pp. 331 (New York: Routledge).

Whitsel, B. (1998). The Turner Diaries and Cosmotheism: William Pierce's Theology of Revolution. *Nova Religio*, **1**(2), 183–197.

Wilson, D. (2007). *Where's My Jetpack?: A Guide to the Amazing Science Fiction Future that Never Arrived* (Bloomsbury).

Wildavsky, A.B. (1987). Choosing preferences by constructing institutions: a cultural theory of preference formation. *Am. Polit. Sci. Rev.*, **81**, 3–21.

Wildavsky, A.B. and Douglas, M. (1982). *Risk and Culture: An Essay on the Selection of Technological and Environmental Dangers* (Berkeley, CA: University of California Press).

Wojcik, D. (1997). *The End of the World as We Know It: Faith, Fatalism, and Apocalypse in America* (New York: New York University Press).

Wojcik, D. (1999). Secular apocalyptic themes in the Nuclear Era. In D. Wojcik (ed.), *The End of the World As We Know It: Faith, Fatalism, and Apocalypse in America* (New York: New York University Press).

Zerzan, J. (2002). *Running on Emptiness: The Pathology of Civilization* (Los Angeles, CA: Feral House).

·5·

Cognitive biases potentially affecting judgement of global risks

Eliezer Yudkowsky

5.1 Introduction

All else being equal, not many people would prefer to destroy the world. Even faceless corporations, meddling governments, reckless scientists, and other agents of doom, require a world in which to achieve their goals of profit, order, tenure, or other villainies. If our extinction proceeds slowly enough to allow a moment of horrified realization, the doers of the deed will likely be quite taken aback on realizing that they have actually destroyed the world. Therefore I suggest that if the Earth is destroyed, it will probably be by mistake.

The systematic experimental study of reproducible errors of human reasoning, and what these errors reveal about underlying mental processes, is known as the heuristics and biases programme in cognitive psychology. This programme has made discoveries *highly* relevant to assessors of global catastrophic risks. Suppose you are worried about the risk of Substance P, an explosive of planet-wrecking potency which will detonate if exposed to a strong radio signal. Luckily there is a famous expert who discovered Substance P, spent the last thirty years working with it, and knows it better than anyone else in the world. You call up the expert and ask how strong the radio signal has to be. The expert replies that the critical threshold is probably around 4000 terawatts. 'Probably?' you query. 'Can you give me a 98% confidence interval?' 'Sure', replies the expert. 'I'm 99% confident that the critical threshold is above 500 terawatts, and 99% confident that the threshold is below 80,000 terawatts.' 'What about 10 terawatts?' you ask. 'Impossible', replies the expert.

The above methodology for expert elicitation looks perfectly reasonable, the sort of thing any competent practitioner might do when faced with such a problem. Indeed, this methodology was used in the Reactor Safety Study (Rasmussen, 1975), now widely regarded as the first major attempt at probabilistic risk assessment. But the student of heuristics and biases will recognize at least two major mistakes in the method – not logical flaws, but conditions extremely susceptible to human error. I shall return to this example in the discussion of anchoring and adjustments biases (Section 5.7).

The heuristics and biases programme has uncovered results that may startle and dismay the unaccustomed scholar. Some readers, first encountering the experimental results cited here, may sit up and say: 'Is that really an experimental result? Are people really such poor guessers? Maybe the experiment was poorly designed, and the result would go away with such-and-such manipulation.' Lacking the space for exposition, I can only plead with the reader to consult the primary literature. The obvious manipulations have already been tried, and the results found to be robust.

5.2 Availability

Suppose you randomly sample a word of three or more letters from an English text. Is it more likely that the word starts with an 'R' ('rope'), or that 'R' is its third letter ('park')?

A general principle underlying the heuristics and biases programme is that human beings use methods of thought – *heuristics* – which quickly return good approximate answers in many cases; but which also give rise to systematic errors called *biases*. An example of a heuristic is to judge the frequency or probability of an event by its *availability*, the ease with which examples of the event come to mind. 'R' appears in the third-letter position of more English words than in the first-letter position, yet it is much easier to recall words that begin with 'R' than words whose third letter is 'R'. Thus, a majority of respondents guess that words beginning with 'R' are more frequent, when the reverse is the case (Tversky and Kahneman, 1973).

Biases implicit in the availability heuristic affect estimates of risk. A pioneering study by Lichtenstein et al. (1978) examined absolute and relative probability judgements of risk. People know in general terms which risks cause large numbers of deaths and which cause few deaths. However, asked to quantify risks more precisely, people severely overestimate the frequency of rare causes of death and severely underestimate the frequency of common causes of death. Other repeated errors were also apparent: accidents were judged to cause as many deaths as disease. (Diseases cause about 16 times as many deaths as accidents.) Homicide was incorrectly judged a more frequent cause of death than diabetes or stomach cancer. A follow-up study by Combs and Slovic (1979) tallied *reporting* of deaths in two newspapers, and found that errors in probability judgements correlated strongly (.85 and .89) with selective reporting in newspapers.

People refuse to buy flood insurance even when it is heavily subsidized and priced far below an actuarially fair value. Kates 1962 suggest that underreaction to threats of flooding may arise from 'the inability of individuals to conceptualize floods that have never occurred . . . Men on flood plains appear to be very much prisoners of their experience . . . Recently experienced floods appear to set an upward bound to the size of loss with which managers believe they ought to be concerned'. Burton et al. (1978) report that when dams and levees are built, they reduce the frequency of floods, and thus apparently create a false sense of security, leading to reduced precautions. While building dams decreases the *frequency* of floods, damage *per flood* is so much greater afterwards that the average yearly damage *increases*.

It seems that most people do not extrapolate from experienced small hazards to a possibility of large risks; rather, the past experience of small hazards sets a perceived upper bound on risks. A society well protected against minor hazards will take no action against major risks (building on flood plains once the regular minor floods are eliminated). A society subject to regular minor hazards will treat those minor hazards as an upper bound on the size of the risks (guarding against regular minor floods but not occasional major floods).

Risks of human extinction may tend to be underestimated since, obviously, humanity has never yet encountered an extinction event.[1]

5.3 Hindsight bias

Hindsight bias is when subjects, after learning the eventual outcome, give a much higher estimate for the *predictability* of that outcome than subjects who predict the outcome without advance knowledge. Hindsight bias is sometimes called the I-knew-it-all-along effect.

Fischhoff and Beyth (1975) presented students with historical accounts of unfamiliar incidents such as a conflict between the Gurkhas and the British in 1814. Given the account as background knowledge, five groups of students were asked what they would have predicted as the *probability* for each of four outcomes: British victory, Gurkha victory, stalemate with a peace settlement, or stalemate with no peace settlement. Four experimental groups were, respectively, told that these four outcomes were the historical outcome. The fifth, control group was not told any historical outcome. In every case, a group told an outcome assigned substantially higher probability to that outcome, than did any other group or the control group.

Hindsight bias is important in legal cases, where a judge or jury must determine whether a defendant was legally negligent in failing to foresee a hazard (Sanchiro, 2003). In an experiment based on an actual legal case, Kamin and Rachlinski (1995) asked two groups to estimate the probability of flood damage caused by blockage of a city-owned drawbridge. The control

[1] Milan M. Ćirković points out that the Toba supereruption (~73,000 BCE) may count as a near-extinction event. The blast and subsequent winter killed off a super majority of humankind; genetic evidence suggests there were only a few thousand survivors, perhaps even less (Ambrose, 1998). Note that this event is not in our *historical* memory – it predates writing.

group was told only the background information known to the city when it decided not to hire a bridge watcher. The experimental group was given this information, plus the fact that a flood had actually occurred. Instructions stated that the city was negligent if the foreseeable probability of flooding was greater than 10%. As many as 76% of the control group concluded the flood was so unlikely that no precautions were necessary; 57% of the experimental group concluded the flood was so likely that failure to take precautions was legally negligent. A third experimental group was told the outcome and also explicitly instructed to avoid hindsight bias, which made no difference: 56% concluded the city was legally negligent. Judges cannot simply instruct juries to avoid hindsight bias; that debiasing manipulation has no significant effect.

When viewing history through the lens of hindsight, we vastly *underestimate* the cost of preventing catastrophe. In 1986, the space shuttle *Challenger* exploded for reasons eventually traced to an O-ring losing flexibility at low temperature (Rogers et al., 1986). There were warning signs of a problem with the O-rings. But preventing the *Challenger* disaster would have required, not attending to the problem with the O-rings, but attending to *every* warning sign which seemed as severe as the O-ring problem, *without benefit of hindsight.*

5.4 Black Swans

Taleb (2005) suggests that hindsight bias and availability bias bear primary responsibility for our failure to guard against what he calls *Black Swans.* Black Swans are an especially difficult version of the problem of the fat tails: sometimes *most of* the variance in a process comes from exceptionally rare, exceptionally huge events. Consider a financial instrument that earns $10 with 98% probability, but loses $1000 with 2% probability; it is a poor net risk, but it looks like a steady winner. Taleb (2001) gives the example of a trader whose strategy worked for 6 years without a single bad quarter, yielding close to $80 million – then lost $300 million in a single catastrophe.

Another example is that of Long-term Capital Management (LTCM), a hedge fund whose founders included two winners of the Nobel Prize in Economics. During the Asian currency crisis and Russian bond default of 1998, the markets behaved in a literally *unprecedented* fashion, assigned a negligible probability by LTCM's historical model. As a result, LTCM began to lose $100 million per day, day after day. On a single day in 1998, LTCM lost more than $500 million (Taleb, 2005).

The founders of LTCM later called the market conditions of 1998 a '10-sigma event'. But obviously it was *not* that improbable. Mistakenly believing that the past was predictable, people conclude that the future is predictable. As

Fischhoff (1982) puts it:

When we attempt to understand past events, we implicitly test the hypotheses or rules we use both to interpret and to anticipate the world around us. If, in hindsight, we systematically underestimate the surprises that the past held and holds for us, we are subjecting those hypotheses to inordinately weak tests and, presumably, finding little reason to change them.

The lesson of history is that swans happen. People are surprised by catastrophes lying outside their anticipation, beyond their historical probability distributions. Then why are we so taken aback when Black Swans occur? Why did LTCM borrow a leverage of $125 billion against $4.72 billion of equity, almost ensuring that *any* Black Swan would destroy them?

Because of hindsight bias, we learn *overly specific* lessons. After September 11, the US Federal Aviation Administration prohibited box-cutters on airplanes. The hindsight bias rendered the event too predictable in retrospect, permitting the angry victims to find it the result of 'negligence' – such as intelligence agencies' failure to distinguish warnings of Al Qaeda activity amid a thousand *other* warnings. We learned not to allow hijacked planes to overfly our cities. We did not learn the lesson: 'Black Swans do occur; do what you can to prepare for the unanticipated.'

Taleb (2005) writes:

It is difficult to motivate people in the prevention of Black Swans ... Prevention is not easily perceived, measured, or rewarded; it is generally a silent and thankless activity. Just consider that a costly measure is taken to stave off such an event. One can easily compute the costs while the results are hard to determine. How can one tell its effectiveness, whether the measure was successful or if it just coincided with no particular accident? ... Job performance assessments in these matters are not just tricky, but may be biased in favor of the observed 'acts of heroism'. History books do not account for heroic preventive measures.

5.5 The conjunction fallacy

Linda is thirty-one year old, single, outspoken, and very bright. She majored in philosophy. As a student, she was deeply concerned with issues of discrimination and social justice, and also participated in anti-nuclear demonstrations.

Rank the following statements from most probable to least probable:

1. Linda is a teacher in an elementary school.
2. Linda works in a bookstore and takes Yoga classes.
3. Linda is active in the feminist movement.
4. Linda is a psychiatric social worker.
5. Linda is a member of the League of Women Voters.

6. Linda is a bank teller.
7. Linda is an insurance salesperson.
8. Linda is a bank teller and is active in the feminist movement.

Among the eighty-eight undergraduate subjects, 89% ranked statement (8) as more probable than (6) (Tversky and Kahneman, 1982). Since the given description of Linda was chosen to be similar to a feminist and dissimilar to a bank teller, (8) is more *representative* of Linda's description. However, ranking (8) as more *probable* than (6) violates the conjunction rule of probability theory which states that $p(A \& B) \leq p(A)$. Imagine a sample of 1000 women; surely more women in this sample are bank tellers than are feminist bank tellers.

Could the conjunction fallacy rest on subjects interpreting the experimental instructions in an unanticipated way? Perhaps subjects think that by 'probable' is meant the probability of Linda's description given statements (6) and (8), rather than the probability of (6) and (8) given Linda's description. It could also be that subjects interpret (6) to mean 'Linda is a bank teller and is not active in the feminist movement'. Although many creative alternative hypotheses have been invented to explain away the conjunction fallacy, the conjunction fallacy has survived all experimental tests meant to disprove it; see, for example, Sides et al. (2002) for a summary. For example, the following experiment excludes both of the alternative hypotheses proposed earlier.

Consider a regular six-sided die with four green faces and two red faces. The die will be rolled 20 times and the sequence of greens (G) and reds (R) will be recorded. You are asked to select one sequence, from a set of three, and you will win $ 25 if the sequence you chose appears on successive rolls of the die. Please check the sequence of greens and reds on which you prefer to bet.

1. RGRRR
2. GRGRRR
3. GRRRRR

A total of 125 undergraduates at University of British Columbia (UBC) and Stanford University played this gamble with real pay-offs. Among them, 65% of subjects chose sequence (2) (Tversky and Kahneman, 1983). Sequence (2) is most *representative* of the die, since the die is mostly green and sequence (2) contains the greatest proportion of green faces. However, sequence (1) *dominates* sequence (2) because (1) is strictly included in (2) – to get (2) you must roll (1) *preceded* by a green face.

In the above-mentioned task, the exact probabilities for each event could in principle have been calculated by the students. However, rather than go to the effort of a numerical calculation, it would seem that (at least 65% of) the students made an intuitive guess, based on which sequence seemed most 'representative' of the die. Calling this 'the representativeness heuristic' does not imply that students deliberately decided that they would estimate probability by estimating similarity. Rather, the representativeness heuristic is what produces the intuitive sense that sequence (2) 'seems more likely' than sequence (1). In other words, the 'representativeness heuristic' is a built-in feature of the brain for producing rapid probability judgements, rather than a consciously adopted procedure. We are not *aware* of substituting judgement of representativeness for judgement of probability.

The conjunction fallacy similarly applies to futurological forecasts. Two independent sets of professional analysts at the Second International Congress on Forecasting were asked to rate, respectively, the probability of 'a complete suspension of diplomatic relations between the United States and the Soviet Union sometime in 1983' or 'a Russian invasion of Poland, and a complete suspension of diplomatic relations between the United States and the Soviet Union sometime in 1983.' The second set of analysts responded with significantly higher probabilities (Tversky and Kahneman 1983).

In a study by Johnson et al. (1993), MBA students at Wharton were scheduled to travel to Bangkok as part of their degree programme. Several groups of students were asked how much they were willing to pay for terrorism insurance. One group of subjects was asked how much they were willing to pay for terrorism insurance covering the flight *from* Thailand *to* the United States. A second group of subjects was asked how much they were willing to pay for terrorism insurance covering the round-trip flight. A third group was asked how much they were willing to pay for terrorism insurance that covered the complete trip to Thailand. These three groups responded with average willingness to pay of $17.19, $13.90, and $7.44, respectively.

According to probability theory, adding a detail to a hypothesis *must* render the hypothesis less probable. It is less probable that Linda is a feminist bank teller than that she is a bank teller, since all feminist bank tellers are necessarily bank tellers. Yet human psychology seems to follow the rule that *adding a detail can make the story more plausible.*

People might pay more for international diplomacy intended to prevent nanotechnological warfare *by China* than for an engineering project to defend against nanotechnological attack *from any source.* The second threat scenario is less vivid and alarming, but the defence is more useful *because* it is more vague. More valuable still would be strategies which make humanity harder to extinguish without being specific to nanotechnological threats – such as colonizing space, or see Chapter 15 (this volume) on Artificial Intelligence. Security expert Bruce Schneier observed (both before and after the 2005 hurricane in New Orleans) that the US government was guarding *specific* domestic targets against 'movie-plot scenarios' of terrorism, at the cost of taking away resources from emergency-response capabilities that could respond to *any* disaster (Schneier, 2005).

Overly detailed reassurances can also create false perceptions of safety: 'X is *not* an existential risk and you don't need to worry about it, because of A, B, C, D, and E'; where the failure of any *one* of propositions A, B, C, D, or E potentially extinguishes the human species. 'We don't need to worry about nanotechnological war, because a UN commission will initially develop the technology and prevent its proliferation until such time as an active shield is developed, capable of defending against all accidental and malicious outbreaks that contemporary nanotechnology is capable of producing, and this condition will persist indefinitely.' Vivid, specific scenarios can inflate our probability estimates of security, as well as misdirecting defensive investments into needlessly narrow or implausibly detailed risk scenarios.

More generally, people tend to overestimate conjunctive probabilities and underestimate disjunctive probabilities (Tversky and Kahneman, 1974). That is, people tend to overestimate the probability that, for example, seven events of 90% probability will *all* occur. Conversely, people tend to underestimate the probability that *at least one* of seven events of 10% probability will occur. Someone judging whether to, for instance, incorporate a new start-up, must evaluate the probability that many individual events will *all* go right (there will be sufficient funding, competent employees, customers will want the product) while also considering the likelihood that *at least one* critical failure will occur (the bank refuses a loan, the biggest project fails, the lead scientist dies). This may help explain why only 44% of entrepreneurial ventures[2] survive after four years (Knaup, 2005).

Dawes (1988, p. 133) observes: 'In their summations lawyers avoid arguing from disjunctions ('either this or that or the other could have occurred, all of which would lead to the same conclusion') in favor of conjunctions. Rationally, of course, disjunctions are *much* more probable than are conjunctions.'

The scenario of humanity going extinct in the next century is a disjunctive event. It could happen as a result of any of the existential risks discussed in this book – or some other cause which none of us foresaw. Yet for a futurist, disjunctions make for an awkward and unpoetic-sounding prophecy.

5.6 Confirmation bias

Peter Wason (1960) conducted a now-classic experiment that became known as the '2–4–6' task. Subjects had to *discover a rule*, known to the experimenter but not to the subject – analogous to scientific research. Subjects wrote three numbers, such as '2–4–6' or '10–12–14', on cards, and the experimenter told them whether the triplet *fit* the rule or *did not fit* the rule. Initially subjects were given the triplet 2–4–6, and told that this triplet fitted the rule. Subjects could

[2] Note that the figure 44% is for all new businesses, including small restaurants, rather than, say, dot-com start-ups.

continue testing triplets until they felt sure they knew the experimenter's rule, at which point the subject announced the rule.

Although subjects typically expressed high confidence in their guesses, only 21% of Wason's subjects guessed the experimenter's rule, and replications of Wason's experiment usually report success rates of around 20%. Contrary to the advice of Karl Popper, subjects in Wason's task try to *confirm* their hypotheses rather than *falsifying* them. Thus, someone who forms the hypothesis 'Numbers increasing by two' will test the triplets 8–10–12 or 20–22–24, hear that they fit, and confidently announce the rule. Someone who forms the hypothesis X–2X–3X will test the triplet 3–6–9, discover that it fits, and then announce that rule. In every case the *actual* rule is the same: the three numbers must be in ascending order. In some cases subjects devise, 'test', and announce rules far more complicated than the actual answer.

Wason's 2–4–6 task is a 'cold' form of *confirmation bias*; people seek confirming but not falsifying evidence. 'Cold' means that the 2–4–6 task is an emotionally neutral case of confirmation bias; the belief held is logical, not emotional. 'Hot' refers to cases where the belief is emotionally charged, such as political argument. Unsurprisingly, 'hot' confirmation biases are stronger – larger in effect and more resistant to change. Active, effortful confirmation biases are labelled *motivated cognition* (more ordinarily known as 'rationalization'). As stated by Brenner et al. (2002, p. 503) in 'Remarks on Support Theory':

> Clearly, in many circumstances, the desirability of believing a hypothesis may markedly influence its perceived support ... Kunda (1990) discusses how people who are motivated to reach certain conclusions attempt to construct (in a biased fashion) a compelling case for their favored hypothesis that would convince an impartial audience. Gilovich (2000) suggests that conclusions a person does not want to believe are held to a higher standard than conclusions a person wants to believe. In the former case, the person asks if the evidence *compels* one to accept the conclusion, whereas in the latter case, the person asks instead if the evidence *allows* one to accept the conclusion.

When people subject disagreeable evidence to more scrutiny than agreeable evidence, this is known as *motivated* scepticism or *disconfirmation bias*. Disconfirmation bias is especially destructive for two reasons: (1) Two biased reasoners considering the *same* stream of evidence can shift their beliefs in *opposite* directions – both sides selectively accepting only favourable evidence. Gathering more evidence may not bring biased reasoners to agreement. (2) People who are more skilled sceptics – who know a larger litany of logical flaws – but apply that skill *selectively*, may change their minds more slowly than *unskilled* reasoners.

Taber and Lodge (2000) examined the prior attitudes and attitude changes of students – when exposed to political literature for and against gun control and affirmative action, two issues of particular salience in the political life of the United States. The study tested six hypotheses using two experiments:

1. *Prior attitude effect.* Subjects who feel strongly about an issue – even when encouraged to be objective – will evaluate supportive arguments more favourably than contrary arguments.
2. *Disconfirmation bias.* Subjects will spend more time and cognitive resources denigrating contrary arguments than supportive arguments.
3. *Confirmation bias.* Subjects free to choose their information sources will seek out supportive rather than contrary sources.
4. *Attitude polarization.* Exposing subjects to an apparently balanced set of pro and con arguments will exaggerate their initial polarization.
5. *Attitude strength effect.* Subjects voicing stronger attitudes will be more prone to the above biases.
6. *Sophistication effect.* Politically knowledgeable subjects, because they possess greater ammunition with which to counter-argue incongruent facts and arguments, will be more prone to the above biases.

Ironically, Taber and Lodge's experiments confirmed all six of the authors' prior hypotheses. Perhaps you will say: 'The experiment only reflects the beliefs the authors started out with – it is just a case of confirmation bias.' If so, then by making you a more sophisticated arguer – by teaching you another bias of which to accuse people – I have actually harmed you; I have made you slower to react to evidence. I have given you another opportunity to fail each time you face the challenge of changing your mind.

Heuristics and biases are widespread in human reasoning. Familiarity with heuristics and biases can enable us to detect a wide variety of logical flaws that might otherwise evade our inspection. But, as with *any* ability to detect flaws in reasoning, this inspection must be applied *even-handedly* – both to our own ideas and the ideas of others; to ideas which discomfort us and also to ideas which comfort us. Awareness of human fallibility is dangerous knowledge, if you remind yourself of the fallibility of those who disagree with you. If I am selective about *which* arguments I inspect for errors, or even *how hard* I inspect for errors, then every new rule of rationality I learn, every new logical flaw I know how to detect, makes me that much stupider. Intelligence, to be useful, must be used for something other than defeating itself.

You cannot 'rationalize' what is not rational to begin with – as if lying were called 'truthization'. There is no way to obtain more truth for a proposition by bribery, flattery, or the most passionate argument – you can make more people *believe* the proposition, but you cannot make it more *true*. To improve the truth of our beliefs we *must* change our beliefs. Not every change is an improvement, but every improvement is necessarily a change.

Our beliefs are more swiftly determined than we think. Griffin and Tversky (1992) discreetly approached twenty-four colleagues faced with a choice between two job offers and asked them to estimate the probability that they would choose each job offer. The average confidence in the choice assigned the greater probability was a modest 66%. Yet only one of twenty-four respondents chose the option initially assigned the lower probability, yielding an overall accuracy of 96% (one of few reported instances of human *under*confidence).

The moral may be that *once you can guess what your answer will be* – once you can assign a greater probability to your answering one way than another – you have, in all probability, already decided. And if you were honest with yourself, you would often be able to guess your final answer within seconds of hearing the question. We change our minds less often than we think. How fleeting is that brief unnoticed moment when we cannot yet guess what our answer will be, the tiny fragile instant when there is a chance for intelligence to act – in questions of choice, as in questions of fact.

Thor Shenkel said: 'It ain't a true crisis of faith unless things could just as easily go either way.'

Norman R.F. Maier said: 'Do not propose solutions until the problem has been discussed as thoroughly as possible without suggesting any.' Robyn Dawes, commenting on Maier, said: 'I have often used this edict with groups I have led – particularly when they face a very tough problem, which is when group members are most apt to propose solutions immediately.'

In computer security, a 'trusted system' is one that you *are in fact trusting*, not one that is in fact trustworthy. A 'trusted system' is a system which, if it is untrustworthy, can cause a failure. When you read a paper which proposes that a potential global catastrophe is impossible, or has a specific annual probability, or can be managed using some specific strategy, then you trust the rationality of the authors. You trust the authors' ability to be driven from a comfortable conclusion to an uncomfortable one, even in the absence of overwhelming experimental evidence to prove a cherished hypothesis wrong. You trust that the authors did not unconsciously look just a little bit harder for mistakes in equations that seemed to be leaning the wrong way, before you ever saw the final paper.

However, if authority legislates that the mere suggestion of an existential risk is enough to shut down a project, or if it becomes a de facto truth of the political process that no possible calculation can overcome the burden of a suggestion once made, no scientist will ever again make a suggestion, which is worse. I do not know how to solve this problem. But I think it would be well for estimators of existential risks to know something about heuristics and biases in general, and disconfirmation bias in particular.

5.7 Anchoring, adjustment, and contamination

An experimenter spins a 'Wheel of Fortune' device as you watch, and the Wheel happens to come up pointing to (version one) the number 65 or (version two) the number 15. The experimenter then asks you whether the percentage of African countries in the United Nations is above or below this number. After you answer, the experimenter asks you your estimate of the percentage of African countries in the United Nations.

Tversky and Kahneman (1974) demonstrated that subjects who were first asked if the number was above or below fifteen, later generated substantially lower percentage estimates than subjects first asked if the percentage was above or below 65. The groups' median estimates of the percentage of African countries in the United Nations were twenty-five and forty-five, respectively. This, even though the subjects had watched the number being generated by an apparently random device, the Wheel of Fortune, and hence believed that the number bore no relation to the actual percentage of African countries in the United Nations. Payoffs for accuracy did not change the magnitude of the effect. Tversky and Kahneman hypothesized that this effect was due to *anchoring and adjustment;* subjects took the initial uninformative number as their starting point, or *anchor,* and then *adjusted* the number up or down until they reached an answer that sounded plausible to them; then they stopped adjusting. The result was under-adjustment from the anchor.

In the example that opens this chapter, we *first* asked the expert on Substance P to guess the actual value for the strength of radio signal that would detonate Substance P, and only *afterwards* asked for confidence bounds around this value. This elicitation method leads people to adjust upwards and downwards *from their starting estimate,* until they reach values that 'sound improbable' and stop adjusting. This leads to under-adjustment and too-narrow confidence bounds.

Following the study by Tversky and Kahneman (1974), continued research showed a wider range of anchoring and pseudo-anchoring effects. Anchoring occurred even when they represented utterly implausible answers to the question; for example, asking subjects to estimate the year Einstein first visited the United States, after considering anchors of 1215 or 1992. These implausible anchors produced anchoring effects just as large as more plausible anchors such as 1905 or 1939 (Strack and Mussweiler, 1997). Walking down the supermarket aisle, you encounter a stack of cans of canned tomato soup, and a sign saying 'Limit 12 per customer'. Does this sign actually prompt people to buy more cans of tomato soup? According to empirical experiment, it does (Wansink et al., 1998).

Such generalized phenomena became known as *contamination* effects, since it turned out that almost *any* information could work its way into a cognitive judgement (Chapman and Johnson, 2002). Attempted manipulations to eliminate contamination include paying subjects for correct answers (Tversky and Kahneman, 1974), instructing subjects to avoid anchoring on the initial quantity (Quattrone et al., 1981), and facing real-world problems (Wansink et al., 1998). These manipulations did not decrease, or only slightly decreased, the magnitude of anchoring and contamination effects. Furthermore, subjects asked whether they had been influenced by the contaminating factor typically did not believe they had been influenced, when experiment showed they had been (Wilson et al., 1996).

A manipulation which consistently *increases* contamination effects is placing the subjects in cognitively 'busy' conditions such as rehearsing a word-string while working (Gilbert et al., 1988) or asking the subjects for quick answers (Gilbert and Osborne, 1989). Gilbert et al. (1988) attribute this effect to the extra task interfering with the ability to *adjust* away from the anchor; that is, less adjustment was performed in the cognitively busy condition. This decreases adjustment, hence increases the under-adjustment effect known as anchoring.

To sum up, information that is *visibly* irrelevant still anchors judgements and contaminates guesses. When people start from information known to be irrelevant and adjust until they reach a plausible-sounding answer, they under-adjust. People under-adjust more severely in cognitively busy situations and other manipulations that make the problem harder. People deny they are anchored or contaminated, even when experiment shows they are. These effects are not diminished or only slightly diminished by financial incentives, explicit instruction to avoid contamination, and real-world situations.

Now consider how many media stories on Artificial Intelligence cite the *Terminator* movies as if they were documentaries, and how many media stories on brain-computer interfaces mention *Star Trek's* Borg.

If briefly presenting an anchor has a substantial effect on subjects' judgements, how much greater an effect should we expect from reading an entire book or watching a live-action television show? In the ancestral environment, there were no moving pictures; whatever you saw with your own eyes was true. People do seem to realize, so far as conscious thoughts are concerned, that fiction is fiction. Media reports that mention *Terminator* do not *usually* treat Cameron's screenplay as a prophecy or a fixed truth. Instead the reporter seems to regard Cameron's vision as something that, having happened before, might well happen again – the movie is recalled (is *available*) as if it were an illustrative historical case. I call this mix of anchoring and availability the *logical fallacy of generalization from fictional evidence.*[3]

Storytellers obey strict rules of narrative unrelated to reality. Dramatic logic is not logic. Aspiring writers are warned that *truth is no excuse*: you may not justify an unbelievable event in your fiction by citing an instance of real life. A good story is painted with bright details, illuminated by glowing metaphors; a storyteller must be concrete, as hard and precise as stone. But in forecasting,

[3] A related concept is the *good-story bias* hypothesized in Bostrom (2001). Fictional evidence usually consists of 'good stories' in Bostrom's sense. Note that not all good stories are presented as fiction.

every added detail is an extra burden! Truth is hard work and not the kind of hard work done by storytellers. We should avoid not only being *duped* by fiction – failing to expend the mental effort necessary to 'unbelieve' it – but also being *contaminated* by fiction, letting it anchor our judgements. And we should be aware that we are not always aware of this contamination. Not uncommonly in a discussion of existential risk, the categories, choices, consequences, and strategies derive from movies, books, and television shows. There are subtler defeats, but this is outright surrender.

5.8 The affect heuristic

The *affect heuristic* refers to the way in which subjective impressions of 'goodness' or 'badness' can act as a heuristic, capable of producing fast perceptual judgements, and also systematic biases.

In a study by Slovic et al. (2002), two groups of subjects evaluated a scenario in which an airport had to decide whether to spend money to purchase new equipment, while critics argued that money should be spent on other aspects of airport safety. The response scale ranged from zero (would not support at all) to twenty (very strong support). A measure that was described as 'Saving 150 lives' had mean a support of 10.4, whereas a measure that was described as 'Saving 98% of 150 lives' had a mean support of 13.6. Even 'Saving 85% of 150 lives' had higher support than simply 'Saving 150 lives'. The hypothesis motivating the experiment was that saving 150 lives sounds diffusely good and is therefore only weakly evaluable, whereas saving 98% of something is clearly very good because it is so close to the upper bound on the percentage scale.

Finucane et al. (2000) wondered if people conflated their assessments of the *possible benefits* of a technology such as nuclear power, and their assessment of *possible risks,* into an overall good or bad feeling about the technology. Finucane et al. tested this hypothesis by providing four kinds of information that would increase or decrease perceived risk or perceived benefit. There was no logical relation between the information provided (e.g., about risks) and the non-manipulated variable (e.g., benefits). In each case, the manipulated information produced an inverse effect on the affectively inverse characteristic. Providing information that increased perception of risk decreased perception of benefit. Similarly, providing information that decreased perception of benefit increased perception of risk. Finucane et al. also found that time pressure greatly *increased* the inverse relationship between perceived risk and perceived benefit – presumably because time pressure increased the dominance of the affect heuristic over analytical reasoning.

Ganzach (2001) found the same effect in the realm of finance: analysts seemed to base their judgements of risk and return for *unfamiliar* stocks upon a global affective attitude. Stocks perceived as 'good' were judged to have low risks and high return; stocks perceived as 'bad' were judged to have low return and high risks. That is, for unfamiliar stocks, perceived risk and perceived return were negatively correlated, as predicted by the affect heuristic.[4] For *familiar* stocks, perceived risk and perceived return were positively correlated; riskier stocks were expected to produce higher returns, as predicted by ordinary economic theory. (If a stock is safe, buyers pay a premium for its safety and it becomes more expensive, driving down the expected return.)

People typically have sparse information in considering future technologies. Thus it is not surprising that their attitudes should exhibit affective polarization. When I first began to think about such matters, I rated biotechnology as having relatively smaller benefits compared to nanotechnology, *and* I worried more about an engineered supervirus than about misuse of nanotechnology. Artificial Intelligence, from which I expected the largest benefits of all, gave me not the least anxiety. Later, after working through the problems in much greater detail, my assessment of relative benefit remained much the same, but my worries had inverted: the more powerful technologies, with greater anticipated benefits, now appeared to have correspondingly more difficult risks. In retrospect this is what one would expect. But analysts with scanty information may rate technologies affectively, so that information about perceived benefit seems to mitigate the force of perceived risk.

5.9 Scope neglect

Migrating birds (2000/20,000/200,000) die each year by drowning in uncovered oil ponds, which the birds mistake for water bodies . These deaths could be prevented by covering the oil ponds with nets. How much money would you be willing to pay to provide the needed nets?

Three groups of subjects considered three versions of the above question, asking them how high a tax increase they would accept to save 2,000, 20,000, or 200,000 birds. The response – known as Stated Willingness-to-Pay (SWTP) – had a mean of $80 for the 2000-bird group, $78 for 20,000 birds, and $88 for 200,000 birds (Desvousges et al., 1993). This phenomenon is known as *scope insensitivity* or *scope neglect.*

Similar studies have shown that Toronto residents would a pay little more to clean up all polluted lakes in Ontario than polluted lakes in a particular region of Ontario (Kahneman, 1986); and that residents of four western US states

[4] Note that in this experiment, *sparse information* played the same role as cognitive business or time pressure in increasing reliance on the affect heuristic.

would pay only 28% more to protect all fifty-seven wilderness areas in those states than to protect a single area (McFadden and Leonard, 1995).

The most widely accepted explanation for scope neglect appeals to the affect heuristic. Kahneman et al. (1999, pp. 212–213) write:

> The story constructed by Desvouges et al. probably evokes for many readers a mental representation of a prototypical incident, perhaps an image of an exhausted bird, its feathers soaked in black oil, unable to escape. The hypothesis of valuation by prototype asserts that the affective value of this image will dominate expressions of the attitude to the problem – including the willingness to pay for a solution. Valuation by prototype implies extension neglect.

Two other hypotheses accounting for scope neglect include *purchase of moral satisfaction* (Kahneman and Knetsch, 1992) and *good cause dump* (Harrison, 1992). 'Purchase of moral satisfaction' suggests that people spend enough money to create a 'warm glow' in themselves, and the amount required is a property of the person's psychology, having nothing to do with birds. 'Good cause dump' suggests that people have some amount of money they are willing to pay for 'the environment', and *any* question about environmental goods elicits this amount.

Scope neglect has been shown to apply to human lives. Carson and Mitchell (1995) report that increasing the alleged risk associated with chlorinated drinking water from 0.004 to 2.43 annual deaths per 1000 (a factor of 600) increased stated willingness to pay from $3.78 to $15.23 (a factor of four). Baron and Greene (1996) found no effect from varying lives saved by a factor of ten.

Fetherstonhaugh et al. (1997), in a paper titled 'Insensitivity to the value of human life: a study of psychophysical numbing', found evidence that our perception of human deaths, and valuation of human lives, obeys Weber's Law – meaning that we use a *logarithmic* scale. And indeed, studies of scope neglect in which the quantitative variations are huge enough to elicit any sensitivity at all, show small *linear* increases in Willingness-to-Pay corresponding to *exponential* increases in scope. Kahneman et al. (1999) interpret this as an additive effect of scope affect and prototype affect – the prototype image elicits most of the emotion, and the scope elicits a smaller amount of emotion which is *added* (not multiplied) with the first amount.

Albert Szent-Györgyi, famous Hungarian physiologist and the discoverer of vitamin C, said: 'I am deeply moved if I see one man suffering and would risk my life for him. Then I talk impersonally about the possible pulverization of our big cities, with a hundred million dead. I am unable to multiply one man's suffering by a 100 million.' Human emotions take place within an analogous brain. The human brain cannot release enough neurotransmitters to feel emotion a 1000 times as strong as the grief of one funeral. A prospective risk going from 10,000,000 deaths to 100,000,000 deaths does not multiply by ten the strength of our determination to stop it. It adds one more zero on paper for our eyes to glaze over, an effect so small that one must usually jump several orders of magnitude to detect the difference experimentally.

5.10 Calibration and overconfidence

What confidence do people place in their erroneous estimates? In Section 5.2 on Availability, I discussed an experiment on perceived risk, in which subjects overestimated the probability of newsworthy causes of death in a way that correlated to their selective reporting in newspapers. Slovic et al. (1982, p. 472) also observed:

> A particularly pernicious aspect of heuristics is that people typically have great confidence in judgments based upon them. In another followup to the study on causes of death, people were asked to indicate the odds that they were correct in choosing the more frequent of two lethal events (Fischhoff, Slovic, and Lichtenstein, 1977) – In Experiment 1, subjects were reasonably well calibrated when they gave odds of 1:1, 1.5:1, 2:1, and 3:1. That is, their percentage of correct answers was close to the appropriate percentage correct, given those odds. However, as odds increased from 3:1 to 100:1, there was little or no increase in accuracy. Only 73% of the answers assigned odds of 100:1 were correct (instead of 99.1%). Accuracy 'jumped' to 81% at 1000:1 and to 87% at 10,000:1. For answers assigned odds of 1,000,000:1 or greater, accuracy was 90%; the appropriate degree of confidence would have been odds of 9:1 – In summary, subjects were frequently wrong at even the highest odds levels. Moreover, they gave many extreme odds responses. More than half of their judgments were greater than 50:1. Almost one-fourth were greater than 100:1 – 30% of the respondents in Experiment 1 gave odds greater than 50:1 to the incorrect assertion that homicides are more frequent than suicides.

This extraordinary-seeming result is quite common within the heuristics and biases literature, where it is known as *overconfidence*. Suppose I ask you for your best guess as to an uncertain quantity, such as the number of 'Physicians and Surgeons' listed in the Yellow Pages of the Boston phone directory, or total US egg production in millions. You will generate some value, which surely will not be *exactly* correct; the true value will be more or less than your guess. Next I ask you to name a *lower bound* such that you are 99% confident that the true value lies *above* this bound and an *upper bound* such that you are 99% confident the true value lies *beneath* this bound. These two bounds form your 98% *confidence interval*. If you are *well calibrated*, then on a test with 100 such questions, around two questions will have answers that fall outside your 98% confidence interval.

Alpert and Raiffa (1982) asked subjects a collective total of 1000 general knowledge questions like those described above; 426 of the true values lay outside the subjects' 98% confidence intervals. If the subjects were properly calibrated there would have been approximately twenty surprises. Put another way: Events to which subjects assigned a probability of 2% happened 42.6% of the time.

Another group of thirty-five subjects was asked to estimate 99.9% confident upper and lower bounds. They received 40% surprises. Another thirty-five subjects were asked for 'minimum' and 'maximum' values and were surprised 47% of the time. Finally, a fourth group of thirty-five subjects were asked for 'astonishingly low' and 'astonishingly high' values; they recorded 38% surprises.

In a second experiment, a new group of subjects was given a first set of questions, scored, provided with feedback, told about the results of previous experiments, had the concept of calibration explained to them at length, and then asked to provide 98% confidence intervals for a new set of questions. The post-training subjects were surprised 19% of the time, a substantial improvement over their pre-training score of 34% surprises, but still a far cry from the well-calibrated value of 2% surprises.

Similar failure rates have been found for experts. Hynes and Vanmarke (1976) asked seven internationally known geotechnical engineers to predict the height of an embankment that would cause a clay foundation to fail and to specify confidence bounds around this estimate that were wide enough to have a 50% chance of enclosing the true height. None of the bounds specified by the engineers enclosed the true failure height. Christensen-Szalanski and Bushyhead (1981) reported physician estimates for the probability of pneumonia for 1531 patients examined because of a cough. At the highest calibrated bracket of stated confidences, with average verbal probabilities of 88%, the proportion of patients actually having pneumonia was less than 20%.

In the words of Alpert and Raiffa (1982, p. 301): 'For heaven's sake, *Spread Those Extreme Fractiles!* Be honest with yourselves! Admit what you don't know!'

Lichtenstein et al. (1982) reviewed the results of fourteen papers on thirty-four experiments performed by twenty-three researchers studying human calibration. The *overwhelmingly* strong result was that people are overconfident. In the modern field, overconfidence is no longer noteworthy; but it continues to show up, in passing, in nearly any experiment where subjects are allowed to assign extreme probabilities.

Overconfidence applies forcefully to the domain of planning, where it is known as the *planning fallacy*. Buehler et al. (1994) asked psychology students to predict an important variable – the delivery time of their psychology honours thesis. They waited until students approached the end of their year-long projects, then asked the students when they realistically expected to submit their thesis and also when they would submit the thesis 'if everything went as poorly as it possibly could'. On average, the students took fifty-five days to complete their thesis, twenty-two days longer than they had anticipated, and seven days longer than their *worst-case* predictions.

Buehler et al. (1995) asked students for times by which they were 50% sure, 75% sure, and 99% sure that they would finish their academic project. Only 13% of the participants finished their project by the time assigned a 50% probability level, only 19% finished by the time assigned a 75% probability, and 45% finished by the time of their 99% probability level. Buehler et al. (2002) wrote: 'The results for the 99% probability level are especially striking: Even when asked to make a highly conservative forecast, a prediction that they felt virtually certain that they would fulfill, students' confidence in their time estimates far exceeded their accomplishments.'

Newby-Clark et al. (2000) found that asking subjects for their predictions based on realistic 'best guess' scenarios and asking subjects for their hoped-for 'best case' scenarios produced indistinguishable results. When asked for their 'most probable' case, people tend to envision everything going exactly as planned, with no unexpected delays or unforeseen catastrophes: the same vision as their 'best case'. *Reality, it turns out, usually delivers results somewhat worse than the 'worst case'.*

This chapter discusses overconfidence *after* discussing the confirmation bias and the sub-problem of the disconfirmation bias. The calibration research is dangerous knowledge – so tempting to apply selectively. 'How foolish my opponent is, to be so certain of his arguments! Doesn't he know how often people are surprised on their certainties?' If you realize that expert opinions have less force than you thought, you had better also realize that your own thoughts have *much* less force than you thought, so that it takes less force to compel you away from your preferred belief. Otherwise you become slower to react to incoming evidence. You are left worse off than if you had never heard of calibration. That is why – despite frequent great temptation – I avoid discussing the research on calibration unless I have previously spoken of the confirmation bias, so that I can deliver this same warning.

Note also that a confidently expressed expert opinion is quite a different matter from a calculation made *strictly* from actuarial data, or *strictly* from a *precise, precisely confirmed* model. Of all the times an expert has ever stated, even from strict calculation, that an event has a probability of 10^{-6}, they have undoubtedly been wrong more often than one time in a million. But if combinatorics could not correctly predict that a lottery ticket has a 10^{-8} chance of winning, ticket sellers would go broke.

5.11 Bystander apathy

My last bias comes, not from the field of heuristics and biases, but from the field of social psychology. A now-famous series of experiments by Latane and Darley (1969) uncovered the *bystander effect*, also known as *bystander apathy*, in which larger numbers of people are less likely to act in emergencies – not only individually, but also collectively. Among subjects alone in a room, on noticing smoke entering from under a door, 75% of them left the room to report it. When three naïve subjects were present, the smoke was reported only 38% of the time. A naive subject in the presence of two confederates who purposely ignored the smoke, even when the room became hazy, left to report the smoke only 10% of the time. A college student apparently having an epileptic seizure was helped 85% of the time by a single bystander and 31% of the time by five bystanders.

The bystander effect is usually explained as resulting from *diffusion of responsibility* and *pluralistic ignorance*. Being part of a group reduces individual responsibility. Everyone hopes that someone else will handle the problem instead, and this reduces the individual pressure to the point that no one does anything. Support for this hypothesis is adduced from manipulations in which subjects believe that the victim is especially dependent on them; this reduces the bystander effect or negates it entirely. Cialdini (2001) recommends that if you are ever in an emergency, you single out *one* bystander, and ask that person to help – thereby overcoming the diffusion.

Pluralistic ignorance is a more subtle effect. Cialdini (2001, p. 114) writes:

> Very often an emergency is not obviously an emergency. Is the man lying in the alley a heart-attack victim or a drunk sleeping one off? ... In times of such uncertainty, the natural tendency is to look around at the actions of others for clues. We can learn from the way the other witnesses are reacting whether the event is or is not an emergency. What is easy to forget, though, is that everybody else observing the event is likely to be looking for social evidence, too. Because we all prefer to appear poised and unflustered among others, we are likely to search for that evidence placidly, with brief, camouflaged glances at those around us. Therefore everyone is likely to see everyone else looking unruffled and failing to act.

The bystander effect is not about individual selfishness or insensitivity to the suffering of others. Alone subjects *do* usually act. Pluralistic ignorance can explain, and individual selfishness cannot explain, subjects failing to react to a room filling up with smoke. In experiments involving apparent dangers to either others or the self, subjects placed with non-reactive confederates frequently glance at the non-reactive confederates.

I am sometimes asked: 'If *existential risk X* is real, why aren't more people doing something about it?' There are many possible answers, a few of which I have touched on here. People may be overconfident and over-optimistic. They may focus on overly specific scenarios for the future, to the exclusion of all others. They may not recall any past extinction events in memory. They may overestimate the predictability of the past and hence underestimate the surprise of the future. They may not realize the difficulty of preparing for emergencies without benefit of hindsight. They may prefer philanthropic gambles with higher pay-off probabilities, neglecting the value of the stakes. They may conflate positive information about the benefits of a technology as negative information about its risks. They may be contaminated by movies, where the world ends up being saved. They may purchase moral satisfaction more easily by giving to other charities. Otherwise, the extremely unpleasant prospect of human extinction may spur them to seek arguments that humanity will *not* go extinct, without an equally frantic search for reasons why we *would*.

But if the question is, specifically, 'Why aren't more people doing something about it?', one possible component is that people are asking that very question – darting their eyes around to see if anyone else is reacting to the emergency, meanwhile trying to appear poised and unflustered. If you want to know why others are not responding to an emergency, before you respond yourself, you may have just answered your own question.

5.12 A final caution

Every true idea which discomforts you will seem to match the pattern of at least one psychological error.

Robert Pirsig said: 'The world's biggest fool can say the sun is shining, but that doesn't make it dark out.' If you believe someone is guilty of a psychological error, then demonstrate your competence by first demolishing their consequential factual errors. If there are no factual errors, then what matters the psychology? The temptation of psychology is that, knowing a little psychology, we can meddle in arguments where we have no technical expertise.

If someone wrote a novel about an asteroid strike destroying modern civilization, then the reader might criticize that novel as dystopian, apocalyptic, and symptomatic of the author's naïve inability to deal with a complex technological society. We should recognize this as a literary criticism and not a scientific one; it is about good or bad novels and not about good or bad hypotheses. To quantify the annual probability of an asteroid strike *in real life*, one *must* study astronomy and the historical record (while avoiding the case-specific biases; see Chapters 1 and 11, this volume). No amount of literary criticism can put a number on it. Garreau (2005) seems to hold that a scenario of a mind slowly increasing in capability, is more *mature* and *sophisticated* than a scenario of extremely rapid intelligence increase. But that is a technical question, not a matter of taste; no amount of psychologizing can tell you the exact slope of that curve.

It is harder to abuse heuristics and biases than psycho-analysis. Accusing someone of conjunction fallacy leads naturally into listing the specific details that you think are burdensome and drive down the joint probability. Even so, do not lose track of the real-world facts of primary interest; do not let the argument become *about* psychology.

Despite all dangers and temptations, it is better to know about psychological biases than not to know. Otherwise we will walk directly into the whirling helicopter blades of life. But be very careful not to have *too much fun* accusing others of biases. That is the road that leads to becoming a sophisticated arguer – someone who, faced with any discomforting argument, finds at once a bias in it; the one whom you must watch above all is yourself.

Jerry Cleaver said: 'What does you in is not failure to apply some high-level, intricate, complicated technique. It's overlooking the basics. Not keeping your eye on the ball.'

Analyses should finally *centre* on testable real-world assertions. Do not take your eyes off the ball.

5.13 Conclusion

Why should there be an organized body of thinking about global catastrophic and existential risks? Falling asteroids are not like engineered superviruses; physics disasters are not like nanotechnological wars. Why not consider each of these problems separately?

If someone proposes a physics disaster, then the committee convened to analyse the problem must obviously include physicists. But someone on that committee should also know how terribly dangerous it is to have an answer in your mind before you finish asking the question. Someone on that committee should remember the reply of Enrico Fermi to Leo Szilard's proposal that a fission chain reaction could be used to build nuclear weapons. (The reply was 'Nuts!' – Fermi considered the possibility so remote as to not be worth investigating.) Someone should remember the history of errors in physics calculations: the Castle Bravo nuclear test that produced a 15-megaton explosion, instead of four to eight, because of an unconsidered reaction in lithium-7: they correctly solved the wrong equation, failed to think of all the terms that needed to be included, and at least one person in the expanded fallout radius died. Someone should remember Lord Kelvin's careful proof, using multiple, independent quantitative calculations from well-established theories, that the Earth could not possibly have existed for as much as 40 million years. Someone should know that when an expert says the probability is 'a million to one' without using actuarial data or calculations from a precise, precisely confirmed model, the calibration is probably more like 20 to 1 (although this is not an exact conversion).

Any existential risk evokes problems that it shares with all other existential risks, *in addition to* the domain-specific expertise required for the *specific* existential risk (see more details in Chapter 1, this book). Someone on the physics-disaster committee should know what the term 'existential risk' means and should possess whatever skills the field of existential risk management has accumulated or borrowed. For maximum safety, that person should also be a physicist. The domain-specific expertise and the expertise pertaining to existential risks should combine in one person. I am sceptical that a scholar of heuristics and biases, unable to read physics equations, could check the work of physicists who knew nothing of heuristics and biases.

Once upon a time I made up overly detailed scenarios, without realizing that *every* additional detail was an extra burden. I really did think that I could say there was a 90% chance of Artificial Intelligence being developed between 2005 and 2025, with the peak in 2018. This statement now seems to me like complete gibberish. Why did I *ever* think I could generate a tight probability distribution over a problem like that? Where did I even get those numbers in the first place?

I once met a lawyer who had made up his own theory of physics. I said to the lawyer: 'You cannot invent your own physics theories without knowing math and studying for years; physics is hard.' He replied: 'But if you really understand physics you can explain it to your grandmother, Richard Feynman told me so.' And I said to him: 'Would you advise a friend to argue his own court case?' At this he fell silent. He knew abstractly that physics was difficult, but I think it had honestly never occurred to him that physics might be as difficult as lawyering.

One of many biases *not* discussed in this chapter describes the biasing effect of not knowing what we do not know. When a company recruiter evaluates his own skill, he recalls in his mind the performance of candidates he hired, many of whom subsequently excelled; therefore the recruiter thinks highly of his skill. But the recruiter never sees the work of candidates *not* hired. Thus I must warn that this paper touches upon only a small subset of heuristics and biases; for when you wonder how much you have already learned, you will recall the few biases this chapter *does* mention, rather than the many biases it does not. Brief summaries cannot convey a sense of the field, the larger understanding which weaves a set of memorable experiments into a unified interpretation. Many highly relevant biases, such as *need for closure*, I have not even mentioned. The purpose of this chapter is not to teach the knowledge needful to a student of existential risks but to intrigue you into learning more.

Thinking about existential risks falls prey to all the same fallacies that prey upon thinking in general. But the stakes are much, much higher. A common result in heuristics and biases is that offering money or other incentives does not eliminate the bias. (Kachelmeier and Shehata [1992] offered subjects living in the People's Republic of China the equivalent of three months' salary.) The subjects in these experiments do not make mistakes on purpose; they make mistakes because they do not know how to do better. Even if you told them the survival of humankind was at stake, they still would not thereby know how to do better. (It might increase their need for closure, causing them to do worse.)

It is a terribly frightening thing, but people do not become any smarter, *just* because the survival of humankind is at stake.

In addition to standard biases, I have personally observed what look like harmful modes of thinking specific to existential risks. The Spanish flu of 1918 killed 25–50 million people. World War II killed 60 million people; 10^7 is the order of the largest catastrophes in humanity's written history. Substantially larger numbers, such as 500 million deaths, and *especially* qualitatively different scenarios such as the extinction of the entire human species, seem to trigger a *different mode of thinking* – enter into a 'separate magisterium'. People who would never dream of hurting a child hear of an existential risk, and say, 'Well, maybe the human species doesn't really deserve to survive.'

There is a saying in heuristics and biases that people do not evaluate events, but descriptions of events – what is called non-extensional reasoning. The *extension* of humanity's extinction includes the death of yourself, your friends, your family, your loved ones, your city, your country, your political fellows. Yet people who would take great offence at a proposal to wipe the country of Britain from the map, to kill every member of the Democratic Party in the United States, to turn the city of Paris to glass – who would feel still greater horror on hearing the doctor say that their child had cancer – these people will discuss the extinction of humanity with perfect calm. The phrase 'extinction of humanity', as words on paper, appears in fictional novels or is discussed in philosophy books – it belongs to a different context compared to the Spanish flu. We evaluate descriptions of events, not extensions of events. The cliché phrase *end of the world* invokes the magisterium of myth and dream, of prophecy and apocalypse, of novels and movies. The challenge of existential risks to rationality is that, the catastrophes being so huge, people snap into a different mode of thinking. Human deaths are suddenly no longer bad, and detailed predictions suddenly no longer require any expertise, and whether the story is told with a happy ending or a sad ending is a matter of personal taste in stories.

But that is only an anecdotal observation of mine. I thought it better that this essay should focus on mistakes well documented in the literature – the general literature of cognitive psychology, because there is not yet experimental literature specific to the psychology of existential risks. There should be.

In the mathematics of Bayesian decision theory there is a concept of *information value* – the expected utility of knowledge. The value of information emerges from the value of whatever it is *about*; if you double the stakes, you double the value of information about the stakes. The value of rational thinking works similarly – the value of performing a computation that integrates the evidence is calculated much the same way as the value of the evidence itself (Good, 1952; Horvitz et al., 1989).

No more than Albert Szent-Györgyi could multiply the suffering of one human by a 100 million can I truly understand the value of clear thinking about global risks. Scope neglect is the hazard of being a biological human, running on an analogous brain; the brain cannot multiply by 6 billion. And the stakes of existential risk extend beyond even the 6 billion humans alive today, to all the stars in all the galaxies that humanity and humanity's descendants may some day touch. All that vast potential hinges on our survival here, now, in the days when the realm of humankind is a single planet orbiting a single star. I cannot feel our future. All I can do is try to defend it.

Acknowledgement

I thank Michael Roy Ames, Olie Lamb, Tamas Martinec, Robin Lee Powell, Christian Rovner, and Michael Wilson for their comments, suggestions, and criticisms. Needless to say, any remaining errors in this chapter are my own.

Suggestions for further reading

Dawes, R. (1988). *Rational Choice in an Uncertain World: The Psychology of Intuitive Judgment* (San Diego, CA: Harcourt, Brace, Jovanovich). First edition 1988 by Dawes and Kagan, second edition 2001 by Dawes and Hastie. This book aims to introduce heuristics and biases to an intelligent general audience. (For example, Bayes's Theorem is explained, rather than assumed, but the explanation is only a few pages.) A good book for quickly picking up a sense of the field.

Kahneman, D., Slovic, P., and Tversky, A. (eds.) (1982). *Judgment Under Uncertainty: Heuristics and Biases* (New York: Cambridge University Press). This is the edited volume that helped establish the field, written with the outside academic reader firmly in mind. Later research has generalized, elaborated, and better explained the phenomena treated in this volume, but the basic results given are still standing strong.

Kahneman, D. and Tversky, A. (eds.) (2000). *Choices, Values, and Frames* (Cambridge: Cambridge University Press). Gilovich, T. Griffin, D. and Kahneman, D. (2003). *Heuristics and Biases.* These two edited volumes overview the field of heuristics and biases in its current form. They are somewhat less accessible to a general audience.

References

Alpert, M. and Raiffa, H. (1982). A progress report on the training of probability assessors. In Kahneman, D., Slovic, P., and Tversky, A. (eds.), *Judgement Under Uncertainty: Heuristics and Biases*, pp. 294–305 (Cambridge: Cambridge University Press).

Ambrose, S.H. (1998). Late Pleistocene human population bottlenecks, volcanic winter, and differentiation of modern humans. *J. Human Evol.*, **34**, 623–651.

Baron, J. and Greene, J. (1996). Determinants of insensitivity to quantity in valuation of public goods: contribution, warm glow, budget constraints, availability, and prominence. *J. Exp. Psychol.: Appl.*, **2**, 107–125.

Bostrom, N. (2001). Existential risks: analyzing human extinction scenarios. *J. Evol. Technol.*, **9**.

Brenner, L.A., Koehler, D.J., and Rottenstreich, Y. (2002). Remarks on support theory: recent advances and future directions. In Gilovich, T., Griffin, D., and Kahneman, D. (eds.), *Heuristics and Biases: The Psychology of Intuitive Judgment*, pp. 489–509 (Cambridge: Cambridge University Press).

Buehler, R., Griffin, D., and Ross, M. (1994). Exploring the 'planning fallacy': why people underestimate their task completion times. *J. Personal. Social Psychol.*, **67**, 366–381.

Buehler, R., Griffin, D., and Ross, M. (1995). It's about time: optimistic predictions in work and love. *Eur. Rev. Social Psychol.*, **6**, 1–32.

Buehler, R., Griffin, D., and Ross, M. (2002). Inside the planning fallacy: the causes and consequences of optimistic time predictions. In Gilovich, T., Griffin, D., and Kahneman, D. (eds.), *Heuristics and Biases: The Psychology of Intuitive Judgment*, pp. 250–270 (Cambridge: Cambridge University Press).

Burton, I., Kates, R., and White, G. (1978). *Environment as Hazard* (New York: Oxford University Press).

Carson, R.T. and Mitchell, R.C. (1995). Sequencing and nesting in contingent valuation surveys. *J. Environ. Econ. Manag.*, **28**(2), 155–173.

Chapman, G.B. and Johnson, E.J. (2002). Incorporating the irrelevant: anchors in judgments of belief and value. In Gilovich, T., Griffin, D., and Kahneman, D. (eds.), *Heuristics and Biases: The Psychology of Intuitive Judgment*, pp. 120–138 (Cambridge: Cambridge University Press.

Christensen-Szalanski, J.J.J. and Bushyhead, J.B. (1981). Physicians' use of probabilistic information in a real clinical setting. *J. Exp. Psychol. Human Percept. Perf.*, **7**, 928–935.

Cialdini, R.B. (2001). *Influence: Science and Practice* (Boston, MA: Allyn and Bacon).

Combs, B. and Slovic, P. (1979). Causes of death: Biased newspaper coverage and biased judgments. *Journalism Quarterly*, **56**, 837–843.

Dawes, R.M. (1988). *Rational Choice in an Uncertain World* (San Diego, CA: Harcourt, Brace, Jovanovich).

Desvousges, W.H., Johnson, F.R., Dunford, R.W., Boyle, K.J., Hudson, S.P., and Wilson, N. (1993). Measuring natural resource damages with contingent valuation: tests of validity and reliability. In Hausman, J.A. (ed.), *Contingent Valuation: A Critical Assessment*, pp. 91–159 (Amsterdam: North Holland).

Fetherstonhaugh, D., Slovic, P., Johnson, S., and Friedrich, J. (1997). Insensitivity to the value of human life: a study of psychophysical numbing. *J. Risk Uncertainty*, **14**, 238–300.

Finucane, M.L., Alhakami, A., Slovic, P., and Johnson, S.M. (2000). The affect heuristic in judgments of risks and benefits. *J. Behav. Decision Making*, **13**(1), 1–17.

Fischhoff, B. (1982). For those condemned to study the past: heuristics and biases in hindsight. In Kahneman, D., Slovic, P., and Tversky, A. (eds.), *Judgement Under*

Uncertainty: Heuristics and Biases, pp. 306–354 (Cambridge: Cambridge University Press).

Fischhoff, B. and Beyth, R. (1975). I knew it would happen: remembered probabilities of once-future things. *Organ. Behav. Human Perf.*, **13**, 1–16.

Fischhoff, B., Slovic, P., and Lichtenstein, S. (1977). Knowing with certainty: the appropriateness of extreme confidence. *J. Exp. Psychol Human Percept. Perf.*, **3**, 522–564.

Ganzach, Y. (2001). Judging risk and return of financial assets. *Organ. Behav. Human Decision Processes*, **83**, 353–370.

Garreau, J. (2005). *Radical Evolution: The Promise and Peril of Enhancing Our Minds, Our Bodies – and What It Means to Be Human* (New York: Doubleday).

Gilbert, D.T. and Osborne, R.E. (1989). Thinking backward: Some curable and incurable consequences of cognitive busyness. *J. Person. Social Psychol.*, **57**, 940–949.

Gilbert, D.T., Pelham, B.W., and Krull, D.S. (1988). On cognitive busyness: when person perceivers meet persons perceived. *J. Person. Social Psychol.*, **54**, 733–740.

Gilovich, T. (2000). *Motivated Skepticism and Motivated Credulity: Differential Standards of Evidence in the Evaluation of Desired and Undesired Propositions.* Presented at the 12th Annual Convention of the American Psychological Society, Miami Beach, FL.

Gilovich, T., Griffin, D., and Kahneman, D. (eds.) (2003). *Heuristics and Biases: The Psychology of Intuitive Judgment* (Cambridge: Cambridge University Press).

Good, I.J. (1952). Rational decisions. *J. Royal Statist. Soc., Series B*, **14**, 107–114.

Griffin, D. and Tversky, A. (1992). The weighing of evidence and the determinants of confidence. *Cogn. Psychol.*, **24**, 411–435.

Harrison, G.W. (1992). Valuing public goods with the contingent valuation method: a critique of Kahneman and Knestch. *J. Environ. Econ. Manag.*, **23**, 248–257.

Horvitz, E.J., Cooper, G.F., and Heckerman, D.E. (1989). Reflection and action under scarce resources: theoretical principles and empirical study. *Proceedings of the Eleventh International Joint Conference on Artificial Intelligence*, pp. 1121–1127 (Detroit, MI).

Hynes, M.E. and Vanmarke, E.K. (1976). Reliability of embankment performance predictions. *Proceedings of the ASCE Engineering Mechanics Division Specialty Conference* (Waterloo, Ontario: University of Waterloo Press).

Johnson, E., Hershey, J., Meszaros, J., and Kunreuther, H. (1993). Framing, probability distortions and insurance decisions. *J. Risk Uncertainty*, **7**, 35–51.

Kachelmeier, S.J. and Shehata, M. (1992). Examining risk preferences under high monetary incentives: experimental evidence from the People's Republic of China. *Am. Econ. Rev.*, **82**, 1120–1141.

Kahneman, D. (1986). Comments on the contingent valuation method. In Cummings, R.G., Brookshire, D.S., and Schulze, W.D. (eds.), *Valuing Environmental Goods: A State of the Arts Assessment of the Contingent Valuation Method*, pp. 185–194 (Totowa, NJ: Roweman and Allanheld).

Kahneman, D. and Knetsch, J.L. (1992). Valuing public goods: the purchase of moral satisfaction. *J. Environ. Econ. Manag.*, **22**, 57–70.

Kahneman, D., Ritov, I., and Schkade, D.A. (1999). Economic preferences or attitude expressions?: An analysis of dollar responses to public issues. *J. Risk Uncertainty*, **19**, 203–235.

Kahneman, D., Slovic, P., and Tversky, A. (eds.) (1982). *Judgment Under Uncertainty: Heuristics and Biases* (New York: Cambridge University Press).

Kahneman, D. and Tversky, A. (eds.) (2000). *Choices, Values, and Frames* (Cambridge: Cambridge University Press).

Kamin, K. and Rachlinski, J. (1995). Ex post ≠ ex ante: determining liability in hindsight. *Law Human Behav.*, **19**(1), 89–104.

Kates, R. (1962). Hazard and choice perception in flood plain management. Research Paper No. 78. Chicago, IL: University of Chicago, Department of Geography.

Knaup, A. (2005). Survival and longevity in the business employment dynamics data. *Monthly Labor Rev.*, **May** 2005, 50–56.

Kunda, Z. (1990). The case for motivated reasoning. *Psychol. Bull.*, **108**(3), 480–498.

Latane, B. and Darley, J. (1969). Bystander 'Apathy'. *Am. Scientist*, **57**, 244–268.

Lichtenstein, S., Fischhoff, B., and Phillips, L.D. (1982). Calibration of probabilities: the state of the art to 1980. In Kahneman, D., Slovic, P., and Tversky, A. (eds.),*Judgment Under Uncertainity: Heuristics and Biases*, pp. 306–334 (New York: Cambridge University Press).

Lichtenstein, S., Slovic, P., Fischhoff, B., Layman, M., and Combs, B. (1978). Judged frequency of lethal events. *J. Exp. Psychol.: Human Learn. Memory*, **4**(6), 551–578.

McFadden, D. and Leonard, G. (1995). Issues in the contingent valuation of environmental goods: methodologies for data collection and analysis. In Hausman, J.A. (ed.), *Contingent Valuation: A Critical Assessment*, pp. 165–208 (Amsterdam: North Holland).

Newby-Clark, I.R., Ross, M., Buehler, R., Koehler, D.J., and Griffin, D. (2000). People focus on optimistic and disregard pessimistic scenarios while predicting their task completion times. *J. Exp. Psychol. Appl.*, **6**, 171–182.

Quattrone, G.A., Lawrence, C.P., Finkel, S.E., and Andrus, D.C. (1981). Explorations in anchoring: the effects of prior range, anchor extremity, and suggestive hints. Manuscript, Stanford University.

Rasmussen, N.C. (1975). *Reactor Safety Study: An Assessment of Accident Risks in U.S. Commercial Nuclear Power Plants.* NUREG-75/014, WASH-1400 (Washington, DC: U.S. Nuclear Regulatory Commission)

Rogers, W.P., Armstrong, N., Acheson, D.C., Covert, E.E., Feynman, R.P., Hotz, R.B., Kutyna, D.J., Ride, S.K., Rummel, R.W., Suffer, D.F., Walker, A.B.C., Jr., Wheelon, A.D., Yeager, C., Keel, A.G., Jr. (1986). Report of the Presidential Commission on the Space Shuttle Challenger Accident. *Presidential Commission on the Space Shuttle Challenger Accident.* Washington, DC.

Sanchiro, C. (2003). Finding Error. *Mich. St. L. Rev.* 1189.

Schneier, B. (2005). Security lessons of the response to hurricane Katrina. http://www.schneier.com/blog/archives/2005/09/security_lesson.html. Viewed on 23 January 2006.

Sides, A., Osherson, D., Bonini, N., and Viale, R. (2002). On the reality of the conjunction fallacy. *Memory Cogn.*, **30**(2), 191–198.

Slovic, P., Finucane, M., Peters, E., and MacGregor, D. (2002). Rational actors or rational fools: implications of the affect heuristic for behavioral economics. *J. Socio-Econ.*, **31**, 329–342.

Slovic, P., Fischoff, B., and Lichtenstein, S. (1982). Facts versus fears: understanding perceived risk. In Kahneman, D., Slovic, P., and Tversky, A. (eds.), *Judgment Under Uncertainity: Heuristics and Biases*, pp. 463–492 (Cambridge: Cambridge University Press).

Strack, F. and Mussweiler, T. (1997). Explaining the enigmatic anchoring effect: mechanisms of selective accessibility. *J. Person. Social Psychol.*, **73**, 437–446.

Taber, C.S. and Lodge, M. (2000). Motivated skepticism in the evaluation of political beliefs. Presented at the 2000 meeting of the American Political Science Association.

Taleb, N. (2001). *Fooled by Randomness: The Hidden Role of Chance in Life and in the Markets*, pp. 81–85 (New York: Textre).

Taleb, N. (2005). *The Black Swan: Why Don't We Learn that We Don't Learn?* (New York: Random House).

Tversky, A. and Kahneman, D. (1973). Availability: a heuristic for judging frequency and probability. *Cogn. Psychol.*, **4**, 207–232.

Tversky, A. and Kahneman, D. (1974). Judgment under uncertainty: heuristics and biases. *Science*, **185**, 251–284.

Tversky, A. and Kahneman, D. (1982). Judgments of and by representativeness. In Kahneman, D., Slovic, P., and Tversky, A. (eds.), *Judgement Under Uncertainty: Heuristics and Biases*, pp. 84–98 (Cambridge: Cambridge University Press).

Tversky, A. and Kahneman, D. (1983). Extensional versus intuitive reasoning: the conjunction fallacy in probability judgment. *Psychol. Rev.*, **90**, 293–315.

Wansink, B., Kent, R.J., and Hoch, S.J. (1998). An anchoring and adjustment model of purchase quantity decisions. *J. Market. Res.*, **35**(February), 71–81.

Wason, P.C. (1960). On the failure to eliminate hypotheses in a conceptual task. *Quarterly J. Exp. Psychol.*, **12**, 129–140.

Wilson, T.D., Houston, C., Etling, K.M., and Brekke, N. (1996). A new look at anchoring effects: basic anchoring and its antecedents. *J. Exp. Psychol.: General*, **4**, 387–402.

·6·

Observation selection effects and global catastrophic risks

Milan M. Ćirković

> Treason doth never prosper: what's the reason?
> Why if it prosper, none dare call it treason.
>
> Sir John Harrington (1561–1612)

6.1 Introduction: anthropic reasoning and global risks

Different types of global catastrophic risks (GCRs) are studied in various chapters of this book by direct analysis. In doing so, researchers benefit from a detailed understanding of the interplay of the underlying causal factors. However, the causal network is often excessively complex and difficult or impossible to disentangle. Here, we would like to consider limitations and theoretical constraints on the risk assessments which are provided by the general properties of the world in which we live, as well as its contingent history. There are only a few of these constraints, but they are important because they do not rely on making a lot of guesses about the details of future technological and social developments. The most important of these are observation selection effects.

Physicists, astronomers, and biologists have been familiar with the observational selection effect for a long time, some aspects of them (e.g., Malmquist bias in astronomy[1] or Signor-Lipps effect in paleontology[2]) being the subject of detailed mathematical modelling. In particular, cosmology is fundamentally incomplete without taking into account the necessary 'anthropic bias': the conditions we observe in fundamental physics, as well as in the universe at large, seem atypical when judged against what one would

[1] The difference between the average absolute magnitudes of stars (or galaxies or any other similar sources) in magnitude- and distance-limited samples, discovered in 1920 by K.G. Malmquist.

[2] The effect by which rare species seemingly disappear earlier than their numerous contemporaries (thus making extinction episodes more prolonged in the fossil record than they were in reality), discovered by P.W. Signor and J.H. Lipps in 1982.

expect as 'natural' according to our best theories, and require an explanation compatible with our existence as intelligent observers at this particular epoch in the history of the universe. In contrast, the observation selection effects are still often overlooked in philosophy and epistemology, and practically completely ignored in risk analysis, since they usually do not apply to conventional categories of risk (such as those used in insurance modelling). Recently, Bostrom (2002a) laid foundations for a detailed theory of observation selection effects, which has applications for both philosophy and several scientific areas including cosmology, evolution theory, thermodynamics, traffic analysis, game theory problems involving imperfect recall, astrobiology, and quantum physics. The theory of observation selection effects can tell us what we should expect to observe, given some hypothesis about the distribution of observers in the world. By comparing such predictions to our actual observations, we get probabilistic evidence for or against various hypotheses.[3]

Many conclusions pertaining to GCRs can be reached by taking into account observation selection effects. For instance, people often erroneously claim that we should not worry too much about existential disasters, since none has happened in the last thousand or even million years. This fallacy needs to be dispelled. Similarly, the conclusion that we are endangered primarily by our own activities and their consequences can be seen most clearly only after we filter out selection effects from our estimates.

In the rest of this chapter, we shall consider several applications of the anthropic reasoning to evaluation of our future prospects: first the anthropic overconfidence argument stemming from the past–future asymmetry in presence of intelligent observers (Section 6.2) and then the (in)famous Doomsday Argument (DA; Section 6.3). We proceed with Fermi's paradox and some specific risks related to the concept of extraterrestrial intelligence (Section 6.4) and give a brief overview of the Simulation Argument in connection with GCRs in Section 6.5 before we pass on to concluding remarks.

6.2 Past–future asymmetry and risk inferences

One important selection effect in the study of GCRs arises from the breakdown of the temporal symmetry between past and future catastrophes when our existence at the present epoch and the necessary conditions for it are taken into account. In particular, some of the predictions derived from past records are unreliable due to observation selection, thus introducing an essential qualification to the general and often uncritically accepted gradualist principle that 'the past is a key to the future'. This resulting *anthropic overconfidence bias*

[3] For a summary of vast literature on observation selection, anthropic principles, and anthropic reasoning in general, see Barrow and Tipler (1986); Balashov (1991); and Bostrom (2002a).

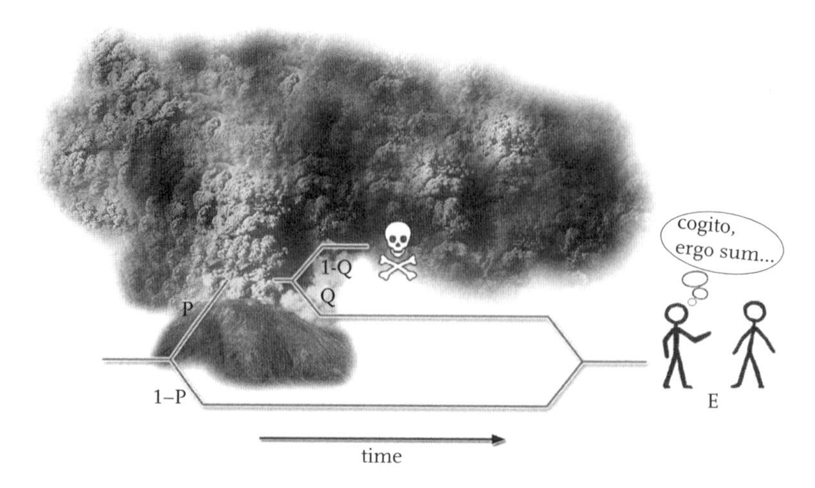

Fig. 6.1 A schematic presentation of the single-event toy model. The evidence E consists of our present-day existence.

is operative in a wide range of catastrophic events, and leads to potentially dangerous underestimates of the corresponding risk probabilities. After we demonstrate the effect on a toy model applied to a single catastrophic event situation in Section 6.2.1, we shall develop the argument in more detail in Section 6.2.2, while considering its applicability conditions for various types of GCRs in Section 6.2.3. Finally, we show that with the help of additional astrobiological information, we may do even better and constrain the probabilities of some very specific exogenous risks in Section 6.2.4.[4]

6.2.1 A simplified model

Consider the simplest case of a single very destructive global catastrophe, for instance, a worse-than-Toba super-volcanic eruption (see Chapter 10, this volume). The evidence we take into account in a Bayesian manner is the fact of our existence at the present epoch; this, in turn, implies the existence of a complicated web of evolutionary processes upon which our emergence is contingent; we shall neglect this complication in the present binary toy model and shall return to it in the next subsection. The situation is schematically shown in Fig. 6.1. The a priori probability of catastrophe is P and the probability of human extinction (or a sufficiently strong perturbation leading to divergence of evolutionary pathways from the morphological subspace containing humans) upon the catastrophic event is Q. We shall suppose that the two probabilities are (1) constant, (2) adequately normalized, and (3) applicable to a particular well-defined interval of past time. Event B_2 is the

[4] Parts of this section are loosely based upon Ćirković (2007).

occurrence of the catastrophe, and by E we denote the evidence of our present existence.

The direct application of the Bayes formula for expressing conditional probabilities in form

$$P(B_2|E) = \frac{P(B_2)\, P(E|B_2)}{P(B_1)\, P(E|B_1) + P(B_2)\, P(E|B_2)}, \tag{6.1}$$

using our notation, yields the a posteriori probability as

$$P(B_2|E) = \frac{PQ}{(1-P)\cdot(1+PQ)} = \frac{PQ}{1-P+PQ}. \tag{6.2}$$

By simple algebraic manipulation, we can show that

$$P\,(B_2|E) \leq P \tag{6.3}$$

that is, we tend to *underestimate* the true catastrophic risk. It is intuitively clear why: the symmetry between past and future is broken by the existence of an evolutionary process leading to our emergence as observers at this particular epoch in time. We can expect a large catastrophe tomorrow, but we cannot – even without any empirical knowledge – expect to find traces of a large catastrophe that occurred yesterday, since it would have pre-empted our existence today.

Note that

$$\lim_{Q\to 0} \frac{P}{P\,(B_2|\,E)} = \infty. \tag{6.4}$$

Very destructive events completely destroy predictability! An obvious consequence is that absolutely destructive events, which humanity has no chance of surviving at all ($Q = 0$), completely annihilate our confidence in predicting from past occurrences. This almost trivial conclusion is not, however, widely appreciated.

The issue at hand is the possibility of *vacuum phase transition* (see Chapter 16, this volume). This is an example par excellence of the $Q = 0$ event: its ecological consequences are such that the extinction not only of humanity but also of the terrestrial biosphere is certain.[5] However, the anthropic bias was noticed neither by Hut and Rees (1983), nor by many of subsequent papers citing it. Instead, these authors suggested that the idea of high-energy experiments triggering vacuum phase transition can be rejected by comparison with the high-energy events occurring in nature. Since the energies of particle collisions taking place, for instance, in interactions between cosmic rays and the Earth's atmosphere or the solid mass of the Moon are still orders of magnitude higher than those achievable in human laboratories in the near future, and

[5] For a more optimistic view of this possibility in a fictional context see Egan (2002).

with plausible general assumptions on the scaling of the relevant reaction cross-sections with energy, Hut and Rees concluded that in view of the fact that the Earth (and the Moon) survived the cosmic-ray bombardment for about 4.5 Gyr, we are safe for the foreseeable future. In other words, their argument consists of the claim that the absence of a catastrophic event of this type in our past light cone gives us the information that the probability P (or its rate per unit time p) has to be so extremely small, that any fractional increase caused by human activities (like the building and operating of a new particle collider) is insignificant. If, for example, p is 10^{-50} per year, then its doubling or even its increasing 1000-fold by deliberate human activities is arguably unimportant. Thus, we can feel safe with respect to the future on the basis of our observations about the past. As we have seen, there is a hole in this argument: all observers everywhere will always find that no such disaster has occurred in their backward light cone – and this is true whether such disasters are common or rare.

6.2.2 Anthropic overconfidence bias

In order to predict the future from records of the past, scientists use a wide variety of methods with one common feature: the construction of an empirical distribution function of events of a specified type (e.g., extraterrestrial impacts, supernova/gamma-burst explosions, or super-volcanic eruptions). In view of the Bayesian nature of our approach, we can dub this distribution function the a posteriori distribution function. Of course, deriving such function from observed traces is often difficult and fraught with uncertainty. For instance, constructing the distribution function of asteroidal/cometary impactors from the sample of impact craters discovered on Earth (Earth Impact Database, 2005) requires making physical assumptions, rarely uncontroversial, about physical parameters of impactors such as their density, as well as astronomical (velocity distribution of Earth-crossing objects) and geological (the response of continental or oceanic crust to a violent impact, formation conditions of impact glasses) input.

However, the *a posteriori distribution* is not the end of the story. What we are interested in is the 'real' distribution of chances of events (or their causes), which is 'given by Nature' but not necessarily completely revealed by the historical record. This underlying objective characteristic of a system can be called its *a priori distribution function.* It reflects the (evolving) state of the system considered without reference to incidental spatio-temporal specifics. Notably, the a priori distribution function describes the stochastic properties of a chance-generating system in nature rather than the contingent outcomes of that generator in the particular history of a particular place (in this case planet Earth). The relationship between a priori and a posteriori distribution functions for several natural catastrophic hazards is shown in a simplified manner in Table 6.1. Only a priori distribution is useful for predicting the future, since it is not constrained by observation selection effects.

Table 6.1 Examples of Natural Hazards Potentially Comprising GCRs and Two Types of Their Distribution Functions

Type of Event	*A Priori* Distribution	Empirical (*A Posteriori*) Distribution
Impacts	Distribution of near-Earth objects and Earth-crossing comets	Distribution of impact craters, shock glasses, and so on
Super-volcanism	Distribution of geophysical 'hot spots' and their activity	Distribution of calderas, volcanic ash, ice cores, and so on
Supernovae and/ or GRBs	Distribution of progenitors and their motions in the Solar neighbourhood	Geochemical trace anomalies, distribution of remnants

Note: Only *a priori* distribution is veritably describing nature and can serve as a source of predictions about the future events.

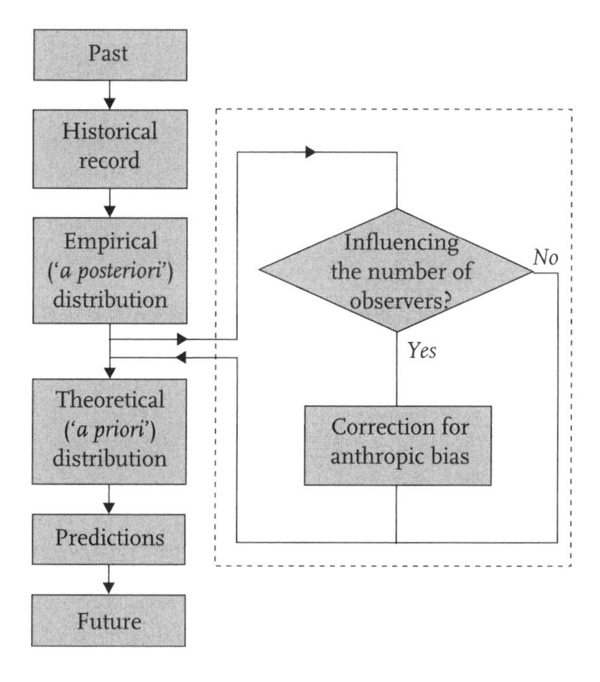

Fig. 6.2 A sketch of the common procedure for deriving predictions about the future from the past records. This applies to quite benign events as well as to GCRs, but only in the latter case do we need to apply the correction symbolically shown in dashed-line box. Steps framed by dashed line are – surprisingly enough – usually *not* performed in the standard risk analysis; they are, however, necessary in order to obtain unbiased estimates of the magnitude of natural GCRs.

The key insight is that the inference to the inherent (a priori) distribution function from the reconstructed empirical (a posteriori) distribution must take account of an observation selection effect (Fig 6.2). Catastrophic events exceeding some severity threshold eliminate all observers and are hence unobservable. Some types of catastrophes may also make the existence of observers on a planet impossible in a subsequent interval of time, the size of which might be correlated with the magnitude of the catastrophe. Because of this observation selection effect, the events reflected in our historical record are not sampled from the full events space but rather from just the part of the events space that lies beneath the 'anthropic compatibility boundary' drawn on the time-severity diagram for each type of catastrophe. This biased sampling effect must be taken into account when we seek to infer the objective chance distribution from the observed empirical distribution of events. Amazingly, it is usually *not* taken into account in most of the real analyses, perhaps 'on naive ergodic grounds'.[6]

This observation selection effect is in addition to what we might call 'classical' selection effects applicable to any sort of event (e.g., removal of traces of events in the distant part by erosion and other instances of the natural entropy increase; see Woo, 1999). Even after these classical selection effects have been taken into account in the construction of an empirical (a posteriori) distribution, the observation selection effects remain to be corrected for in order to derive the *a priori* distribution function.

6.2.3 Applicability class of risks

It seems obvious that the reasoning sketched above applies to GCRs of natural origin since, with one partial exception, there is no unambiguous way of treating major anthropogenic hazards (like global nuclear war or misuse of biotechnology) statistically. This is a necessary, but not yet sufficient, condition for the application of this argument. In order to establish the latter, we need natural catastrophic phenomena which are

- sufficiently destructive (at least in a part of the severity spectrum)
- sufficiently random (in the epistemic sense) and
- leaving traces in the terrestrial (or in general local) record allowing statistical inference.

There are many conceivable threats satisfying these broad desiderata. Some examples mentioned in the literature comprise the following:

1. asteroidal/cometary impacts (severity gauged by the Turin scale or the impact crater size)
2. super-volcanism episodes (severity gauged by the so-called volcanic explosivity index (VEI) or a similar measure)

[6] I thank C.R. Shalizi for this excellent formulation.

3. supernovae/gamma-ray bursts (severity gauged by the distance/intrinsic power)
4. superstrong solar flares (severity gauged by the spectrum/intrinsic power of electromagnetic and corpuscular emissions).

The crucial point here is to have events sufficiently influencing our past, but without too much information which can be obtained externally to the terrestrial biosphere. Thus, there are differences between kinds of catastrophic events in this regard. For instance, the impact history of the Solar System (or at least the part where the Earth is located) is, in theory, easier to be obtained for the Moon, where erosion is orders of magnitude weaker than on Earth. In practice, in the current debates about the rates of cometary impacts, it is precisely the terrestrial cratering rates that are used as an argument for or against existence of a large dark impactor population (see Napier, 2006; Chapter 11 in this volume), thus offering a good model on which the anthropic bias can, at least potentially, be tested. In addition to the impact craters, there is a host of other traces one attempts to find in field work which contribute to the building of the empirical distribution function of impacts, notably searching for chemical anomalies or shocked glasses (e.g., Schultz et al., 2004).

Supernovae/gamma-ray bursts distribution frequencies are also inferred (albeit much less confidently!) from observations of distant regions, notably external galaxies similar to the Milky Way. On one hand, finding local traces of such events in the form of geochemical anomalies (Dreschhoff and Laird, 2006) is excessively difficult and still very uncertain. This external evidence decreases the importance of the Bayesian probability shift. On the other hand, the destructive capacities of such events have been known and discussed for quite some time (see Chapter 12, this volume; Hunt, 1978; Ruderman, 1974; Schindewolf, 1962), and have been particularly enhanced recently by successful explanation of hitherto mysterious gamma-ray bursts as explosions occurring in distant galaxies (Scalo and Wheeler, 2002). The possibility of such cosmic explosions causing a biotic crisis and possibly even a mass extinction episode has returned with a vengeance (Dar et al., 1998; Melott et al., 2004).

Super-volcanic episodes (see Chapter 10, this volume) – both explosive pyroclastic and non-explosive basaltic eruptions of longer duration – are perhaps the best example of global terrestrial catastrophes (which is the rationale for choosing it in the toy model above). They are interesting for two additional recently discovered reasons: (1) Super-volcanism creating Siberian basaltic traps almost certainly triggered the end-Permian mass extinction (251.4 $\pm$ 0.7 Myr before present), killing up to 96% of the terrestrial non-bacterial species (e.g., Benton, 2003; White, 2002). Thus, its global destructive potential is today beyond doubt. (2) Super-volcanism is perhaps the single almost-realized existential catastrophe: the Toba super-eruption probably reduced human population to approximately 1000 individuals, nearly causing the extinction of humanity (Ambrose, 1998; Rampino and Self, 1992). In that light, we would do very well to consider seriously this threat which, ironically in view of historically well-known calamities like the distruction of Santorini, Pompeii, or Tambora, has become an object of concern only very recently (e.g., McGuire, 2002; Roscoe, 2001).

As we have seen, one frequently cited argument in the debate on GCRs, the one of Hut and Rees (1983), actually demonstrates how misleading (but comforting!) conclusions about risk probabilities can be reached when anthropic overconfidence bias is not taken into account.

6.2.4 Additional astrobiological information

The bias affecting the conclusions of Hut and Rees (1983) can be at least partially corrected by using the additional information coming from astrobiology, which has been recently done by Tegmark and Bostrom (2005). Astrobiology is the nascent and explosively developing discipline that deals with three canonical questions: How does life begin and develop? Does life exist elsewhere in the universe? What is the future of life on Earth and in space? One of the most interesting of many astrobiological results of recent years has been the study by Lineweaver (2001), showing that the Earth-like planets around other stars in the Galactic Habitable Zone (GHZ; Gonzalez et al., 2001) are, on average, 1.8 ± 0.9 Gyr older than our planet (see also the extension of this study by Lineweaver et al., 2004). His calculations are based on the tempo of chemical enrichment as the basic precondition for the existence of terrestrial planets. Moreover, Lineweaver's results enable constructing a planetary age distribution, which can be used to constrain the rate of particularly destructive catastrophes, like the vacuum decay or a strangelet catastrophe.

The central idea of the Tegmark and Bostrom study is that planetary age distribution, as compared to the Earth's age, bounds the rate for many doomsday scenarios. If catastrophes that permanently destroy or sterilize a cosmic neighbourhood were very frequent, then almost all intelligent observers would arise much earlier than we did, since the Earth is a latecomer within the habitable planet set. Using the Lineweaver data on planetary formation rates, it is possible to calculate the distribution of birth rates for intelligent species under different assumptions about the rate of sterilization by catastrophic events. Combining this with the information about our own temporal location enables the rather optimistic conclusion that the cosmic (permanent) sterilization rate is, at most, of the order of one per 10^9 years.

How about catastrophes that do not permanently sterilize a cosmic neighbourhood (preventing habitable planets from surviving and forming in that neighbourhood)? Most catastrophes are obviously in this category. Is biological evolution on the other habitable planets in the Milky Way influenced more or less by catastrophes when compared to the Earth? We cannot easily say, because the stronger the catastrophic stress is (the larger analogue of our probability $1 - Q$ is on average), the less useful information can we extract about the proximity – or else – of our particular historical experience to what is generally to be expected. However, future astrobiological studies could help us to resolve this conundrum. Some data already exist. For instance, one recently well-studied case is the system of the famous nearby Sun-like star Tau Ceti which contains both planets and a massive debris disc, analogous to the Solar System Kuiper Belt. Modelling of Tau Ceti's dust disc observations indicate, however, that the mass of the colliding bodies up to 10 km in size may total around 1.2 Earth-masses, compared with 0.1 Earth-masses estimated to be in the Solar System's Kuiper Belt (Greaves et al., 2004). Thus, Tau Ceti's dust disc may have around ten times more cometary and asteroidal material than is currently found in the Solar System – in spite of the fact that Tau Ceti seems to be about twice as old as the Sun (and it is conventionally expected for the amount of such material to decrease with time). Why the Tau Ceti System would have a more massive cometary disc than the Solar System is not fully understood, but it is reasonable to conjecture that any hypothetical terrestrial planet of this extrasolar planetary system has been subjected to much more severe impact stress than the Earth has been during the course of its geological and biological history.[7]

6.3 Doomsday Argument

The Doomsday Argument (DA) is an anthropic argument purporting to show that we have systematically underestimated the probability that humankind will become extinct relatively soon. Originated by the astrophysicist Brandon Carter and developed at length by the philosopher John Leslie,[8] DA purports to show that we have neglected to fully take into account the indexical information residing in the fact about *when* in the history of the human species we exist. Leslie (1996) – in what can be considered the first serious study of GCRs facing humanity and their philosophical aspects – gives a substantial weight to DA, arguing that it prompts immediate re-evaluation of probabilities of extinction obtained through direct analysis of particular risks and their causal mechanisms.

The core idea of DA can be expressed through the following thought experiment. Place two large urns in front of you, one of which you know contains ten balls, the other a million, but you do not know which is which.

[7] Earth-like planets have not been discovered yet around Tau Ceti, but in view of the crude observational techniques employed so far, it has not been expected; the new generation of planet-searching instruments currently in preparation (Darwin, Gaia, TPF, etc.) will settle this problem.

[8] Originally in Leslie (1989); for his most comprehensive treatment, see Leslie (1996). Carter did not publish on DA.

The balls in each urn are numbered 1, 2, 3, 4, ... Now take one ball at random from the left urn; it shows the number 7. This clearly is a strong indication that the left urn contains only ten balls. If the odds originally were 50:50 (identically looking urns), an application of Bayes' theorem gives the posterior probability that the left urn is the one with only ten balls as P_{post} $(n = 10) = 0.99999$. Now consider the case where instead of two urns you have two possible models of humanity's future, and instead of balls you have human individuals, ranked according to birth order. One model suggests that the human race will soon become extinct (or at least that the number of individuals will be greatly reduced), and as a consequence the total number of humans that ever will have existed is about 100 billion. Even the vociferous optimists would not put the prior probability of such a development excessively low – certainly not lower than the probability of the largest certified natural disaster (so-called 'asteroid test') of about 10^{-8} per year. The other model indicates that humans will colonize other planets, spread through the Galaxy, and continue to exist for many future millennia; we consequently can take the number of humans in this model to be of the order of, say, 10^{18}. As a matter of fact, you happen to find that your rank is about 60 billion. According to Carter and Leslie, we should reason in the same way as we did with the urn balls. That you should have a rank of 60 billion is much more likely if only 100 billion humans ever will have lived than if the number was 10^{18}. Therefore, by Bayes' theorem, you should update your beliefs about mankind's prospects and realize that an impending doomsday is much more probable than you thought previously.[9]

Its underlying idea is formalized by Bostrom (1999, 2002a) as the Self-sampling Assumption (SSA):

SSA: One should reason as if one were a random sample from the set of all observers in one's reference class.

In effect, it tells us that there is no structural difference between doing statistics with urn balls and doing it with intelligent observers. SSA has several seemingly paradoxical consequences, which are readily admitted by its supporters; for a detailed discussion, see Bostrom (2001). In particular, the *reference class problem* ('what counts as an observer?') has been plaguing the entire field of anthropic reasoning. A possible response to it is an improved version of SSA, 'Strong SSA' (SSSA):

[9] This is the original, Carter–Leslie version of DA. The version of Gott (1993) is somewhat different, since it does not deal with the number of observers, but with intervals of time characterizing any phenomena (including humanity's existence). Where Gott does consider the number of observers, his argument is essentially temporal, depending on (obviously quite speculative) choice of particular population model for future humanity. It seems that a gradual consensus has been reached about inferiority of this version compared to Leslie–Carter's (see especially Caves, 2000; Olum, 2002), so we shall concentrate on the latter.

SSSA: One should reason as if one's present observer-moment were a random sample from the set of all observer-moments in its reference class.

It can be shown that by taking *more* indexical information into account than SSA does (SSA considers only information about which observer you are, but you also have information about, for example, *which temporal part of this observer* = observer-moment you are at the current moment), it is possible to relativize your reference class so that it may contain different observers at different times, depending partly on your epistemic situation on the occasion. SSA, therefore, describes the correct way of assigning probabilities only in certain special cases; and revisiting the existing arguments for SSA, we find that this is all they establish. In particular, DA is inconclusive. It is shown to depend on particular assumptions about the part of one's subjective prior probability distribution that has to do with indexical information – assumptions that one is free to reject, and indeed, arguably, ought to reject in light of their strongly counterintuitive consequences. Thus, applying the argument to our actual case may be a mistake; at least, a serious methodological criticism could be made of such an inference.

6.4 Fermi's paradox

Fermi's paradox (also known as the 'Great Silence' problem) consists in the tension between (1) naturalistic origin of life and intelligence, as well as astrophysical sizes and ages of our Galaxy and (2) the absence of extraterrestrials in the Solar System, or any other traces of extraterrestrial intelligent activities in the universe.[10] In particular, the lack of macro-engineering (or astroengineering) activities observable from interstellar distances tells us that it is not the case that life evolves on a significant fraction of Earth-like planets and proceeds to develop advanced technology, using it to colonize the universe or perform astroengineering feats in ways that would have been detected with our current instrumentation. The characteristic time for colonization of the Galaxy, according to Fermi's argument, is 10^6–10^8 years, making the fact that the Solar System is not colonized hard to explain, if not for the absence of extraterrestrial cultures. There must be (at least) one Great Filter – an evolutionary step that is extremely improbable – somewhere on the line between Earth-like planet and colonizing-in-detectable-ways civilization (Hanson, 1999). If the Great Filter is not in our past, we must fear it in our

[10] It would be more appropriate to call it the Tsiolkovsky–Fermi–Viewing–Hart–Tipler Paradox (for more history, see Brin, 1983; Kuiper and Brin, 1989; Webb, 2002, and references therein). We shall use the locution 'Fermi's Paradox' for the sake of brevity, and with full respect for the contributions of the other authors.

(near) future. Maybe almost every civilization that develops a certain level of technology causes its own extinction.

Fermi's paradox has become significantly more serious, even disturbing, of late. This is due to several independent lines of scientific and technological advances occurring during the last two decades:

- The discovery of 496 extrasolar planets, as of 13th Novemer 2010 (for regular updates see http://exoplanet.eu/). Although most of them are 'hot Jupiters' and not suitable for life as we know it (some of their satellites could still be habitable, however; see Williams et al., 1997), many other exoworlds are reported to be parts of systems with stable circumstellar habitable zones (Asghari et al., 2004; Beaugé et al., 2005; Noble et al., 2002). It seems that only the selection effects and capacity of present-day instruments stand between us and the discovery of Earth-like extrasolar planets, envisioned by the new generation of orbital observatories.
- Improved understanding of the details of chemical and dynamical structure of the Milky Way and its GHZ. In particular, the already mentioned calculations of Lineweaver (2001; Lineweaver et al., 2004) on the histories of Earth-like planet formation show their median age as 6.4 ± 0.7 Gyr, significantly larger than the Earth's age.
- Confirmation of the relatively *rapid* origination of life on early Earth (e.g., Mojzsis et al., 1996); this rapidity, in turn, offers weak probabilistic support to the idea of many planets in the Milky Way inhabited by at least simple life forms (Lineweaver and Davis, 2002).
- Discovery of extremophiles and the general resistance of simple life forms to much more severe environmental stresses than was hitherto thought possible (Cavicchioli, 2002). These include representatives of all three great domains of terrestrial life (*Bacteria*, *Archaea*, and *Eukarya*), showing that the number and variety of cosmic habitats for life are probably much larger than conventionally imagined.
- Our improved understanding in molecular biology and biochemistry leading to heightened confidence in the theories of naturalistic origin of life (Bada, 2004; Ehrenfreund et al., 2002; Lahav et al., 2001). The same can be said, to a lesser degree, for our understanding of the origin of intelligence and technological civilization (e.g., Chernavskii, 2000).
- Exponential growth of the technological civilization on Earth, especially manifested through Moore's Law and other advances in information technologies (see, for instance, Bostrom, 2000; Schaller, 1997).
- Improved understanding of the feasibility of interstellar travel in both the classical sense (e.g., Andrews, 2003) and in the more efficient form of

sending inscribed matter packages over interstellar distances (Rose and Wright, 2004).

- Theoretical grounding for various astroengineering/macroengineering projects (Badescu and Cathcart, 2000, 2006; Korycansky et al., 2001) potentially detectable over interstellar distances. Especially important in this respect is the possible synergistic combination of astroengineering and computation projects of advanced civilizations, like those envisaged by Sandberg (1999).

Although admittedly uneven and partially conjectural, this list of advances and developments (entirely unknown at the time of Tsiolkovsky's and Fermi's original remarks and even Viewing's, Hart's and Tipler's later re-issues) testifies that Fermi's paradox is not only still with us more than half a century later, but that it is more puzzling and disturbing than ever.

There is a tendency to interpret Fermi's paradox as an argument against contemporary Search for Extra-Terrestrial Intelligence (SETI) projects (e.g., Tipler, 1980). However, this is wrong, since the argument is at best inconclusive – there are many solutions which retain both the observed 'Great Silence' and the rationale for engaging in vigorous SETI research (Gould 1987; Webb, 2002). Furthermore, it is possible that the question is wrongly posed; in an important recent paper, the distinguished historian of science Steven J. Dick argued that there is a tension between SETI, as conventionally understood, and prospects following exponential growth of technology as perceived in recent times on Earth (Dick, 2003, p. 66):

> [I]f there is a flaw in the logic of the Fermi paradox and extraterrestrials *are* a natural outcome of cosmic evolution, then cultural evolution may have resulted in a postbiological universe in which *machines* are the predominant intelligence. This is more than mere conjecture; it is a recognition of the fact that cultural evolution – the final frontier of the Drake Equation – needs to be taken into account no less than the astronomical and biological components of cosmic evolution. [emphasis in the original]

It is easy to understand the necessity of redefining SETI studies in general and our view of Fermi's paradox in particular in this context. For example, post-biological evolution makes those behavioural and social traits like territoriality or expansion drive (to fill the available ecological nïche) which are – more or less successfully – 'derived from nature', lose their relevance. Other important guidelines must be derived which will encompass the vast realm of possibilities stemming from the concept of post-biological evolution. In addition, we have witnessed substantial research leading to a decrease in confidence in the so-called Carter's (1983) 'anthropic' argument, the other mainstay of SETI scepticism (Ćirković et al., 2007; Livio, 1999; Wilson, 1994). All this is accompanied by an increased public interest in astrobiology and related issues (such as Cohen and Stewart, 2002; Grinspoon, 2003; Ward and Brownlee, 2000).

6.4.1 Fermi's paradox and GCRs

Faced with the aggravated situation vis-à-vis Fermi's paradox the solution is usually sought in either (1) some version of the 'rare Earth' hypothesis (i.e., the picture which emphasizes the inherent uniqueness of evolution on our planet and hence uniqueness of human intelligence and technological civilization in the Galactic context), or (2) 'neo-catastrophic' explanations (ranging from the classical 'mandatory self-destruction' explanation, championed for instance by disenchanted SETI pioneers from the Cold War epoch like Sebastian von Hoerner or Iosif Shklovsky, to the modern emphasis on mass extinctions in the history of life and the role of catastrophic impacts, gamma-ray bursts, and similar dramatic events). Both these broad classes of hypotheses are unsatisfactory on several counts: for instance, the 'rare Earth' hypotheses reject the usual Copernican assumption (the Earth is a typical member of the planetary set), and neo-catastrophic explanations usually fail to pass the non-exclusivity requirement[11] (but see Ćirković, 2004, 2006). None of these is a clear, straightforward solution. It is quite possible that a 'patchwork solution', comprised of a combination of suggested and other solutions, remains our best option for solving this deep astrobiological problem. This motivates the continuation of the search for plausible explanations of Fermi's paradox. It should be emphasized that even the founders of 'rare Earth' picture readily admit that simple life forms are ubiquitous throughout the universe (Ward and Brownlee, 2000). It is clear that with the explosive development of astrobiological techniques, very soon we shall be able to directly test this default conjecture.

On the other hand, neo-catastrophic explanations pose important dilemmas related to GCRs – if the 'astrobiological clock' is quasiperiodically reset by exogenous events (like Galactic gamma-ray bursts; Annis, 1999; Ćirković, 2004, 2006), how dangerous is it to be living at present? Seemingly paradoxically, our prospects are quite bright under this hypothesis, since (1) the frequency of forcing events decreases in time and (2) exogenous forcing implies 'astrobiological phase transition' – namely that we are currently located in the temporal window enabling emergence and expansion of intelligence throughout the Galaxy. This would give a strong justification to our present and future SETI projects (Ćirković, 2003). Moreover, this class of solutions of Fermi's paradox does not suffer from usual problems like assuming something about arguably nebulous extraterrestrial sociology in contrast to

[11] The requirement that any process preventing formation of a large and detectable interstellar civilization operates over large spatial (millions of habitable planets in the Milky Way) and temporal (billions of years of the Milky Way history) scales. For more details, see Brin (1983).

solutions such as the classical 'Zoo' or 'Interdict' hypotheses (Ball, 1973; Fogg, 1987).

Somewhat related to this issue is Olum's anthropic argument dealing with the recognition that, if large interstellar civilizations are physically possible, they should, in an infinite universe strongly suggested by modern cosmology, predominate in the total tally of observers (Olum, 2004). As shown by Ćirković (2006), neo-catastrophic solution based on the GRB-forcing of astrobiological timescales can successfully resolve this problem which, as many other problems in astrobiology, including Carter's argument, is based upon implicit acceptance of insidious gradualist assumptions. In particular, while in the equilibrium state most of observers would indeed belong to large (in an appropriately loose sense) civilizations, it is quite reasonable to assume that such an equilibrium has not been established yet. On the contrary, we are located in the phase-transition epoch, in which *all* civilizations are experiencing rapid growth and complexification. Again, neo-catastrophic scenarios offer a reasonable hope for the future of humanity, in agreement with all our empirical evidence.

The relevance of some of particular GCRs discussed in this book to Fermi's paradox has been repeatedly addressed in recent years (e.g., Chapter 10, this volume; Rampino, 2002). It seems that the promising way for future investigations is formulation of joint 'risk function' describing all (both local and correlated) risks facing a habitable planet; such a multi-component function will act as a constraint on the emergence of intelligence and in conjunction with the planetary formation rates, this should give us specific predictions on the number and spatiotemporal distribution of SETI targets.

6.4.2 Risks following from the presence of extraterrestrial intelligence

A particular GCR not covered elsewhere in this book is the one of which humans have been at least vaguely aware since 1898 and the publication of H.G. Wells' *The War of the Worlds* (Wells, 1898) – conflict with hostile extraterrestrial intelligent beings. The famous Orson Welles radio broadcast for Halloween on 30 October 1938 just reiterated the presence of this threat in the mind of humanity. The phenomenon of the mass hysteria displayed on that occasion has proved a goldmine for psychologists and social scientists (e.g., Bulgatz, 1992; Cantril, 1947) and the lessons are still with us. However, we need to recognize that analysing various social and psychological reactions to such bizarre events could induce disconfirmation bias (see Chapter 5 on cognitive biases in this volume) in the rational consideration of the probability, no matter how minuscule, of this and related risks.

The probability of this kind of GCR obviously depends on how frequent extraterrestrial life is in our astrophysical environment. As discussed in the preceding section, opinions wildly differ on this issue.[12] Apart from a couple of 'exotic' hypotheses ('Zoo', 'Interdict', but also the simulation hypotheses below), most researchers would agree that the average distance between planets inhabited by technological civilizations in the Milky Way is *at least* of the order of 10^2 parsecs.[13] This directs us to the second relevant issue for this particular threat: apart from the frequency of extraterrestrial intelligence (which is a necessary, but not sufficient condition for this GCR), the reality of the risk depends on the following:

1. the feasibility of conflict over huge interstellar distances
2. the magnitude of threat such a conflict would present for humanity in the sense of general definition of GCRs, and
3. motivation and willingness of intelligent communities to engage in this form of conflict.

Item (1) seems doubtful, to say the least, if the currently known laws of physics hold without exception; in particular, the velocity limit ensures that such conflict would necessary take place over timescales measured by at least centuries and more probably millennia or longer (compare the timescales of wars between terrestrial nations with the transportation timescales on Earth!). The limitations of computability in chaotic systems would obviate the detailed strategic thinking and planning on such long timescales even for superintelligences employed by the combatants. In addition, the nature of clumpy astronomical distribution of matter and resources, which are tightly clustered around the central star(s) of planetary systems, ensures that a takeover of an inhabited and industrialized planetary system would be possible only in the case of large technological asymmetry between the parties in the conflict. We have seen, in the discussion of Fermi's paradox that, given observable absence of astroengineering activities, such an asymmetry seems unlikely. This means, among other things, that even if we encounter hostile extraterrestrials, the conflict need not jeopardize the existence of human (or post-human) civilization in the Solar System and elsewhere. Finally, factor (3) is even more unlikely and not only for noble, but at present hardly conceivable, ethical reasons. If we take seriously the lessons of sociobiology that suggest that historical human warfare is part of the 'Darwinian baggage' inherited by human cultures, an obvious consequence is that, with the transition to

[12] In the pioneering paper on GCRs/existential risks Bostrom (2002b) has put this risk in the 'whimpers' column, meaning that it is an exceedingly slow and temporally protracted possibility (and the one assigned low probability anyway). Such a conclusion depends on the specific assumptions about extraterrestrial life and intelligence, as well as on the particular model of future humanity and thus is of rather narrow value. We would like to generalize that treatment here while pointing out that still further generalization is desirable.

[13] Parsec, or paralactic second is a standard unit in astronomy: 1 pc = 3.086×10^{16} m. One parsec is, for instance, the average distance between stars in the solar neighbourhood.

a post-biological phase of our evolution, any such archaic impulses will be obviated. *Per analogiam,* this will apply to other intelligent communities in the Milky Way. On the other hand, the resources of even our close astronomical environment are so vast, as is the space of efficiency-improving technologies, that no real ecological pressures can arise to prompt imperial-style expansion and massive colonization over interstellar distances. Even if such unlikely pressures arise, it seems clear that the capacities of seizing defended resources would always (lacking the already mentioned excessive technological asymmetry) be far less cost-effective than expansion into the empty parts of the Milky Way and the wider universe.

There is one particular exception to this generally optimistic view on the (im)possibilities of interstellar warfare which can be worrying: the so-called 'deadly probes' scenario for explaining Fermi's paradox (e.g., Brin, 1983; for fictional treatments of this idea see Benford, 1984; Schroeder, 2002). If the first or one of the first sets of self-replicating von Neumann probes to be released by Galactic civilizations was either programmed to destroy other civilizations or mutated to the same effect (see Benford, 1981), this would explain the 'Great Silence' by another non-exclusive risk. In the words of Brin (1983) '[i]ts logic is compellingly self-consistent'. The 'deadly probes' scenario seems to be particularly disturbing in conjunction with the basic theme of this book, since it shares some of the features of conventional technological optimism vis-à-vis the future of humanity: capacity of making self-replicating probes, AI, advanced spaceflight propulsion, probably also nanotechnology.

It is unfortunate that the 'deadly probes' scenario has not to date been numerically modelled. It is to be hoped that future astrobiological and SETI research will explore these possibilities in a more serious and quantitative manner. In the same time, our astronomical SETI efforts, especially those aimed at the discovery of astroengineering projects (Freitas, 1985; Ćirković and Bradbury, 2006) should be intensified. The discovery of any such project or artefact (see Arnold, 2005) could, in fact, gives us strong probabilistic argument against the 'deadly probes' risk and thus be of long-term assuring comfort.[14]

A related, but distinct, set of threats follows from the possible inadvertent activities of extraterrestrial civilizations that can bring destruction to humanity. A clear example of such activities are quantum field theory-related risks (see Chapter 16, this volume), especially the vacuum decay triggering.

[14] A version of the 'deadly probes' scenario is a purely informatics concept of Moravec (1988), where the computer viruses roam the Galaxy using whatever physical carrier available and replicating at the expense of resources of any receiving civilization. This, however, hinges on the obviously limited capacity to pack sufficiently sophisticated self-replicating algorithm in the bit-string of size small enough to be received non-deformed often enough – which raises some interesting issues from the point of view of algorithmic information theory (e.g., Chaitin, 1977). It seems almost certain that the rapidly occurring improvements in information security will be able to clear this possible threat in check.

A 'new vacuum' bubble produced anywhere in the visible universe – say by powerful alien particle accelerators – would expand at the speed of light, possibly encompassing the Earth and humanity at some point. Clearly, such an event could, in principle, have happened somewhere within our cosmological horizon long ago, the expanding bubble not yet having reached our planet. Fortunately, at least with a set of rather uncontroversial assumptions, the reasoning of Tegmark and Bostrom explained in Section 6.2.4 above applies to this class of events, and the relevant probabilities can be rather tightly constrained by using additional astrobiological information. The conclusion is optimistic since it gives a very small probability that humanity will be destroyed in this manner in the next billion years.

6.5 The Simulation Argument

A particular speculative application of the theory of observation selection leads to the so-called Simulation Argument of Bostrom (2003). If we accept the possibility that a future advanced human (post-human) civilization might have the technological capability of running 'ancestor-simulations' – computer simulations of people like our historical predecessors sufficiently detailed for the simulated people to be conscious – we run into an interesting consequence illuminated by Bostrom (2003). Starting from a rather simple reasoning for the fraction of observers living in simulations (f_{sim})

$$f_{\text{sim}} = \frac{\text{number of observers in simulations}}{\text{total number of observers}} = \frac{\text{number of observers in simulations}}{\begin{array}{l}\text{number of observers in simulations +}\\ \text{number of observers outside of simulations}\end{array}}, \tag{6.5}$$

Bostrom reaches the intriguing conclusion that this commits one to the belief that either (1) we are living in the simulation, or (2) we are almost certain never to reach the post-human stage, or (3) almost all post-human civilizations lack individuals who run significant numbers of ancestor-simulations, that is, computer-emulations of the sort of human-like creatures from which they evolved. Disjunct (3) looks at first glance most promising, but it should be clear that it suggests a quite uniform or monolithic social organization of the future, which could be hallmark of totalitarianism, a GCR in its own right (see Chapter 22, this volume). The conclusion of the Simulation Argument appears to be a pessimistic one, for it narrows down quite substantially the range of positive future scenarios that are tenable in light of the empirical information we now have. The Simulation Argument increases the probability that we are living in a simulation (which may in many subtle ways affect our estimates of how likely various outcomes are) and it decreases the probability that the post-human world would contain lots of free individuals who have large computational resources and human-like motives. But how does it threaten us right now?

In a nutshell, the simulation risk lies in disjunct (1), which implies the possibility that the simulation we inhabit could be shut down. As Bostrom (2002b, p. 7) writes: 'While to some it may seem frivolous to list such a radical or "philosophical" hypothesis next to the concrete threat of nuclear holocaust, we must seek to base these evaluations on reasons rather than untutored intuition.' Until a refutation appears of the argument presented in Bostrom (2003), it would be intellectually dishonest to neglect to mention simulation-shutdown as a potential extinction mode.

A decision to terminate our simulation, taken on the part of the post-human director (under which we shall subsume any relevant agency), may be prompted by our actions or by any number of exogenous factors. Such exogenous factors may include generic properties of such ancestor-simulations such as fixed temporal window or fixed amount of allocated computational resources, or emergent issues such as a realization of a GCR in the director's world. Since we cannot know much about these hypothetical possibilities, let us pick one that is rather straightforward to illustrate how a risk could emerge: the energy cost of running an ancestor-simulation.

From the human experience thus far, especially in sciences such as physics and astronomy, the cost of running large simulations may be very high, though it is still dominated by the capital cost of computer processors and human personnel, not the energy cost. However, as the hardware becomes cheaper and more powerful and the simulating tasks more complex, we may expect that at some point in future the energy cost will become dominant. Computers necessarily dissipate energy as heat, as shown in classical studies of Landauer (1961) and Brillouin (1962) with the finite minimum amount of heat dissipation required per processing of 1 bit of information.[15] Since the simulation of complex human society will require processing a huge amount of information, the accompanying energy cost is necessarily huge. This could imply that the running of ancestor-simulations is, even in advanced technological societies, expensive and/or subject to strict regulation. This makes the scenario in which a simulation runs until it dissipates a fixed amount of energy allocated in advance (similar to the way supercomputer or telescope resources are allocated for today's research) more plausible. Under this assumption, the simulation must necessarily either end abruptly or enter a prolonged phase of gradual simplification and asymptotic dying-out. In the best possible case,

[15] There may be exceptions to this related to the complex issue of reversible computing. In addition, if the Landauer–Brillouin bound holds, this may have important consequences for the evolution of advanced intelligent communities, as well as for our current SETI efforts, as shown by Ćirković and Bradbury (2006).

the simulation is allocated a fixed *fraction* of energy resources of the director's civilization. In such case, it is, in principle, possible to have a simulation of indefinite duration, linked only to the (much more remote) options for ending of the director's world. On the other hand, our activities may make the simulation shorter by increasing the complexity of present entities and thus increasing the running cost of the simulation.[16]

6.6 Making progress in studying observation selection effects

Identifying and correcting for observation selection biases plays the same role as correcting for other biases in risk analysis, evaluation, and mitigation: we need to know the underlying mechanisms of risk very precisely in order to make any progress towards making humanity safe. Although we are not very well prepared for any of the emergencies discussed in various chapters of this book, a general tendency easily emerges: some steps in mitigation have been made only for risks that are rationally sufficiently understood in as objective manner as possible: pandemics, nuclear warfare, impacts, and so on. We have shown that the observation selection acts to decrease the perceived probability of future risks in several wide classes, giving us a false sense of security.

The main lesson is that we should be careful not to use the fact that life on Earth has survived up to this day and that our humanoid ancestors did not go extinct in some sudden disaster to infer that the Earth-bound life and humanoid ancestors are highly resilient. Even if on the vast majority of Earth-like planets life goes extinct before intelligent life forms evolve, we should still expect to find ourselves on one of the exceptional planets that were lucky enough to escape devastation. In particular, the case of Tau Ceti offers a glimpse of what situation in many other places throughout the universe may be like. With regard to some existential risks, our past success provides no ground for expecting success in the future.

Acknowledgement

I thank Rafaela Hildebrand, Cosma R. Shalizi, Alexei Turchin, Slobodan Popović, Zoran Knežević, Nikola Božić, Momčilo Jovanović, Robert J. Bradbury, Danica Ćirković, Zoran Živković and Irena Diklić for useful discussion and comments.

[16] The 'planetarium hypothesis' advanced by Baxter (2000) as a possible solution to Fermi's paradox, is actually very similar to the general simulation hypothesis; however, Baxter suggests exactly 'risky' behaviour in order to try to force the contact between us and the director(s)!

Suggestions for further reading

Bostrom, N. (2002). *Anthropic Bias: Observation Selection Effects* (New York: Routledge). A comprehensive summary of the theory of observation selection effects with some effective examples and a chapter specifically devoted to the Doomsday Argument.

Bostrom, N. (2003). Are you living in a computer simulation? *Philosophical Quarterly*, 53, 243–255. Quite accessible original presentation of the Simulation Argument.

Grinspoon, D. (2003). *Lonely Planets: The Natural Philosophy of Alien Life* (New York: HarperCollins). The most comprehensive and lively written treatment of various issues related to extraterrestrial life and intelligence, as well as an excellent introduction into astrobiology. Contains a popular-level discussion of physical, chemical, and biological preconditions for observership.

Webb, S. (2002). *Where Is Everybody? Fifty Solutions to the Fermi's Paradox* (New York: Copernicus). The most accessible and comprehensive introduction to the various proposed solutions of Fermi's paradox.

References

Ambrose, S.H. (1998). Late Pleistocene human population bottlenecks, volcanic winter, and differentiation of modern humans. *J. Human Evol.*, **34**, 623–651.

Andrews, D.G. (2003). Interstellar Transportation using Today's Physics. *AIAA Paper* 2003-4691. Report to 39th Joint Propulsion Conference & Exhibit.

Annis, J. (1999). An astrophysical explanation for the great silence. *J. Brit. Interplan. Soc.*, **52**, 19–22.

Arnold, L.F.A. (2005). Transit lightcurve signatures of artificial objects. *Astrophys. J.*, **627**, 534–539.

Asghari, N. et al. (2004). Stability of terrestrial planets in the habitable zone of Gl 777 A, HD 72659, Gl 614, 47 UMa and HD 4208. *Astron. Astrophys.*, **426**, 353–365.

Bada, J.L. (2004). How life began on Earth: a status report. *Earth Planet. Sci. Lett.*, **226**, 1–15.

Badescu, V. and Cathcart, R.B. (2000). Stellar engines for Kardashev's type II civilisations. *J. Brit. Interplan. Soc.*, **53**, 297–306.

Badescu, V. and Cathcart, R.B. (2006). Use of class A and class C stellar engines to control Sun's movement in the galaxy. *Acta Astronautica*, **58**, 119–129.

Balashov, Yu. (1991). Resource Letter AP-1: The Anthropic Principle. *Am. J. Phys.*, **59**, 1069–1076.

Ball, J.A. (1973). The Zoo hypothesis. *Icarus*, **19**, 347–349.

Barrow, J.D. and Tipler, F.J. (1986). *The Anthropic Cosmological Principle* (New York: Oxford University Press).

Baxter, S. (2000). The planetarium hypothesis: a resolution of the Fermi paradox. *J. Brit. Interplan. Soc.*, **54**, 210–216.

Beaugé, C., Callegari, N., Ferraz-Mello, S., and Michtchenko, T.A. (2005). Resonance and stability of extra-solar planetary systems. In Knežević, Z. and Milani, A. (eds.),

Dynamics of Populations of Planetary Systems. Proceedings of the IAU Colloquium No. 197, pp. 3–18 (Cambridge: Cambridge University Press).
Benford, G. (1981). Extraterrestrial intelligence? *Quarterly J. Royal Astron. Soc.*, **22**, 217.
Benford, G. (1984). *Across the Sea of Suns* (New York: Simon & Schuster).
Benton, M.J. (2003). *When Life Nearly Died: The Greatest Mass Extinction of All Time* (London: Thames and Hudson).
Bostrom, N. (1999). The Doomsday Argument is alive and kicking. *Mind*, **108**, 539–550.
Bostrom, N. (2000). When machines outsmart humans. *Futures*, **35**, 759–764.
Bostrom, N. (2001). The Doomsday Argument, Adam & Eve, UN^{++} and Quantum Joe. *Synthese*, **127**, 359.
Bostrom, N. (2002a). *Anthropic Bias: Observation Selection Effects in Science and Philosophy* (New York: Routledge).
Bostrom, N. (2002b). Existential risks. *J. Evol. Technol.*, http://www.jetpress.org/volume9/risks.html
Bostrom, N. (2003). Are you living in a computer simulation? *Philos. Quarterly*, **53**, 243–255.
Brillouin, L. (1962). *Science and Information Theory* (New York: Academic Press).
Brin, G.D. (1983). The Great Silence – the controversy concerning extraterrestrial intelligent life. *Quarterly J. Royal Astron. Soc.*, **24**, 283.
Bulgatz, J. (1992). *Ponzi Schemes, Invaders from Mars and More Extraordinary Popular Delusions and the Madness of Crowds* (New York: Harmony Books).
Cantril, H. (1947). *The Invasion from Mars: A Study in the Psychology of Panic* (Princeton, NJ: Princeton University Press).
Carter, B. (1983). The anthropic principle and its implications for biological evolution. *Philos. Trans. Royal Soc. London A*, **310**, 347–363.
Caves, C. (2000). Predicting future duration from present age: a critical assessment. *Contemp. Phys.*, **41**, 143.
Cavicchioli, R. (2002). Extremophiles and the search for extraterrestrial life. *Astrobiology*, **2**, 281–292.
Chaitin, G.J. (1977). Algorithmic information theory. *IBM J. Res. Develop.*, **21**, 350.
Chernavskii, D.S. (2000). The origin of life and thinking from the viewpoint of modern physics. *Physics-Uspekhi*, **43**, 151–176.
Ćirković, M.M. (2003). On the importance of SETI for transhumanism. *J. Evol. Technol.*, **13**. http://www.jetpress.org/volume13/cirkovic.html
Ćirković, M.M. (2004). On the temporal aspect of the Drake Equation and SETI. *Astrobiology*, **4**, 225–231.
Ćirković, M.M. (2006). Too early? On the apparent conflict of astrobiology and cosmology. *Biol. Philos.*, **21**, 369–379.
Ćirković, M.M. (2007). Evolutionary catastrophes and the Goldilocks problem. *Int. J. Astrobiol.* **6**, 325–329.
Ćirković, M.M. and Bradbury, R.J. (2006). Galactic gradients, postbiological evolution and the apparent failure of SETI. *New Astronomy*, **11**, 628–639.
Ćirković, M.M., Dragićević, I., and Vukotić, B. (2008). Galactic punctuated equilibrium: how to undermine Carter's anthropic argument in astrobiology. *Astrobiology*, in press.
Cohen, J. and Stewart, I. (2002). *What Does a Martian Look Like?* (Hoboken, NJ: John Wiley & Sons).

Dar, A., Laor, A., and Shaviv, N.J. (1998). Life extinctions by cosmic ray jets. *Phys. Rev. Lett.*, **80**, 5813–5816.

Dick, S.J. (2003) Cultural evolution, the postbiological universe and SETI. *Int. J. Astrobiol.*, **2**, 65–74.

Dreschhoff, G.A.M. and Laird, C.M. (2006). Evidence for a stratigraphic record of supernovae in polar ice. *Adv. Space Res.*, **38**, 1307–1311.

Earth Impact Database. (2005). http://www.unb.ca/passc/ImpactDatabase/

Egan, G. (2002). *Schild's Ladder* (HarperCollins, New York).

Ehrenfreund, P. et al. (2002). Astrophysical and astrochemical insights into the origin of life. *Rep. Prog. Phys.*, **65**, 1427–1487.

Fogg, M.J. (1987). Temporal aspects of the interaction among the First Galactic Civilizations: The 'Interdict Hypothesis'. *Icarus*, **69**, 370–384.

Freitas, R.A. Jr (1985). Observable characteristics of extraterrestrial technological civilizations. *J. Brit. Interplanet. Soc.*, **38**, 106–112.

Gonzalez, G., Brownlee, D., and Ward, P. (2001). The Galactic Habitable Zone: galactic chemical evolution. *Icarus*, **152**, 185–200.

Gott, J.R. (1993). Implications of the Copernican principle for our future prospects. *Nature*, **363**, 315–319.

Gould, S.J. (1987). SETI and the Wisdom of Casey Stengel. In *The Flamingo's Smile: Reflections in Natural History*, pp. 403–413 (New York: W. W. Norton & Company).

Greaves, J.S., Wyatt, M.C., Holland, W.S., and Dent, W.R.F. (2004). The debris disc around τ Ceti: a massive analogue to the Kuiper Belt. *MNRAS*, **351**, L54–L58.

Grinspoon, D. (2003). *Lonely Planets: The Natural Philosophy of Alien Life* (New York: HarperCollins).

Hanson, R. (1999). Great Filter. Preprint at http://hanson.berkeley.edu/greatfilter.html

Hanson, R. (2001). How to live in a simulation. *J. Evol. Technol.*, **7**, http://www.jetpress.org/volume7/simulation.html

Hunt, G.E. (1978). Possible climatic and biological impact of nearby supernovae. *Nature*, **271**, 430–431.

Hut, P. and Rees, M.J. (1983). How stable is our vacuum? *Nature*, **302**, 508–509.

Korycansky, D.G., Laughlin, G., and Adams, F.C. (2001). Astronomical engineering: a strategy for modifying planetary orbits. *Astrophys. Space Sci.*, **275**, 349–366.

Kuiper, T.B.H. and Brin, G.D. (1989). Resource Letter ETC-1: Extraterrestrial civilization. *Am. J. Phys.*, **57**, 12–18.

Lahav, N., Nir, S., and Elitzur, A.C. (2001). The emergence of life on Earth. *Prog. Biophys. Mol. Biol.*, **75**, 75–120.

Landauer, R. (1961). Irreversibility and heat generation in the computing process. *IBM J. Res. Develop.*, **5**, 183–191.

Leslie, J. (1989), Risking the World's End. *Bull. Canadian Nucl. Soc.*, **May**, 10–15.

Leslie, J. (1996). *The End of the World: The Ethics and Science of Human Extinction* (London: Routledge).

Lineweaver, C.H. (2001). An estimate of the age distribution of terrestrial planets in the Universe: quantifying metallicity as a selection effect. *Icarus*, **151**, 307–313.

Lineweaver, C.H. and Davis, T.M. (2002). Does the rapid appearance of life on earth suggest that life is common in the Universe? *Astrobiology*, **2**, 293–304.

Lineweaver, C.H., Fenner, Y., and Gibson, B.K. (2004). The Galactic Habitable Zone and the age distribution of complex life in the Milky Way. *Science*, **303**, 59–62.

Livio, M. (1999). How rare are extraterrestrial civilizations, and when did they emerge? *Astrophys. J.*, **511**, 429–431.

McGuire, B. (2002). *A Guide to the End of the World: Everything You Never Wanted to Know* (Oxford: Oxford University Press).

Melott, A.L., Lieberman, B.S., Laird, C.M., Martin, L.D., Medvedev, M.V., Thomas, B.C., Cannizzo, J.K., Gehrels, N., and Jackman, C.H. (2004). Did a gamma-ray burst initiate the late Ordovician mass extinction? *Int. J. Astrobiol.*, **3**, 55–61.

Mojzsis, S.J., Arrhenius, G., McKeegan, K.D., Harrison, T.M., Nutman, A.P., and Friend, C.R.L. (1996). Evidence for life on Earth before 3800 million years ago. *Nature*, **384**, 55–59.

Moravec, H.P. (1988). *Mind Children: The Future of Robot and Human Intelligence* (Cambridge, MA: Harvard University Press).

Napier, W.M. (2006). Evidence for cometary bombardment episodes. *MNRAS*, **366**, 977–982.

Noble, M., Musielak, Z.E., and Cuntz, M. (2002). Orbital stability of terrestrial planets inside the Habitable Zones of extrasolar planetary systems. *Astrophys. J.*, **572**, 1024–1030.

Olum, K. (2002). The doomsday argument and the number of possible observers. *Philos. Quarterly*, **52**, 164–184.

Olum, K. (2004). Conflict between anthropic reasoning and observation. *Analysis*, **64**, 1–8.

Rampino, M.R. (2002). Supereruptions as a threat to civilizations on earth-like planets. *Icarus*, **156**, 562–569.

Rampino, M.R. and Self, S. (1992). Volcanic winter and accelerated glaciation following the Toba super-eruption. *Nature*, **359**, 50–52.

Roscoe, H.K. (2001). The risk of large volcanic eruptions and the impact of this risk on future ozone depletion. *Nat. Haz.*, **23**, 231–246.

Rose, C. and Wright, G. (2004). Inscribed matter as an energy-efficient means of communication with an extraterrestrial civilization. *Nature*, **431**, 47–49.

Ruderman, M.A. (1974). Possible consequences of nearby supernova explosions for atmospheric ozone and terrestrial life. *Science*, **184**, 1079–1081.

Sandberg, A. (1999). The physics of information processing superobjects: daily life among the Jupiter brains. *J. Evol. Tech.*, **5**, http://www.jetpress.org/volume5/Brains2.pdf

Scalo, J. and Wheeler, J.C. (2002). Astrophysical and astrobiological implications of gamma-ray burst properties. *Astrophys. J.*, **566**, 723–737.

Schaller, R.R. (1997). Moore's law: past, present, and future. *IEEE Spectrum*, **June**, 53–59.

Schindewolf, O. (1962). Neokatastrophismus? *Deutsch Geologische Gesellschaft Zeitschrift Jahrgang*, **114**, 430–445.

Schroeder, K. (2002). *Permanence* (New York: Tor Books).

Schultz, P.H. et al. (2004). The Quaternary impact record from the Pampas, Argentina. *Earth Planetary Sci. Lett.*, **219**, 221–238.

Tegmark, M. and Bostrom, N. (2005). Is a doomsday catastrophe likely? *Nature,* **438**, 754.

Tipler, F.J. (1980). Extraterrestrial intelligent beings do not exist. *Quarterly J. Royal. Astron. Soc.,* **21**, 267–281.

Ward, P.D. and Brownlee, D. (2000). *Rare Earth: Why Complex Life Is Uncommon in the Universe* (New York: Springer).

Webb, S. (2002). *Where Is Everybody? Fifty Solutions to the Fermi's Paradox* (New York: Copernicus).

Wells, H.G. (1898). *The War of the Worlds* (London: Heinemann).

White, R.V. (2002). Earth's biggest 'whodunnit': unravelling the clues in the case of the end-Permian mass extinction. *Philos. Trans. Royal Soc. Lond. A,* **360**, 2963–2985.

Williams, D.M., Kasting, J.F., and Wade, R.A. (1997). Habitable moons around extrasolar giant planets. *Nature,* **385**, 234–236.

Wilson, P.A. (1994). Carter on Anthropic Principle Predictions. *Brit. J. Phil. Sci.,* **45**, 241–253.

Woo, G. (1999). *The Mathematics of Natural Catastrophes* (Singapore: World Scientific).

·7·

Systems-based risk analysis

Yacov Y. Haimes

7.1 Introduction

Risk models provide the roadmaps that guide the analyst throughout the journey of risk assessment, if the adage 'To manage risk, one must measure it' constitutes the compass for risk management. The process of risk assessment and management may be viewed through many lenses, depending on the perspective, vision, values, and circumstances. This chapter addresses the complex problem of coping with catastrophic risks by taking a systems engineering perspective. Systems engineering is a multidisciplinary approach distinguished by a practical philosophy that advocates holism in cognition and decision making. The ultimate purposes of systems engineering are to (1) build an understanding of the system's nature, functional behaviour, and interaction with its environment, (2) improve the decision-making process (e.g., in planning, design, development, operation, and management), and (3) identify, quantify, and evaluate risks, uncertainties, and variability within the decision-making process.

Engineering systems are almost always designed, constructed, and operated under unavoidable conditions of risk and uncertainty and are often expected to achieve multiple and conflicting objectives. The overall process of identifying, quantifying, evaluating, and trading-off risks, benefits, and costs should be neither a separate, cosmetic afterthought nor a gratuitous add-on technical analysis. Rather, it should constitute an integral and explicit component of the overall managerial decision-making process. In risk assessment, the analyst often attempts to answer the following set of three questions (Kaplan and Garrick, 1981): 'What can go wrong?', 'What is the likelihood that it would go wrong?', and 'What are the consequences?' Answers to these questions help risk analysts identify, measure, quantify, and evaluate risks and their consequences and impacts.

Risk management builds on the risk assessment process by seeking answers to a second set of three questions (Haimes, 1991): 'What can be done and what options are available?', 'What are their associated trade-offs in terms of all costs, benefits, and risks?', and 'What are the impacts of current management decisions on future options?' Note that the last question is the most critical one for any managerial decision-making. This is so because unless the negative and positive impacts of current decisions on future options are assessed and evaluated (to the extent possible), these policy decisions cannot be deemed to be 'optimal' in any sense of the word. Indeed, the assessment and management of risk is essentially a synthesis and amalgamation of the empirical and normative, the quantitative and qualitative, and the objective and subjective efforts. Total risk management can be realized only when these questions are addressed in the broader context of management, where all options and their associated trade-offs are considered within the hierarchical organizational structure. Evaluating the total trade-offs among all important and related system objectives in terms of costs, benefits, and risks cannot be done seriously and meaningfully in isolation from the modelling of the system and from considering the prospective resource allocations of the overall organization.

Theory, methodology, and computational tools drawn primarily from systems engineering provide the technical foundations upon which the above two set of three questions are addressed quantitatively. Good management must thus incorporate and address risk management within a holistic, systemic, and all-encompassing framework and address the following four sources of failure: hardware, software, organizational, and human. This set of sources is intended to be internally comprehensive (i.e., comprehensive within the system's own internal environment. External sources of failure are not discussed here because they are commonly system-dependent.) However, the above four failure elements are not necessarily independent of each other. The distinction between software and hardware is not always straightforward, and separating human and organizational failure often is not an easy task. Nevertheless, these four categories provide a meaningful foundation upon which to build a total risk management framework.

In many respects, systems engineering and risk analysis are intertwined, and only together do they make a complete process. To paraphrase Albert Einstein's comment about the laws of mathematics and reality, we say: 'To the extent to which risk analysis is real, it is not precise; to the extent to which risk analysis is precise, it is not real'. The same can be applied to systems engineering, since modelling constitutes the foundations for both quantitative risk analysis and systems engineering, and the reality is that no single model can precisely represent large-scale and complex systems.

7.2 Risk to interdependent infrastructure and sectors of the economy

The myriad economic, organizational, and institutional sectors, among others, that characterize countries in the developed world can be viewed as a complex large-scale system of systems. (In a similar way, albeit on an entirely different scale, this may apply to the terrorist networks and to the global socio-economic and political environment.) Each system is composed of numerous interconnected and interdependent cyber, physical, social, and organizational infrastructures (subsystems), whose relationships are dynamic (i.e., ever changing with time), non-linear (defeating a simplistic modelling schema), probabilistic (fraught with uncertainty), and spatially distributed (agents and infrastructures with possible overlapping characteristics are spread all over the continent(s)). These systems are managed or coordinated by multiple government agencies, corporate divisions, and decision makers, with diverse missions, resources, timetables, and agendas that are often in competition and conflict. Because of the above characteristics, failures due to human and organizational errors are common. Risks of extreme and catastrophic events facing this complex enterprise cannot and should not be addressed using the conventional expected value of risk as the sole measure for risk. Indeed, assessing and managing the myriad sources of risks facing many countries around the world will be fraught with difficult challenges. However, systems modelling can greatly enhance the likelihood of successfully managing such risks. Although we recognize the difficulties in developing, sustaining, and applying the necessary models in our quest to capture the essence of the multiple perspectives and aspects of these complex systems, there is no other viable alternative but to meet this worthy challenge.

The literature of risk analysis is replete with misleading definitions of vulnerability. Of particular concern is the definition of risk as the product of impact, vulnerability, and threat. This means that in the parlance of systems engineering we must rely on the building blocks of mathematical models, focusing on the use of state variables. For example, to control the production of steel, one must have an understanding of the states of the steel at any instant – its temperature and other physical and chemical properties. To know when to irrigate and fertilize a farm to maximize crop yield, a farmer must assess the soil moisture and the level of nutrients in the soil. To treat a patient, a physician must first know the temperature, blood pressure, and other states of the patient's physical health.

State variables, which constitute the building blocks for representing and measuring risk to infrastructure and economic systems, are used to define the following terms (Haimes, 2004, 2006, Haimes et al. 2002):

1. *Vulnerability* is the manifestation of the inherent states of the system (e.g., physical, technical, organizational, cultural) that can be exploited to adversely affect (cause harm or damage to) that system.
2. *Intent* is the desire or motivation to attack a target and cause adverse effects.
3. *Capability* is the ability and capacity to attack a target and cause adverse effects.
4. *Threat* is the *intent* and *capability* to adversely affect (cause harm or damage to) the system by adversely changing its states.
5. *Risk* (when viewed from the perspectives of terrorism) can be considered qualitatively as the result of a threat with adverse effects to a vulnerable system. Quantitatively, however, risk is a measure of the probability and severity of adverse effects.

Thus, it is clear that modelling risk as the probability and severity of adverse effects requires knowledge of the vulnerabilities, intents, capabilities, and threats to the infrastructure system. Threats to a vulnerable system include terrorist networks whose purposes are to change some of the fundamental states of a country: from a stable to an unstable government, from operable to inoperable infrastructures, and from a trustworthy to an untrustworthy cyber system. These terrorist networks that threaten a country have the same goals as those commissioned to protect its safety, albeit in opposite directions – both want to control the states of the systems in order to achieve their objectives.

Note that the vulnerability of a system is multidimensional, namely, a *vector* in mathematical terms. For example, suppose we consider the risk of hurricane to a major hospital. The states of the hospital, which represent vulnerabilities, are functionality/availability of the electric power, water supply, telecommunications, and intensive care and other emergency units, which are critical to the overall functionality of the hospital. Furthermore, each one of these state variables is not static in its operations and functionality – its levels of functionality change and evolve continuously. In addition, each is a system of its own and has its own sub-state variables. For example, the water supply system consists of the main pipes, distribution system, and pumps, among other elements, each with its own attributes. Therefore, to use or oversimplify the multidimensional vulnerability to a scalar quantity in representing risk could mask the underlying causes of risk and lead to results that are not useful.

The significance of understanding the systems-based nature of a system's vulnerability through its essential state variables manifests itself in both the risk assessment process (*the problem*) and the risk management process (*the remedy*). As noted in Section 7.1, in *risk assessment* we ask: What can go wrong? What is the likelihood? What might be the consequences? (Kaplan and Garrick, 1981). In *risk management* we ask: What can be done and what options are available? What are the trade-offs in terms of all relevant costs, benefits, and risks? What are the impacts of current decisions on future options? (Haimes, 1991, 2004) This significance is also evident in the interplay between vulnerability and threat. (Recall that a threat to a vulnerable system, with adverse effects, yields risk.) Indeed, to answer the three questions in risk assessment, it is imperative to have knowledge of those states that represent the essence of the system under study and of their levels of functionality and security.

7.3 Hierarchical holographic modelling and the theory of scenario structuring

7.3.1 Philosophy and methodology of hierarchical holographic modelling

Hierarchical holographic modelling (HHM) is a holistic philosophy/ methodology aimed at capturing and representing the essence of the inherent diverse characteristics and attributes of a system – its multiple aspects, perspectives, facets, views, dimensions, and hierarchies. Central to the mathematical and systems basis of holographic modelling is the overlapping among various holographic models with respect to the objective functions, constraints, decision variables, and input–output relationships of the basic system. The term *holographic* refers to the desire to have a multi-view image of a system when identifying vulnerabilities (as opposed to a single view, or a flat image of the system). Views of risk can include but are not limited to (1) economic, (2) health, (3) technical, (4) political, and (5) social systems. In addition, risks can be geography related and time related. In order to capture a holographic outcome, the team that performs the analysis must provide a broad array of experience and knowledge.

The term *hierarchical* refers to the desire to understand what can go wrong at many different levels of the system hierarchy. HHM recognizes that for the risk assessment to be complete, one must recognize that there are macroscopic risks that are understood at the upper management level of an organization that are very different from the microscopic risks observed at lower levels. In a particular situation, a microscopic risk can become a critical factor in making things go wrong. To carry out a complete HHM analysis, the team that performs the analysis must include people who bring knowledge up and down the hierarchy.

HHM has turned out to be particularly useful in modelling large-scale, complex, and hierarchical systems, such as defence and civilian infrastructure systems. The multiple visions and perspectives of HHM add strength to risk analysis. It has been extensively and successfully deployed to study risks for government agencies such as the President's Commission on Critical Infrastructure Protection (PCCIP), the FBI, NASA, the Virginia Department of Transportation (VDOT), and the National Ground Intelligence Center, among others. (These cases are discussed as examples in Haimes [2004].) The HHM methodology/philosophy is grounded on the premise that in the process of modelling large-scale and complex systems, more than one mathematical or conceptual model is likely to emerge. Each of these models may adopt a specific point of view, yet all may be regarded as acceptable representations of the infrastructure system. Through HHM, multiple models can be developed and coordinated to capture the essence of many dimensions, visions, and perspectives of infrastructure systems.

7.3.2 The definition of risk

In the first issue of *Risk Analysis,* Kaplan and Garrick (1981) set forth the following 'set of triplets' definition of risk, R:

$$R = \{< S_i, L_i, X_i >\} \tag{7.1}$$

where S_i here denotes the i-th 'risk scenario', L_i denotes the likelihood of that scenario, and X_i the 'damage vector' or resulting consequences. This definition has served the field of risk analysis well since then, and much early debate has been thoroughly resolved about how to quantify the L_i and X_i, and the meaning of 'probability', 'frequency', and 'probability of frequency' in this connection (Kaplan, 1993, 1996).

In Kaplan and Garrick (1981) the S_i themselves were defined, somewhat informally, as answers to the question, 'What can go wrong?' with the system or process being analysed. Subsequently, a subscript 'c' was added to the set of triplets by Kaplan (1991, 1993):

$$R = \{< S_i, L_i, X_i >\}_c. \tag{7.2}$$

This denotes that the set of scenarios $\{S_i\}$ should be 'complete', meaning it should include 'all the possible scenarios, or at least all the important ones'.

7.3.3 Historical perspectives

At about the same time that Kaplan and Garrick's (1981) definition of risk was published, so too was the first article on HHM (Haimes, 1981, 2004). Central to the HHM method is a particular diagram (see, for example, Fig. 7.1). This is particularly useful for analysing systems with multiple, interacting (perhaps overlapping) subsystems, such as a regional transportation or water supply system. The different columns in the diagram reflect different 'perspectives' on the overall system.

HHM can be seen as part of the theory of scenario structuring (TSS) and vice versa, that is, TSS as part of HHM. Under the sweeping generalization of the HHM method, the different methods of scenario structuring can lead to

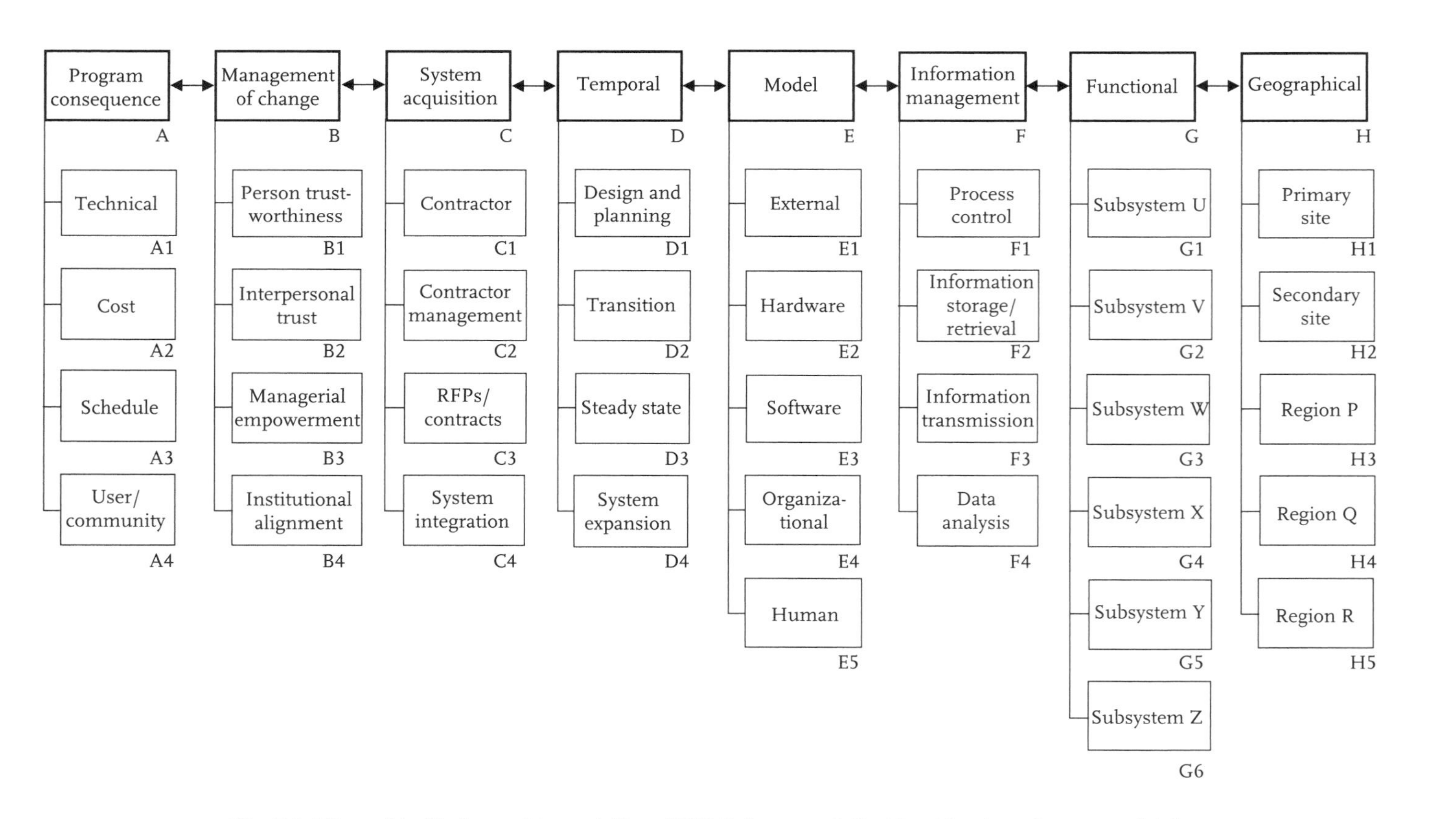

Fig. 7.1 Hierarchical holographic modelling (HHM) framework for identification of sources of risk.

seemingly different sets of scenarios for the same underlying problem. This fact is a bit awkward from the standpoint of the 'set of triplets' definition of risk (Kaplan and Garrick, 1981).

The HHM approach divides the continuum but does not necessarily partition it. In other words, it allows the set of subsets to be overlapping, that is, non-disjoint. It argues that disjointedness is required only when we are going to quantify the likelihood of the scenarios, and even then, only if we are going to add up these likelihoods (in which case the overlapping areas would end up counted twice). Thus, if the risk analysis seeks mainly to identify scenarios rather than to quantify their likelihood, the disjointedness requirement can be relaxed somewhat, so that it becomes a preference rather than a necessity.

To see how HHM and TSS fit within each other (Kaplan et al., 2001), one key idea is to view the HHM diagram as a depiction of the success scenario S_0. Each box in Fig. 7.1 may then be viewed as defining a set of actions or results required of the system, as part of the definition of 'success'. Conversely then, each box also defines a set of risk scenarios in which there is failure to accomplish one or more of the actions or results defined by that box. The union of all these sets contains all possible risk scenarios and is then 'complete'.

This completeness is, of course, a very desirable feature. However, the intersection of two of our risk scenario sets, corresponding to two different HHM boxes, may not be empty. In other words, our scenario sets may not be 'disjoint'.

7.4 Phantom system models for risk management of emergent multi-scale systems

No single model can capture all the dimensions necessary to adequately evaluate the efficacy of risk assessment and management activities of emergent multi-scale systems. This is because it is impossible to identify all relevant state variables and their sub-states that adequately represent large and multi-scale systems (Haimes, 1977, 1981, 2004). Indeed, there is a need for theory and methodology that will enable analysts to appropriately rationalize risk management decisions through a process that

1. identifies existing and potential emergent risks systemically,
2. evaluates, prioritizes, and filters these risks based on justifiable selection criteria,
3. collects, integrates, and develops appropriate metrics and a collection of models to understand the critical aspects of regions,
4. recognizes emergent risks that produce large impacts and risk management strategies that potentially reduce those impacts for various time frames,
5. optimally learns from implementing risk management strategies, and

6. adheres to an adaptive risk management process that is responsive to dynamic, internal, and external forced changes. *To do so effectively, models must be developed to periodically quantify, to the extent possible, the efficacy of risk management options in terms of their costs, benefits, and remaining risks.*

A risk-based, multi-model, systems-driven approach can effectively address these emergent challenges. Such an approach must be capable of maximally utilizing what is known now and optimally learn, update, and adapt through time as decisions are made and more information becomes available at various regional levels. The methodology must quantify risks as well as measure the extent of learning to quantify adaptability. This learn-as-you-go tactic will result in re-evaluation and evolving/learning risk management over time.

Phantom system models (PSMs) (Haimes, 2007) enable research teams to effectively analyse major forced (contextual) changes on the characteristics and performance of emergent multi-scale systems, such as cyber and physical infrastructure systems, or major socio-economic systems. The PSM is aimed at providing a reasoned virtual-to-real experimental modelling framework with which to explore and thus understand the relationships that characterize the nature of emergent multi-scale systems. The PSM philosophy rejects a dogmatic approach to problem-solving that relies on a modelling approach structured exclusively on a single school of thinking. Rather, PSM attempts to draw on a pluri-modelling schema that builds on the multiple perspectives gained through generating multiple models. This leads to the construction of appropriate complementary models on which to deduce logical conclusions for future actions in risk management and systems engineering. Thus, we shift *from* only deciding what is optimal, given what we know, *to* answering questions such as (1) What do we need to know? (2) What are the impacts of having more precise and updated knowledge about complex systems from a risk reduction standpoint? and (3) What knowledge is needed for acceptable risk management decision-making? Answering this mandates seeking the 'truth' about the unknowable complex nature of emergent systems; it requires intellectually bias-free modellers and thinkers who are empowered to experiment with a multitude of modelling and simulation approaches and to collaborate for appropriate solutions.

The PSM has three important functions: (1) identify the states that would characterize the system, (2) enable modellers and analysts to explore cause-and-effect relationships in virtual-to-real laboratory settings, and (3) develop modelling and analyses capabilities to assess irreversible extreme risks; anticipate and understand the likelihoods and consequences of the forced changes around and within these risks; and design and build reconfigured systems to be sufficiently resilient under forced changes, and at acceptable recovery time and costs.

The PSM builds on and incorporates input from HHM, and by doing so seeks to develop causal relationships through various modelling and simulation tools; it imbues life and realism into phantom ideas for emergent systems that otherwise would never have been realized. In other words, with different modelling and simulation tools, PSM legitimizes the exploration and experimentation of out-of-the-box and seemingly 'crazy' ideas and ultimately discovers insightful implications that otherwise would have been completely missed and dismissed.

7.5 Risk of extreme and catastrophic events

7.5.1 The limitations of the expected value of risk

One of the most dominant steps in the risk assessment process is the quantification of risk, yet the validity of the approach most commonly used to quantify risk – its expected value – has received neither the broad professional scrutiny it deserves nor the hoped-for wider mathematical challenge that it mandates. One of the few exceptions is the conditional expected value of the risk of extreme events (among other conditional expected values of risks) generated by the partitioned multi-objective risk method (PMRM) (Asbeck and Haimes, 1984; Haimes, 2004).

Let $p_x(x)$ denote the probability density function of the random variable X, where, for example, X is the concentration of the contaminant trichloroethylene (TCE) in a groundwater system, measured in parts per billion (ppb). The expected value of the concentration (the risk of the groundwater being contaminated by an average concentration of TCE), is $E(X)$ ppb. If the probability density function is discretized to n regions over the entire universe of contaminant concentrations, then $E(X)$ equals the sum of the product of p_i and x_i, where p_i is the probability that the i-th segment of the probability regime has a TCE concentration of x_i. Integration (instead of summation) can be used for the continuous case. Note, however, that the expected-value operation commensurates contaminations (events) of low concentration and high frequency with contaminations of high concentration and low frequency. For example, events $x_1 = 2$ pbb and $x_2 = 20,000$ ppb that have the probabilities $p_1 = 0.1$ and $p_2 = 0.00001$, respectively, yield the same contribution to the overall expected value: (0.1) (2) + (0.00001) (20,000) = 0.2+0.2. However, to the decision maker in charge, the relatively low likelihood of a disastrous contamination of the groundwater system with 20,000 ppb of TCE cannot be equivalent to the contamination at a low concentration of 0.2 ppb, even with a very high likelihood of such contamination. Owing to the nature of mathematical smoothing, the averaging function of the contaminant concentration in this example does not lend itself to prudent management decisions. This is because the expected value of risk does not accentuate catastrophic events and their consequences, thus misrepresenting what would be perceived as an unacceptable risk.

7.5.2 The partitioned multi-objective risk method

Before the partitioned multi-objective risk method (PMRM) was developed, problems with at least one random variable were solved by computing and minimizing the unconditional expectation of the random variable representing damage. In contrast, the PMRM isolates a number of damage ranges (by specifying so-called partitioning probabilities) and generates conditional expectations of damage, given that the damage falls within a particular range. A *conditional expectation* is defined as the expected value of a random variable, given that this value lies within some pre-specified probability range. Clearly, the values of conditional expectations depend on where the probability axis is partitioned. The analyst subjectively chooses where to partition in response to the extreme characteristics of the decision-making problem. For example, if the decision-maker is concerned about the once-in-a-million-years catastrophe, the partitioning should be such that the expected catastrophic risk is emphasized.

The ultimate aim of good risk assessment and management is to suggest some theoretically sound and defensible foundations for regulatory agency guidelines for the selection of probability distributions. Such guidelines should help incorporate meaningful decision criteria, accurate assessments of risk in regulatory problems, and reproducible and persuasive analyses. Since these risk evaluations are often tied to highly infrequent or low-probability catastrophic events, it is imperative that these guidelines consider and build on the statistics of extreme events. Selecting probability distributions to characterize the risk of extreme events is a subject of emerging studies in risk management (Bier and Abhichandani, 2003; Haimes, 2004; Lambert et al., 1994; Leemis, 1995).

There is abundant literature that reviews the methods of approximating probability distributions from empirical data. Goodness-of-fit tests determine whether hypothesized distributions should be rejected as representations of empirical data. Approaches such as the *method of moments* and *maximum likelihood* are used to estimate distribution parameters. The caveat in directly applying accepted methods to natural hazards and environmental scenarios is that most deal with selecting the best matches for the 'entire' distribution. The problem is that these assessments and decisions typically address worst-case scenarios on the tails of distributions. The differences in distribution tails can be very significant even if the parameters that characterize the central tendency of the distribution are similar. A normal and a uniform distribution that have similar expected values can markedly differ on the tails. The possibility of significantly misrepresenting the tails, which are potentially the most relevant portion of the distribution, highlights the importance of considering extreme events when selecting probability distributions (see also Chapter 8, this volume).

More time and effort should be spent to characterize the tails of distributions when modelling the entire distribution. Improved matching between extreme events and distribution tails provides policymakers with more accurate and relevant information. Major factors to consider when developing distributions that account for tail behaviours include (1) the availability of data, (2) the characteristics of the distribution tail, such as shape and rate of decay, and (3) the value of additional information in assessment.

The conditional expectations of a problem are found by partitioning the problem's probability axis and mapping these partitions onto the damage axis. Consequently, the damage axis is partitioned into corresponding ranges. Clearly, the values of conditional expectations are dependent on where the probability axis is partitioned. The choice of where to partition is made subjectively by the analyst in response to the extreme characteristics of the problem. If, for example, the analyst is concerned about the once-in-a-million-years catastrophe, the partitioning should be such that the expected catastrophic risk is emphasized. Although no general rule exists to guide the partitioning, Asbeck and Haimes (1984) suggest that if three damage ranges are considered for a normal distribution, then the $+1s$ and $+4s$ partitioning values provide an effective rule of thumb. These values correspond to partitioning the probability axis at 0.84 and 0.99968; that is, the low-damage range would contain 84% of the damage events, the intermediate range would contain just under 16%, and the catastrophic range would contain about 0.032% (probability of 0.00032). In the literature, catastrophic events are generally said to be those with a probability of exceedance of 10^{-5} (see, for instance, the National Research Council report on dam safety [National Research Council, 1985]). This probability corresponds to events exceeding $+4s$.

A continuous random variable X of damages has a cumulative distribution function (cdf) $P(x)$ and a probability density function (pdf) $p(x)$, which are defined by the following relationships:

$$P(x) = \text{Prob}[X \leq x] \tag{7.3}$$

$$p(x) = \frac{\mathrm{d}P(x)}{\mathrm{d}x}. \tag{7.4}$$

The cdf represents the *non-exceedance probability* of x. The *exceedance probability* of x is defined as the probability that X is observed to be greater than x and is equal to one minus the cdf evaluated at x. The expected value, average, or mean value of the random variable X is defined as

$$E[X] = \int_0^{\infty} xp(x)\, dx. \tag{7.5}$$

For the discrete case, where the universe of events (sample space) of the random variable X is discretized into I segments, the expected value of damage $E[X]$ can be written as

$$E[X] = \sum_{i=1}^{I} p_i x_i \quad p_i \geq 0, \sum p_i = 1, \tag{7.6}$$

where x_i is the i-th segment of the damage.

In the PMRM, the concept of the expected value of damage is extended to generate multiple *conditional expected-value functions*, each associated with a particular range of exceedance probabilities or their corresponding range of damage severities. The resulting conditional expected-value functions, in conjunction with the traditional expected value, provide a family of risk measures associated with a particular policy.

Let $1 - \alpha_1$ and $1 - \alpha_2$, where $0 < \alpha_1 < \alpha_2 < 1$, denote exceedance probabilities that partition the domain of X into three ranges, as follows. On a plot of exceedance probability, there is a unique damage β_1 on the damage axis that corresponds to the exceedance probability $1 - \alpha_1$ on the probability axis. Similarly, there is a unique damage β_2 that corresponds to the exceedance probability $1-\alpha_2$. Damages less than β_1 are considered to be of low severity, and damages greater than β_2 of high severity. Similarly, damages of a magnitude between β_1 and β_2 are considered to be of moderate severity. The partitioning of risk into three severity ranges is illustrated in Fig. 7.2. For example, if the partitioning probability α_1 is specified to be 0.05, then β_1 is the 5th exceedance percentile. Similarly, if α_2 is 0.95 (i.e., $1 - \alpha_2$ is equal to 0.05), then β_2 is the 95th exceedance percentile.

For each of the three ranges, the conditional expected damage (given that the damage is within that particular range) provides a measure of the risk associated with the range. These measures are obtained by defining the *conditional expected value*. Consequently, the new measures of risk are $f_2(\cdot)$, of high exceedance probability and low severity; $f_3(\cdot)$, of medium exceedance probability and moderate severity; and $f_4(\cdot)$, of low exceedance probability and high severity. The function $f_2(\cdot)$ is the conditional expected value of X, given that x is less than or equal to β_1:

$$f_2(\cdot) = E[X \mid X \leq \beta_1]$$

$$f_2(\cdot) = \frac{\int_0^{\beta_1} xp(x)\, dx}{\int_0^{\beta_1} p(x)\, dx}$$

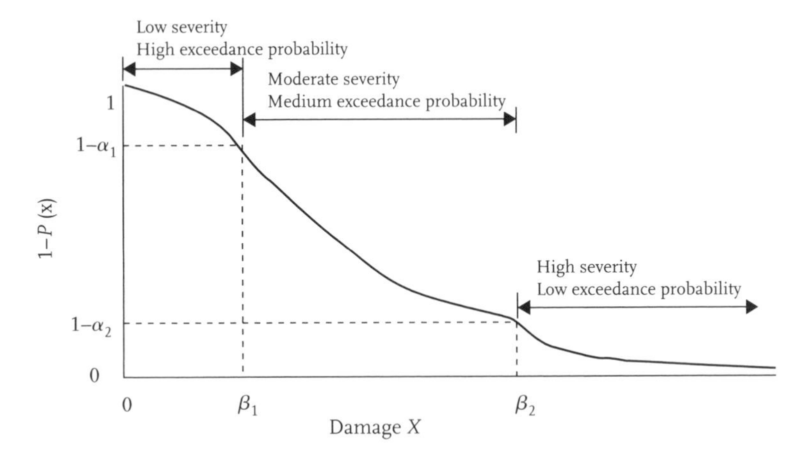

Fig. 7.2 PDF of failure rate distributions for four designs.

$$f_3(\cdot) = E[X \mid \beta_1 \leq X \leq \beta_2]$$

$$f_3(\cdot) = \frac{\int_{\beta_1}^{\beta_2} xp(x)\,\mathrm{d}x}{\int_{\beta_1}^{\beta_2} p(x)\,\mathrm{d}x}$$

$$f_4(\cdot) = E[X|X > \beta_2]$$

$$f_4(\cdot) = \frac{\int_{\beta_1}^{\infty} xp(x)\mathrm{d}x}{\int_{\beta_2}^{\infty} p(x)\mathrm{d}x}. \tag{7.7}$$

Thus, for a particular policy option, there are three measures of risk,$f_2(\cdot)$, $f_3(\cdot)$, and $f_4(\cdot)$, in addition to the traditional expected value denoted by $f_5(\cdot)$. The function $f_1(\cdot)$ is reserved for the cost associated with the management of risk. Note that

$$f_5(\cdot) = \frac{\int_0^{\infty} xp(x)\,\mathrm{d}x}{\int_0^{\infty} p(x)\,\mathrm{d}x} = \int_0^{\infty} xp(x)\,\mathrm{d}x \tag{7.8}$$

since the total probability of the sample space of X is necessarily equal to one. In the PMRM, all or some subset of these five measures are balanced in a multi-objective formulation. The details are made more explicit in the next two sections.

7.5.3 Risk versus reliability analysis

Over time, most, if not all, man-made products and structures ultimately fail. Reliability is commonly used to quantify this time-dependent failure of a system. Indeed, the concept of reliability plays a major role in engineering planning, design, development, construction, operation, maintenance, and replacement.

The distinction between reliability and risk is not merely a semantic issue; rather, it is a major element in resource allocation throughout the life cycle of a product (whether in design, construction, operation, maintenance, or replacement). The distinction between risk and safety, well articulated over two decades ago by Lowrance (1976), is vital when addressing the design, construction, and maintenance of physical systems, since by their nature such systems are built of materials that are susceptible to failure. The probability of such a failure and its associated consequences constitute the measure of risk. Safety manifests itself in the level of risk that is acceptable to those in charge of the system. For instance, the selected strength of chosen materials, and their resistance to the loads and demands placed on them, is a manifestation of the level of acceptable safety. The ability of materials to sustain loads and avoid failures is best viewed as a random process – a process characterized by two random variables: (1) the load (demand) and (2) the resistance (supply or capacity).

Unreliability, as a measure of the probability that the system does not meet its intended functions, does not include the consequences of failures. On the other hand, as a measure of the probability (i.e., unreliability) and severity (consequences) of the adverse effects, risk is inclusive and thus more representative. Clearly, not all failures can justifiably be prevented at all costs. Thus, system reliability cannot constitute a viable metric for resource allocation unless an a priori level of reliability has been determined. This brings us to the duality between risk and reliability on the one hand, and multiple objectives and a single objective optimization on the other.

In the multiple-objective model, the level of acceptable reliability is associated with the corresponding consequences (i.e., constituting a risk measure) and is thus traded off with the associated cost that would reduce the risk (i.e., improve the reliability). In the simple-objective model, on the other hand, the level of acceptable reliability is not explicitly associated with the corresponding consequences; rather it is predetermined (or parametrically evaluated) and thus is considered as a constraint in the model.

There are, of course, both historical and evolutionary reasons for the more common use of reliability analysis rather than risk analysis, as well as substantive and functional justifications. Historically, engineers have always been concerned with strength of materials, durability of product, safety, surety, and operability of various systems. The concept of risk as a quantitative measure of both the probability and consequences (or an adverse effect) of a failure has evolved relatively recently. From the substantive–functional perspective, however, many engineers or decision-makers cannot relate to the amalgamation of two diverse concepts with different units – probabilities and consequences – into one concept termed *risk*. Nor do they accept the metric with which risk is commonly measured. The common metric for risk – the expected value of adverse outcome – essentially commensurates events of low probability and high consequences with those of high probability and low consequences. In this sense, one may find basic philosophical justifications for engineers to avoid using the risk metric and instead work with reliability. Furthermore and most important, dealing with reliability does not require the engineer to make explicit trade-offs between cost and the outcome resulting from product failure. Thus, design engineers isolate themselves from the social consequences that are by-products of the trade-offs between reliability and cost. The design of levees for flood protection may clarify this point.

Designating a 'one-hundred-year return period' means that the engineer will design a flood protection levee for a predetermined water level that on average is not expected to be exceeded more than once every hundred years. Here, ignoring the socio-economic consequences, such as loss of lives and property damage due to a high water level that would most likely exceed the one-hundred-year return period, the design engineers shield themselves from the broader issues of consequences, that is, risk to the population's social well-being. On the other hand, addressing the multi-objective dimension that the risk metric brings requires much closer interaction and coordination between the design engineers and the decision makers. In this case, an interactive process is required to reach acceptable levels of risks, costs, and benefits. In a nutshell, complex issues, especially those involving public policy with health and socio-economic dimensions, should not be addressed through overly simplified models and tools. As the demarcation line between hardware and software slowly but surely fades away, and with the ever-evolving and increasing role of design engineers and systems analysts in technology-based decision-making, a new paradigm shift is emerging. This shift is characterized by a strong overlapping of the responsibilities of engineers, executives, and less technically trained managers.

The likelihood of multiple or compound failure modes in infrastructure systems (as well as in other physical systems) adds another dimension to the limitations of a single reliability metric for such infrastructures (Park et al., 1998; Schneiter et al., 1996). Indeed, because the multiple reliabilities of a system must be addressed, the need for explicit trade-offs among risks and costs becomes more critical. Compound failure modes are defined as two or more paths to failure with consequences that depend on the occurrence of combinations of failure paths. Consider the following examples: (1) a water distribution system, which can fail to provide adequate pressure, flow volume, water quality, and other needs; (2) the navigation channel of an inland waterway, which can fail by exceeding the dredge capacity and by closure to barge traffic; and (3) highway bridges, where failure can occur from deterioration of the bridge deck, corrosion or fatigue of structural elements, or an external loading such as floodwater. None of these failure modes is independent of the others in probability or consequence. For example, in the case of the bridge, deck cracking can contribute to structural corrosion and structural deterioration in turn can increase the vulnerability of the bridge to floods. Nevertheless, the individual failure modes of bridges are typically analysed independently of one another. Acknowledging the need for multiple metrics of reliability of an infrastructure could markedly improve decisions regarding maintenance and rehabilitation, especially when these multiple reliabilities are augmented with risk metrics.

Suggestions for further reading

Apgar, D. (2006). *Risk Intelligence: Learning to Manage What We Don't Know* (Boston, MA: Harvard Business School Press). This book is to help business managers deal more effectively with risk.

Levitt, S.D. and Dubner, S.J. (2005). *Freakonomics: A Rogue Economist Explores the Hidden Side of Everything* (New York: HarperCollins). A popular book that adapts insights from economics to understand a variety of everyday phenomena.

Taleb, N.N. (2007). *The Black Swan: The Impact of the Highly Improbable* (Random House: New York). An engagingly written book, from the perspective of an investor, about risk, especially long-shot risk and how people fail to take them into account.

References

Asbeck, E.L. and Haimes, Y.Y. (1984). The partitioned multiobjective risk method (PMRM). *Large Scale Syst.*, **6**(1), 13–38.

Bier, V.M. and Abhichandani, V. (2003). Optimal allocation of resources for defense of simple series and parallel systems from determined adversaries. In Haimes, Y.Y., Moser, D.A., and Stakhiv, E.Z. (eds.), *Risk-based Decision Making in Water Resources X* (Reston, VA: ASCE).

Haimes, Y.Y. (1977), Hierarchical Analyses of Water Resources Systems, Modeling & Optimization of Large Scale Systems, McGraw Hill, New York.

Haimes, Y.Y. (1981). Hierarchical holographic modeling. *IEEE Trans. Syst. Man Cybernet.*, **11**(9), 606–617.

Haimes, Y.Y. (1991). Total risk management. *Risk Anal.*, **11**(2), 169–171.

Haimes, Y.Y. (2004). *Risk Modeling, Assessment, and Management.* 2nd edition (New York: John Wiley).

Haimes, Y.Y. (2006). On the definition of vulnerabilities in measuring risks to infrastructures. *Risk Anal.*, **26**(2), 293–296.

Haimes, Y.Y. (2007). Phantom system models for emergent multiscale systems. *J. Infrastruct. Syst.*, **13**, 81–87.

Haimes, Y.Y., Kaplan, S., and Lambert, J.H. (2002). Risk filtering, ranking, and management framework using hierarchical holographic modeling. *Risk Anal.*, **22**(2), 383–397.

Kaplan, S. (1991). The general theory of quantitative risk assessment. In Haimes, Y., Moser, D., and Stakhiv, E. (eds.), *Risk-based Decision Making in Water Resources V*, pp. 11–39 (New York: American Society of Civil Engineers).

Kaplan, S. (1993). The general theory of quantitative risk assessment – its role in the regulation of agricultural pests. *Proc. APHIS/NAPPO Int. Workshop Ident. Assess. Manag. Risks Exotic Agric. Pests,* **11**(1), 123–126.

Kaplan, S. (1996). *An Introduction to TRIZ, The Russian Theory of Inventive Problem Solving* (Southfield, MI: Ideation International).

Kaplan, S. and Garrick, B.J. (1981). On the quantitative definition of risk. *Risk Anal.,* **1**(1), 11–27.

Kaplan, S., Haimes, Y.Y., and Garrick, B.J. (2001). Fitting hierarchical holographic modeling (HHM) into the theory of scenario structuring, and a refinement to the quantitative definition of risk. *Risk Anal.,* **21**(5), 807–819.

Kaplan, S., Zlotin, B., Zussman, A., and Vishnipolski, S. (1999). *New Tools for Failure and Risk Analysis – Anticipatory Failure Determination and the Theory of Scenario Structuring* (Southfield, MI: Ideation).

Lambert, J.H., Matalas, N.C., Ling, C.W., Haimes, Y.Y., and Li, D. (1994). Selection of probability distributions in characterizing risk of extreme events. *Risk Anal.,* **149**(5), 731–742.

Leemis, M.L. (1995). *Reliability: Probabilistic Models and Statistical Methods* (Englewood Cliffs, NJ: Prentice-Hall).

Lowrance, W.W. (1976). *Of Acceptable Risk* (Los Altos, CA: William Kaufmann).

National Research Council (NRC), Committee on Safety Criteria for Dams. (1985). *Safety of Dams – Flood and Earthquake Criteria* (Washington, DC: National Academy Press).

Park, J.I., Lambert, J.H., and Haimes, Y.Y. (1998). Hydraulic power capacity of water distribution networks in uncertain conditions of deterioration. *Water Resources Res.,* **34**(2), 3605–3614.

Schneiter, C.D. Li, Haimes, Y.Y., and Lambert, J.H. (1996). Capacity reliability and optimum rehabilitation decision making for water distribution networks. *Water Resources Res.,* **32**(7), 2271–2278.

·8·

Catastrophes and insurance

Peter Taylor

This chapter explores the way financial losses associated with catastrophes can be mitigated by insurance. It covers what insurers mean by catastrophe and risk, and how computer modelling techniques have tamed the problem of quantitative estimation of many hitherto intractable extreme risks. Having assessed where these techniques work well, it explains why they can be expected to fall short in describing emerging global catastrophic risks such as threats from biotechnology. The chapter ends with some pointers to new techniques, which offer some promise in assessing such emerging risks.

8.1 Introduction

Catastrophic risks annually cause tens of thousands of deaths and tens of billions of dollars worth of losses. The figures available from the insurance industry (see, for instance, the Swiss Re [2007] Sigma report) show that mortality has been fairly consistent, whilst the number of recognized catastrophic events, and even more, the size of financial losses, has increased. The excessive rise in financial losses, and with this the number of recognized 'catastrophes', primarily comes from the increase in asset values in areas exposed to natural catastrophe. However, the figures disguise the size of losses affecting those unable to buy insurance and the relative size of losses in developing countries. For instance, Swiss Re estimated that of the estimated $46 billion losses due to catastrophe in 2006, which was a very mild year for catastrophe losses, only some $16 billion was covered by insurance. In 2005, a much heavier year for losses, Swiss Re estimated catastrophe losses at $230 billion, of which $83 billion was insured. Of the $230 billion, Swiss Re estimated that $210 billion was due to natural catastrophes and, of this, some $173 billion was due to the US hurricanes, notably Katrina ($135 billion). The huge damage from the Pakistan earthquake, though, caused relatively low losses in monetary terms (around $5 billion mostly uninsured), reflecting the low asset values in less-developed countries.

In capitalist economies, insurance is the principal method of mitigating potential financial loss from external events in capitalist economies. However, in most cases, insurance does not directly mitigate the underlying causes and risks themselves, unlike, say, a flood prevention scheme. Huge losses in recent years from asbestos, from the collapse of share prices in 2000/2001, the 9/11 terrorist attack, and then the 2004/2005 US hurricanes have tested the global insurance industry to the limit. But disasters cause premiums to rise, and where premiums rise capital follows.

Losses from hurricanes, though, pale besides the potential losses from risks that are now emerging in the world as technological, industrial, and social changes accelerate. Whether the well-publicized risks of global warming, the misunderstood risks of genetic engineering, the largely unrecognized risks of nanotechnology and machine intelligence, or the risks brought about by the fragility to shocks of our connected society, we are voyaging into a new era of risk management. Financial loss will, as ever, be an important consequence of these risks, and we can expect insurance to continue to play a role in mitigating these losses alongside capital markets and governments. Indeed, the responsiveness of the global insurance industry to rapid change in risks may well prove more effective than regulation, international cooperation, or legislation.

Insurance against catastrophes has been available for many years – we need to only think of the San Francisco 1906 earthquake when Cuthbert Heath sent the telegram 'Pay all our policyholders in full irrespective of the terms of their policies' back to Lloyd's of London, an act that created long-standing confidence in the insurance markets as providers of catastrophe cover. For much of this time, assessing the risks from natural hazards such as earthquakes and hurricanes was largely guesswork and based on market shares of historic worst losses rather than any independent assessment of the chance of a catastrophe and its financial consequence. In recent years, though, catastrophe risk management has come of age with major investments in computer-based modelling. Through the use of these models, the insurance industry now understands the effects of many natural catastrophe perils to within an order of magnitude. The recent book by Eric Banks (see Suggestions for further reading) offers a thorough, up-to-date reference on the insurance of property against natural catastrophe. Whatever doubts exist concerning the accuracy of these models – and many in the industry do have concerns as we shall see – there is no questioning that models are now an essential part of the armoury of any carrier of catastrophe risk.

Models notwithstanding, there is still a swathe of risks that commercial insurers will not carry. They fall into two types (1) where the risk is uneconomic, such as houses on a flood plain and (2) where the uncertainty of the outcomes is too great, such as terrorism. In these cases, governments in developed countries may step in to underwrite the risk as we saw with TRIA (Terrorism

Risk Insurance Act) in the United States following 9/11. An analysis[1] of uninsured risks revealed that in some cases risks remain uninsured for a further reason – that the government will bail them out! There are also cases where underwriters will carry the risk, but policyholders find them too expensive. In these cases, people will go without insurance even if insurance is a legal requirement, as with young male UK drivers.

Another concern is whether the insurance industry is able to cope with the sheer size of the catastrophes. Following the huge losses of 9/11 a major earthquake or windstorm would have caused collapse of many re-insurers and threatened the entire industry. However, this did not occur and some loss-free years built up balance sheets to a respectable level. But then we had the reminders of the multiple Florida hurricanes in 2004, and hurricane Katrina (and others!) in 2005, after which the high prices for hurricane insurance have attracted capital market money to bolster traditional re-insurance funds. So we are already seeing financial markets merging to underwrite these extreme risks – albeit 'at a price'. With the doom-mongering of increased weather volatility due to global warming, we can expect to see inter-governmental action, such as the Ethiopian drought insurance bond, governments taking on the role of insurers of the last resort, as we saw with the UK Pool Re-arrangement, bearing the risk themselves through schemes, such as the US FEMA flood scheme, or indeed stepping in with relief when a disaster occurs.

8.2 Catastrophes

What are catastrophic events? A catastrophe to an individual is not necessarily a catastrophe to a company and thus unlikely to be a catastrophe for society. In insurance, for instance, a nominal threshold of $5 million is used by the Property Claims Service (PCS) in the United States to define a catastrophe. It would be a remarkable for a loss of $5 million to constitute a 'global catastrophe'!

We can map the semantic minefield by characterizing three types of catastrophic risk as treated in insurance (see Table 8.1): *physical* catastrophes, such as windstorm and earthquake, whether due to natural hazards or man-made accidental or intentional cause; *liability* catastrophes, whether intentional such as terrorism or accidental such as asbestosis; and *systemic* underlying causes leading to large-scale losses, such as the dotcom stock market collapse.

Although many of these catastrophes are insured today, some are not, notably emerging risks from technology and socio-economic collapse. These types of risk present huge challenges to insurers as they are potentially catastrophic losses and yet lack an evidential loss history.

[1] Uninsured Losses, report for the Tsunami Consortium, November 2000.

Table 8.1 Three Types of Catastrophic Risk as Treated in Insurance

	Natural	Man-made	
		Intentional	Accidental
Physical (property)	Earthquake, windstorm, volcano, flood, tsunami, wildfire, landslip, space storm, asteroid	Nuclear bomb (war), nuclear action (terrorism), arson (property terrorism) (e.g., 9/11)	Climate change, nuclear accident
Liability	Pandemic	War (conventional, nano-, bio-, nuclear), terrorism (nano, bio-, nuclear)	Product liability (e.g., asbestos), environmental pollution (chemical, bio-, nuclear), bio-accident, nano-accident, nuclear accident
Systemic		Social failure (e.g., war, genocide), economic failure (e.g., energy embargo, starvation)	Technology failure (e.g., computer network failure), financial dislocation (e.g., 2000 stock market crash)

Catastrophe risks can occur in unrelated combination within a year or in clusters, such as a series of earthquakes and even the series of Florida hurricanes seen in 2004. Multiple catastrophic events in a year would seem to be exceptionally rare until we consider that the more extreme an event the more likely it is to trigger another event. This can happen, for example, in natural catastrophes where an earthquake could trigger a submarine slide, which causes a tsunami or triggers a landslip, which destroys a dam, which in turn floods a city. Such high-end correlations are particularly worrying when they might induce man-made catastrophes such as financial collapse, infrastructure failure, or terrorist attack. We return to this question of high-end correlations later in the chapter.

You might think that events are less predictable the more extreme they become. Bizarre as it is, this is not necessarily the case. It is known from statistics that a wide class of systems show, as we look at the extreme tail, a regular 'extreme value' behaviour. This has, understandably, been particularly important in Holland (de Haan, 1990), where tide level statistics along the Dutch coast since 1880 were used to set the dike height to a 1 in 10,000-year exceedance level. This compares to the general 1 in 30-year exceedance level for most New Orleans dikes prior to Hurricane Katrina (Kabat et al., 2005)!

You might also have thought that the more extreme an event is the more obvious must be its cause, but this does not seem to be true in general either. Earthquakes, stock market crashes, and avalanches all exhibit sudden large failures without clear 'exogenous' (external) causes. Indeed, it is characteristic of many complex systems to exhibit 'endogenous' failures following from their intrinsic structure (see, for instance, Sornette et al., 2003).

In a wider sense, there is the problem of predictability. Many large insurance losses have come from 'nowhere' – they simply were not recognized in advance as realistic threats. For instance, despite the UK experience with IRA bombing in the 1990s, and sporadic terrorist attacks around the world, no one in the insurance industry foresaw concerted attacks on the World Trade Center and the Pentagon on 11 September 2001.

Then there is the problem of latency. Asbestos was considered for years to be a wonder material[2] whose benefits were thought to outweigh any health concerns. Although recognized early on, the 'latent' health hazards of asbestos did not receive serious attention until studies of its long-term consequences emerged in the 1970s. For drugs, we now have clinical trials to protect people from unforeseen consequences, yet material science is largely unregulated. Amongst the many new developments in nanotechnology, could there be latent modern versions of asbestosis?

8.3 What the business world thinks

You would expect the business world to be keen to minimize financial adversity, so it is of interest to know what business sees as the big risks.

A recent survey of perceived risk by Swiss Re (see Swiss Re, 2006, based on interviews in late 2005) of global corporate executives across a wide range of industries identified computer-based risk the highest priority risk in all major countries by level of concern and second in priority as an emerging risk. Also, perhaps surprisingly, terrorism came tenth, and even natural disasters only made seventh. However, the bulk of the recognized risks were well within the traditional zones of business discomfort such as corporate governance, regulatory regimes, and accounting rules.

The World Economic Forum (WEF) solicits expert opinion from business leaders, economists, and academics to maintain a finger on the pulse of risk and trends. For instance, the 2006 WEF Global Risks report (World Economic Forum, 2006) classified risks by likelihood and severity with the most severe risks being those with losses greater than $1 trillion or mortality greater than one million deaths or adverse growth impact greater than 2%. They were as follows.

[2] See, for example, http://environmentalchemistry.com/yogi/environmental/asbestoshistory 2004.html

1. US current account deficit was considered a severe threat to the world economy in both short (1–10% chance) and long term (<1% chance).
2. Oil price shock was considered a short-term severe threat of low likelihood (<1%).
3. Japan earthquake was rated as a 1–10% likelihood. No other natural hazards were considered sufficiently severe.
4. Pandemics, with avian flu as an example, was rated as a 1–10% chance.
5. Developing world disease: spread of HIV/AIDS and TB epidemics were similarly considered a severe and high likelihood threat (1–20%).
6. Organized crime counterfeiting was considered to offer severe outcomes (long term) due to vulnerability of IT networks, but rated low frequency (<1%).
7. International terrorism considered potentially severe, through a conventional simultaneous attack (short term estimated at <1%) or a non-conventional attack on a major city in longer term (1–10%).

No technological risks were considered severe, nor was climate change. Most of the risks classified as severe were considered of low likelihood (<1%) and all were based on subjective consensual estimates.

The more recent 2007 WEF Global Risks report (World Economic Forum, 2007) shows a somewhat different complexion with risk potential generally increased, most notably the uncertainty in the global economy from trade protectionism and over-inflated asset values (see Fig. 8.1). The report also takes a stronger line on the need for intergovernmental action and awareness.

It seems that many of the risks coming over the next five to twenty years from advances in biotechnology, nanotechnology, machine intelligence, the resurgence of nuclear, and socio-economic fragility, all sit beyond the radar of the business world today. Those that are in their sights, such as nanotechnology, are assessed subjectively and largely disregarded.

And that is one of the key problems when looking at global catastrophic risk and business. These risks are too big and too remote to be treated seriously.

8.4 Insurance

Insurance is about one party taking on another's financial risk. Given what we have just seen of our inability to predict losses, and given the potential for dispute over claims, it is remarkable that insurance even exists, yet it does! Through the protection offered by insurance, people can take on the risks of ownership of property and the creation of businesses. The principles of ownership and its financial protection that we have in the capitalist West, though, do not apply to many countries; so, for instance, commercial insurance

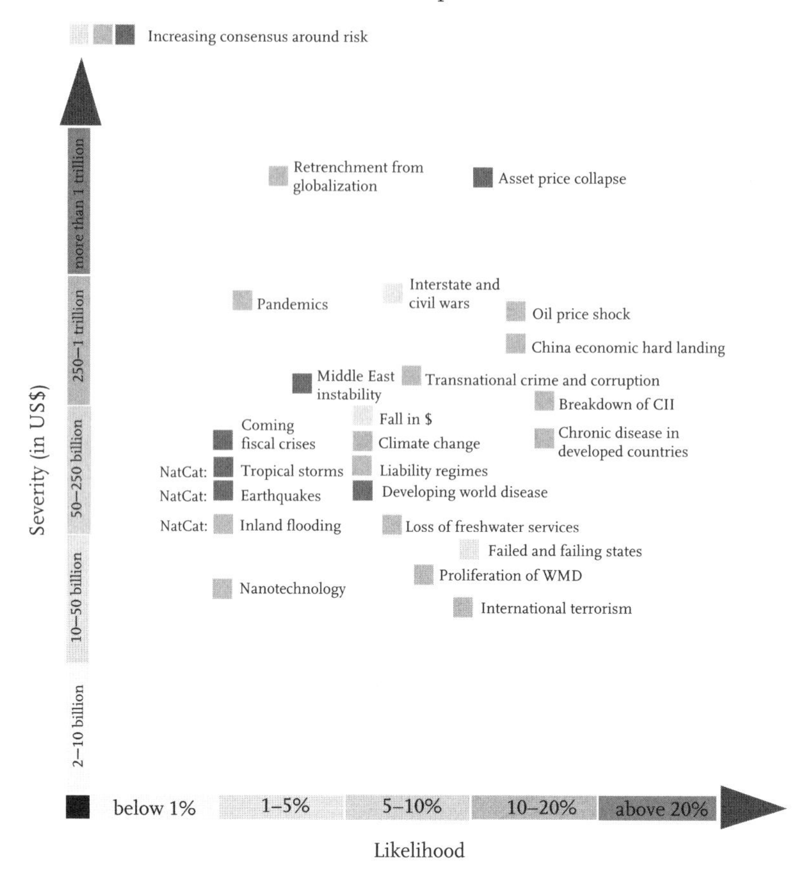

Fig. 8.1 World Economic Forum 2007 – The 23 core global risks: likelihood with severity by economic loss.

did not exist in Soviet Russia. Groups with a common interest, such as farmers, can share their common risks either implicitly by membership of a collective, as in Soviet Russia, or explicitly by contributing premiums to a mutual fund. Although mutuals were historically of importance as they often initiated insurance companies, insurance is now almost entirely dominated by commercial risk-taking.

The principles of insurance were set down over 300 years ago in London by shipowners at the same time as the theory of probability was being formulated to respond to the financial demands of the Parisian gaming tables. Over these years a legal, accounting, regulatory, and expert infrastructure has built up to make insurance an efficient and effective form of financial risk transfer.

To see how insurance works, let us start with a person or company owning property or having a legal liability in respect of others. They may choose to take their chances of avoiding losses by luck but will generally prefer to protect against the consequences of any financial losses due to a peril such as fire or accident. In some cases, such as employer's liability, governments require by law that insurance be bought. Looking to help out are insurers who promise (backed normally by capital or a pledge of capital) to pay for these losses in return for a payment of money called a 'premium'. The way this deal is formulated is through a contract of insurance that describes what risks are covered. Insurers would only continue to stay in business over a period of years if premiums exceed claims plus expenses. Insurers will nonetheless try and run their businesses with as little capital as they can get away with, so government regulators exist to ensure they have sufficient funds. In recent years, regulators such as the Financial Services Authority in the United Kingdom have put in place stringent quantitative tests on the full range of risk within an insurer, which include underwriting risks, such as the chance of losing a lot of money in one year due to a catastrophe, financial risks such as risk of defaulting creditors, market risks such as failure of the market to provide profitable business, and operational risks from poor systems and controls.

Let us take a simple example: your house. In deciding the premium to insure your house for a year, an underwriter will apply a 'buildings rate' for your type of house and location to the rebuild cost of the house, and then add on an amount for 'contents rate' for your home's location and safety features against fire and burglary multiplied by the value of contents. The rate is the underwriter's estimate of the chance of loss – in simple terms, a rate of 0.2% is equivalent to expecting a total loss once in 500 years. So, for example, the insurer might think you live in a particularly safe area and have good fire and burglary protection, and so charge you, say, 0.1% rate on buildings and 0.5% on contents. Thus, if your house's rebuild cost was estimated at £500,000 and your contents at £100,000, then you would pay £500 for buildings and £500 for contents, a total of £1000 a year.

Most insurance works this way. A set of 'risk factors' such as exposure to fire, subsidence, flood, or burglary are combined – typically by addition – in the construction of the premium. The rate for each of these factors comes primarily from claims experience – this type of property in this type of area has this proportion of losses over the years. That yields an *average* price. Insurers, though, need to guard against bad years and to do this they will try to underwrite enough of these types of risk, so that a 'law of large numbers' or 'regression to the mean' reduces the volatility of the losses in relation to the total premium received. Better still, they can diversify their portfolio of risks so that any correlations of losses within a particular class (e.g., a dry winter causes earth shrinkage and subsidence of properties built on clay soils) can be counteracted by uncorrelated classes.

If only all risks were like this, but they are not. There may be no decent claims history, the conditions in the future may not resemble the past, there may be a possibility of a few rare but extremely large losses, it may not be possible to reduce volatility by writing a lot of the same type of risk, and it may not be possible to diversify the risk portfolio. One or more of these circumstances can apply. For example, lines of business where we have low claims experience and doubt over the future include 'political risks' (protecting a financial asset against a political action such as confiscation). The examples most relevant to this chapter are 'catastrophe' risks, which typically have low claims experience, large losses, and limited ability to reduce volatility. To understand how underwriters deal with these, we need to revisit what the pricing of risk and indeed risk itself are all about.

8.5 Pricing the risk

The primary challenge for underwriters is to set the premium to charge the customer – the price of the risk. In constructing this premium, an underwriter will usually consider the following elements:

1. Loss costs, being the expected cost of claims to the policy.
2. Acquisition costs, such as brokerage and profit commissions.
3. Expenses, being what it costs to run the underwriting operation.
4. Capital costs, being the cost of supplying the capital required by regulators to cover the possible losses according their criterion (e.g., the United Kingdom's FSA currently requires capital to meet a 1-in-200-year or more annual chance of loss).
5. Uncertainty cost, being an additional subjective charge in respect of the uncertainty of this line of business. This can, in some lines of business, such as political risk, be the dominant factor.
6. Profit, being the profit margin required of the business. This can sometimes be set net of expected investment income from the cash flow of receiving premiums before having to pay out claims, which for 'liability' contracts can be many years.

Usually the biggest element of price is the loss cost or 'pure technical rate'. We saw above how this was set for household building cover. The traditional method is to model the history of claims, suitably adjusted to current prices, with frequency/severity probability distribution combinations, and then trend these in time into the future, essentially a model of the past playing forward. The claims can be those either on the particular contract or on a large set of contracts with similar characteristics.

Setting prices from rates is – like much of the basic mathematics of insurance – essentially a linear model even though non-linearity appears pronounced when large losses happen. As an illustration of non-linearity, insurers made allowance for some inflation of rebuild/repair costs when underwriting for US windstorms, yet the actual 'loss amplification' in Hurricane Katrina was far greater than had been anticipated. Another popular use of linearity has been linear regression in modelling risk correlations. In extreme cases, though, this assumption, too, can fail. A recent and expensive example was when the dotcom bubble burst in April 2000. The huge loss of stock value to millions of Americans triggered allegations of impropriety and legal actions against investment banks, class actions against the directors and officers of many high technology companies whose stock price had collapsed, for example, the collapse of Enron and WorldCom and Global Crossing, and the demise of the accountants Arthur Andersen discredited when the Enron story came to light. Each of these events has led to massive claims against the insurance industry. Instead of linear correlations, we need now to deploy the mathematics of copulas.[3] This phenomenon is also familiar from physical damage where a damaged asset can in turn enhance the damage to another asset, either directly, such as when debris from a collapsed building creates havoc on its neighbours ('collateral damage'), or indirectly, such as when loss of power exacerbates communication functions and recovery efforts ('dependency damage').

As well as pricing risk, underwriters have to guard against accumulation of risk. In catastrophe risks the simplest measure of accumulations is called the 'aggregate' – the cost of total ruin when everything is destroyed. Aggregates represent the worst possible outcome and are an upper limit on an underwriter's exposure, but have unusual arithmetic properties. (As an example, the aggregate exposure of California is typically lower than the sum of the aggregate exposures in each of the Cresta zones into which California is divided for earthquake assessment. The reason for this is that many insurance policies cover property in more than one zone but have a limit of loss across the zones. Conversely, for fine-grained geographical partitions, such as postcodes, the sum across two postal codes can be higher than the aggregate of each. The reason for this is that risks typically have a per location [per policy when dealing with re-insurance] deductible!)

8.6 Catastrophe loss models

For infrequent and large catastrophe perils such as earthquakes and severe windstorms, the claims history is sparse and, whilst useful for checking the results of models, offers insufficient data to support reliable claims analysis.

[3] A copula is a functional, whose unique existence is guaranteed from Sklar's theorem, which says that a multivariate probability distribution can be represented uniquely by a functional of the marginal probability functions.

Instead, underwriters have adopted computer-based catastrophe loss models, typically from proprietary expert suppliers such as RMS (Risk Management Solutions), AIR (Applied Insurance Research), and EQECAT.

The way these loss models work is well-described in several books and papers, such as the recent UK actuarial report on loss models (GIRO, 2006). From there we present Fig. 8.2, which shows the steps involved.

Quoting directly from that report:

Catastrophe models have a number of basic modules:

- Event module 1em

  A database of stochastic events (the event set) with each event defined by its physical parameters, location, and annual probability/frequency of occurrence.
- Hazard module 1em

  This module determines the hazard of each event at each location. The hazard is the consequence of the event that causes damage – for a hurricane it is the wind at ground level, for an earthquake, the ground shaking.
- Inventory (or exposure) module 1em

  A detailed exposure database of the insured systems and structures. As well as location this will include further details such as age, occupancy, and construction.
- Vulnerability module 1em

  Vulnerability can be defined as the degree of loss to a particular system or structure resulting from exposure to a given hazard (often expressed as a percentage of sum insured).
- Financial analysis module 1em

  This module uses a database of policy conditions (limits, excess, sub limits, coverage terms) to translate this loss into an insured loss.

Of these modules, two, the inventory and financial analysis modules, rely primarily on data input by the user of the models. The other three modules represent the engine of the catastrophe model, with the event and hazard modules being based on seismological and meteorological assessment and the vulnerability module on engineering assessment. (GIRO, 2006, p. 6)

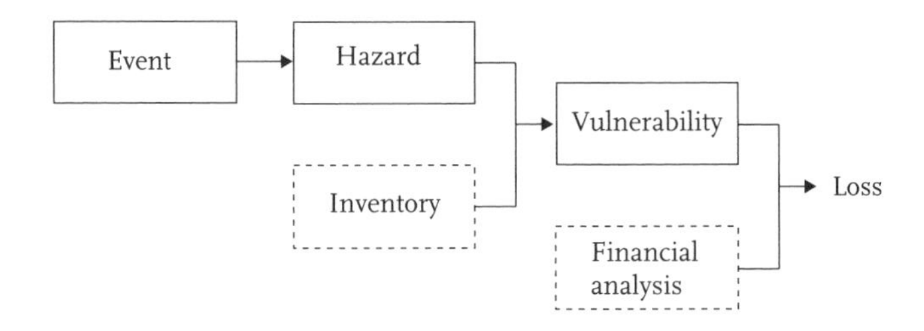

Fig. 8.2 Generic components of a loss model.

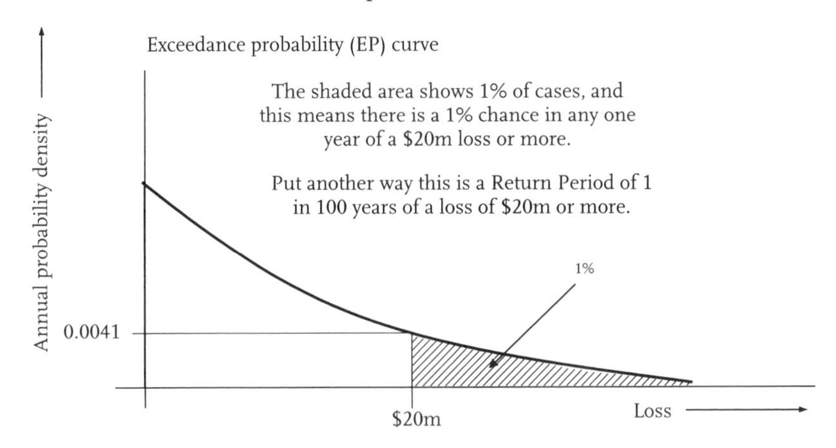

Fig. 8.3 Exceedance probability loss curve.

The model simulates a catastrophic event such as a hurricane by giving it a geographical extent and peril characteristics so that it 'damages' – as would a real hurricane – the buildings according to 'damageability' profiles for occupancy, construction, and location. This causes losses, which are then applied to the insurance policies in order to calculate the accumulated loss to the insurer. The aim is to produce an estimate of the probability of loss in a year called the occurrence exceedance probability (OEP), which estimates the chance of exceeding a given level of loss in any one year, as shown in Fig. 8.3. When the probability is with respect to all possible losses in a given year, then the graph is called an aggregate exceedance probability curve (AEP).

You will have worked out that just calculating the losses on a set of events does not yield a smooth curve. You might also have asked yourself how the 'annual' bit gets in. You might even have wondered how the damage is chosen because surely in real life there is a range of damage even for otherwise similar buildings. Well, it turns out that different loss modelling companies have different ways of choosing the damage percentages and of combining these events, which determine the way the exceedance probability distributions are calculated.[4] Whatever their particular solutions, though, we end up with a two-dimensional estimate of risk through the 'exceedance probability (EP) curve'.

[4] Applied insurance research (AIR), for example, stochastically samples the damage for each event on each property and the simulation is for a series of years. Risk management solutions (RMS), on the other hand, take each event as independent, described by a Poisson arrival rate and treat the range of damage as a parameterized beta function ('secondary uncertainty') in order to come up with the OEP curve, and then some fancy mathematics for the AEP curve. The AIR method is the more general and conceptually the simplest, as it allows for non-independence of events in the construction of the events hitting a given year, and has built-in damage variability (the so-called secondary uncertainty).

8.7 What is risk?

[R]isk is either a condition of, or a measure of, exposure to misfortune – more concretely, exposure to unpredictable losses. However, as a measure, risk is not one-dimensional – it has three distinct aspects or 'facets' related to the anticipated values of unpredictable losses. The three facets are Expected Loss, Variability of Loss Values, and Uncertainty about the Accuracy of Mental Models intended to predict losses.

Ted Yellman, 2000

Although none of us can be sure whether tomorrow will be like the past or whether a particular insurable interest will respond to a peril as a representative of its type, the assumption of such external consistencies underlies the construction of rating models used by insurers. The parameters of these models can be influenced by past claims, by additional information on safety factors and construction, and by views on future medical costs and court judgments. Put together, these are big assumptions to make, and with catastrophe risks the level of uncertainty about the chance and size of loss is of primary importance, to the extent that such risks can be deemed uninsurable. Is there a way to represent this further level of uncertainty? How does it relate to the 'EP curves' we have just seen?

Kaplan and Garrick (1981) defined quantitative risk in terms of three elements – probability for likelihood, evaluation measure for consequence, and 'level 2 risk' for the uncertainty in the curves representing the first two elements. Yellman (see quote in this section) has taken this further by elaborating the 'level 2 risk' as uncertainty of the likelihood and adversity relationships.

We might represent these ideas by the EP curve in Fig. 8.4. When dealing with insurance, 'Likelihood' is taken as probability density and 'Adversity' as loss. Jumping further up the abstraction scale, these ideas can be extended to qualitative assessments, where instead of defined numerical measures of probability and loss we look at categoric (low, medium, high) measures of Likelihood and Adversity. The loss curves now look like those shown in Figs. 8.5 and 8.6. Putting these ideas together, we can represent these elements of risk in terms of fuzzy exceedance probability curves as shown in Fig. 8.7.

Another related distinction is made in many texts on risk (e.g., see Woo, 1999) between intrinsic or 'aleatory' (from the Greek for dice) uncertainty and avoidable or 'epistemic' (implying it follows from our lack of knowledge) risk. The classification of risk we are following looks at the way models predict outcomes in the form of a relationship between chance and loss. We can have many different parameterizations of a model and, indeed, many different models. The latter types of risk are known in insurance as 'process risk' and 'model risk', respectively.

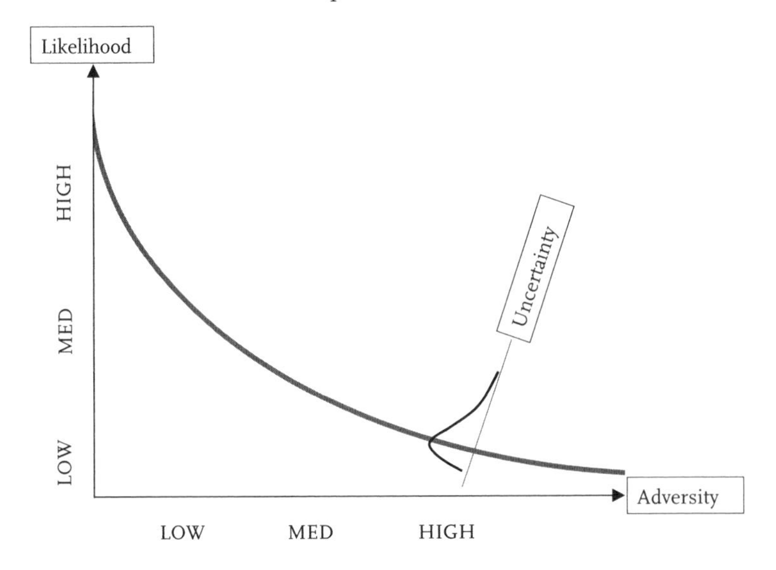

Fig. 8.4 Qualitative loss curve.

Impact	Risk distribution		
Significant	S2, E2	E1, L3, T1	T2, F1
Moderate	L2	T3, F3, S3	
Minor	L1, F2	S1	E3
	Low	Medium	High
	Likelihood		

Economic and financial
F1 Interest rate
F2 Securities
F3 Cost of insurance

Environmental
E1 Climate change
E2 Pollution
E3 Ozone depletion

Legal
L1 Liabilities
L2 Human rights
L3 International agreements

Technological
T1 Nuclear power
T2 Biotechnology
T3 Genetic engineering

Safety and security
S1 Invasion
S2 Terrorism
S3 Organized crime

Fig. 8.5 Qualitative risk assessment chart – Treasury Board of Canada.

These distinctions chime very much with the way underwriters in practice perceive risk and set premiums. There is a saying in catastrophe re-insurance that 'nothing is less than one on line', meaning the vagaries of life are such that you should never price high-level risk at less than the chance of a total

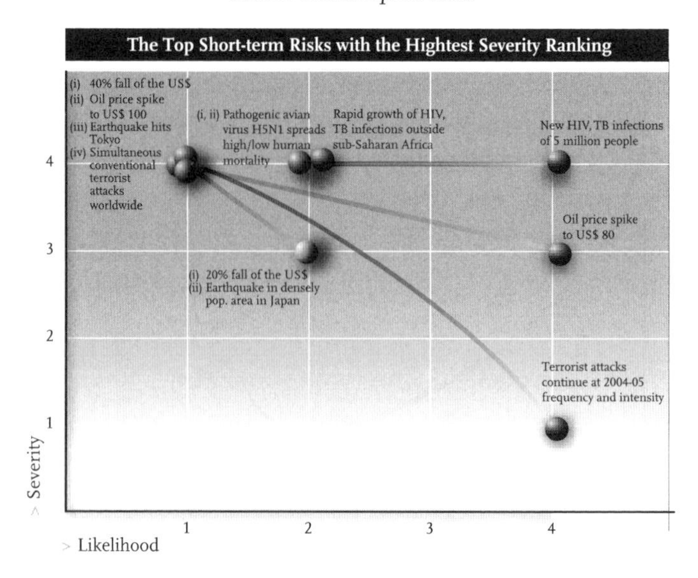

Fig. 8.6 Qualitative risk assessment chart – World Economic Forum 2006.

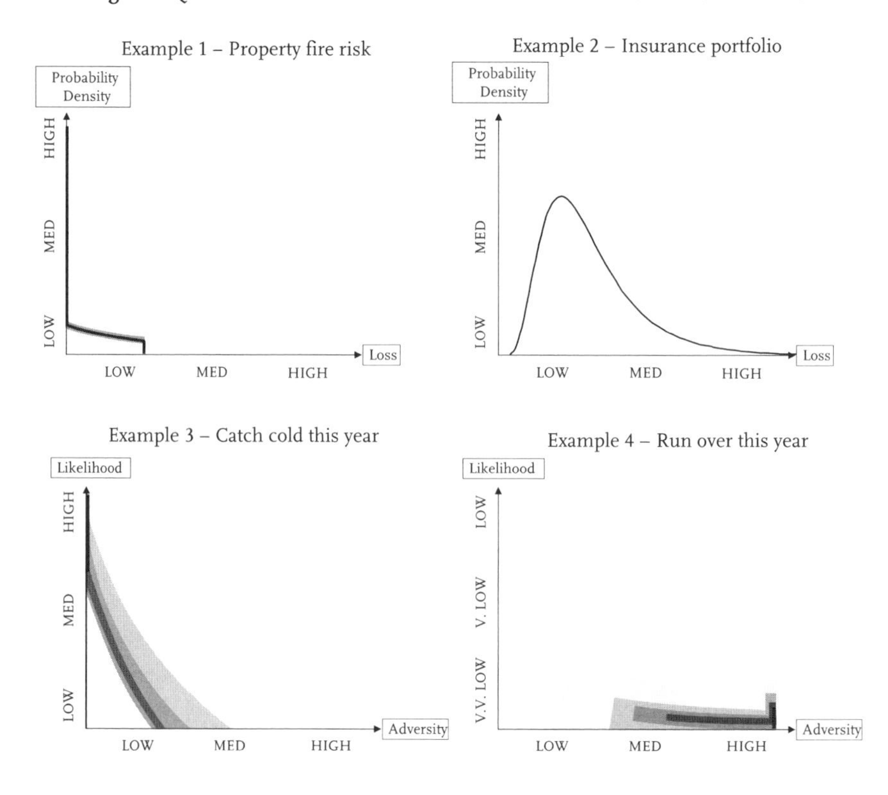

Fig. 8.7 Illustrative qualitative loss curves.

loss once in a hundred years (1%). So, whatever the computer models might tell the underwriter, the underwriter will typically allow for the 'uncertainty' dimension of risk. In commercial property insurance this add-on factor has taken on a pseudo-scientific flavour, which well illustrates how intuition may find an expression with whatever tools are available.

8.8 Price and probability

Armed with a recognition of the three dimensions of risk – chance, loss, and uncertainty – the question arises as to whether the price of an insurance contract, or indeed some other financial instrument related to the future, is indicative of the probability of a particular outcome. In insurance it is common to 'layer' risks as 'excess of loss' to demarcate the various parts of the EP curve. When this is done, then we can indeed generally say that the element of price due to loss costs (see aforementioned) represents the mean of the losses to that layer and that, for a given shape of curve, tells us the probability under the curve. The problem is whether that separation into 'pure technical' price can be made, and generally it cannot be as we move into the extreme tail because the third dimension – uncertainty – dominates the price. For some financial instruments such as weather futures, this probability prediction is much easier to make as the price is directly related to the chance of exceedance of some measure (such as degree days). For commodity prices, though, the relationship is generally too opaque to draw any such direct relationships of price to probability of event.

8.9 The age of uncertainty

We have seen that catastrophe insurance is expressed in a firmly probabilistic way through the EP curve, yet we have also seen that this misses many of the most important aspects of uncertainty.

Choice of model and choice of parameters can make a big difference to the probabilistic predictions of loss we use in insurance. In a game of chance, the only risk is process risk, so that the uncertainty resides solely with the probability distribution describing the process. It is often thought that insurance is like this, but it is not: it deals with the vagaries of the real world. We attempt to approach an understanding of that real world with models, and so for insurance there is additional uncertainty from incorrect or incorrectly configured models.

In practice, though, incorporating uncertainty will not be that easy to achieve. Modellers may not wish to move from the certainties of a single EP curve to the demands of sensitivity testing and the subjectivities of qualitative risk assessment. Underwriters in turn may find adding further levels of explicit uncertainty uncomfortable. Regulators, too, may not wish to have the apparent scientific purity of the loss curve cast into more doubt, giving insurers more not less latitude! On the plus side, though, this approach will align the tradition of expert underwriting, which allows for many risk factors and uncertainties, with the rigour of analytical models such as modern catastrophe loss models.

One way to deal with 'process' risk is to find the dependence of the model on the source assumptions of damage and cost, and the chance of events. Sensitivity testing and subjective parameterizations would allow for a diffuse but more realistic EP curve.

This leaves 'model' risk – what can we do about this? The common solution is to try multiple models and compare the results to get a feel for the spread caused by assumptions. The other way is to make an adjustment to reflect our opinion of the adequacy or coverage of the model, but this is today largely a subjective assessment.

There is a way we can consider treating parameter and model risk, and that is to construct adjusted EP curves to represent the parameter and model risk. Suppose that we could run several different models and got several different EP curves? Suppose, moreover, that we could rank these different models with different weightings. Well, that would allow us to create a revised EP curve, which is the 'convolution' of the various models.

In the areas of emerging risk, parameter and model risk, not process risk, play a central role in the risk assessment as we have little or no evidential basis on which to decide between models or parameterizations.

But are we going far enough? Can we be so sure the future will be a repeat of the present? What about factors outside our domain of experience? Is it possible that for many risks we are unable to produce a probability distribution even allowing for model and parameter risk? Is insurance really faced with 'black swan' phenomena ('black swan' refers to the failure of the inductive principle that all swans were white when black swans were discovered in Australia), where factors outside our models are the prime driver of risk?

What techniques can we call upon to deal with these further levels of uncertainty?

8.10 New techniques

We have some tools at our disposal to deal with these challenges.

8.10.1 Qualitative risk assessment

Qualitative risk assessment, as shown in the figures in the chapter, is the primary way in which most risks are initially assessed. Connecting these qualitative tools to probability and loss estimates is a way in which we can couple the intuitions and judgements of everyday sense with the analytical techniques used in probabilistic loss modelling.

8.10.2 Complexity science

Complexity science is revealing surprising order in what was hitherto the most intractable of systems. Consider, for instance, wildfires in California, which have caused big losses to the insurance industry in recent years. An analysis of wildfires in different parts of the world (Malamud et al., 1998) shows several remarkable phenomena at work: first, that wildfires exhibit negative linear behaviour on a log–log graph of frequency and severity; second, that quite different parts of the world have comparable gradients for these lines; and third, that where humans interfere, they can create unintended consequences and actually increase the risk, as it appears that forest management by stamping out small fires has actually made large fires more severe in southern California. Such log–log negative linear plots correspond to inverse power probability density functions (pdfs) (Sornette, 2004), and this behaviour is quite typical of many complex systems as popularized in the book *Ubiquity* by Mark Buchanan (see Suggestions for further reading).

8.10.3 Extreme value statistics

In extreme value statistics similar regularities have emerged in the most surprising of areas – the extreme values we might have historically treated as awkward outliers. Can it be coincidence that complexity theory predicts inverse power law behaviour, extreme value theory predicts an inverse power pdf, and that empirically we find physical extremes of tides, rainfall, wind, and large losses in insurance showing pareto (inverse power pdf) distribution behaviour?

8.11 Conclusion: against the gods?

Global catastrophic risks are extensive, severe, and unprecedented. Insurance and business generally are not geared up to handling risks of this scale or type. Insurance can handle natural catastrophes such as earthquakes and windstorms, financial catastrophes such as stock market failures to some extent, and political catastrophes to a marginal extent. Insurance is best when there is an evidential basis and precedent for legal coverage. Business is best when the capital available matches the capital at risk and the return reflects the risk of loss of this capital. Global catastrophic risks unfortunately fail to meet any of these criteria. Nonetheless, the loss modelling techniques developed for the insurance industry coupled with our deeper understanding of uncertainty and new techniques give good reason to suppose we can deal with these risks as we have with others in the past. Do we believe the fatalist cliché that 'risk is the currency of the gods' or can we go 'against the gods' by thinking the causes and consequences of these emerging risks through, and then estimating their chances, magnitudes, and uncertainties? The history of insurance indicates that we should have a go!

Acknowledgement

I thank Ian Nicol for his careful reading of the text and identification and correction of many errors.

Suggestions for further reading

Banks, E. (2006). *Catastrophic Risk* (New York: John Wiley). Wiley Finance Series. This is a thorough and up-to-date text on the insurance and re-insurance of catastrophic risk. It explains clearly and simply the way computer models generate exceedance probability curves to estimate the chance of loss for such risks.

Buchanan, M. (2001). *Ubiquity* (London: Phoenix). This is a popular account – one of several now available including the same author's *Small Worlds* – of the 'inverse power' regularities somewhat surprisingly found to exist widely in complex systems. This is of particular interest to insurers as the long-tail probability distribution most often found for catastrophe risks is the pareto distribution which is 'inverse power'.

GIRO (2006). Report of the Catastrophe Modelling Working Party (London: Institute of Actuaries). This specialist publication provides a critical survey of the modelling methodology and commercially available models used in the insurance industry.

References

De Haan, L. (1990). Fighting the arch enemy with mathematics. *Statistica Neerlandica*, **44**, 45–68.

Kabat, P., van Vierssen, W., Veraart, J., Vellinga, P., and Aerts, J. (2005). Climate proofing the Netherlands. *Nature*, **438**, 283–284.

Kaplan, S. and Garrick, B.J. (1981). On the quantitative definition of risk. *Risk Anal.*, **1**(**1**), 11.

Malamud, B.D., Morein, G., and Turcotte, D.L. (1998). Forest fires – an example of self-organised critical behaviour. *Science*, **281**, 1840–1842.

Sornette, D. (2004). *Critical Phenomena in Natural Sciences – Chaos, Fractals, Selforganization and Disorder: Concepts and Tools*, 2nd edition (Berlin: Springer).

Sornette, D., Malevergne, Y., and Muzy, J.F. (2003). Volatility fingerprints of large shocks: endogeneous versus exogeneous. *Risk Magazine*.

Swiss Re. (2006). Swiss Re corporate survey 2006 report. Zurich: Swiss Re.

Swiss Re. (2007). Natural catastrophes and man-made disasters 2006. Sigma report no 2/2007. Zurich: Swiss Re.

Woo, G. (1999). *The Mathematics of Natural Catastrophes* (London: Imperial College Press).

World Economic Forum. *Global Risks* 2006 (Geneva: World Economic Forum).

World Economic Forum. *Global Risks* 2007 (Geneva: World Economic Forum).

Yellman, T.W. (2000). The three facets of risk (Boeing Commercial Airplane Group, Seattle, WA) AIAA-2000-5594 2000. In World Aviation Conference, San Diego, CA, 10–12 October, 2000.

·9·

Public policy towards catastrophe

Richard A. Posner

The Indian Ocean tsunami of December 2004 focused attention on a type of disaster to which policymakers pay too little attention – a disaster that has a very low or unknown probability of occurring, but that if it does occur creates enormous losses. The flooding of New Orleans in the late summer of 2005 was a comparable event, although the probability of the event was known to be high; the Corps of Engineers estimated its annual probability as 0.33% (Schleifstein and McQuaid, 2002), which implies a cumulative probability of almost 10% over a thirty-year span. The particular significance of the New Orleans flood for catastrophic-risk analysis lies in showing that an event can inflict enormous loss even if the death toll is small – approximately 1/250 of the death toll from the tsunami.

Great as that toll was, together with the physical and emotional suffering of survivors, and property damage, even greater losses could be inflicted by other disasters of low (but not negligible) or unknown probability. The asteroid that exploded above Siberia in 1908 with the force of a hydrogen bomb might have killed millions of people had it exploded above a major city. Yet that asteroid was only about 200 feet in diameter, and a much larger one (among the thousands of dangerously large asteroids in orbits that intersect the earth's orbit) could strike the earth and cause the total extinction of the human race through a combination of shock waves, fire, tsunamis, and blockage of sunlight, wherever it struck.[1] Another catastrophic risk is that of abrupt global warming, discussed later in this chapter.

Oddly, with the exception of global warming (and hence the New Orleans flood, to which global warming may have contributed, along with man-made destruction of wetlands and barrier islands that formerly provided some protection for New Orleans against hurricane winds), none of the catastrophes mentioned above, including the tsunami, is generally considered an 'environmental' catastrophe. This is odd, since, for example, abrupt catastrophic global change would be a likely consequence of a major asteroid

[1] That cosmic impacts (whether from asteroids or comets) of modest magnitude can cause very destructive tsunamis is shown in Ward and Asphaug (2000) and Chesley and Ward (2006); see also the Chapter 11 in this volume.

strike. The reason non-asteroid-induced global warming is classified as an environmental disaster but the other disasters are not is that environmentalists are concerned with human activities that cause environmental harm but not with natural activities that do so. This is an arbitrary separation because the analytical issues presented by natural and human-induced environmental catastrophes are very similar.

To begin the policy analysis, suppose that a tsunami as destructive as the Indian Ocean one occurs on average once a century and kills 250,000 people. That is an average of 2500 deaths per year. Even without attempting a sophisticated estimate of the value of life to the people exposed to the risk, one can say with some confidence that if an annual death toll of 2500 could be substantially reduced at moderate cost, the investment would be worthwhile. A combination of educating the residents of low-lying coastal areas about the warning signs of a tsunami (tremors and a sudden recession in the ocean), establishing a warning system involving emergency broadcasts, telephoned warnings, and air-raid-type sirens, and improving emergency response systems would have saved many of the people killed by the Indian Ocean tsunami, probably at a total cost less than any reasonable estimate of the average losses that can be expected from tsunamis. Relocating people away from coasts would be even more efficacious, but except in the most vulnerable areas or in areas in which residential or commercial uses have only marginal value, the costs would probably exceed the benefits – for annual costs of protection must be matched with annual, not total, expected costs of tsunamis. In contrast, the New Orleans flood might have been prevented by flood-control measures such as strengthening the levees that protect the city from the waters of the Mississippi River and the Gulf of Mexico, and in any event, the costs inflicted by the flood could have been reduced at little cost simply by a better evacuation plan.

The basic tool for analysing efficient policy towards catastrophe is cost-benefit analysis. Where, as in the case of the New Orleans flood, the main costs, both of catastrophe and of avoiding catastrophe, are fairly readily monetizable and the probability of the catastrophe if avoidance measures are not taken is known with reasonable confidence, analysis is straightforward In the case of the tsunami, however, and of many other possible catastrophes, the main costs are not readily monetizable and the probability of the catastrophe may not be calculable.[2] Regarding the first problem, however, there is now a substantial economic literature inferring the value of life from the costs people are willing to incur to avoid small risks of death; if from behaviour towards risk one infers that a person would pay $70 to avoid a 1 in 100,000 risk of death, his value of

[2] Deaths caused by Hurricane Katrina were a small fraction of overall loss relative to property damage, lost earnings, and other readily monetizable costs. The ratio was reversed in the case of the Indian Ocean tsunami, where 300,000 people were killed versus only 1200 from Katrina.

life would be estimated at $7 million ($70/.00001), which is in fact the median estimate of the value of life of a 'prime-aged US worker' today (Viscusi and Aldy, 2003, pp. 18, 63).[3] Because value of life is positively correlated with income, this figure cannot be used to estimate the value of life of most of the people killed by the Indian Ocean tsunami. A further complication is that the studies may not be robust with respect to risks of death much smaller than the 1 in 10,000 to 1 in 100,000 range of most of the studies (Posner, 2004, pp. 165–171); we do not know what the risk of death from a tsunami was to the people killed. Additional complications come from the fact that the deaths were only a part of the cost inflicted by the disaster – injuries, suffering, and property damage also need to be estimated, along with the efficacy and expense of precautionary measures that would have been feasible. The risks of smaller but still destructive tsunamis that such measures might protect against must also be factored in; nor can there be much confidence about the 'once a century' risk estimate. Nevertheless, it is apparent that the total cost of the recent tsunami was high enough to indicate that precautionary measures would have been cost-justified, even though they would have been of limited benefit because, unlike the New Orleans flood, there was no possible measure for preventing the tsunami.

So why were such measures not taken in anticipation of a tsunami on the scale that occurred? Tsunamis are a common consequence of earthquakes, which themselves are common; and tsunamis can have other causes besides earthquakes – a major asteroid strike in an ocean would create a tsunami that could dwarf the Indian Ocean one. A combination of factors provides a plausible answer. First, although a once-in-a-century event is as likely to occur at the beginning of the century as at any other time, it is much less likely to occur in the first decade of the century than later. That is, probability is relative to the span over which it is computed; if the annual probability of some event is 1%, the probability that it will occur in 10 years is just a shade under 10%. Politicians with limited terms of office and thus foreshortened political horizons are likely to discount low-risk disaster possibilities, since the risk of damage to their careers from failing to take precautionary measures is truncated. Second, to the extent that effective precautions require governmental action, the fact that government is a centralized system of control makes it difficult for officials to respond to the full spectrum of possible risks against which cost-justified measures might be taken. The officials, given the variety of matters to which they must attend, are likely to have a high threshold of attention below which

[3] Of course, not all Americans are 'prime-aged workers', but it is not clear that others have lower values of life. Economists compute value of life by dividing how much a person is willing to pay to avoid a risk of death (or insists on being paid to take the risk) by the risk itself. Elderly people, for example, are not noted for being risk takers, despite the shortened span of life that remains to them; they would probably demand as high a price to bear a risk of death as a prime-aged worker – indeed, possibly more.

risks are simply ignored. Third, where risks are regional or global rather than local, many national governments, especially in the poorer and smaller countries, may drag their heels in the hope of taking a free ride on the larger and richer countries. Knowing this, the latter countries may be reluctant to take precautionary measures and by doing so reward and thus encourage free riding. (Of course, if the large countries are adamant, this tactic will fail.) Fourth, often countries are poor because of weak, inefficient, or corrupt government, characteristics that may disable poor nations from taking cost-justified precautions. Fifth, because of the positive relation between value of life and per capita income, even well-governed poor countries will spend less per capita on disaster avoidance than rich countries will.

An even more dramatic example of neglect of low-probability/high-cost risks concerns the asteroid menace, which is analytically similar to the menace of tsunamis. NASA, with an annual budget of more than $10 billion, spends only $4 million a year on mapping dangerously close large asteroids, and at that rate may not complete the task for another decade, even though such mapping is the key to an asteroid defence because it may give us years of warning. Deflecting an asteroid from its orbit when it is still millions of miles from the earth appears to be a feasible undertaking. Although asteroid strikes are less frequent than tsunamis, there have been enough of them to enable the annual probabilities of various magnitudes of such strikes to be estimated, and from these estimates, an expected cost of asteroid damage can be calculated (Posner, 2004, pp. 24–29, 180).

As in the case of tsunamis, if there are measures beyond those being taken already that can reduce the expected cost of asteroid damage at a lower cost, thus yielding a net benefit, the measures should be taken, or at least seriously considered.

Often it is not possible to estimate the probability or magnitude of a possible catastrophe, and so the question arises whether or how cost-benefit analysis, or other techniques of economic analysis, can be helpful in devising responses to such a possibility. One answer is what can be called 'inverse cost-benefit analysis' (Posner, 2004, pp. 176–184). Analogous to extracting probability estimates from insurance premiums, it involves dividing what the government is spending to prevent a particular catastrophic risk from materializing by what the social cost of the catastrophe would be if it did materialize. The result is an approximation of the implied probability of the catastrophe. Expected cost is the product of probability and consequence (loss): $C = PL$. If P and L are known, C can be calculated. If instead C and L are known, P can be calculated: if $1 billion ($C$) is being spent to avert a disaster, which, if it occurs, will impose a loss (L) of $100 billion, then $P = C/L = .01$.

If P so calculated diverges sharply from independent estimates of it, this is a clue that society may be spending too much or too little on avoiding L. It is just a clue, because of the distinction between marginal and total costs and benefits. The optimal expenditure on a measure is the expenditure that equates marginal cost to marginal benefit. Suppose we happen to know that P is not .01 but .1, so that the expected cost of the catastrophe is not $1 billion but $10 billion. It does not follow that we should be spending $10 billion, or indeed anything more than $1 billion, to avert the catastrophe. Maybe spending just $1 billion would reduce the expected cost of catastrophe from $10 billion all the way down to $500 million and no further expenditure would bring about a further reduction, or at least a cost-justified reduction. For example, if spending another $1 billion would reduce the expected cost from $500 million to zero, that would be a bad investment, at least if risk aversion is ignored.

The federal government is spending about $2 billion a year to prevent a bioterrorist attack (raised to $2.5 billion for 2005, however, under the rubric of 'Project BioShield') (U.S. Department of Homeland Security, 2004; U.S. Office of Management and Budget, 2003). The goal is to protect Americans, so in assessing the benefits of this expenditure casualties in other countries can be ignored. Suppose the most destructive biological attack that seems reasonably possible on the basis of what little we now know about terrorist intentions and capabilities would kill 100 million Americans. We know that value-of-life estimates may have to be radically discounted when the probability of death is exceedingly slight. However, there is no convincing reason for supposing the probability of such an attack less than, say, one in 100,000; and the value of life that is derived by dividing the cost that Americans will incur to avoid a risk of death of that magnitude by the risk is about $7 million. Then if the attack occurred, the total costs would be $700 trillion – and that is actually too low an estimate because the death of a third of the population would have all sorts of collateral consequences, mainly negative. Let us, still conservatively however, refigure the total costs as $1 quadrillion. The result of dividing the money being spent to prevent such an attack, $2 billion, by $1 quadrillion is 1/500,000. Is there only a 1 in 500,000 probability of a bioterrorist attack of that magnitude in the next year? One does not know, but the figure seems too low.

It does not follow that $2 billion a year is too little to be spending to prevent a bioterrorist attack; one must not forget the distinction between total and marginal costs. Suppose that the $2 billion expenditure reduces the probability of such an attack from .01 to .0001. The expected cost of the attack would still be very high – $1 quadrillion multiplied by .0001 is $100 billion – but spending more than $2 billion might not reduce the residual probability of .0001 at all. For there might be no feasible further measures to take to combat bioterrorism, especially when we remember that increasing the number of people involved in defending against bioterrorism, including not only scientific and technical personnel but also security guards in laboratories where lethal pathogens are stored, also increases the number of people capable, alone or in conjunction with others, of mounting biological attacks. But there *are* other response measures that should be considered seriously, such as investing in developing and stockpiling broad-spectrum vaccines, establishing international controls over biological research, and limiting publication of bioterror 'recipes'. One must also bear in mind that expenditures on combating bioterrorism do more than prevent mega-attacks; the lesser attacks, which would still be very costly both singly and cumulatively, would also be prevented.

Costs, moreover, tend to be inverse to time. It would cost a great deal more to build an asteroid defence in one year than in ten years because of the extra costs that would be required for a hasty reallocation of the required labour and capital from the current projects in which they are employed; so would other crash efforts to prevent catastrophes. Placing a lid on current expenditures would have the incidental benefit of enabling additional expenditures to be deferred to a time when, because more will be known about both the catastrophic risks and the optimal responses to them, considerable cost savings may be possible. The case for such a ceiling derives from comparing marginal benefits to marginal costs; the latter may be sharply increasing in the short run.[4]

A couple of examples will help to show the utility of cost-benefit analytical techniques even under conditions of profound uncertainty. The first example involves the Relativistic Heavy Ion Collider (RHIC), an advanced research particle accelerator that went into operation at Brookhaven National Laboratory in Long Island in 2000. As explained by the distinguished English physicist Sir Martin Rees (2003, pp. 120–121), the collisions in RHIC might conceivably produce a shower of quarks that would 'reassemble themselves into a very compressed object called a strangelet. ... A strangelet could, by contagion, convert anything else it encountered into a strange new form of matter. ... A hypothetical strangelet disaster could transform the entire planet Earth into an inert hyperdense sphere about one hundred metres across'. Rees (2003, p. 125) considers this 'hypothetical scenario' exceedingly unlikely, yet points out that even an annual probability of 1 in 500 million is not wholly negligible when the result, should the improbable materialize, would be so total a disaster.

Concern with such a possibility led John Marburger, the director of the Brookhaven National Laboratory and now the President's science advisor, to commission a risk assessment by a committee of physicists chaired by Robert Jaffe before authorizing RHIC to begin operating. Jaffe's committee concluded that the risk was slight, but did not conduct a cost-benefit analysis.

RHIC cost $600 million to build and its annual operating costs were expected to be $130 million. No attempt was made to monetize the benefits that the experiments conducted in it were expected to yield but we can get the analysis going by making a wild guess (to be examined critically later) that the benefits

[4] The 'wait and see' approach is discussed further below, in the context of responses to global warming.

can be valued at $250 million per year. An extremely conservative estimate, which biases the analysis in favour of RHIC's passing a cost-benefit test, of the cost of the extinction of the human race is $600 trillion.[5] The final estimate needed to conduct a cost-benefit analysis is the annual probability of a strangelet disaster in RHIC: here a 'best guess' is 1 in 10 million. (See also Chapter 16 in this volume.)

Granted, this really is a guess. The physicist Arnon Dar and his colleagues estimated the probability of a strangelet disaster during RHIC's planned period of 10-year life as no more than 1 in 50 million, which on an annual basis would mean roughly 1 in 500 million. Robert Jaffe and his colleagues, the official risk-assessment team for RHIC, offered a series of upper-bound estimates, including a 1 in 500,000 probability of a strangelet disaster over the ten-year period, which translates into an annual probability of such a disaster of approximately 1 in 5 million.

A 1 in 10 million estimate yields an annual expected extinction cost of $60 million for 10 years to add to the $130 million in annual operating costs and the initial investment of $600 million – and with the addition of that expected cost, it is easily shown that the total costs of the project exceed its benefits if the benefits are only $250 million a year. Of course this conclusion could easily be reversed by raising the estimate of the project's benefits above my 'wild guess' figure of $250 million. But probably the estimate should be lowered rather than raised. For, from the standpoint of economic policy, it is unclear whether RHIC could be expected to yield any social benefits and whether, if it did, the federal government should subsidize particle-accelerator research. The purpose of RHIC is not to produce useful products, as earlier such research undoubtedly did, but to yield insights into the earliest history of the universe. In other words, the purpose is to quench scientific curiosity. Obviously, that is a benefit to scientists, or at least to high-energy physicists. But it is unclear why it should be thought a benefit to society as a whole, or in any event why it should be paid for by the taxpayer, rather than financed by the universities that employ the physicists who are interested in conducting such research. The same question can be asked concerning other government subsidies for other types of purely academic research but with less urgency for research that is harmless. If there is no good answer to the general question, the fact that particular research poses even a slight risk of global catastrophe becomes a compelling argument against its continued subsidization.

The second example, which will occupy much of the remaining part of this chapter, involves global warming. The Kyoto Protocol, which recently came into effect by its terms when Russia signed it, though the United States has not, requires the signatory nations to reduce their carbon dioxide emissions to a level 7–10% below what they were in the late 1990s, but exempts developing

[5] This calculation is explained in Posner (2004, pp. 167–70).

countries, such as China, a large and growing emitter, and Brazil, which is destroying large reaches of the Amazon rain forest, much of it by burning. The effect of carbon dioxide emissions on the atmospheric concentration of the gas is cumulative, because carbon dioxide leaves the atmosphere (by being absorbed into the oceans) at a much lower rate than it enters it, and therefore the concentration will continue to grow even if the annual rate of emission is cut down substantially. Between this phenomenon and the exemptions, it is feared that the Kyoto Protocol will have only a slight effect in arresting global warming. Yet the tax or other regulatory measures required to reduce emissions below their level of 6 years ago will be very costly.

The Protocol's supporters are content to slow the rate of global warming by encouraging, through heavy taxes (e.g., on gasoline or coal) or other measures (such as quotas) that will make fossil fuels more expensive to consumers, conservation measures such as driving less or driving more fuel-efficient cars that will reduce the consumption of these fuels. This is either too much or too little. It is too much if, as most scientists believe, global warming will continue to be a gradual process, producing really serious effects – the destruction of tropical agriculture, the spread of tropical diseases such as malaria to currently temperate zones, dramatic increases in violent storm activity (increased atmospheric temperatures, by increasing the amount of water vapour in the atmosphere, increase precipitation),[6] and a rise in sea levels (eventually to the point of inundating most coastal cities) – only towards the end of the century. For by that time science, without prodding by governments, is likely to have developed economical 'clean' substitutes for fossil fuels (we already *have* a clean substitute – nuclear power) and even economic technology for either preventing carbon dioxide from being emitted into the atmosphere by the burning of fossil fuels or for removing it from the atmosphere.[7] However, the Protocol, at least without the participation of the United States and China, the two largest emitters, is too limited a response to global warming if the focus is changed from gradual to abrupt global warming. Because of the cumulative effect of carbon-dioxide emissions on the atmospheric concentration of the gas, a modest reduction in emissions will not reduce that concentration, but merely modestly reduce its rate of growth.

At various times in the earth's history, drastic temperature changes have occurred in the course of just a few years. In the most recent of these periods, which geologists call the 'Younger Dryas' and date to about 11,000 years ago,

[6] There is evidence that global warming is responsible in part at least for the increasing intensity of hurricanes (Emanuel, 2005; Trenberth, 2005).

[7] For an optimistic discussion of the scientific and economic feasibility of trapping carbon dioxide before it can be released into the atmosphere and capturing it after it has been released, see Socolow (2005).

shortly after the end of the last ice age, global temperatures soared by about 14°F in about a decade (Mithin, 2003). Because the earth was still cool from the ice age, the effect of the increased warmth on the human population was positive. However, a similar increase in a modern decade would have devastating effects on agriculture and on coastal cities, and might even cause a shift in the Gulf Stream that would result in giving all of Europe a Siberian climate. Recent dramatic shrinking of the north polar icecap, ferocious hurricane activity, and a small westward shift of the Gulf Stream are convincing many scientists that global warming is proceeding much more rapidly than expected just a few years ago.

Because of the enormous complexity of the forces that determine climate, and the historically unprecedented magnitude of human effects on the concentration of greenhouse gases, the possibility that continued growth in that concentration could precipitate – and within the near rather than the distant future – a sudden warming similar to that of the Younger Dryas cannot be excluded. Indeed, no probability, high or low, can be assigned to such a catastrophe. But it may be significant that, while dissent continues, many climate scientists are now predicting dramatic effects from global warming within the next twenty to forty years, rather than just by the end of the century (Lempinen, 2005).[8] It may be prudent, therefore, to try to stimulate the rate at which economical substitutes for fossil fuels, and technology both for limiting the emission of carbon dioxide by those fuels when they are burned in internal-combustion engines or electrical generating plants, and for removing carbon dioxide from the atmosphere, are developed.

Switching focus from gradual to abrupt global warming has two advantages from the standpoint of analytical tractability. The first is that, given the rapid pace of scientific progress, if disastrous effects from global warming can safely be assumed to lie at least fifty years in the future, it makes sense not to incur heavy costs now but instead to wait for science to offer a low-cost solution of the problem. Second, to compare the costs of remote future harms with the costs of remedial measures taken in the present presents baffling issues concerning the choice of a discount rate. Baffling need not mean insoluble; the 'time horizons' approach to discounting offers a possible solution (Fearnside, 2002). A discounted present value can be equated to an undiscounted present value simply by shortening the time horizon for the consideration of costs and benefits. For example, the present value of an infinite stream of costs discounted at 4% is equal to the undiscounted sum of those costs for twenty-five years, while the present value of an infinite stream of costs discounted at 1% is equal to the undiscounted sum of those costs for 100 years. The formula for the present value of \$1 per year forever is $\$1/r$, where r is the discount rate.

[8] In fact, scientists have already reported dramatic effects from global warming in melting Arctic glaciers and sea ice (Hassol, 2004).

So if r is 4%, the present value is \$25, and this is equal to an undiscounted stream of \$1 per year for twenty-five years. If r is 1%, the undiscounted equivalent is 100 years.

One way to argue for the 4% rate (i.e., for truncating our concern for future welfare at 25 years) is to say that people are willing to weight the welfare of the next generation as heavily as our own welfare but that's the extent of our regard for the future. One way to argue for the 1% rate is to say that they are willing to give equal weight to the welfare of everyone living in this century, which will include us, our children, and our grandchildren, but beyond that we do not care. Looking at future welfare in this way, one may be inclined towards the lower rate – which would have dramatic implications for willingness to invest today in limiting gradual global warming. The lower rate could even be regarded as a ceiling. Most people have some regard for human welfare, or at least the survival of some human civilization, in future centuries. We are grateful that the Romans did not exterminate the human race in chagrin at the impending collapse of their empire.

Another way to bring future consequences into focus without conventional discounting is by aggregating risks over time rather than expressing them in annualized terms. If we are concerned about what may happen over the next century, then instead of asking what the annual probability of a collision with a 10 km asteroid is, we might ask what the probability is that such a collision will occur within the next 100 years. An annual probability of 1 in 75 million translates into a century probability of roughly 1 in 750,000. That may be high enough – considering the consequences if the risk materializes – to justify spending several hundred million dollars, perhaps even several billion dollars to avert it.

The choice of a discount rate can be elided altogether if the focus of concern is *abrupt* global warming, which could happen at any time and thus constitutes a present rather than merely a remote future danger. Because it is a present danger, gradual changes in energy use that promise merely to reduce the rate of emissions are not an adequate response. What is needed is some way of accelerating the search for a technological response that will drive the annual emissions to zero or even below. Yet the Kyoto Protocol might actually do this by impelling the signatory nations to impose stiff taxes on carbon dioxide emissions in order to bring themselves into compliance with the Protocol. The taxes would give the energy industries, along with business customers of them such as airlines and manufacturers of motor vehicles, a strong incentive to finance R&D designed to create economical clean substitutes for such fuels and devices to 'trap' emissions at the source, before they enter the atmosphere, or even to remove carbon dioxide from the atmosphere. Given the technological predominance of the United States, it is important that these taxes be imposed on US firms, which they would be if the United States ratified the Kyoto Protocol and by doing so became bound by it.

One advantage of the technology-forcing tax approach over public subsidies for R&D is that the government would not be in the business of picking winners – the affected industries would decide what R&D to support – and another is that the brunt of the taxes could be partly offset by reducing other taxes, since emission taxes would raise revenue as well as inducing greater R&D expenditures.

It might seem that subsidies would be necessary for technologies that would have no market, such as technologies for removing carbon dioxide from the atmosphere. There would be no private demand for such technologies because, in contrast to ones that reduce emissions, technologies that remove already emitted carbon dioxide from the atmosphere would not reduce any emitter's tax burden. This problem is, however, easily solved by making the tax a tax on *net* emissions. Then an electrical generating plant or other emitter could reduce its tax burden by removing carbon dioxide from the atmosphere as well as by reducing its own emissions of carbon dioxide into the atmosphere.

The conventional assumption about the way that taxes, tradable permits, or other methods of capping emissions of greenhouse gases work is that they induce substitution away from activities that burn fossil fuels and encourage more economical use of such fuels. To examine this assumption, imagine (unrealistically) that the demand for fossil fuels is completely inelastic in the short run.[9] Then even a very heavy tax on carbon dioxide emissions would have no short-run effect on the level of emissions, and one's first reaction is likely to be that, if so, the tax would be ineffectual. Actually it would be a highly efficient tax from the standpoint of generating government revenues (the basic function of taxation); it would not distort the allocation of resources, and therefore its imposition could be coupled with a reduction in less efficient taxes without reducing government revenues, although the substitution would be unlikely to be complete because, by reducing taxpayer resistance, more efficient taxes facilitate the expansion of government.

More important, such a tax might – paradoxically – have an even greater impact on emissions, *precisely* because of the inelasticity of short-run demand, than a tax that induced substitution away from activities involving the burning of fossil fuels or that induced a more economical use of such fuels. With immediate substitution of alternative fuels impossible and the price of fossil fuels soaring because of the tax, there would be powerful market pressures both to speed the development of economical alternatives to fossil fuels as energy sources and to reduce emissions, and the atmospheric concentration, of carbon dioxide directly.

[9] The length of the 'short run' is, unfortunately, difficult to specify. It depends, in the present instance, on how long it would take for producers and consumers of energy to minimize the impact of the price increase by changes in production (increasing output in response to the higher price) and consumption (reducing consumption in response to the higher price).

From this standpoint a tax on emissions would be superior to a tax on the fossil fuels themselves (e.g., a gasoline tax, or a gas on B.T.U. content). Although an energy tax is cheaper to enforce because there is no need to monitor emissions, only an emissions tax would be effective in inducing carbon sequestration, because sequestration reduces the amount of atmospheric carbon dioxide without curtailing the demand for fossil fuels. A tax on gasoline will reduce the demand for gasoline but will not induce efforts to prevent the carbon dioxide emitted by the burning of the gasoline that continues to be produced from entering the atmosphere.

Dramatic long-run declines in emissions are likely to result *only* from technological breakthroughs that steeply reduce the cost of both clean fuels and carbon sequestration, rather than from insulation, less driving, lower thermostat settings, and other energy-economizing moves; and it is dramatic declines that we need. Even if the short-run elasticity of demand for activities that produce carbon dioxide emissions were −1 (i.e., if a small increase in the price of the activity resulted in a proportionately equal reduction in the scale of the activity), a 20% tax on emissions would reduce their amount by only 20% (this is on the assumption that emissions are produced in fixed proportions with the activities generating them). Because of the cumulative effect of emissions on atmospheric concentrations of greenhouse gases, those concentrations would continue to grow, albeit at a 20% lower rate; thus although emissions might be elastic with respect to the tax, the actual atmospheric concentrations, which are the ultimate concern, would not be. In contrast, a stiff emissions tax might precipitate within a decade or two technological breakthroughs that would enable a drastic reduction of emissions, perhaps to zero. If so, the effect of the tax would be much greater than would be implied by estimates of the elasticity of demand that ignored such possibilities. The possibilities are masked by the fact that because greenhouse-gas emissions are not taxed (or classified as pollutants), the private incentives to reduce them are meagre.

Subsidizing research on measures to control global warming might seem more efficient than a technology-forcing tax because it would create a direct rather than merely an indirect incentive to develop new technology. But the money to finance the subsidy would have to come out of tax revenues, and the tax (whether an explicit tax, or inflation, which is a tax on cash balances) that generated these revenues might be less efficient than a tax on emissions if the latter taxed less elastic activities, as it might. A subsidy, moreover, might induce overinvestment. A problem may be serious and amenable to solution through an expenditure of resources, but above a certain level additional expenditures may contribute less to the solution than they cost. An emissions tax set equal to the social cost of emissions will not induce overinvestment, as industry will have no incentive to incur a greater cost to avoid the tax. If the social cost of emitting a specified quantity of carbon dioxide is $1 and the tax therefore is $1, industry will spend up to $1, but not more, to avoid the tax. If it can avoid the tax only by spending $1.01 on emission-reduction measures, it will forgo the expenditure and pay the tax.

Furthermore, although new technology is likely to be the ultimate solution to the problem of global warming, methods for reducing carbon dioxide emissions that do not depend on new technology, such as switching to more fuel-efficient cars, may have a significant role to play, and the use of such methods would be encouraged by a tax on emissions but not by a subsidy for novel technologies, at least until those technologies yielded cheap clean fuels.

The case for subsidy would be compelling only if inventors of new technologies for combating global emissions could not appropriate the benefits of the technologies and therefore lacked incentives to develop them. But given patents, trade secrets, trademarks, the learning curve (which implies that the first firm in a new market will have lower production costs than latecomers), and other methods of internalizing the benefits of inventions, appropriability should not be a serious problem, with the exception of basic research, including research in climate science.

A superficially appealing alternative to the Kyoto Protocol would be to adopt a 'wait and see' approach – the approach of doing nothing at all about greenhouse-gas emissions in the hope that a few more years of normal (as distinct from tax-impelled) research in climatology will clarify the true nature and dimensions of the threat of global warming, and then we can decide what if any measures to take to reduce emissions. This probably would be the right approach were it not for the practically irreversible effect of greenhouse-gas emissions on the atmospheric concentration of those gases. Because of that irreversibility, stabilizing the atmospheric concentration of greenhouse gases at some future date might require far deeper cuts in emissions then than if the process of stabilization begins now. Making shallower cuts now can be thought of as purchasing an option to enable global warming to be stopped or slowed at some future time at a lower cost. Should further research show that the problem of global warming is not a serious one, the option would not be exercised.

To illustrate, suppose there is a 70% probability that in 2024 global warming will cause a social loss of $1 trillion (present value) and a 30% probability that it will cause no loss, and that the possible loss can be averted by imposing emission controls now that will cost the society $500 billion (for simplicity's sake, the entire cost is assumed to be borne this year). In the simplest form of cost-benefit analysis, since the discounted loss from global warming in 2024 is $700 billion, imposing the emission controls now is cost-justified. But suppose that in 2014 we will learn for certain whether there is going to be the bad ($1 trillion) outcome in 2024. Suppose further that if we postpone imposing the emission controls until 2014, we can still avert the $1 trillion loss. Then clearly we should wait, not only for the obvious reason that the present value of $500 billion to be spent in ten years is less than $500 billion (at a discount rate of 3% it is approximately $425 billion) but also and more interestingly because there is a 30% chance that we will not have to incur *any* cost of emission controls. As a result, the expected cost of the postponed controls is not $425 billion, but only 70% of that amount, or $297.5 billion, which is a lot less than $500 billion. The difference is the value of waiting.

Now suppose that if today emission controls are imposed that cost society $100 billion, this will, by forcing the pace of technological advance (assume for simplicity that this is their only effect – that there is no effect in reducing emissions), reduce the cost of averting in 2014 the global-warming loss of $1 trillion in 2024 from $500 billion to $250 billion. After discounting to present value at 3% and by 70% to reflect the 30% probability that we will learn in 2014 that emission controls are not needed, the $250 billion figure shrinks to $170 billion. This is $127.5 billion less than the superficially attractive pure wait-and-see approach ($297.5 billion minus $170 billion). Of course, there is a price for the modified wait-and-see option – $100 billion. But the value is greater than the price.

This is an example of how imposing today emissions limits more modest than those of the Kyoto Protocol might be a cost-justified measure even if the limits had no direct effect on atmospheric concentrations of greenhouse gases. Global warming could be abrupt without being catastrophic and catastrophic without being abrupt. But abrupt global warming is more likely to be catastrophic than gradual global warming, because it would deny or curtail opportunities for adaptive responses, such as switching to heat-resistant agriculture or relocating population away from coastal regions. The numerical example shows that the option approach is attractive even if the possibility of abrupt global warming is ignored; in the example, we know that we are safe until 2024. However, the possibility of abrupt warming should not be ignored. Suppose there is some unknown but not wholly negligible probability that the $1 trillion global-warming loss will hit in 2014 and that it will be too late then to do anything to avert it. That would be a ground for imposing stringent emissions controls earlier even though by doing so we would lose the opportunity to avoid their cost by waiting to see whether they would actually be needed. Since we do not know the point at which atmospheric concentrations of greenhouse gases would trigger abrupt global warming, the imposition of emissions limits now may, given risk aversion, be an attractive insurance policy. An emissions tax that did not bring about an immediate reduction in the level of emissions might still be beneficial by accelerating technological breakthroughs that would result in zero emissions before the trigger point was reached.

The risk of abrupt global warming is not only an important consideration in deciding what to do about global warming; unless it is given significant weight, the political prospects for strong controls on greenhouse-gas emissions are poor. The reason can be seen in a graph that has been used without much success to galvanize public concern about global warming (IPCC, 2001; Fig. 9.1). The shaded area is the distribution of predictions of global temperature changes over the course of the century, and is at first glance alarming. However, a closer look reveals that the highest curve, which is based on the assumption that nothing will be done to curb global warming, shows a temperature increase of only about 10° Fahrenheit over the course of the century. Such an increase would be catastrophic if it occurred in a decade, but it is much less alarming when spread out over a century, as that is plenty of time for a combination of clean fuels and cheap carbon-sequestration methods to reduce carbon dioxide emissions to zero or even (through carbon sequestration) below zero without prodding by governments. Given such an outlook, convincing governments to incur heavy costs now to reduce the century increase from ten to say five degrees is distinctly an uphill fight. There is also a natural scepticism about any attempt to predict what is going to happen a hundred years in the future, and a belief that since future generations will be wealthier than our generation they will find it less burdensome to incur large costs to deal with serious environmental problems.

Nevertheless, once abrupt global warming is brought into the picture, any complacency induced by the graph is quickly dispelled. For we then understand that the band of curves in the graph is arbitrarily truncated; that we could have a vertical takeoff say in 2020 that within a decade would bring us to the highest point in the graph. Moreover, against *that* risk, a technology-forcing tax on emissions might well be effective even if only the major emitting countries imposed substantial emission taxes. If manufacturers of automobiles sold in North America, the European Union, and Japan were hit with a heavy tax on carbon dioxide emissions from their automobiles, the fact that China was not taxing automobiles sold in its country would not substantially erode the incentive of the worldwide automobile industry to develop effective methods for reducing the carbon dioxide produced by their automobiles.

It is tempting to suppose that measures to deal with long-run catastrophic threats can safely be deferred to the future because the world will be richer and therefore abler to afford costly measures to deal with catastrophe. However, such complacency is unwarranted. Catastrophes can strike at any time and if they are major could make the world significantly poorer. Abrupt climate change is a perfect example. Change on the order of the Younger Dryas might make future generations markedly poorer than we are rather than wealthier, as might nuclear or biological attacks, cosmic impacts, or super-volcanic eruptions. These possibilities might actually argue for using a negative rather

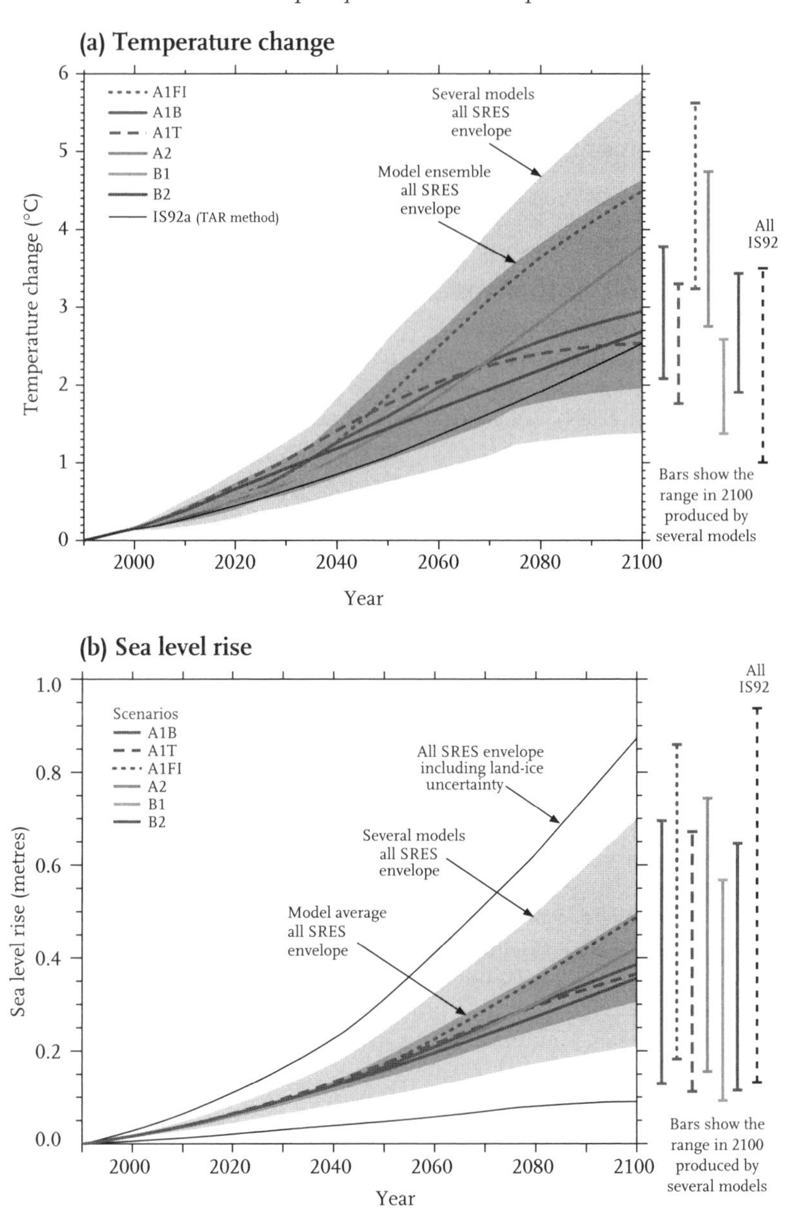

Fig. 9.1 The global climate of the twenty-first century will depend on natural changes and the response of the climate system to human activities.
Credit: IPCC, 2001: *Climate Change 2001: Scientific Basis. Contribution of Working Group I to the Third Assessment Report of the Intergovernmental Panel on Climate Change* [Houghton, J.T., Y. Ding, D.J. Griggs, M. Noguer, P.J. van der Linden, X. Dai, K. Maskell, and C.A. Johnson (eds.)]. Figure 5, p 14. Cambridge University Press, Cambridge, United Kingdom and New York, NY, USA.

than positive discount rate to determine the present-value cost of a future climate disaster.

Acknowledgement

I thank Megan Maloney for her helpful research.

Suggestions for further reading

Grossi, P. and Kunreuther, H. (2005). *Catastrophe Modeling: A New Approach to Managing Risk* (New York: Springer).

Belton, M.J.S. et al. (2004). *Mitigation of Hazardous Comets and Asteroids.* (Cambridge: Cambridge University Press).

Nickerson, R.S. (2004). *Cognition and Chance: The Psychology of Probabilistic Reasoning* (New Jersey: Lawrence Erlbaum Associates).

OECD (2004). *Large-scale Disasters: Lessons Learned* (Organisation for Economic Co-operation and Development).

Posner, R.A. (2004). *Catastrophe: Risk and Response* (Oxford: Oxford University Press).

Smith, K. (2004). *Environmental Hazards: Assessing Risk and Reducing Disaster,* 4th ed. (Oxford: Routledge).

Rees, M. (2003). *Our Final Hour: A Scientist's Warning: How Terror, Error, and Environmental Disaster Threaten Humankind's Future in this Century—On Earth and Beyond* (New York: Basic Books).

References

Chesley, S.R. and Ward, S.N. (2006). A quantitative assessment of the human hazard from impact-generated tsunami. *J. Nat. Haz.*, **38**, 355–374.

Emanuel, K. (2005). Increasing destructiveness of tropical cyclones over the past 30 years. *Nature,* **436**, 686–688.

Fearnside, P.M. (2002). Time preference in global warming calculations: a proposal for a unified index. *Ecol. Econ.*, **41**, 21–31.

Hassol, S.J. (2004). *Impacts of a Warming Arctic: Arctic Climate Impact Assessment* (Cambridge: Cambridge University Press). Available online at http://amap.no/acia/

IPCC (Houghton, J.T., Ding, Y., Griggs, D.J., Noguer, M., van der Linden, P.J., and Xiaosu, D. (eds.)) (2001). *Climate Change 2001: The Scientific Basis.* Contribution of Working Group I to the Third Assessment Report of the Intergovernmental Panel on Climate Change (IPCC) (Cambridge: Cambridge University Press).

Lempinen, E.W. (2005). Scientists on AAAS panel warn that ocean warming is having dramatic impact (AAAS news release 17 Feb 2005) http://www.aaas.org/news/releases/2005/0217warmingwarning.shtml

Mithin, S. (2003). *After the Ice : A Global Human History, 20,000–5,000 BC* (Cambridge, MA: Harvard University Press).

Posner, R.A. (2004). *Catastrophe: Risk and Response* (New York: Oxford University Press).

Rees, M.J. (2003). *Our Final Hour: A Scientist's Warning; How Terror, Error, and Environmental Disaster Threaten Humankind's Future in this Century – on Earth and Beyond* (New York: Basic Books).

Schleifstein, M. and McQuaid, J. (3 July 2002). The big easy is unprepared for the big one, experts say. *Newhouse News Service.* http://www.newhouse.com/archive/story-1b070502.html

Socolow, R.H. (July 2005). Can we bury global warming? *Scientific Am.*, **293**, 49–55.

Trenberth, K. (2005). Uncertainty in hurricanes and global warming. *Science*, **308**, 1753–1754.

U.S. Department of Homeland Security (2004). Fact sheet: Department of Homeland Security Appropriations Act of 2005 (Press release 18 October 2004) http://www.dhs.gov/dhspublic/interapp/press_release/press_release_0541.xml

U.S. Office of Management and Budget (2003). 2003 report to Congress on combating terrorism (O.M.B. report Sept 2003). http://www.whitehouse.gov/omb/inforeg/2003_combat_terr.pdf

Viscusi, W.K. and Aldy, J.E. (2003). The value of a statistical life: a critical review of market estimates throughout the world. *J. Risk Uncertainty*, **27**, 5–76.

Ward, S.N. and Asphaug, E. (2000). Asteroid impact tsunami: a probabilistic hazard assessment. *Icarus*, **145**, 64–78.

PART II
Risks from nature

·10·

Super-volcanism and other geophysical processes of catastrophic import

Michael R. Rampino

10.1 Introduction

In order to classify volcanic eruptions and their potential effects on the atmosphere, Newhall and Self (1982) proposed a scale of explosive magnitude, the Volcanic Explosivity Index (VEI), based mainly on the volume of the erupted products (and the height of the volcanic eruption column). VEI's range varies from VEI = 0 (for strictly non-explosive eruptions) to VEI = 8 (for explosive eruptions producing $\sim 10^{12}$ m^3 bulk volume of tephra). Eruption rates for VEI = 8 eruptions may be greater than 10^6 $m^3 s^{-1}$ (Ninkovich et al., 1978a, 1978b).

Eruptions also differ in the amounts of sulphur-rich gases released to form stratospheric aerosols. Therefore, the sulphur content of the magma, the efficiency of degassing, and the heights reached by the eruption column are important factors in the climatic effects of eruptions (Palais and Sigurdsson, 1989; Rampino and Self, 1984). Historic eruptions of VEI ranging from three to six (volume of ejecta from <1 km^3 to a few tens of km^3) have produced stratospheric aerosol clouds up to a few tens of Mt. These eruptions, including Tambora 1815 and Krakatau 1883, have caused cooling of the Earth's global climate of a few tenths of a degree Centigrade (Rampino and Self, 1984). The most recent example is the Pinatubo (Philippines) eruption of 1991 (Graf et al., 1993; Hansen et al., 1996).

Volcanic super-eruptions are defined as eruptions that are tens to hundreds of times larger than historic eruptions, attaining a VEI of 8 (Mason et al., 2004; Rampino, 2002; Rampino et al., 1988; Sparks et al., 2005). Super-eruptions are usually caldera-forming events and more than twenty super-eruption sites for the last 2 million years have been identified in North America, South America, Italy, Indonesia, the Philippines, Japan, Kamchatka, and New Zealand. No doubt additional super-eruption sites for the last few million years exist (Sparks et al., 2005).

The Late Pleistocene eruption of Toba in Sumatra, Indonesia was one of the greatest known volcanic events in the geologic record (Ninkovich et al., 1978a, 1978b; Rampino and Self, 1993a; Rose and Chesner, 1990). The relatively recent age and the exceptional size of the Toba eruption make it an important test case of the possible effects of explosive volcanism on the global atmosphere and climate (Oppenheimer, 2002; Rampino and Self, 1992, 1993a; Rampino et al., 1988; Sparks et al., 2005). For the Toba event, we have data on intercaldera fill, outflow sheets produced by pyroclastic flows and tephra fallout. Recent information on the environmental effects of super-eruptions supports the exceptional climatic impact of the Toba eruption, with significant effects on the environment and human population.

10.2 Atmospheric impact of a super-eruption

The Toba eruption has been dated by various methods K/Ar method at 73,500 $\pm$ 3500 yr BP (Chesner et al., 1991). The Toba ash layer occurs in deep-sea cores from the Indian Ocean and South China Sea (Huang et al., 2001; Shultz et al., 2002; Song et al., 2000). The widespread ash layer has a dense rock equivalent volume (DRE) of approximately 800 km^3 (Chesner et al., 1991). The pyroclastic flow deposits on Sumatra have a volume of approximately 2000 km^3 DRE (Chesner et al., 1991; Rose and Chesner, 1990), for a total eruption volume of approximately 2800 km^3 (DRE). Woods and Wohletz (1991) estimated Toba eruption cloud heights of 32 $\pm$ 5 km, and the duration of continuous fallout of Toba ash over the Indian Ocean has been estimated at two weeks or less (Ledbetter and Sparks, 1979).

Release of sulphur volatiles is especially important for the climatic impact of an eruption, as these form sulphuric acid aerosols in the stratosphere (Rampino and Self, 1984). Although the intrinsic sulphur content of rhyolite magmas is generally low, the great volume erupted is sufficient to give an enormous volatile release. Based on studies of the sulphur content of the Toba deposits, Rose and Chesner (1990) estimated that approximately 3×10^{15} g of H_2S/SO_2 (equivalent to $\sim 1 \times 10^{16}$ g of H_2SO_4 aerosols) could have been released from the erupted magma. The amounts of fine ash and sulphuric acid aerosols that could have been generated by Toba was estimated independently using data from smaller historical rhyolitic eruptions (Rampino and Self, 1992). By this simple extrapolation, the Toba super-eruption could have produced up to 2×10^{16} g of fine ($<2\ \mu$) dust and approximately 1.5×10^{15} g of sulphuric acid aerosols.

Physical and chemical processes in dense aerosol clouds may act in a 'self-limiting' manner, significantly reducing the amount of long-lived H_2SO_4 aerosols (Rampino and Self, 1982; Pinto et al., 1989). Using one-dimensional aerosol microphysical and photochemical models, Pinto and others (1989) showed that for an aerosol cloud of approximately 10^{14} g of SO_2, condensation and coagulation are important in producing larger-sized particles, which have a smaller optical effect per unit mass, and settle out of the stratosphere faster than smaller particles. However, the maximum sulphur volatile emission that they modelled was 2×10^{14} g of SO_2, and no data exist on the behaviour of H_2SO_4 aerosols in more than 10 times denser clouds.

Another possible limitation on aerosol loading is the amount of water in the stratosphere available to convert SO_2 to H_2SO_4. Stothers et al. (1986) calculated that approximately 4×10^{15} g of water might be available in the ambient stratosphere, and injection into the stratosphere of up to 5.4×10^{17} g of H_2O from Toba is possible (Rose and Chesner, 1990), more than enough water to convert the sulphur gases emitted by Toba into H_2SO_4 aerosols.

The exceptional magnitude of the Toba eruption makes it a natural target in the studies of large volcanic events preserved in polar ice cores. Work on the GISP2 ice core from Summit, Greenland, revealed an approximately 6-year long period of enhanced volcanic sulphate dated at $71,100 \pm 5000$ years ago identified with the Toba eruption (Zielinski et al., 1996a, 1996b). The magnitude of this sulphate signal is the largest in the entire 110,000 years of the GISP2 record.

Zielinski and others (1996a) estimated that the total atmospheric loading of H_2SO_4 for the approximately 6-year period of the ice-core peak ranged from approximately 0.7 to 4.4×10^{15} g, in general agreement with the above estimates derived from volcano-logical techniques and scaling from smaller eruptions (Rampino and Self, 1992, 1993a; Rose and Chesner, 1990). Estimates of aerosol loadings range from approximately 150 to 1000 Mt per year, over the approximately 6-year period of the ice-core peak.

The SO_4^{2-} signal identified with Toba coincides with the beginning of an approximately 1000-year cooling event seen in the ice-core record between brief warm periods (interstadials), but is separated from the most recent major approximately 9000-year glacial period by the approximately 2000-year-long warmer period. A similar cool pulse between interstadials is seen in the pollen record of the Grande Pile in northeastern France, dated as approximately 70,000 years BP (Woillard and Mook, 1982).

Thus, the ice-core evidence suggests that the Toba signal occurred during the transition from a warm interglacial climate and was preceded and followed by abrupt climate oscillations that preceded the start of the most recent major early glaciation (Zielinski et al., 1996a, 1996b).

10.3 Volcanic winter

Since Toba is a low-latitude volcano, dust and volatiles would have been injected efficiently into both Northern and Southern Hemispheres (Rampino et al., 1988), although the season of the eruption is unknown. These estimated aerosol optical effects are roughly equivalent in visible opacity to smoke-clouds (Turco et al., 1990), which is within the range used in nuclear-winter scenarios of massive emissions of soot emanating from burning urban and industrial areas in the aftermath of nuclear war.

Although the climate conditions and duration of a nuclear winter have been much debated, simulations by Turco and others (1990) predicted that land temperatures in the 30°–70°N latitude zone could range from approximately 5°C to approximately 15°C colder than normal, with freezing events in mid-latitudes during the first few months. At lower latitudes, model simulations suggest cooling of 10°C or more, with drastic decreases in precipitation in the first few months. Ocean-surface cooling of approximately 2–6°C might extend for several years, and persistence of significant soot for 1–3 years might lead to longer term (decadal) climatic cooling, primarily through climate feedbacks including increased snow cover and sea ice, changes in land surface albedo, and perturbed sea-surface temperatures (Rampino and Ambrose, 2000).

The injection of massive amounts of volcanic dust into the stratosphere by a super-eruption such as Toba might be expected to lead to similar immediate surface cooling, creating a 'volcanic winter' (Rampino and Self, 1992; Rampino et al., 1988). Volcanic dust probably has a relatively shorter residence time in the atmosphere (3–6 months) than soot (Turco et al., 1990) and spreads from a point source, but volcanic dust is injected much higher into the stratosphere, and hence Toba ash could have had a wide global coverage despite its short lifetime. Evidence of the wide dispersal of the dust and ash from Toba can be seen from lake deposits in India, where the reworked Toba ash forms a layer up to 3 m thick, and from the widespread ash layer in the Indian Ocean and South China Sea (Acharya and Basu, 1993; Huang et al., 2001; Shane et al., 1995).

Evidence for rapid and severe cooling from the direct effects of volcanic ash clouds comes from the aftermath of the 1815 Tambora eruption. Madras, India experienced a dramatic cooling during the last week of April 1815, a time when the relatively fresh ash and aerosol cloud from Tambora (10–11 April) would have been overhead. Morning temperatures dropped from 11°C on Monday to −3°C on Friday (Stothers, 1984a). A similar, but much smaller effect, occurred as the dust cloud from the 1980 Mt St Helens eruption passed over downwind areas (Robock and Mass, 1982).

The stratospheric injection of sulphur volatiles ($\geq 10^{15}$ g), and the time required for the formation and spread of volcanic H_2SO_4 aerosols in the stratosphere should lead to an extended period of increased atmospheric opacity and surface cooling. The ice-core record, however, indicates stratospheric loadings of 10^{14} to 10^{15} g of H_2SO_4 aerosols for up to 6 years after the eruption (Zielinski et al., 1996a).

This agrees with model calculations by Pope and others (1994) that predict oxidation lifetimes (time required to convert a given mass of sulphur into

H_2SO_4 aerosols) of between 4 and 17 years, and diffusion lifetimes (time required to remove unoxidized SO_2 by diffusion to the troposphere) of between four and seven years for total sulphur masses between 10^{15} and 10^{16} g. For atmospheric injection in this range, the diffusion lifetime is the effective lifetime of the cloud because the SO_2 reservoir is depleted before oxidation is completed.

If the relationship between Northern Hemisphere cooling and aerosol loading from large eruptions is approximately linear, then scaling up from the 1815 AD. Tambora eruption would lead to an approximately 3.5°C hemispheric cooling after Toba (Rampino and Self, 1993a). Similarly, empirical relationships between SO_2 released and climate response (Palais and Sigurdsson, 1989) suggested a hemispheric surface-temperature decrease of about 4 ± 1°C. The eruption clouds of individual historic eruptions have been too short-lived to drive lower tropospheric temperatures to their steady-state values (Pollack et al., 1993), but the apparently long-lasting Toba aerosols may mean that the temperature changes in the troposphere attained a larger fraction of their steady-state values. Huang et al. (2001) were able to correlate the Toba ash in the South China Sea with a 1°C cooling of surface waters that lasted about 1000 years.

Considering a somewhat smaller super-eruption, the Campanian eruption of approximately 37,000 cal yr BP in Italy (150 km^3 of magma discharged) was coincident with Late Pleistocene bio-cultural changes that occurred within and outside the Mediterranean region. These included the Middle to Upper Paleolithic cultural transition and the replacement of Neanderthals by 'modern' *Homo sapiens* (Fedele et al., 2002).

10.4 Possible environmental effects of a super-eruption

The climatic and environmental impacts of the Toba super-eruption are potentially so much greater than that of recent historical eruptions (e.g., Hansen et al., 1992; Stothers, 1996) that instrumental records, anecdotal information, and climate-model studies of the effects of these eruptions may not be relevant in scaling up to the unique Toba event (Rampino and Self, 1993a; Rampino et al., 1988). Various studies on the effects of extremes of atmospheric opacity and climate cooling on the environment and life have been carried out, however, in connection with studies of nuclear winter and the effects of asteroid impacts on the earth (e.g., Green et al., 1985; Harwell, 1984; Tinus and Roddy, 1990), and some of these may be relevant to the Toba situation.

Two major effects on plant life from high atmospheric opacity are reduction of light levels and cold temperatures. Reduction in light levels expected from the Toba eruption would range from dim-sun conditions (~75% sunlight transmitted) like those seen after the 1815 Tambora eruption, to that of an overcast day (~10% sunlight transmitted). Experiments with young grass plants have shown how net photosynthesis varies with light intensity. For a decrease to 10% of the noon value for a sunny summer day, photosynthesis was reduced by about 85% (van Kuelan et al., 1975), and photosynthesis also drops with decreasing temperatures (Redman, 1974).

Resistance of plants to unusual cold conditions varies somewhat. Conditions in the tropical zone are most relevant to possible impacts on early human populations in Africa. Tropical forests are very vulnerable to chilling, and Harwell and others (1985) argue that for freezing events in evergreen tropical forests, essentially all aboveground plant tissues would be killed rapidly.

Average surface temperatures in the tropics today range from approximately 16–24°C. Nuclear winter scenarios predict prolonged temperature decreases of 3–7°C in Equatorial Africa, and short-term temperature decreases of up to 10°C. Many tropical plants are severely damaged by chilling to below 10–15°C for a few days (Greene et al., 1985; Leavitt, 1980). Most tropical forest plants have limited seed banks, and the seeds typically lack a dormant phase. Furthermore, regrowth tends to produce forests of limited diversity, capable of supporting much less biomass (Harwell et al., 1985).

Even for temperate forests, destruction could be very severe (Harwell, 1984; Harwell et al., 1985). In general, the ability of well-adapted trees to withstand low temperatures (cold hardiness) is much greater than that needed at any single time of the year, but forests can be severely damaged by unusual or sustained low temperatures during certain times of the year. A simulation of a 10°C decrease in temperatures during winter shows a minimal effect on the cold-hardy and dormant trees, whereas a similar 10°C drop in temperature during the growing season (when cold hardiness is decreased) leads to a 50% dieback, and severe damage to surviving trees, resulting in the loss of at least a year's growth.

The situation for deciduous forest trees would be even worse than that for the evergreens, as their entire foliage would be new and therefore lost. For example, Larcher and Bauer (1981) determined that cold limits of photosynthesis of various temperate zone plants range from −1.3 to −3.9°C, approximately the same range as the tissue-freezing temperatures for these plants. Lacking adequate food reserves, most temperate forest trees would not be able to cold harden in a timely manner, and would die or suffer additional damage during early freezes in the Fall (Tinus and Roddy, 1990).

The effect of the Toba super-eruption on the oceans is more difficult to estimate. Regionally, the effect on ocean biota of the fallout of approximately 4 g/cm^2 of Toba ash over an area of 5×10^6 km^2 in the Indian Ocean must have been considerable. Deposition rates of N, organic C, and $CaCO_3$ all rise sharply in the first few centimetres of the Toba ash layer, indicating that the ash fallout swept the water column of most of its particulate organic carbon and calcium carbonate (Gilmour et al., 1990).

Another possible effect of a dense aerosol cloud is decreased ocean productivity. For example, satellite observations after the 1982 El Chichón eruption showed high aerosol concentrations over the Arabian Sea, and these values were associated with low surface productivity (as indicated by phytoplankton concentrations) from May through October of that year (Strong, 1993). Brock and McClain (1992) suggested that the low productivity was related to weaker-than-normal monsoon winds, and independent evidence suggests that the southwest monsoon in the area arrived later and withdrew earlier than usual, and that the wind-driven Somali current was anomalously weak. Conversely, Genin and others (1995) reported enhanced vertical mixing of cooled surface waters in weakly stratified areas of the Red Sea following the Pinatubo eruption, which resulted in algal and phytoplankton blooms that precipitated widespread coral death.

Studies following the 1991 Pinatubo eruption provide evidence that aerosol-induced cooling of the southwestern Pacific could lead to significant weakening of Hadley Cell circulation and rainfall, and might precipitate long-term El Niño-like anomalies with extensive drought in many tropical areas (Gagan and Chivas, 1995). Some climate-model simulations predict significant drought in tropical areas from weakening of the trade winds/Hadley circulation and from reduction in the strength of the summer monsoon (e.g., Pittock et al., 1986, 1989; Turco et al., 1990). For example, Pittock and others (1989) presented GCM results that showed a 50% reduction in convective rainfall in the tropics and monsoonal regions.

10.5 Super-eruptions and human population

Recent debate about the origin of modern humans has focused on two competing hypotheses: (1) the 'multiregional' hypothesis, in which the major subdivisions of our species evolved slowly and in situ, with gene flow accounting for the similarities now observed among groups, and (2) the 'replacement' hypothesis, in which earlier populations were replaced 30,000 to 100,000 years ago by modern humans that originated in Africa (Hewitt, 2000; Rogers and Joude, 1995).

Genetic studies have been used in attempts to test these two hypotheses. Studies of nuclear and mitochondrial DNA from present human populations led to the conclusion that the modern populations originated in Africa and spread to the rest of the Old World approximately 50,000 ± 20,000 years ago (Harpending et al., 1993; Jones and Rouhani, 1986; Wainscoat et al., 1986). This population explosion apparently followed a severe population bottleneck, estimated by Harpending and others (1993) to have reduced the human population to approximately 500 breeding females, or a total population as small as 4000 for approximately 20,000 years. At the same time, Neanderthals who were probably better adapted for cold climate moved into the Levant region when modern humans vacated it (Hewitt, 2000).

Harpending and others (1993) proposed that the evidence may fit an intermediate 'Weak Garden of Eden' hypothesis that a small ancestral human population separated into partially isolated groups about 100,000 years ago, and about 30,000 years later these populations underwent either simultaneous bottlenecks or simultaneous expansions in size. Sherry and others (1994) estimated mean population expansions times ranging from approximately 65,000 to 30,000 years ago, with the African expansion possibly being the earliest.

Ambrose (1998, 2003; see Gibbons, 1993) pointed out that the timing of the Toba super-eruption roughly matched the inferred timing of the bottleneck and release, and surmised that the environmental after-effects of the Toba eruption might have been so severe as to lead to a precipitous decline in the population of human ancestors (but see Gathorne-Hardy and Harcourt-Smith, 2003 for opposing views). Rampino and Self (1993a) and Rampino and Ambrose (2000) concurred that the climatic effects of Toba could have constituted a true 'volcanic winter', and could have caused severe environmental damage. It may be significant that analysis of mtDNA of Eastern Chimpanzee (*Pan troglodytes schweinfurthii*) shows a similar pattern to human DNA, suggesting a severe reduction in population at about the same time as in the human population (see Rogers and Jorde, 1995).

10.6 Frequency of super-eruptions

Decker (1990) proposed that if all magnitude 8 eruptions in the recent past left caldera structures that have been recognized, then the frequency of VEI 8 eruptions would be approximately 2×10^{-5} eruptions per year, or roughly one VEI 8 eruption every 50,000 years.

The timing and magnitude of volcanic eruptions, however, are difficult to predict. Prediction strategies have included (1) recognition of patterns of eruptions at specific volcanoes (e.g., Godano and Civetta, 1996; Klein, 1982), (2) precursor activity of various kinds (e.g., Chouet, 1996; Nazzaro, 1998), (3) regional and global distribution of eruptions in space and time (Carr, 1977; Mason et al., 2004; Pyle, 1995), and (4) theoretical predictions based on behaviour of materials (Voight, 1988; Voight and Cornelius, 1991). Although significant progress has been made in short-term prediction of eruptions, no method has proven successful in consistently predicting the timing, and more importantly, the magnitude of the resulting eruption or its magmatic sulphur content and release characteristics.

State-of-the-art technologies involving continuous satellite monitoring of gas emissions, thermal anomalies and ground deformation (e.g., Alexander, 1991; Walter, 1990) promise improved forecasting and specific prediction of volcanic events, but these technologies are thus far largely unproven.

For example, although we have 2000 years of observations for the Italian volcano Vesuvius (Nazzaro, 1998), and a long history of monitoring and scientific study, prediction of the timing and magnitude of the next Vesuvian eruption remains a problem (Dobran et al., 1994; Lirer et al., 1997). For large caldera-forming super-eruptions, which have not taken place in historic times, we have little in the way of meaningful observations on which to base prediction or even long-range forecasts.

10.7 Effects of a super-eruptions on civilization

The regional and global effects of the ash fallout and aerosol clouds on climate, agriculture, health, and transportation would present a severe challenge to modern civilization. The major effect on civilization would be through collapse of agriculture as a result of the loss of one or more growing seasons (Toon et al., 1997). This would be followed by famine, the spread of infectious diseases, breakdown of infrastructure, social and political unrest, and conflict. Volcanic winter predictions are for global cooling of 3–5°C for several years, and regional cooling up to 15°C (Rampino and Self, 1992; Rampino and Ambrose, 2000). This could devastate the major food-growing areas of the world. For example, the Asian rice crop could be destroyed by a single night of below-freezing temperatures during the growing season. In the temperate grain-growing areas, similar drastic effects could occur. In Canada, a 2–3°C average local temperature drop would destroy wheat production, and 3–4°C would halt all Canadian grain production. Crops in the American Midwest and the Ukraine could be severely injured by a 3–4°C temperature decrease (Harwell and Hutchinson, 1985; Pittock et al., 1986). Severe climate would also interfere with global transportation of foodstuffs and other goods. Thus, a super-eruption could compromise global agriculture, leading to famine and possible disease pandemics (Stothers, 2000).

Furthermore, large volcanic eruptions might lead to longer term climatic change through positive feedback effects on climate such as cooling the surface oceans, formation of sea-ice, or increased land ice (Rampino and Self, 1992, 1993a, 1993b), prolonging recovery from the 'volcanic winter'. The result could be widespread starvation, famine, disease, social unrest, financial collapse, and severe damage to the underpinnings of civilization (Sagan and Turco, 1990; Sparks et al., 2005).

The location of a super-eruption can also be an important factor in its regional and global effects. Eruptions from the Yellowstone Caldera over the last 2 million years have included three super-eruptions. Each of these produced thick ash deposits over the western and central United States (compacted ash thicknesses of 0.2 m occur ~1500 km from the source; Wood and Kienle, 1990).

One mitigation strategy could involve the stockpiling of global food reserves. In considering the vagaries of normal climatic change, when grain stocks dip below about 15% of utilization, local scarcities, worldwide price jumps and sporadic famine were more likely to occur. Thus a minimum world level of accessible grain stocks near 15% of global utilization should be maintained as a hedge against year-to-year production fluctuations due to climatic and socio-economic disruptions. This does not take into account social and economic factors that could severely limit rapid and complete distribution of food reserves.

At present, a global stockpile equivalent to a two-month global supply of grain exists, which is about 15% of annual consumption. For a super-volcanic catastrophe, however, several years of growing season might be curtailed, and hence a much larger stockpile of grain and other foodstuffs would have to be maintained, along with the means for rapid global distribution.

10.8 Super-eruptions and life in the universe

The chances for communicative intelligence in the Galaxy is commonly represented by a combination of the relevant factors called the Drake Equation, which can be written as

$$N = R * f_p n_e f_l f_i f_c L \qquad (10.1)$$

where N is the number of intelligent communicative civilizations in the Galaxy; R^* is the rate of star formation averaged over the lifetime of the Galaxy; f_p is the fraction of stars with planetary systems; n_e is the mean number of planets within such systems that are suitable for life; f_l is the fraction of such planets on which life actually occurs; f_i is the fraction of planets on which intelligence arises; f_c is the fraction of planets on which intelligent life develops a communicative phase; and L is the mean lifetime of such technological civilizations (Sagan, 1973).

Although the Drake Equation is useful in organizing the factors that are thought to be important for the occurrence of extraterrestrial intelligence, the actual assessment of the values of the terms in the equation is difficult. The only well-known number is R^*, which is commonly taken as 10 yr^{-1}. Estimates for N have varied widely from approximately 0 to >108 civilizations (Sagan, 1973).

It has been pointed out recently that f_c and L are limited in part by the occurrence of asteroid and comet impacts that could prove catastrophic to technological civilizations (Sagan and Ostro, 1994; Chyba, 1997). Present human civilization, dependent largely on annual crop yields, is vulnerable to an 'impact winter' that would result from dust lofted into the stratosphere by the impact of objects ≥ 1 km in diameter (Chapman and Morrison, 1994; Toon et al., 1997). Such an impact would release approximately 10^5–10^6 Mt (TNT equivalent) of energy, produce a crater approximately 20–40 km in diameter, and is calculated to generate a global cloud consisting of approximately 1000 Mt of submicron dust (Toon et al., 1997). Covey et al. (1990) performed 3-D climate-model simulations for a global dust cloud containing submicron particles with a mass corresponding to that produced by an impact of 6×10^5 Mt (TNT). In this model, global temperatures dropped by approximately 8°C during the first few weeks. Chapman and Morrison (1994) estimated that an impact of this size would kill more than 1.5 billion people through direct and non-direct effects.

Impacts of this magnitude are expected to occur on average about every 100,000 years (Chapman and Morrison, 1994). Thus, a civilization must develop science and technology sufficient to detect and deflect such threatening asteroids and comets on a time scale shorter than the typical times between catastrophic impacts. Recent awareness of the impact threat to civilization has led to investigations of the possibilities of detection, and deflection or destruction of asteroids and comets that threaten the Earth (e.g., Gehrels, 1994; Remo, 1997). Planetary protection technology has been described as essential for the long-term survival of human civilization on the Earth.

The drastic climatic and ecological effects predicted for explosive super-eruptions leads to the question of the consequences for civilization here on Earth, and on other earth-like planets that might harbour intelligent life (Rampino, 2002; Sparks et al., 2005). Chapman and Morrison (1994) suggested that the global climatic effects of super-eruptions such as Toba might be equivalent to the effects of an approximately 1 km diameter asteroid. Fine volcanic dust and sulphuric acid aerosols have optical properties similar to the submicron dust produced by impacts (Toon et al., 1997), and the effects on atmospheric opacity should be similar. Volcanic aerosols, however, have a longer residence time of several years (Bekki et al., 1996) compared to a few months for fine dust, so a huge eruption might be expected to have a longer lasting effect on global climate than an impact producing a comparable amount of atmospheric loading.

Estimates of the frequency of large volcanic eruptions that could cause 'volcanic winter' conditions suggest that they should occur about once every 50,000 years. This is approximately a factor of two more frequent than asteroid or comet collisions that might cause climate cooling of similar severity (Rampino, 2002). Moreover, predicting or preventing a volcanic climatic disaster might be more difficult than tracking and diverting incoming asteroids and comets. These considerations suggest that volcanic super-eruptions pose a real threat to civilization, and efforts to predict and mitigate volcanic climatic disasters should be contemplated seriously (Rampino, 2002; Sparks et al., 2005).

Acknowledgement

I thank S. Ambrose, S. Self, R. Stothers, and G. Zielinski for the information provided.

Suggestions for further reading

Bindeman, I.N. (2006). The Secrets of Supervolcanoes. *Scientific American Magazine* (June 2006). A well-written popular introduction in the rapidly expanding field of super-volcanism.

Mason, B.G., Pyle, D.M., and Oppenheimer, C. (2004). The size and frequency of the largest explosive eruptions on Earth. *Bull. Volcanol.*, 66, 735–748. The best modern treatment of statistics of potential globally catastrophic volcanic eruptions. It includes a comparison of impact and super-volcanism threats and concludes that super-eruptions present a significantly higher risk per unit energy yield.

Rampino, M.R. (2002). Super-eruptions as a threat to civilizations on Earth-like planets. *Icarus*, 156, 562–569. Puts super-volcanism into a broader context of evolution of intelligence in the universe.

Rampino, M.R., Self, S., and Stothers, R.B. (1988). Volcanic winters. *Annu. Rev. Earth Planet. Sci.*, 16, 73–99. Detailed discussion of climatic consequences of volcanism and other potentially catastrophic geophysical processes.

References

Acharya, S.K. and Basu, P.K. (1993). Toba ash on the Indian subcontinent and its implications for correlations of Late Pleistocene alluvium. *Quat. Res.*, **40**, 10–19.

Alexander, D. (1991). Information technology in real-time for monitoring and managing natural disasters. *Prog. Human Geogr.*, **15**, 238–260.

Ambrose, S.H. (1998). Late Pleistocene human population bottlenecks, volcanic winter, and the differentiation of modern humans. *J. Human Evol.*, **34**, 623–651.

Ambrose, S.H. (2003). Did the super-eruption of Toba cause a human population bottleneck? Reply to Gathorne-Hardy and Harcourt-Smith. *J. Human Evol.*, **45**, 231–237.

Bekki, S., Pype, J.A., Zhong, W., Toumi, R., Haigh, J.D., and Pyle, D.M. (1996). The role of microphysical and chemical processes in prolonging the climate forcing of the Toba eruption. *Geophys. Res. Lett.*, **23**, 2669–2672.

Bischoff, J.L., Solar, N., Maroto, J., and Julia, R. (1989). Abrupt Mousterian-Aurignacian boundaries at c. 40 ka bp: Accelerator ^{14}C dates from l'Arbreda Cave (Catalunya, Spain). *J. Archaeol. Sci.*, **16**, 563–576.

Brock, J.C. and McClain, C.R. (1992). Interannual variability in phytoplankton blooms observed in the northwestern Arabian Sea during the southwest monsoon. *J. Geophys. Res.*, **97**, 733–750.

Carr, M.J. (1977). Volcanic activity and great earthquakes at convergent plate margins. *Science*, **197**, 655–657.

Chapman, C.R. and Morrison, D. (1994). Impacts on the Earth by asteroids and comets: assessing the hazards. *Nature*, **367**, 33–40.

Chesner, C.A., Rose, W.I., Deino, A., Drake, R. and Westgate, J.A. (1991). Eruptive history of the earth's largest Quaternary caldera (Toba, Indonesia) clarified. *Geology*, **19**, 200–203.

Chouet, B.A. (1996). Long-period volcano seismicity: Its source and use in eruption forecasting. *Nature*, **380**, 316.

Chyba, C.F. (1997). Catastrophic impacts and the Drake Equation. In Cosmovici, C.B., Bowyer, S. and Werthimer, D. (eds.), *Astronomical and Biochemical Origins and the Search for Life in the Universe*, pp. 157–164 (Bologna: Editrice Compositori).

Covey, C., Ghan, S.J., Walton, J.J., and Weissman, P.R. (1990). Global environmental effects of impact-generated aerosols: Results from a general circulation model. *Geol. Soc. Am. Spl. Paper*, **247**, 263–270.

De La Cruz-reyna, S. (1991). Poisson-distributed patterns of explosive eruptive activity. *Bull. Volcanol.*, **54**, 57–67.

Decker, R.W. (1990). How often does a Minoan eruption occur? In Hardy, D.A. (ed.), *Thera and the Aegean World III* 2, pp. 444–452 (London: Thera Foundation).

Dobran, F., Neri, A., and Tedesco, M. (1994). Assessing the pyroclastic flow hazard at Vesuvius. *Nature*, **367**, 551–554.

Fedele, F.G., Giaccio, B., Isaia, R., and Orsi, G. (2002). Ecosystem impact of the Campanian ignimbrite eruption in Late Pleistocene Europe. *Quat. Res.*, **57**, 420–424.

Gagan, M.K. and Chivas, A.R. (1995). Oxygen isotopes in western Australian coral reveal Pinatubo aerosol-induced cooling in the Western Pacific Warm Pool. *Geophys. Res. Lett.*, **22**, 1069–1072.

Gathorne-Hardy, F.J. and Harcourt-Smith, W.E.H. (2003). The super-eruption of Toba, did it cause a human population bottleneck? *J. Human Evol.*, **45**, 227–230.

Genin, A., Lazar, B., and Brenner, S. (1995). Vertical mixing and coral death in the Red Sea following the eruption of Mount Pinatubo. *Nature*, **377**, 507–510.

Gehrels, T. (ed.) (1994). *Hazards Due to Comets & Asteroids*, 1300 p (Tucson: University of Arizona Press).

Gibbons, A. (1993). Pleistocene population explosions. *Science*, **262**, 27–28.

Gilmour, I., Wolbach, W.S., and Anders, E. (1990). Early environmental effects of the terminal Cretaceous impact. *Geol. Soc. Am. Spl. Paper*, **247**, 383–390.

Godano, C. and Civetta, L. (1996). Multifractal analysis of Vesuvius volcano eruptions. *Geophys. Res. Lett.*, **23**, 1167–1170.

Graf, H.-F., Kirschner, I., Robbock, A., and Schult, I. (1993). Pinatubo eruption winter climate effects: Model versus observations. *Clim. Dynam.*, **9**, 81–93.

Green, O., Percival, I., and Ridge, I. (1985). *Nuclear Winter, the Evidence and the Risks*, 216 p (Cambridge: Polity Press).

Hansen, J., Lacis, A., Ruedy, R., and Sato, M. (1992). Potential climate impact of the Mount Pinatubo eruption. *Geophys. Res. Lett.*, **19**, 215–218.

Hansen, J.E., Sato, M., Ruedy, R., Lacis, A., Asamoah, K, Borenstein, S., Brown, E., Cairns, B., Caliri, G., Campbell. M., Curran, B., De Castrow, S., Druyan, L., Fox, M., Johnson, C., Lerner, J., Mscormick, M.P., Miller, R., Minnis, P., Morrison, A., Palndolfo, L., Ramberran, I., Zaucker, F., Robinson, M., Russell, P., Shah, K., Stone, P., Tegen, I., Thomason, L., Wilder, J., and Wilson, H. (1996). A Pinatubo modeling investigation. In Fiocco, G., Fua, D., and Visconti, G. (eds.), *The Mount Pinatubo Eruption: Effects on the Atmosphere and Climate* NATO ASI Series Volume 142, pp. 233–272 (Heidelberg: Springer Verlag).

Harpending, H.C., Sherry, S.T., Rogers, A.L., and Stoneking, M. (1993). The genetic structure of ancient human populations. *Curr. Anthropol.*, **34**, 483–496.

Harwell, M.A. (1984). *The Human and Environmental Consequences of Nuclear War*, 179 p (New York: Springer-Verlag).

Harwell, M.A. and Hutchinson, T.C. (eds.) (1985). *Environmental Consequences of Nuclear War, Volume II Ecological and Agricultural Effects*, 523 p (New York: Wiley).

Harwell, M.A., Hutchinson, T.C., Cropper, W.P., Jr, and Harwell, C.C. (1985). Vulnerability of ecological systems to climatic effects of nuclear war. In Harwell, M.A. and Hutchinson, T.C. (eds.), *Environmental Consequences of Nuclear War, Volume II Ecological and Agricultural Effects*, pp. 81–171 (New York: Wiley).

Hewitt, G. (2000). The genetic legacy of the Quaternary ice ages. *Nature*, **405**, 907–913.

Huang, C.-H., Zhao, M., Wang, C.-C., and Wei, G. (2001). Cooling of the South China Sea by the Toba eruption and other proxies ~71,000 years ago. *Geophys. Res. Lett.*, **28**, 3915–3918.

Jones, J.S. and Rouhani, S. (1986). How small was the bottleneck? *Nature*, **319**, 449–450.

Klein, F.W. (1982). Patterns of historical eruptions at Hawaiian volcanoes. *J. Volcanol. Geotherm. Res.*, **12**, 1–35.

Larcher, W. and Bauer, H. (1981). Ecological significance of resistance to low temperature. In Lange, O.S., Nobel, P.S., Osmond, C.B., and Zeigler, H. (eds.), *Encyclopedia of Plant Physiology, Volume 12A: Physiological Plant Ecology I*, pp. 403–37 (Berlin: Springer-Verlag).

Leavitt, J. (1980). *Responses of Plants to Environmental Stresses. I. Chilling, Freezing and High Temperature Stresses*, 2nd edition (New York: Academic Press).

Ledbetter, M.T. and Sparks, R.S.J. (1979). Duration of large-magnitude explosive eruptions deduced from graded bedding in deep-sea ash layers. *Geology*, **7**, 240–244.

Lirer, L., Munno, R., Postiglione, I., Vinci, A., and Vitelli, L. (1997). The A.D. 79 eruption as a future explosive scenario in the Vesuvian area: Evaluation of associated risk. *Bull. Volcanol.*, **59**, 112–124.

Mason, B.G., Pyle, D.M., and Oppenheimer, C. (2004). The size and frequency of the largest explosive eruptions on Earth. *Bull. Volcanol.*, **66**, 735–748.

Nazzaro, A. (1998). Some considerations on the state of Vesuvius in the Middle Ages and the precursors of the 1631 eruption. *Annali di Geofisica*, **41**, 555–565.

Newhall, C.A. and Self, S. (1982). The volcanic explosivity index (VEI): an estimate of the explosive magnitude for historical volcanism. *J. Geophys. Res.*, **87**, 1231–1238.

Ninkovich, D., Shackleton, N.J., Abdel-Monem, A.A., Obradovich, J.A., and Izett, G. (1978a). K-Ar age of the late Pleistocene eruption of Toba, north Sumatra. *Nature*, **276**, 574–577.

Ninkovich, D., Sparks, R.S.J., and Ledbetter, M.T. (1978b). The exceptional magnitude and intensity of the Toba eruption: an example of the use of deep-sea tephra layers as a geological tool. *Bull. Volcanol.*, **41**, 1–13.

Oppenheimer, C. (2002). Limited global change due to the largest known Quaternary eruption, Toba = 74 kyr BP? *Quat. Sci. Rev.*, **21**, 1593–1609.

Palais, J.M. and Sigurdsson, H. (1989). Petrologic evidence of volatile emissions from major historic and pre-historic volcanic eruptions. *Am. Geophys. Union, Geophys. Monogr.*, **52**, 32–53.

Pinto, J.P., Turco, R.P., and Toon, O.B. (1989). Self-limiting physical and chemical effects in volcanic eruption clouds. *J. Geophys. Res.*, **94**, 11165–11174.

Pittock, A.B., Ackerman, T.P., Crutzen, P.J., MacCracken, M.C., Shapiro, C.S., and Turco, R.P. (eds.) (1986). *Environmental Consequences of Nuclear war, Volume I Physical and Atmospheric Effects*, 359 p (New York: Wiley).

Pittock, A.B., Walsh, K., and Frederiksen, J.S. (1989). General circulation model simulation of mild nuclear winter effects. *Clim. Dynam.*, **3**, 191–206.

Pollack, J.B., Rind, D., Lacis, A., Hansen, J.E., Sato, M., and Ruedy, R. (1993). GCM simulations of volcanic aerosol forcing. Part 1: climate changes induced by steady-state perturbations. *J. Clim.*, **6**, 1719–1742.

Pope, K.O., Baines, K.H., Ocampo, A.C., and Ivanov, B.A. (1994). Impact winter and the Cretaceous/Tertiary extinctions: Results of a Chicxulub asteroid impact model. *Earth Planet. Sci. Lett.*, **128**, 719–725.

Pyle, D.M. (1995). Mass and energy budgets of explosive volcanic eruptions. *Geophys. Res. Lett.*, **22**, 563–566.

Rampino, M.R. (2002). Supereruptions as a threat to civilizations on Earth-like planets. *Icarus*, **156**, 562–569.

Rampino, M.R. and Ambrose, S.H. (2000). Volcanic winter in the Garden of Eden: the Toba super-eruption and the Late Pleistocene human population crash. In McCoy, F.W. and Heiken, G. (eds.), *Volcanic Hazards and Disasters in Human Antiquity*. Special paper 345, pp. 71–82 (Boulder, CO: Geological Society of America).

Rampino, M.R. and Self, S. (1982). Historic eruptions of Tambora (1815), Krakatau (1883), and Agung (1963), their stratospheric aerosols and climatic impact. *Quat. Res.*, **18**, 127–143.

Rampino, M.R. and Self, S. (1984). Sulphur-rich volcanism and stratospheric aerosols. *Nature*, **310**, 677–679.

Rampino, M.R. and Self, S. (1992). Volcanic winter and accelerated glaciation following the Toba super-eruption. *Nature*, **359**, 50–52.

Rampino, M.R. and Self, S. (1993a). Climate-volcanism feedback and the Toba eruption of ~74,000 years ago. *Quat. Res.*, **40**, 269–280.

Rampino, M.R. and Self, S. (1993b). Bottleneck in human evolution and the Toba eruption: *Science*, **262**, 1955.

Rampino, M.R., Self, S., and Stothers, R.B. (1988). Volcanic winters. *Annu. Rev. Earth Planet. Sci.*, **16**, 73–99.

Redman, R.E. (1974). Photosynthesis, plant respiration, and soil respiration measured with controlled environmental chambers in the field: Canadian Committee, IBP Technical Report 49 (Saskatoon: University of Saskatchewan).

Remo, J.L. (ed.) (1997). *Near Earth Objects: The United Nations International Conference*, Volume 822, 623 p (New York Academy of Sciences Annals). New York

Robock, A. and Mass, C. (1982). The Mount St. Helens volcanic eruption of 18 May 1980: large short-term surface temperature effects. *Science*, **216**, 628–630.

Rogers, A.R. and Jorde, L.B. (1995). Genetic evidence on modern human origins. *Human Biol.*, **67**, 1–36.

Rose, W.I. and Chesner, C.A. (1990). Worldwide dispersal of ash and gases from earth's largest known eruption: Toba, Sumatra, 75 Ka. *Global Planet. Change*, **89**, 269–275.

Sagan, C. (ed.) (1973). *Communication with Extraterrestrial Intelligence* (Cambridge, MA: MIT Press).

Sagan, C. and Ostro, S. (1994). Long-range consequences of interplanetary collisions. *Issues Sci. Technol.*, **10**, 67–72.

Sagan C. and Turco, R. (1990). *A Path Where No Man Thought*, 499 p (New York: Random House).

Shane, P., Westgate, J., Williams, M., and Korisettar, R. (1995). New geochemical evidence for the Youngest Toba Tuff in India. *Quat. Res.*, **44**, 200–204.

Sherry, S.T., Rogers, A.R., Harpending, H., Soodyall, H., Jenkins, T., and Stoneking, M. (1994). Mismatch distributions of mtDNA reveal recent human population expansions. *Human Biol.*, **66**, 761–775.

Shultz, H., Emeis, K.-C., Erlenkeuser, H., von Rad, U., and Rolf, C. (2002). The Toba volcanic event and interstadial/stadial climates at the marine isotopic stage 5 to 4 transition in the northern Indian Ocean. *Quat. Res.*, **57**, 22–31.

Song, S.-R., Chen, C.-H., Lee, M.-Y., Yang, T.F., and Wei, K.-Y. (2000). Newly discovered eastern dispersal of the youngest Toba tuff. *Marine Geol.*, **167**, 303–312.

Sparks, S., Self, S., Grattan, J., Oppenheimer, C., Pyle, D., and Rymer, H. (2005). Super-eruptions global effects and future threats. Report of a Geological Society of London Working Group, The Geological Society, London, pp. 1–25.

Stothers, R.B. (1984a). The great Tambora eruption of 1815 and its aftermath. *Science*, **224**, 1191–1198.

Stothers, R.B. (1984b). The mystery cloud of AD 536. *Nature*, **307**, 344–345.

Stothers, R.B. (1996). Major optical depth perturbations to the stratosphere from volcanic eruptions: Pyrheliometric period, 1881–1960. *J. Geophys. Res.*, **101**, 3901–3920.

Stothers, R.B. (2000). Climatic and demographic consequences of the massive volcanic eruption of 1258. *Clim. Change*, **45**, 361–374.

Strong, A.E. (1993). A note on the possible connection between the El Chichón eruption and ocean production in the northwest Arabian Sea during 1982. *J. Geophys. Res.*, **98**, 985–987.

Tinus, R.W. and Roddy, D.J. (1990). Effects of global atmospheric perturbations on forest ecosystems in the Northern Temperate Zone; Predictions of seasonal depressed-temperature kill mechanisms, biomass production, and wildfire soot emissions. *Geol. Soc. Am. Spl. Paper*, **247**, 77–86.

Toon, O.B., Turco, R.P., and Covey, C. (1997). Environmental perturbations caused by the impacts of asteroids and comets. *Rev. Geophys.*, **35**, 41–78.

Turco, R.P., Toon, O.B., Ackerman, T.P., Pollack, J.B., and Sagan, C., (1990). Climate and smoke: An appraisal of nuclear winter. *Science*, **247**, 166–176.

van Keulan, H., Lowerse, W., Sibma, L., and Alberda, M. (1975). Crop simulation and experimental evaluation – a case study, In Cooper, J.P. (ed.), *Photosynthesis and Productivity in Different Environments*, pp. 623–643 (Cambridge: Cambridge University Press).

Voight, B. (1988). A method for prediction of volcanic eruptions. *Nature*, **332**, 125–130.

Voight, B. and Cornelius, R.R. (1991). Prospects for eruption prediction in near real-time. *Nature*, **350**, 695–697.

Wainscoat, J.S., Hill, A.V.S., Thein, S.L., Clegg, J.J. (1986). Evolutionary relationships of human populations from an analysis of nuclear DNA polymorphisms. *Nature*, **319**, 491–493 .

Walter, L.S. (1990). The uses of satellite technology in disaster management. *Disasters*, **14**, 20–35.

Woillard, G. and Mook, W.G. (1982). Carbon-14 dates at Grande Pile: Correlation of land and sea chronologies. *Science*, **215**, 159–161.

Wood, C.A. and Kienle, J. (eds.) (1990). *Volcanoes of North America*, 354 p (Cambridge: Cambridge University Press).

Woods, A.W.M. and Wohletz, K.H. (1991). Dimensions and dynamics of co-ignimbrite eruption columns. *Nature*, **350**, 225–227.

Zielinski, G.A., Mayewski, P.A., Meeker, L.D., Whitlow, S., Twickler, M.S., and Taylor, K., (1996a). Potential atmospheric impact of the Toba mega-eruption ~71,000 years ago. *Geophys. Res. Lett.*, **23**, 837–840.

Zielinski, G.A., Mayewski, P.A., Meeker, L.D., Whitlow, S., and Twickler, M.S. (1996b). An 110,000-year record of explosive volcanism from the GISP2 (Greenland) ice core. *Quat. Res.*, **45**, 109–118.

·11·

Hazards from comets and asteroids

William Napier

There are risks everywhere. Even heaven is a stage of risk.

Wilson Harris, Carnival (1985)

11.1 Something like a huge mountain

The first angel sounded his trumpet, and there came hail and fire mixed with blood, and it was hurled down upon the earth. A third of the earth was burned up, a third of the trees were burned up, and all the green grass was burned up. The second angel sounded his trumpet, and something like a huge mountain, all ablaze, was thrown into the sea ... The third angel sounded his trumpet, and a great star, blazing like a torch, fell from the sky on a third of the rivers ... a third of the sun was struck, a third of the moon, and a third of the stars, so that a third of them turned dark ... and I saw a star that had fallen from the sky to the earth. The star was given the key to the shaft of the Abyss. When he opened the Abyss, smoke rose from it like the smoke from a gigantic furnace. The sun and sky were darkened by the smoke from the Abyss ....

The Revelation of St John was probably written around 100 AD, but is part of a very much older 'Star Wars' literature, going back to the very earliest writings and probably based on pre-literate oral traditions. Common threads in these tales are often hot blast, hurricane winds, flattened forests, tsunami and cataclysmic floods, associated with blazing thunderbolts from the sky, a darkened sun, a great, red-tailed comet and what appears to be a meteor storm. Even without benefit of the twentieth century Tunguska impact, which destroyed 2000 square kilometres of Siberian forest in 1908, classical scholars have long regarded the stories as descriptions of a cosmic impact. Myth was a vehicle for transmitting astronomical and cosmological information through the generations, and it is surely a seductive proposition to see these tales of celestial catastrophe – which are found worldwide – as prehistoric descriptions of cosmic cataclysm, one-off or recurrent, local or global. Inevitably, this is a contentious area – only qualitative statements can be made, and one individual's unifying hypothesis is another's Velikovskian fantasy.

A great earthquake or tsunami may take 100,000 lives; a great impact could take 1000 or 10,000 times as many, and bring civilization to an abrupt halt. In assessing the hazard posed by the stray celestial bodies in our celestial environment, we want to get it right! To do this we must leave the realm of myth and enter that of science. Three quantitative lines of attack are available – impact crater studies, telescopic searches, and dynamical analysis. Each throws light on different aspects of the problem.

11.2 How often are we struck?

11.2.1 Impact craters

The number of impact craters known on the surface of the Earth has increased substantially over the last few decades. Along with this, improved dating techniques have allowed the ages of many of them to be quite well determined. By the end of 2004, about 170 terrestrial impact structures were known. Of these, only forty are useful for statistical purposes: they have diameters over 3 km, are less than 250 million years old, and have been dated with precision better than 10 million years. Most of the forty are dated to precision better than 5 million years, and about half to better than one million years. This is a small database, but it is just enough to search for trends, periodicities, impact episodes, and the like.

Their age distribution is shown in Fig. 11.1. Structure appears on several levels. First, the record gives the appearance that, going into the increasingly remote past, there has been a steep decline in the impact cratering rate. This can only be an illusion, due to the cumulative overlaying of impact craters by sediments, or their erosion by the elements. Even if all craters say less than 10 million years old have been found – an unlikely assumption – the figure reveals that only 40% of impact craters formed 100 million years ago have been discovered, and only about 10% of those formed 200 million years ago.

A closer examination of Fig. 11.1 gives the impression that some craters are bunched in time, and statistical scrutiny bears this out (Napier, 2006). The largest craters all occur within these bombardment episodes: the mean diameter of craters inside an episode is about 50 km, that of craters outside is approximately 20 km. The history of bombardment, then, is not one of random arrivals. Rather, it seems to have the character of 'impact epochs', during which the Earth is heavily bombarded, interspersed with relatively quiescent periods.

A third feature of the cratering record is not obvious to the eye, but can be teased out with detailed analysis: there is a weak periodicity. Strong random surges are also present which make it impossible to determine which of several possible periodic solutions is real, and which are harmonics (Fig. 11.1). The best we can do is say that over the last 250 Myr periodic cratering episodes have recurred at intervals of 24, 30, 36, or 42 million years, interspersed with a handful of random episodes of comparable strength. The peak-to-trough ratios are uncertain, but may be in the range 2:1 to 5:1.

It follows that crater counts on ancient surfaces are of limited use in inferring the contemporary impact rate. The lunar cratering record in particular has

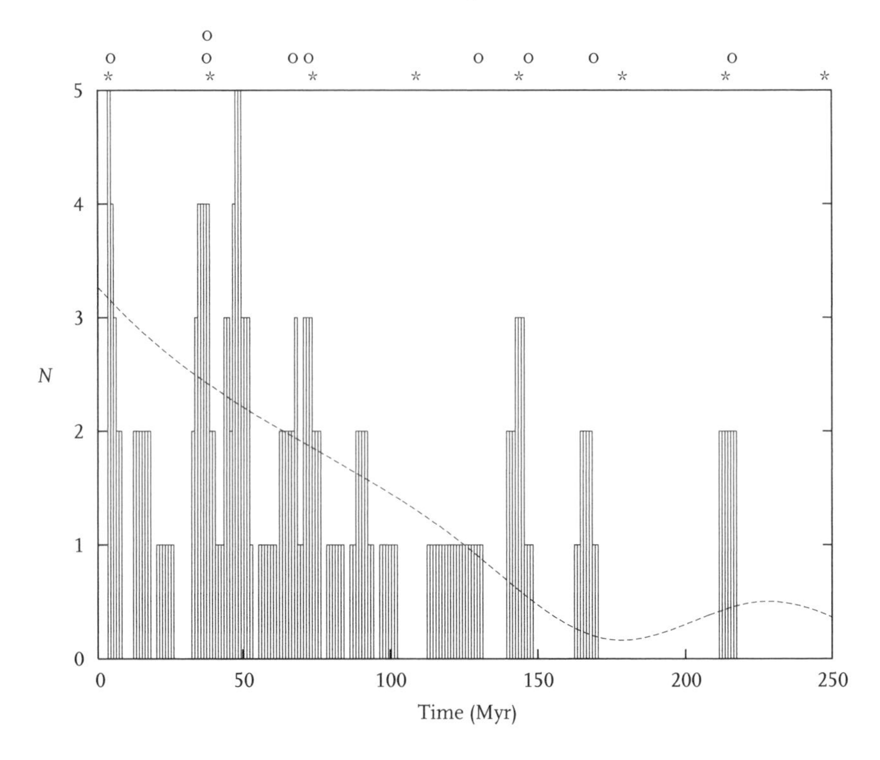

Fig. 11.1 The age distribution of 40 impact craters 3 km or more in diameter, with ages less than 250 Myr, known to precision better than 10 Myr. A rectangular window of width 8 Myr corresponding roughly to the mean uncertainty in ages has been passed over the data, and the smooth curve is a fit to the overall trend. The impression is that impacts occur in discrete episodes of bombardment and is confirmed by detailed statistical analysis (Napier 2005).

been used to infer that an impact capable of yielding a 100 km diameter terrestrial crater (of species-destroying energy) happens once per 27 million years (Neukum and Ivanov, 1994). However the Moon's surface is ancient and fluctuations in impact rates on it cannot be determined with resolving power much better than a billion years. Figure 11.1 implies that the current impact rate may be several times higher than estimates based on the lunar record, and there are indications that coherent structure exists at even higher temporal resolution (Steel et al., 1990).

One conceivable source of bombardment episodes is the collisional break-up of a large asteroid in the main asteroid belt, followed by the feeding of its debris into Earth-crossing orbits. This process may well cause the Earth's cratering rate to fluctuate by about an order of magnitude over 0.1–1 million years timescales, in the case of fragments about a kilometre across

(Menichella et al., 1996). However the largest craters require the disruption of correspondingly large asteroids, and these break up too infrequently by a factor of 10–100 to supply the extinction-level surges of bombardment (Napier, 2006).

Probably, the surges are mostly due to disturbances of the Oort comet cloud. This is a major reservoir of about 100 billion long-period comets, comprising a roughly spherical swarm orbiting the sun out to 50,000 astronomical units, a quarter of the way to the nearest star. The influx of long-period comets from the Oort cloud is mainly due to the perturbing action of the Galactic tide on their orbits. This tide varies cyclically due to the out-of-plane oscillations of the Sun as it orbits the Galaxy, and a periodicity in the long-period comet flux in the range 35–42 million years is predicted (Clube and Napier, 1996; Matese et al., 1995). The sun passed through the plane of the Galaxy 3 or 4 million years ago, and since the infall time of a long-period comet is 2 or 3 million years, we should be close to the peak of a bombardment episode now. There is thus a pleasing consistency with the cratering data (Fig. 11.1). On this evidence the largest impactors at least are more likely to be comets than asteroids, and their current impact rate is likely to be higher than the long-term average.

Crudely speaking, we can think of a cosmic impactor as generating a crater about twenty times its own size. Given the cosmic velocities involved – approximately 20 km per second for an asteroid impact and 55 km per second for a comet – the energies involved in creating the biggest terrestrial craters are characteristically equivalent to the explosion of 100 million megatons TNT, or about ten atomic bombs of Hiroshima size on each square kilometre of the Earth's surface (McCrea, 1981). This enormous energy is released in a fraction of a second and spreads around the globe on a timescale of an hour or so.

Estimates based on the mean impact cratering rate indicate that, on the long-term, a 1 km impactor might be expected every half a million years or so. Again, modelling uncertainties to do with both excavation mechanics and the erratic replenishment of the near-Earth object (NEO) population yield an overall uncertainty of a factor of a few. A rate of one such impact every 100,000 years cannot be excluded by the cratering evidence.

On the sub-kilometre bolide scale the impact cratering record does not strongly constrain the current rate. The Earth's atmosphere acts as a barrier, tending to break up bodies much less than 100 or 200 m across. On airless bodies such as the Moon or the Galilean satellites the situation is unclear since the small crater population appears to be dominated by secondary impacts, arising when hunks of native terrain are thrown out by large impactors (Bierhaus et al., 2005). On this scale – important for assessing the tsunami hazard – we must either extrapolate from the size distribution of larger craters or turn to telescopic surveys.

11.2.2 Near-Earth object searches

Until the 1970s, only a handful of Earth-crossers were known. The subject was of little interest to most astronomers, whose telescopes were (and still are) mainly directed towards stellar, galactic and extragalactic realms. However, following the pioneering work of Helin and Shoemaker (1979), who searched for Earth-crossing bodies using a small, wide-angled telescope on Mount Palomar, it became clear that there is indeed a significant impact hazard out there. This was becoming clear, too, from the increasing number of terrestrial impact craters being discovered. Search programmes got underway in the early 1990s and small bodies in hazardous, Earth-crossing orbits (NEOs) began to be found in serious numbers.

The rate of discovery of Earth-crossers has been impressive, going from 350 in 1995 to 3400 a decade later – of which about 800 are thought to be a kilometre or more in diameter. Most of these small bodies have orbital periods of a few years. It is generally thought that the total population of near-Earth asteroids over a kilometre across is about 1100. If correct, this leads to an expected impact frequency of about one such body every 500,000 years. There is thought to be a sharp threshold between regional and global effects around this size range: civilization ends with an impactor bigger than a kilometre or two! But there is a caveat: extremely dark objects would go undiscovered and not be entered in the inventory of global hazards. The population of sub-kilometre bodies is almost entirely unexplored; but it is this population which may give damaging tsunamis and possible short-lived climatic coolings on timescales of historical interest.

11.2.3 Dynamical analysis

The known population of Earth-crossing objects is transient, with a median lifetime of only about 2 million years. Without replenishment, it would rapidly vanish, most NEOs falling into the sun. This short-lived population must therefore be supplied from other sources. Both asteroidal and cometary reservoirs are available, and the supply from both is likely to be erratic.

A comet is a conglomerate of ice and dust, which, on approaching the sun to within about the orbit of Mars, may grow one or more tails, that can be tens or hundreds of millions of kilometres long. It will die when its volatiles are exhausted. There are several documented cases of comets whose activity has died, leaving a dark, inert body of asteroidal appearance. It is plausible to think that an accumulation of dust on the surface eventually chokes off or insulates underlying ice from solar heating. Equally, numerous comets have been seen to split and a few have disintegrated altogether.

A typical orbital period of a comet arriving from the Oort cloud is a few million years. Recent years have seen the discovery of other major cometary reservoirs on the fringes of the planetary system and these probably help to replenish the Earth-crossing comets. In their active form, long-period comets may amount to only about a percent of the total impact hazard. About one in a hundred, however, are perturbed by the giant planets into Halley-type orbits (high-eccentricity orbits with periods of less than 200 years), whence a single such comet has typically a thousand opportunities to strike the Earth before it falls into the Sun or is ejected from the solar system. This ought to make them a substantial risk, and one, furthermore, difficult to handle because of the high speeds and short warning times involved. There is, however, a paradox: we don't see them! Knowing the rate at which bright comets arrive from the Oort cloud, and the fraction which are captured into the Halley system, it turns out that there should be about 3000 active comets over 5 km or so across in such orbits. And yet only a couple of dozen are observed.

It could be that, after their first passage or two through the inner planetary system, active comets simply become dormant, fading into dark, asteroid-like bodies (Emel'yanenko and Bailey, 1998). The hazard posed by these unseen, dormant bodies would be comparable with that of the observed near-Earth asteroids, in line with other investigators who have likewise concluded that active and dormant comets together 'yield a large, perhaps dominant, contribution to kilometre-sized terrestrial impactors' (Rickman et al., 2001; see also Nurmi et al., 2001). The problem is that, even although very dark, about 400 such bodies – in highly eccentric orbits, with orbital periods up to 200 years – should by now have been discovered, whereas only about 25 are currently known, forming a system known as Damocloids. This is assuming reflectivities 0.04 comparable with the surfaces of the known dormant comets.

Another possibility is that comets thrown into Halley-type orbits disintegrate altogether, turning completely to dust after one or two perihelion passages (Levison et al., 2002). The hypothesis was adopted by NASA's Near-Earth Object Science Definition Team (Stokes et al., 2003) and is the basis of their claim that comets, active and dormant together, constitute no more than 1% of the impact hazard. However, it turns out that this hypothesis also has problems (Napier et al., 2004; Rickman; 2005). For example, for the process to work, almost 99% of incoming Halley-type comets would have to disintegrate in this way. But such complete and rapid disintegration does not seem to be the normal fate of comets: nearly all of the strongest annual meteor showers have, orbiting within them, either an active comet or a large dormant body, presumably a defunct parent (Table 11.1).

One may question the adopted reflectivity of 0.04 for dormant comets, based on the commonly observed surface properties of active comets. A flyby of Comet Borrelly revealed the presence of spots with reflectivities 0.008 (Nelson et al., 2004): if all comet surfaces darken to this extent as they become inactive, then the paradox is solved. Standard radiation theory reveals that a comet which becomes completely inactive, losing its icy interstitial volatiles and leaving a 'bird's nest' structure of organic grains on its surface, may indeed

Table 11.1 Major Annual Meteor Streams

Stream	Peak	Range	Parent	Lunar Swarm	Twentieth Century Impact
Quadrantids	03 January	01–04 January	2003 EH1[a]		
Lyrids	21 April	19–24 April	1861 I		
η Aquarids	04 May	21 April–12 May	Halley		
β Taurids	30 June	24 June–06 July	Encke	22–26 June 1975	30 June 1908
δ Aquarids	29 July	21 July–15 August	—		
Perseids	12 August	25 July–17 August	1862 III	13–14 August 1975	13 August 1930
Orionids	22 October	18–26 October	Halley		
Andromedids	14 November	03–22 November	Biela		
Leonids	17 November	14–20 November	Temple	16–17 November 1974	
Geminids	14 December	07–15 December 15	Phaethon[a]	13–14 December 1974	11 December 1935
Ursids	22 December	16–24 December	Tuttle		

Note: Superscript a indicates an asteroid, otherwise the assumed parent bodies are active comets. Complete disintegration into dust within one or two orbits is apparently not a normal evolutionary route for comets. The dates of prominent meteoroid showers, measured by lunar seismometers over 1971–1975, are listed, along with three twentieth century airburst events in the megaton ballpark; they appear to be associated with major meteor streams.

develop a vanishingly small reflectivity (Napier et al., 2004). But if that is the resolution of the fading paradox, then there is a significant population of high-speed hazards, undiscoverable because they are too dark and spend about 99% of their time beyond the orbit of Mars.

About 25 Damocloids are known at the time of writing. Their average radius is 8 km, which would yield an impact of 60 million megatons, with a mean impact speed of 58 km per second. The reflectivities of six Damocloids have been measured to date, and they are amongst the darkest known objects in the solar system. In general, the more comet-like the orbit of an asteroid, the darker its surface is found to be (Fernandez et al., 2005).

Whether small, dark Damocloids, of, for example, 1 km diameter exist in abundance is unknown – they are in essence undiscoverable with current search programmes. The magnitude of the hazard they present is likewise unknown; it could be negligible, it could more than double the risk assessments based on the objects we see. Crater counts are again of little help, since even the youngest surfaces – such as that of the icy satellite Europa which orbits Jupiter – are older than the probable duration of a cometary bombardment episode and do not strongly constrain contemporary impact rates. The best chance for discovery of such bodies would be through their thermal radiation around perihelion, using infrared instrumentation on the ground (Rivkin et al., 2005) or in satellites.

For a threat object discovered in a short period, Earth-crossing orbit, decades or centuries of advance warning will probably be available. For a comet, the warning time is measured in months. In the case of a dark Damocloid, there will generally be no warning at all.

11.3 The effects of impact

The Tunguska impact of 30 June 1908, in the central Siberian plateau, was an airburst with an energy approximately 10 to 30 megatons, that of a very large hydrogen bomb. It destroyed approximately 2000 km^2 of forest, knocking trees over and charring the barks of trees on one side. Such impacts are local in effect (unless, perhaps, mistaken for a hydrogen bomb explosion in time of crisis). Estimates of their recurrence time range from 200 to about 2000 years.

At 10,000 megatons – comparable to the energy unleashed in a full-scale nuclear war – the area of devastation approaches 100,000 km^2 (Table 11.2). Flying shards of glass in urban areas would cause substantial injury far beyond this area. The rising fireball from such an impact could cause serious burns and extensive conflagration along its line of sight, while earthquakes at the extreme end of human experience occur within a few 100 km of the impact site. There is uncertainty, too, about the recurrence time of impacts in this

Table 11.2 Possible Impact Effects

Megatons	10,000	1 million	100 million
Impactor	500 m	1–2 km	10 km
Scope	Regional	Civilization-destroying	Species-destroying
Land	Fires, blast and earthquake over 250–1000 km	Destructive blast, quake and possibly fire over continental dimensions	Global conflagration and destructive earthquake
Sea	Uncertain	(Mega)tsunamis around ocean rims	Ocean rim devastation; cities replaced by mudflats; oceans acidified
Air	Sun obscured; possible 536 AD event	Agriculture collapses; ozone depletion; acid rain; sky darkened for years	Land and sea ecologies collapse; ozone depletion; acid rain; sky black for years
Climate	Possible brief cooling	Global warming followed by sharp cooling	Global warming followed by cosmic winter

Note: Uncertainties attend all these thumbnail descriptions, and are discussed in the text. Timescales are also open to debate, but a recurrence time of one megaton per annum is a reasonable rule of thumb to within a factor of a few.

energy range, estimates ranging from typically 10,000 years to an order of magnitude more.

There is some disagreement about the likely effects of an ocean impact in this energy range. A large earthquake-generated tsunami will carry an energy of perhaps 5 megatons, and even an inefficient coupling of say a 10,000 megaton impact to wave energy clearly has the potential to cause an immensely damaging tsunami. However, the huge waves generated in the water crater may be so steep that they break up in the open ocean. If, as suggested by some analyses, they generate tsunamis a few metres high on reaching land, then tsunami damage around ocean rims is likely the greatest single hazard of these small, relatively common impacts. A 10,000-megaton Pacific impact would on the more pessimistic analyses lead to waves 4–7 m high all around the rim, presumably with the loss of millions of lives (over 100 million people live within 20 m of sea level and 2 km from the ocean). In other studies, the wave energy dissipates relatively harmlessly before it reaches distant shores. If the pessimistic studies are correct, ocean impacts may peak, in terms of loss of life, at about this level, representing a trade-off between frequency of impact and extent of inundation of coastal lands from the resulting tsunami. The giant wave of the Hollywood movie is thus less to be feared than the more frequent few-metre wave which runs inland for a few kilometres. In terms of species extinction, however, these small impacts are not a problem.

Impacts approaching end-times ferocity probably begin at a million or 2 megatons TNT equivalent (Chapman and Morrison, 1994), corresponding to the impact of bodies a kilometre or so across. Blast and earthquake devastation are now at least continental in scale. While direct radiation from the rising fireball, peaking at 100 km altitude, is limited to a range of 1000 km by the curvature of the Earth, ballistic energy would throw hot ash to the top of the atmosphere, whence it would spread globally. Sunlight would be cut off, and food chains would collapse. The settling time of fine dust is measured in years, and commercial agriculture could not be sustained (Engvild, 2003). Lacking sunlight, continental temperatures would plummet, and heat would flow from the warmer oceans onto the cooled land masses, resulting in violent, freezing winds blowing from sea to land as long as the imbalance persisted. At these higher energies, an ocean impact yields water waves whose dimensions are comparable with the span of underwater earthquakes, and so the transport of the wave energy over global distances seems more assured, as does the hydraulic bore which could create a deep and catastrophic inundation of land.

From 10 million megatons upwards, we may be approaching the mass extinctions of species from a cocktail of prompt and prolonged effects. A land impact of this order could conceivably exterminate humanity and would surely leave signatures in the evolutionary record for future intelligent species to detect. Regionally, the local atmosphere might simply be blown into space. A rain of perhaps 10 million boulders, metre sized and upwards, would be expected over at least continental dimensions if analogy with the Martian impact crater distribution holds (McEwen et al., 2005). Major global effects include wildfires through the incinerating effect of dust thrown around the Earth; poisoning of the atmosphere and ocean by dioxins, acid rain, sulphates and heavy metals; global warming due to water and carbon dioxide injections; followed some years later by global cooling through drastically reduced insolation, all of this happening in pitch black. The dust settling process might last a year to a decade with catastrophic effects on the land and sea food chains (Alvarez et al., 1980; Napier and Clube 1979; Toon et al. 1990). At these extreme energies, multiple bombardments may be involved over several hundred thousand years or more, and in addition prolonged trauma are likely due to dustings from large, disintegrating comets. However this aspect of the hazard is less well understood and the timescales involved are more of geological than societal concern.

11.4 The role of dust

Collisions between asteroids in the main belt may on occasion produce an upsurge of dust on to the Earth (Parkin, 1985), and it has been suggested that climatic and biological effects would follow (Kortenkamp and Dermott, 1998).

There is evidence from seafloor sediments that dust showers of duration about 1.5 million years occurred about 8 and 36 million years ago (Farley et al., 2006). The former event is coincident in time with the known break-up of a large asteroid in the main belt, but the enhancement of dust was modest. The provenance of the 36 Myr dust shower is uncertain: no asteroid breakup capable of yielding this bombardment episode has been identified in the main belt.

Brief episodes (millennia rather than megayears) of cosmic dusting must occur and at some level play a role in modifying the terrestrial climate. The most massive objects to enter the near-Earth environment are rare, giant comets, 100–200 km across, at characteristic intervals of 100,000 years during a bombardment episode. Thrown into a short period, Earth-crossing orbit, such a body will disintegrate under the influence of sunlight and may generate a mass of dust equal that from the simultaneous disintegration of 10,000 Halley comets. Its emplacement could increase the annual flux of cometary dust into the Earth's atmosphere from its present 40,000 tonnes per annum to a million or more tonnes per annum, over its active lifetime of a few millennia. Meteoroids swept up by the atmosphere will ablate to smoke in the form of micron-sized particles (Klekociuk et al., 2005), which are efficient scatterers of sunlight, and whose settling time is characteristically 3–10 years. The atmospheric disintegration of incoming meteoroids into micron-sized aerosols will yield a significant reduction of sunlight reaching the surface of the Earth. Climatic effects would seem unavoidable (Clube et al 1996; Hoyle and Wickramasinghe, 1978; Napier, 2001). Thus mass extinction need not be a single, large impact: multiple bombardments and a series of sharp cooling episodes also provide a reasonable astronomical framework.

The fossil remains of a past large comet can still be discerned in the inner interplanetary environment: it has long been recognised that the progenitor of Comet Encke, and the associated Taurid meteor stream, was such a body (Whipple, 1967). From reconstruction of the initial orbits (Steel and Asher, 1998) it appears that the original comet must have been at least 10 km in diameter, and may have come to the end of its principal activity some 10,000 years ago, with continuing significant activity until about 5000 years BP. With an orbital period of 3.3 ± 0.2 years, this large, disintegrating comet must have been a brilliant object in the Neolithic night sky. Orbital precession would ensure that, at intervals of about 2500 years, the Earth would be subjected to close encounters and annual meteor storms of an intensity outwith modern experience. There may after all be a scientific underpinning to the myths of celestial apocalypse! This Taurid progenitor may have been the offshoot of a much larger body: the zodiacal cloud, a disc of interplanetary dust particles which includes the Taurids, is about two orders of magnitude too massive in relation to current replenishing sources (Hughes, 1996). A recent injection of mass amounting to at least 2000 times the present mass of Comet Halley is implied, corresponding to a comet 150 km in diameter (Hughes, 1996).

The question arises whether material within this complex still constitutes a global risk significantly above background. That there are still concentrations of material within it is established by occasional fireball swarms, by mediaeval records and by seismically detected impacts on the Moon (Table 11.1). Over the period 1971–1975 during which lunar seismometers operated, strong boulder swarms were detected, of a few days duration, coincident with the maxima of the daytime β Taurids, the Perseids, the Leonids and the Geminids. But (see Table 11.1) also close to the peaks of these annual showers were the 1908 Tunguska impact (30 June), a 1930 fall in the megaton range in the Amazon forest (13 August) and a similarly energetic impact in British Guiana in 1935 (11 December 1935). It seems to be stretching coincidence too far to assume that these are chance juxtapositions. Whether meteoroids in say the 1000-megaton range might be concentrated within meteor showers in significant numbers is as yet an unanswered question.

The existence of this material has led to the suggestion that rapid, short-lived, otherwise unexplained coolings of the Earth, taking place at characteristic intervals of a few millennia, might be due to interactions with sub-kilometre meteoroids within the Taurid complex. The most recent of these occurred over 536–545 AD. Tree ring data point to a sudden climatic change, apparently of global reach, accompanied by a dry fog of duration 12–18 months. No chemical signature diagnostic of a volcano is recorded in ice cores over this period. This event – and a similar one at 2345 BC – suggested to Baillie (1994, 1999) that a comet impact was involved. This was modelled by Rigby et al. (2004) as an airburst, wherein vapourized comet material is ejected as a plume and falls back on to the top of the atmosphere as small, condensed particles. They found that a comet of radius of 300 m could reduce sunlight by about 4% which, they considered, might account for both the historical descriptions and the sudden global cooling. The effects of such a cosmic winter can be summarized as crop failures followed by famine. This would be most immediate in the third world, but a cooling event lasting more than a year would also hit the developed countries (Engvild, 2003).

11.5 Ground truth?

There is as yet no clear consensus on the relative importance of dark, dormant comets versus stony asteroids in the overall impact rate. Neither, lacking full exploration of the sub-kilometre regime, can one be confident about the total contemporary hazard presented by bodies in this range. Attempts have been made to correct for the discovery bias against very dark objects, but almost by definition such corrections are most uncertain where they are most needed.

Such models have been used to suggest that impacts of 10 megatons upwards occur on Earth every 2000–3000 years (Stuart and Binzel, 2004), and yet the Tunguska impact took place only 100 years ago. Likewise an impact of energy 1000 megatons or more was predicted to occur every 60,000 years (Morbidelli et al., 2002), and yet within 2 years a 1000 megaton asteroid was discovered which will pass within six Earth radii in 2029. In the same vein active comets more than 7 km across are sometimes estimated to hit the Earth once every 3 billion years, and yet a low-activity comet of this size, IRAS-Araki-Alcock, passed within 750 Earth radii in 1983, consistent with an impact rate 200 times higher. Hughes (2003) examined the distribution of near-miss distances of known NEOs passing the Earth, carrying out this exercise for all known 2002 close encounters. He found that impacts in the Tunguska class are expected every 300 years, and in the 1000-megaton range upwards every 500–5000 years. This 'ground truth' is based on a small number of recent close encounters but does indicate that there is still considerable uncertainty in estimates of impact rate (Asher et al., 2005).

The ultimate ground truth is, of course, to be found in the ground. Courty et al. (2005), in a series of detailed sedimentological studies, finds evidence for a widespread spray of hot, fine impact ejecta throughout tropical Africa, the near East and West Asia, which she dates to around 2600–2300 BC and associates with an abrupt environmental change at that period. Abbott et al. (2005), in a study of West Antarctic Siple Dome ice cores, obtained a variety of data which they find to be consistent with impact ejecta from a 24 km crater, Mahuika, on the southern New Zealand shelf. These latter anomalies are dated at around 1443 AD and a large impact at so recent a date seems very unlikely, if only because its effects would be widely felt; on the other hand deposits from a 130 m high megatsunami in Jervis Bay, Australia, have been dated to 1450 ± 50 AD. These lines of work are quite new and still have to go through the full critical mill; if they continue to hold up, the case for a high current risk will in effect have been made.

11.6 Uncertainties

Multiplying the low probability of an impact by its large consequences, one finds that the per capita impact hazard is at the level associated with the hazards of air travel and the like. Unlike these more mundane risks, however, the impact hazard is unbounded: a big one could end civilization. The effects of cosmic dusting are much less well understood and have hardly been studied; they lack the drama of a great, incinerating impact, and are difficult to assess and handle; but it is not obvious that in terms of frequency or consequences they matter less.

Although the concept of the Earth as a bombarded planet is very old, it is only in the last twenty years that it has become anything like 'mainstream', and there remain qualitative uncertainties and areas of disagreement. We do not know where the bulk of the mass coming into the Halley system from the Oort cloud is going and it remains possible that a large population of unstoppable dark bodies is being missed. A significant population of such 'stealth comets' would be difficult to map out since the bodies are in high eccentricity orbits and spend most of their time beyond Mars. There are uncertainties too about the role of the residual debris from the large, short-period comet which was active as recently as 5000 yr BP. This debris contains concentrations of material which the Earth encounters on various timescales from annual to millennial; whether they include bodies large enough to constitute, say, a significant tsunami hazard, or a short-lived global cooling, cannot be determined from the available evidence.

Progress in this field has been impressively rapid since the 1970s, however, and will no doubt continue apace. There is little doubt that, especially with the mapping out of interplanetary bodies down to the sub-kilometre level, many of these uncertainties will be reduced or disappear.

Suggestions for further reading

Baillie, M.G.L. (1999). *Exodus to Arthur* (London: Batsford). A popular-level discussion of dendrochronological evidence for impacts in recent geological and historical times.

Clube, S.V.M. and Napier, W.M. (1990). *The Cosmic Winter* (Oxford: Basil Blackwell Ltd.). A popular exposition of the theory of coherent catastrophism, arguably the 'worst case' scenario from the point of view of this book.

Gehrels, T. (ed.) (1994). *Hazards Due to Comets and Asteroids* (Tucson: University of Arizona Press). The most general treatment of the impact hazard thus far. While a bit out of date, this comprehensive volume is still the best point-of-entry for the present chapter's concerns.

Kneević, Z. and Milani, A. (eds.) (2005). *Dynamics of Populations of Planetary Systems*, Proc. IAU colloq. 197 (Cambridge: Cambridge University Press). A recent overview of the dynamical side of the impactor populations.

References

Abbott, D., Biscaye, P., Cole-Dai, J., and Breger, D. (2005). Evidence from an ice core of a large impact circa 1443 AD. *EOS Trans. AGU*, **86**, 52, *Fall Meet. Suppl.*, Abstract PP31C-05.

Alvarez, L.W., Alvarez, W., Asaro, F., and Michel, H.V. (1980). Extraterrestrial cause for the Cretaceous-Tertiary extinction. *Science*, **208**, 1095–1108.

Asher, D.J., Bailey, M.E., Emel'yanenko, V., and Napier, W.M. (2005). Earth in the cosmic shooting gallery. *The Observatory*, **125**, 319–322.

Baillie, M.G.L. (1994). Dendrochronology raises questions about the nature of the a.d. 536 dust-veil event. *Holocene,* **4**, 212–217.

Baillie, M.G.L. (1999). *Exodus to Arthur* (London: Batsford).

Bierhaus, E.B., Chapman, C.R., and Merline, W.J. (2005). Secondary craters on Europa and implications for crater surfaces. *Nature,* **437**, 1125–1127.

Chapman, C. and Morrison, D. (1994). Impacts on the Earth by asteroids and Comets. Assessing the hazard. *Nature,* **367**, 33–39.

Clube, S.V.M., Hoyle, F., Napier, W.M., and Wickramasinghe, N.C. (1996). Giant comets, evolution and civilisation. *Astrophys. Space Sci.,* **245**, 43–60.

Clube, S.V.M. and Napier, W.M. (1996). Galactic dark matter and terrestrial periodicities. *Quarterly J. Royal Astron. Soc.,* **37**, 617–642.

Courty, M.-A., et al. (2005). Sequels on humans, lands and climate of the 4-kyr BP impact across the Near East. Presented at European Geosciences Union, Symposium. CL18, Vienna, Austria.

Emel'yanenko, V.V. and Bailey, M.E. (1998). Capture of Halley-type comets from the near-parabolic flux. *MNRAS,* **298**, 212–222.

Engvild, K.C. (2003). A review of the risks of sudden global cooling and its effects on agriculture. *Agric. Forest Meteorol.,* **115**, 127–137.

Farley, K.A., Vokrouhlicky, D., Bottke, W.F., and Nesvorny, D. (2006). A late Miocene dust shower from the break-up of an asteroid in the main belt. *Nature,* **439**, 295–297.

Fernandez, Y.R., Jewitt, D.C., and Sheppard, S.S. (2005). Albedos of asteroids in Comet-like orbits. *Astrophys. J.,* **130**, 308–318.

Helin, E.F. and Shoemaker, E.M. (1979). Palomar planet-crossing asteroid survey, 1973–1978. *Icarus,* **40**, 321–328.

Hoyle, F. and Wickramasinghe, N.C. (1978). Comets, ice ages, and ecological catastrophes. *Astrophys. Space Sci.,* **53**, 523–526.

Hughes, D.W. (1996). The size, mass and evolution of the Solar System dust cloud. *Quarterly J. Royal Astron Soc.,* **37**, 593–604.

Hughes, D.W. (2003). The approximate ratios between the diameters of terrestrial impact craters and the causative incident asteroids. *MNRAS,* **338**, 999–1003.

Klekociuk, A.R., Brown, P.G., Pack, D.W., Revelle, D.O., Edwards, W.N., Spalding, R.E., Tagliaferri, E., Yoo, B.B., and Zagevi, J. (2005). Meteoritic dust from the atmospheric disintegration of a large meteoroid. *Nature,* **436**, 1132–1135.

Kortenkamp, S.J. and Dermott, S.F.A. (1998). A 100,000-year periodicity in the accretion rate of interplanetary dust. *Science,* **280**, 874–876.

Levison, H.F. et al. (2002). The mass disruption of Oort cloud comets. *Science,* **296**, 2212–2215.

Matese, J.J. et al. (1995). Periodic modulation of the Oort cloud comet flux by the adiabatically changing tide. *Icarus,* **116**, 255–268.

McCrea, W.H. (1981). Long time-scale fluctuations in the evolution of the Earth. *Proc. Royal Soc. London,* **A**375, 1–41.

McEwen, A.S., Preblich, B.S., Turke, E.P., Artemieva, N.A., Golombek, M.P., Hurst, M., Kirk, R.L., Burr, D.M., and Christensen, P.R. (2005). The rayed crater Zunil and interpretations of small impact craters on Mars. *Icarus,* **176**, 351–381.

Menichella, M., Paolicci, P., and Farinella, P. (1996). The main belt as a source of near-Earth asteroids. *Earth, Moon, Planets,* **72**, 133–149.

Morbidelli, A., Jedicke, R., Bottke, W.F., Michel, P., and Tedesco, E.F. (2002). From magnitudes to diameters: the albedo distribution of near Earth objects and the Earth collision hazard. *Icarus*, **158**, 329–342.

Napier, W.M. (2001). Temporal variation of the zodiacal dust cloud. *MNRAS*, **321**, 463.

Napier, W.M. (2006). Evidence for cometary bombardment episodes. *MNRAS*, **366**(3), 977–982.

Napier, W.M. and Clube, S.V.M. (1979). A theory of terrestrial catastrophism. *Nature*, **282**, 455–459.

Napier, W.M., Wickramasinghe, J.T., and Wickramasinghe, N.C. (2004). Extreme albedo comets and the impact hazard. *MNRAS*, **355**, 191–195.

Nelson, R.M., Soderblom, L.A., and Hapke, B.W. (2004). Are the circular, dark features on Comet Borrelly's surface albedo variations or pits? *Icarus*, **167**, 37–44.

Neukum, G. and Ivanov, B.A. (1994). Crater size distributions and impact probabilities on Earth from lunar, terrestrial-planet, and asteroid cratering data. In Gehrels, T. (ed.), *Hazards Due to Comets and Asteroids*, p. 359 (Tucson: University of Arizona Press).

Nurmi, P., Valtonen, M.J., and Zheng, J.Q. (2001). Periodic variation of Oort Cloud flux and cometary impacts on the Earth and Jupiter. *MNRAS*, **327**, 1367–1376.

Parkin, D.W. (1985). Cosmic spherules, asteroidal collisions and the possibility of detecting changes in the solar constant. *Geophys. J.*, **83**, 683–698.

Rickman, H. (2005). Transport of comets to the inner solar system. In Knesevic, Z. and Milani, A. (eds.), *Dynamics of Populations of Planetary Systems*, Proc. IAU colloq. no. 197.

Rickman, H., et al. (2001). The cometary contribution to planetary impact rates. In Marov, M. and Rickman, H. (eds.), *Collisional Processes in the Solar System* (Kluwer).

Rigby, E., Symonds, M., and Ward-Thompson, D. (2004). A comet impact in AD 536? *Astron. Geophys.*, **45**, 23–26.

Rivkin, A.S., Binzel, R.P., and Bus, S.J. (2005). Constraining near-Earth object albedos using neo-infrared spectroscopy. *Icarus*, **175**, 175–180.

Steel, D.I., Asher, D.J., Napier, W.M., and Clube, S.V.M. (1990). Are impacts correlated in time? In Gehrels, T. (ed.), *Hazards Due to Comets and Asteroids*, p. 463 (Tucson: University of Arizona).

Steel, D.I., Asher, D.J. 1998: On the possible relation between the Tunguska bolide & comet encke, in: planetary & space science, **46**, pp 205–211.

Stokes, G.H. et. al. (2003). Report of the Near-Earth Object Science Definition Team. NASA, http://neo.jpl.nasa.gov/neo/neoreport030825.pdf

Stuart, J.S. and Binzel, R.P. (2004). NEO impact risk overrated? Tunguska events once every 2000–3000 years? *Icarus*, **170**, 295–311.

Toon, O.B. et al. (1990). Environmental perturbations caused by asteroid impacts. In Gehrels, T. (ed.), *Hazards Due to Comets and Asteroids*, p. 791 (Tucson: University of Arizona).

Whipple, F.L. (1967). The Zodiacal Light and the Interplanetary Medium. NASA SP-150, p. 409, Washington, DC.

·12·

Influence of supernovae, gamma-ray bursts, solar flares, and cosmic rays on the terrestrial environment

Arnon Dar

12.1 Introduction

Changes in the solar neighbourhood due to the motion of the sun in the Galaxy, solar evolution, and Galactic stellar evolution influence the terrestrial environment and expose life on the Earth to cosmic hazards. Such cosmic hazards include impact of near-Earth objects (NEOs), global climatic changes due to variations in solar activity and exposure of the Earth to very large fluxes of radiations and cosmic rays from Galactic supernova (SN) explosions and gamma-ray bursts (GRBs). Such cosmic hazards are of low probability, but their influence on the terrestrial environment and their catastrophic consequences, as evident from geological records, justify their detailed study, and the development of rational strategies, which may minimize their threat to life and to the survival of the human race on this planet. In this chapter I shall concentrate on threats to life from increased levels of radiation and cosmic ray (CR) flux that reach the atmosphere as a result of (1) changes in solar luminosity, (2) changes in the solar environment owing to the motion of the sun around the Galactic centre and in particular, owing to its passage through the spiral arms of the Galaxy, (3) the oscillatory displacement of the solar system perpendicular to the Galactic plane, (4) solar activity, (5) Galactic SN explosions, (6) GRBs, and (7) cosmic ray bursts (CRBs). The credibility of various cosmic threats will be tested by examining whether such events could have caused some of the major mass extinctions that took place on planet Earth and were documented relatively well in the geological records of the past 500 million years (Myr).

12.2 Radiation threats

12.2.1 Credible threats

A credible claim of a global threat to life from a change in global irradiation must first demonstrate that the anticipated change is larger than the periodical changes in irradiation caused by the motions of the Earth, to which terrestrial life has adjusted itself. Most of the energy of the sun is radiated in the visible range. The atmosphere is highly transparent to this visible light but is very opaque to almost all other bands of the electromagnetic spectrum except radio waves, whose production by the sun is rather small. The atmosphere protects the biota at ground level from over-exposure to high fluxes of extraterrestrial gamma-rays, X-ray and UV light. Because of this atmospheric protection, life has not developed immunity to these radiations (except species that perhaps were exposed elsewhere during their evolution to different conditions, such as *Deinoccocus radiodurance*), but has adapted itself to the normal flux levels of radiations that penetrate the atmosphere. In particular, it has adapted itself to the ground level solar irradiance, whose latitudinal and seasonal redistribution undergoes long-term quasi-periodical changes, the so-called Milankovitch cycles, due to quasi-periodical variations in the motions and orientation of the Earth. These include variation in the Earth's eccentricity, tilt of the Earth's axis relative to the normal to the plane of the ecliptic and precession of the Earth's axis. Milutin Milankovitch, the Serbian astronomer, is generally credited with

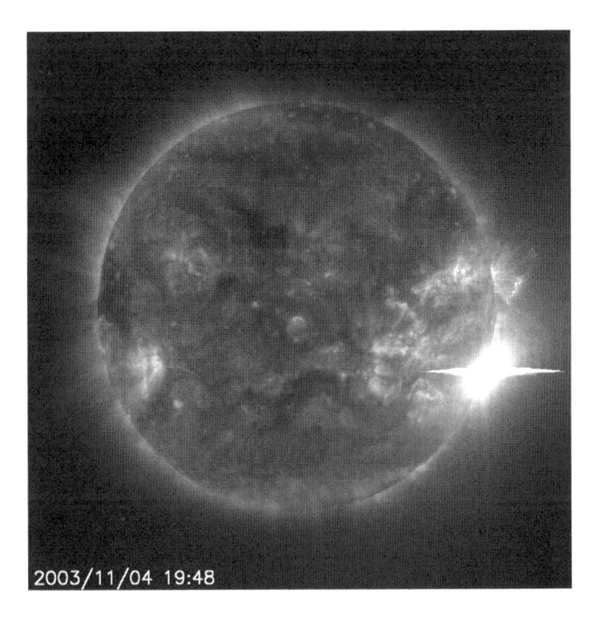

Fig. 12.1 The intense solar flare of 4 November 2003. A giant sun spot region lashed out with an intense solar flare followed by a large coronal mass ejection (CME) on 4 November 2003. The flare itself is seen here at the lower right in an extreme ultraviolet image from the sun-staring SOHO spacecraft's camera. This giant flare was among the most powerful ever recorded since the 1970s, the third such historic blast from AR10486 within two weeks. The energetic particle radiation from the flare did cause substantial radio interference.
Credit: SOHO-EIT Consortium, ESA, NASA

calculating their magnitude and the times of increased or decreased solar radiation, which directly influence the Earth's climate system, thus impacting the advance and retreat of the Earth's glaciers. The climate change, and subsequent periods of glaciation resulting from these variables, is not due to the total amount of solar energy reaching Earth. The three Milankovitch Cycles impact the seasonality and location of solar energy around the Earth, thus affecting contrasts between the seasons. These are important only because the Earth has an asymmetric distribution of land masses, with virtually all (except Antarctica) located in/near the Northern Hemisphere.

Even when all of the orbital parameters favour glaciation, the increase in winter snowfall and decrease in summer melt would barely suffice to trigger glaciation. Snow and ice have a much larger albedo (i.e., the ratio of reflected

Fig. 12.2 Comet Shoemaker-Levy 9 Collision with Jupiter. From 16 through 22 July 1994, pieces of the Comet Shoemaker-Levy 9 collided with Jupiter. The comet consisted of at least twenty-one discernable fragments with diameters estimated at up to two kilometres. The four frames show the impact of the first of the twenty odd fragments of Comet Shoemaker-Levy 9 into Jupiter. The upper left frame shows Jupiter just before impact. The bright object to the right is its closest satellite Io, and the fainter oval structure in the southern hemisphere is the Great Red Spot. The polar caps appear bright at the wavelength of the observations, 2.3 μm, which was selected to maximize contrast between the fireball and the jovian atmosphere. In the second frame, the fireball appears above the southeast (lower left) limb of the planet. The fireball flared to maximum brightness within a few minutes, at which time its flux surpassed that of ten. The final frame shows Jupiter approximately twenty minutes later when the impact zone had faded somewhat.
Credit: Dr. David R. Williams, NASA Goddard Space Flight Center

Fig. 12.3 The supernova remnant Cassiopeia A. Cas A is the 300-year-old remnant created by the SN explosion of a massive star. Each Great Observatory image highlights different characteristics of the remnant. Spitzer Space Telescope reveals warm dust in the outer shell with temperatures of about 10°C (50°F), and Hubble Space Telescope sees the delicate filamentary structures of warmer gases about 10,000°C. Chandra X-ray observatory shows hot gases at about 10 million degrees Celsius. This hot gas was created when ejected material from the SN smashed into surrounding gas and dust at speeds of about 10 million miles per hour.
Credit: NASA/CXC/MIT/UMass Amherst/M.D.Stage et al.

to incident electromagnetic radiation) than ground and vegetation (if the Earth was covered in ice like a giant snowball, its albedo would be approximately 0.84). Snow cover and ice masses tend to reflect more radiation back into space, thus cooling the climate and allowing glaciers to expand. Likewise supernovae (SNe), GRBs, solar flares, and cosmic rays had large influence on the terrestrial environment.

The 1912 Milankovitch theory of glaciation cycles is widely accepted since paleoclimatic archives contain strong spectral components that match the Milankovitch cycles. However, it was recently argued that high precision paleoclimatic data have revealed serious discrepancies with the Milankovitch model that fundamentally challenge its validity and reopen the question of what causes the glacial cycles. For instance, Kirkby et al. (2004) proposed that the ice ages are initially driven not by insolation cycles but by cosmic ray changes, probably through their effect on clouds. Even if the cause of the glacial cycles is still debated, changes in global irradiation of astronomical origin must be larger than the orbital modulation of the solar irradiation in order to pose a credible threat to terrestrial life.

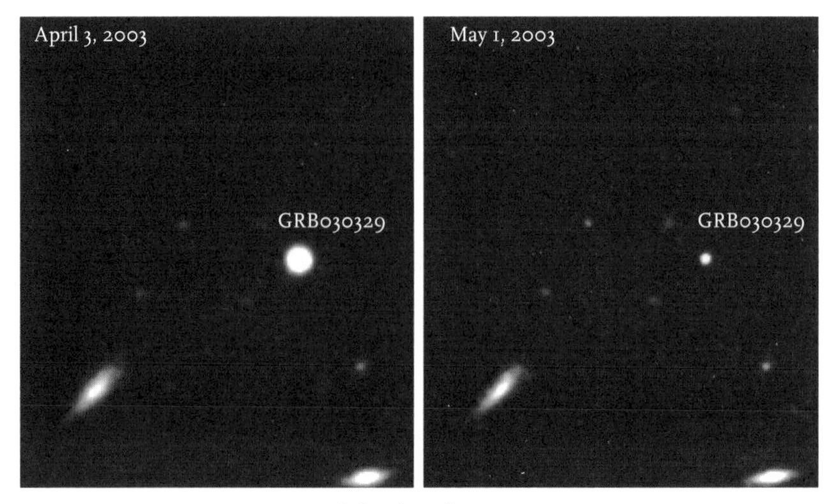

Fig. 12.4 The afterglow of the gamma-ray burst (GRB) 030329: Images of the fading optical afterglow of the GRB 030329 that took place on 29 March 2003 taken by the very large telescope (VLT) of the European Southern Observatory (ESO) in Chile on 3 April 2003 and 1 May 2003. The image taken on 1 May is dominated by an underlying supernova that produced the GRB. The discovery of the underlying supernova SN203dh convinced the majority of the astrophysicists community that 'long-duration' GRBs are produced by highly relativistic jets as long advocated by the Cannonball model of GRBs. The underlying supernova was first discovered spectroscopically in the fading afterglow of GRB 0302329 10 days after the GRB took place, as predicted by Dado et al. (2003) from their study of the early afterglow of GRB 030329.
Credit: European Southern Observatory.

12.2.2 Solar flares

Solar flares are the most energetic explosions in the solar system. They occur in the solar atmosphere. The first solar flare recorded in astronomical literature, by the British astronomer Richard C. Carrington, occurred on 1 September 1859. Solar flares lead to the emission of electromagnetic radiation, energetic electrons, protons and atomic nuclei (solar cosmic rays) and a magnetized plasma from a localized region on the sun. A solar flare occurs when magnetic energy that has built up in the solar atmosphere is suddenly released. The emitted electromagnetic radiation is spread across the entire electromagnetic spectrum, from radio waves at the long wavelength end, through optical emission to X-rays and gamma-rays at the short wavelength end. The energies of solar cosmic rays reach a few giga electron volts = 10^9 ev [1 ev = 1 6021753(14).10^{-13} J]. The frequency of solar flares varies, from several per day when the sun is particularly active to less than one per week when the sun is quiet. Solar flares may take several hours or even days to build up, but the actual flare takes only a matter of minutes to release its energy.

The total energy released during a flare is typically of the order 10^{27} erg s^{-1}. Large flares can emit up to 10^{32} erg. This energy is less than one-tenth of the total energy emitted by the sun every second ($L_{.} = 3.84 \times 10^{33}$ erg s^{-1}). In the unlikely event that all the magnetic field energy in the solar atmosphere is radiated in a single solar flare, the solar flare energy cannot exceed $\sim B^2 R^3/12 \sim 1.4 \times 10^{33}$ erg where $B \sim 50$ Gauss is the strength of the sun's dipole surface magnetic field and $R_{.} = 7 \times 10^{10}$ cm is the solar radius. Even this energy is only approximately one-third of the total energy emitted by the sun every second. Thus, individual solar flares are not energetic enough to cause global catastrophes on planet Earth. However, solar flares and associated coronal mass ejections strongly influence our local space weather. They produce streams of highly energetic particles in the solar wind and the Earth's magnetosphere that can present radiation hazards to spacecraft and astronauts. The soft X-ray flux from solar flares increases the ionization of the upper atmosphere, which can interfere with short-wave radio communication, and can increase the drag on low orbiting satellites, leading to orbital decay. Cosmic rays that pass through living bodies do biochemical damage. The large number of solar cosmic rays and the magnetic storms that are produced by large solar flares are hazardous to unprotected astronauts in interplanetary space. The Earth's atmosphere and magnetosphere protect people on the ground.

12.2.3 Solar activity and global warming

Global temperature has increased over the twentieth century by approximately 0.75°C relative to the period 1860–1900. Land and sea measurements independently show much the same warming since 1860. During this period, the concentration of CO_2 in the Earth's atmosphere has increased by approximately 27% from 290 to 370 parts per million (ppm). This level is considerably higher than at any time during the last 800,000 years, the period for which reliable data has been extracted from ice cores. This increase in the CO_2 concentration in the atmosphere is widely believed to be anthropogenic in origin, that is, derived from human activities, mainly fossil fuel burning and deforestation. Recently, the Intergovernmental Panel on Climate Change (IPCC) concluded that 'most of the observed increase in globally averaged temperatures since the mid-twentieth century is very likely due to the observed increase in anthropogenic greenhouse gas concentrations' via the greenhouse effect (the process in which the emission of infrared radiation by the atmosphere warms a planet's surface, such as that of Earth, Mars and especially Venus, which was discovered by Joseph Fourier in 1829). Relying on climate models, scientists project that global surface temperatures are likely to increase by 1.1–6.4°C until 2100. An increase in global temperatures is expected to cause other changes, including sea level rise, increased intensity of extreme weather events, and changes in the amount and pattern of precipitation. Other effects of global warming include changes in agricultural yields, glacier retreat, species extinctions and increases in the ranges of disease vectors.

However, although scientists essentially agree that mankind should drastically reduce the emission of greenhouse gases and other pollutants, there are scientists who disagree with the IPCC conclusion that there is substantial evidence to prove that the increase of CO_2 concentration in the atmosphere from anthropogenic sources and of other greenhouse gases are the primary cause for global warming. They point out that half of the increase in global temperature over the twentieth century took place in the beginning of the century, long before the bulk of the human influence took place. Moreover, the Earth has experienced pre-human large warming and cooling many times in the past as inferred from geological records and global temperature proxies (variables used to infer past global temperature), such as the concentration of heavy water molecules (D_2O and $H_2{}^{18}O$) in ice cores: The relative rate of evaporation of these molecules from seawater compared to light water molecules (H_2O), increases with temperature. This increases the concentration of the heavy water molecules in precipitation, which solidify into ice over the north and south poles of the Earth. In particular, the ice cores of the Vostok and Epicaant arctic sites, which date back by 740,000 years, reveal eight previous glacial cycles with strong variation in temperature, up to a decrease by −8°C (relative to the present temperature) during the coldest ice ages and an increase by +3°C during the warmest periods. The change in the concentration of atmospheric CO_2 has followed closely the change in global temperature. Supporters of the anthropogenic origin of global warming argue that dramatic increase in greenhouse gases from natural sources was responsible for past global warming, while others suggest that the release of large quantities of CO_2 by the oceans was caused by the increase in their temperature by global warming. Unfortunately, so far no precise data have been presented that allow us to determine which event preceded the other, the increase in atmospheric CO_2 or the global warming.

Global warming remains an active field of research, although the scientific consensus is that greenhouse gases produced by human activity are responsible for it. However, a consensus is not a proper substitute for a scientific proof. Other hypotheses have been suggested to explain the observed increase in mean global temperature and should be examined scientifically. Perhaps the one that looks most credible (to the author) is the hypothesis that the current global warming is largely the result of the reduction in the flux of cosmic rays that reach the atmosphere by an increased solar activity (e.g., Shaviv, 2005; Svensmark, 1998). This possibility is discussed shortly in Section 12.3.

12.2.4 Solar extinction

The sun is about 4.5 billion years old. It will continue to shine for another 5 billion years. But it will start to run out of hydrogen fuel in its core in less than 2 billion years from now. Then its core will contract and become hot enough for helium fusion to occur in it and hydrogen fusion in a shell around its growing helium core. Owing to the growing radiation pressure in the burning shell, the sun will begin expanding into a red giant. This fast phase will last less than 10 million years. When the sun becomes a red giant, Mercury and Venus will be swallowed up by the sun and perhaps the Earth will be too. Even if the Earth will not be swallowed up, conditions on its surface will make it impossible for life to exist. The sun's increased luminosity will heat the Earth's surface so much that the water of oceans and atmosphere will evaporate away. In fact, in only 1 or 2 billion years, prior to the red giant phase, the energy output of the sun will increase to a point where the Earth will probably become too hot to support life.

12.2.5 Radiation from supernova explosions

The most violent events likely to have occurred in the solar neighbourhood during geologic and biological history are SN explosions. Such explosions are the violent death of either massive stars following gravitational collapse of their core (core-collapse supernovae) or white dwarfs in binary systems whose mass increases by accretion beyond the Chandrasekhar mass limit (thermonuclear SN).

Core-collapse supernova takes place when the nuclear fuel in the core of a massive star of more than eight solar masses ($M > 8M_{\odot}$) has been exhausted and can no longer produce the thermal pressure that balances the gravitational pressure of the outlying layers. Then the core collapses into a neutron star or a stellar black hole and releases a huge amount of gravitational energy ($\sim 3 \times 10^{53}$ erg), most of which is converted to neutrinos and only a few percent into kinetic energy of the ejected stellar envelope, which contains radio isotopes whose decay powers most of its radiation.

Thermonuclear supernovae involve the thermonuclear explosion of a white dwarf star in a binary star system. A white dwarf is the end point in the evolution for stars with mass less than eight solar masses ($M < 8M_{\odot}$). It is usually made of carbon or oxygen. Its mass cannot exceed 1.4 times the mass of the sun. A white dwarf in a binary star system can accrete material off its companion star if they are close to each other because of its strong gravitational pull. The in-falling matter from the companion star causes the white dwarf to cross the 1.4 solar-mass limit (a mass called the Chandrasekhar limit after its discoverer) and collapse gravitationally. The gravitational energy release increases the temperature to a level where the carbon and oxygen nuclei fuse uncontrollably. This results in a thermonuclear explosion that disrupts the entire star.

If a supernova explosion occurred sufficiently close to the Earth it could have dramatic effects on the biosphere. Potential implications of a nearby SN explosion for the Earth's biosphere have been considered by a number of authors (Ellis and Schramm, 1995; Ellis et al., 1996; Ruderman, 1979) and later work has suggested that the most important effects might be induced by their cosmic rays. In particular, their possible role in destroying the Earth's ozone layer and opening the biosphere to the extent of irradiation by solar UV radiation has been emphasized (Ellis and Schramm, 1995; Ellis et al., 1996). We shall first consider the direct radiation threats from SN explosions.

Among the new elements produced in core-collapse and thermonuclear SN explosions is radioactive nickel, which liberates huge amounts of energy inside the debris. Most of this energy is absorbed within the debris and radiated as visible light. However, the SN light does not constitute a high risk hazard. The brightest supernovae reach a peak luminosity of approximately 10^{43} erg s^{-1} within a couple of weeks after the explosion, which then declines roughly exponentially with a half-life time of 77 days (the half-life time of the radioactive cobalt that is produced by the decay of nickel). Such a luminosity at a distance of 5 parsecs from the Earth over a couple of weeks adds approximately 1% to the solar radiation that reaches the Earth and has no catastrophic consequences whatsoever. Moreover, the mean rate of Galactic SN explosions is approximately 1 in 50 years (van den Bergh and Tammann, 1991). Most of the SN explosions occur at distances from the Galactic centre much smaller than the radius of the solar orbit. Using the observed distribution of Galactic SN remnants, and the mean rate of SN explosions, the chance probability that during the next 2 billion years (before the red giant phase of the sun) the solar system in its Galactic motion will pass within 15 light years (LY) from an SN explosion is less than 10^{-2}.

The direct threats to life on the Earth from the UV, X-ray and gamma-ray emission from SN explosions and their remnants are even smaller because the atmosphere is opaque to these radiations. The only significant threat is from the possible stripping of the Earth's ozone layer followed by the penetration of UV radiation and absorption of visible sunlight by NO_2 in the atmosphere. However, the threat from supernovae more distant than 30 LY is not larger than that from solar flares. The ozone layer has been frequently damaged by large solar flares and apparently has recovered in relatively short times.

12.2.6 Gamma-ray bursts

Gamma-ray bursts are short-duration flares of MeV gamma-rays that occur in the observable universe at a mean rate of approximately 2–3 per day (e.g., Meegan and Fishman, 1995). They divide into two distinct classes. Nearly 75% are long-duration soft-spectrum bursts which last more than two seconds; the rest are short-duration hard-spectrum bursts (SHBs) that last less than two seconds. There is mounting evidence from observations of the optical afterglows of long-duration GRBs that long bursts are produced by highly relativistic jets ejected during the death of massive stars in SN explosions (e.g., Dar, 2004 and references therein). The origin of SHBs is only partially known. They are not produced in SN explosion of any known type and their energy is typically three orders of magnitude smaller.

Thorsett (1995) was the first to discuss the potential effects on the atmosphere of the Earth and the damage to the biota from the hard X-rays and gamma-rays from a Galactic GRB pointing towards the Earth, while Dar et al. (1998) suggested that the main damage from Galactic GRBs arises from the cosmic rays accelerated by the jets that produce the GRBs (Shaviv and Dar, 1995).Whereas the fluxes of gamma-rays and X-rays from Galactic GRBs that illuminate the Earth and their frequency can be reliably estimated from GRB observations and their association with SN explosions, this is not the case for cosmic rays whose radiation must be estimated from debatable models. Consequently, although the effects of cosmic ray illumination may be much more devastating than those of the gamma-rays and X-rays from the same event, other authors (e.g., Galante and Horvath, 2005; Melott et al., 2004; Scalo and Wheeler, 2002; Smith et al., 2004; Thomas et al., 2005) preferred to concentrate mainly on the effects of gamma-ray and X-ray illumination.

The Galactic distribution of SN explosions is known from the distribution of their remnants. Most of these SN explosions take place in the Galactic disk at Galactocentric distances much shorter than the distance of the Earth from the Galactic centre. Their mean distance from the Earth is roughly 25,000 LY. From the measured energy fluence of GRBs (energy that reaches Earth per unit area) of known red shift it was found that the mean radiation energy emitted in long GRBs is approximately $5 \times 10^{53}/\Delta\Omega/4\pi$ erg, where $\Delta\Omega$ is the solid angle illuminated by the GRB (the beaming angle). The radiation energy per solid angle of short GRBs is smaller by approximately two to three orders of magnitude.

If GRBs in our Galaxy and in external galaxies are not different, then the ratio of their fluences scale like the inverse of their distance ratio squared. Should a typical Galactic GRB at a distance of $d = 25,000$ LY point in the direction of the Earth, its hemisphere facing the GRB will be illuminated by gamma-rays with a total fluence $F_\gamma \sim 5 \times 10^{53}/4\pi\, d^2 \sim 4 \times 10^7$ erg s^{-1} within typically 30s. The gamma-rays' energy and momentum will be deposited within typically 70 g cm^{-2} of the upper atmosphere (the total column density of the atmosphere at sea level is $\sim$1000 g cm^{-2}). Such fluxes would destroy the ozone layer and create enormous shocks going through the atmosphere, provoke giant global storms, and ignite huge fires. Smith et al. (2004) estimated that a fraction between 2×10^{-3} and $4 \times 10^{-2}$of the Gamma-ray fluence will be converted in the atmosphere to an UV fluence at ground level. The UV radiation is mainly harmful to DNA and RNA molecules that absorb this radiation.

The lethal dose for a UV exposure, approximately 10^4 erg cm^{-2}, makes a GRB at 25,000 LY potentially highly lethal (e.g., Galante and Horvath, 2005) to the hemisphere illuminated by the GRB. But, protection can be provided by habitat (underwater, underground, under-roof, shaded areas) or by effective skin covers such as furs of animals or clothing of human beings. The short duration of GRBs and the lack of an nearly warning signal, however, make protection by moving into the shade or a shelter, or by covering up quickly, unrealistic for most species.

It should be noted that the MeV gamma-ray emission from GRBs may be accompanied by a short burst of very high energy gamma-rays, which, so far, could not be detected, either by the gamma-ray and X-ray satellites (CGRO, BeppoSAX, HETE, Chandra, XMMNewton, Integral, SWIFT and the interplanetary network), or by ground-level high energy gamma-ray telescopes such as HESS and Magic (due to timelag in response). Such bursts of GeV and TeV gamma-rays, if produced by GRBs, might be detected by the Gamma-ray Large Area Space Telescope (GLAST), which will be launched into space in 16.V.2008. GeV-TeV gamma-rays from relatively nearby Galactic GRBs may produce lethal doses of atmospheric muons.

12.3 Cosmic ray threats

The mean energy density of Galactic cosmic rays is similar to that of starlight, the cosmic microwave background radiation and the Galactic magnetic field, which all happen to be of the order of approximately 1 eV cm^{-3}. This energy density is approximately eight orders of magnitude smaller than that of solar light at a distance of one astronomical unit, that is, that of the Earth, from the sun. Moreover, cosmic rays interact at the top of the atmosphere and their energy is converted to atmospheric showers. Most of the particles and gamma-rays in the atmospheric showers are stopped in the atmosphere before reaching ground level, and almost only secondary muons and neutrinos, which carry a small fraction of their energy, reach the ground. Thus, at first it appears that Galactic cosmic rays cannot affect life on the Earth significantly. However, this is not the case. There is accumulating evidence that even moderate variations in the flux of cosmic rays that reach the atmosphere have significant climatic effects, despite their low energy density. The evidence comes mainly from two sources:

1. The interaction of cosmic rays with nuclei in the upper atmosphere generates showers of secondary particles, some of which produce the radioisotopes ^{14}C and ^{10}Be that reach the surface of the Earth either via the carbon cycle ($^{14}CO_2$) or in rain and snow (^{10}Be). Since this is the only terrestrial source, their concentration in tree rings, ice cores, and

marine sediments provides a good record of the intensity of Galactic cosmic rays that have reached the atmosphere in the past. They show clear correlation between climate changes and variation of the cosmic ray flux in the Holocene era.

2. The ions produced in cosmic ray showers increase the production of low altitude clouds (e.g., Carslaw et al. 2002). In data collected in the past 20 years, by satellites and neutron monitors, there is a clear correlation between global cloud cover and cosmic ray flux above 10 GeV that penetrate the geomagnetic field. Cloud cover reduces ground level radiation by a global average of 30 Wm^{-2}, which are 13% of the ground level solar irradiance. An increased flux of Galactic cosmic rays is associated with an increase in low cloud cover, which increases the reflectivity of the atmosphere and produces a cooler temperature.

Cosmic rays affect life in other ways:

1. Cosmic ray-produced atmospheric showers of ionized particles trigger lightening discharges in the atmosphere (Gurevich and Zybin, 2005). These showers produce NO and NO_2 by direct ionization of molecules, which destroy ozone at a rate faster than its production in the discharges. The depletion of the ozone in the atmosphere leads to an increased UV flux at the surface.
2. The decay of secondary mesons produced in showers yields high energy penetrating muons, which reach the ground and penetrate deep underground and deep underwater. A small fraction of energetic protons and neutrons from the shower that increases with the energy of the primary cosmic ray particle also reaches the surface. Overall, the very penetrating secondary muons are responsible for approximately 85% of the total equivalent dose delivered by cosmic rays at ground level. Their interactions, and the interactions of their products, with electrons and nuclei in living cells, ionize atoms and break molecules and damage DNA and RNA by displacing electrons, atoms and nuclei from their sites. The total energy deposition dose from penetrating muons resulting in 50% mortality in thirty days is between 2.5 and 3.0 Gy (1 Gy $= 10^4$, erg g^{-1}). A cosmic ray muon deposits approximately 4 Mev g^{-1} in living cells, and thus the lethal flux of cosmic ray muons is 5×10^9 cm^{-2} if delivered in a short time (less than a month). In order to deliver such a dose within a month, the normal cosmic ray flux has to increase by nearly a factor of a thousand during a whole month).

A large increase in cosmic ray flux over extended periods may produce global climatic catastrophes and expose life on the ground, underground, and underwater to hazardous levels of radiation, which result in cancer and leukaemia. However, the bulk of Galactic cosmic rays have energies below

10 GeV. Such cosmic rays that enter the heliosphere are deflected by the magnetic field of the solar wind before they reach the Earth's neighbourhood and by its geomagnetic field before they reach the Earth's atmosphere. Consequently, the Galactic cosmic ray flux that reaches the Earth's atmosphere is modulated by variations in the solar wind and in the Earth's magnetic field.

Life on the Earth has adjusted itself to the normal cosmic ray flux that reaches its atmosphere. Perhaps, cosmic ray-induced mutations in living cells played a major role in the evolution and diversification of life from a single cell to the present millions of species. Any credible claim for a global threat to life from increasing fluxes of cosmic rays must demonstrate that the anticipated increase is larger than the periodical changes in the cosmic ray flux that reaches the Earth, which result from the periodic changes in solar activity, in the geomagnetic field and in the motions of the Earth, and to which terrestrial life has adjusted itself.

12.3.1 Earth magnetic field reversals

The Earth's magnetic field reverses polarity every few hundred thousand years, and is almost non-existent for perhaps a century during the transition. The last reversal was 780 Ky ago, and the magnetic field's strength decreased by 5% during the twentieth century! During the reversals, the ozone layer becomes unprotected from charged solar particles, which weakens its ability to protect humans from UV radiation. However, past reversals were not associated with any major extinction according to the fossil record, and thus are not likely to affect humanity in a catastrophic way.

12.3.2 Solar activity, cosmic rays, and global warming

Cosmic rays are the main physical mechanism controlling the amount of ionization in the troposphere (the lower 10 km or so of the atmosphere). The amount of ionization affects the formation of condensation nuclei required for the formation of clouds in clean marine environment. The solar wind – the outflow of energetic particles and entangled magnetic field from the sun – is stronger and reaches larger distances during strong magnetic solar activity. The magnetic field carried by the wind deflects the Galactic cosmic rays and prevents the bulk of them from reaching the Earth's atmosphere. A more active sun therefore inhibits the formation of condensation nuclei, and the resulting low-altitude marine clouds have larger drops, which are less reflective and live shorter. This decrease in cloud coverage and in cloud reflectivity reduces the Earth's albedo. Consequently, more solar light reaches the surface and warms it.

Cosmic ray collisions in the atmosphere produce ^{14}C, which is converted to $^{14}CO_2$ and incorporated into the tree rings as they form; the year of growth can be precisely determined from dendrochronology. Production of ^{14}C is high during periods of low solar magnetic activity and low during high magnetic activity. This has been used to reconstruct solar activity during the past 8000 years, after verifying that it correctly reproduces the number of sunspots during the past 400 years (Solanki et al., 2004). The number of such spots, which are areas of intense magnetic field in the solar photosphere, is proportional to the solar activity. Their construction demonstrates that the current episode of high sunspot number and very high average level of solar activity, which has lasted for the past seventy years, has been the most intense and has had the longest duration of any in the past 8000 years. Moreover, the solar activity correlates very well with palaeoclimate data, supporting a major solar activity effect on the global climate.

Using historic variations in climate and the cosmic ray flux, Shaviv (2005) could actually quantify empirically the relation between cosmic ray flux variations and global temperature change, and estimated that the solar contribution to the twentieth century warming has been $0.50 \pm 0.20°C$ out of the observed increase of $0.75 \pm 0.15°C$, suggesting that approximately two-thirds of global warming resulted from solar activity while perhaps only approximately one-third came from the greenhouse effect. Moreover, it is quite possible that solar activity coupled with the emission of greenhouse gases is a stronger driver of global warming than just the sum of these two climatic drivers.

12.3.3 Passage through the Galactic spiral arms

Radio emission from the Galactic spiral arms provides evidence for their enhanced CR density. Diffuse radio emission from the interstellar medium is mainly synchrotron radiation emitted by CR electrons moving in the interstellar magnetic fields. High contrasts in radio emission are observed between the spiral arms and the discs of external spiral galaxies. Assuming equipartition between the CR energy density and the magnetic field density, as observed in many astrophysical systems, CR energy density in the spiral arms should be higher than in the disk by a factor of a few. Indeed, there is mounting evidence from radio and X-ray observations that low energy CRs are accelerated by the debris of core-collapse SNe. Most of the supernovae in spiral galaxies like our own are core-collapse SNe. They predominantly occur in spiral arms where most massive stars are born and shortly thereafter die. Thus, Shaviv (2002) has proposed that when the solar system passes through the Galactic spiral arms, the heliosphere is exposed to a much higher cosmic ray flux, which increases the average low-altitude cloud cover and reduces the average global temperature. Coupled with the periodic variations in the geomagnetic field and in the motion of the Earth around the sun, this may have caused the extended periods of glaciations and ice ages, which, in the Phanerozoicera, have typically lasted approximately 30 Myr with a mean separation of approximately 140 Myr.

Indeed, Shaviv (2002) has presented supportive evidence from geological and meteoritic records for a correlation between the extended ice ages and the periods of an increased cosmic ray flux. Also, the duration of the extended ice ages is in agreement with the typical crossing time of spiral arms (typical width of 100 LY divided by a relative velocity of approximately 10 km s^{-1} yields approximately 30 Myr crossing time). Note also that passage of the heliosphere through spiral arms, which contain a larger density of dust grains produced by SN explosions, can enter the heliosphere, reach the atmosphere, scatter away sunlight, reduce the surface temperature and cause an ice age.

12.3.4 Cosmic rays from nearby supernovae

The cosmic ray flux from a supernova remnant (SNR) has been estimated to produce a fluence $F \sim 7.4 \times 10^6\ (30\ \mathrm{LY}/d)^2$ erg cm^{-2} at a distance d from the remnant. The active acceleration time of an SNR is roughly 10^4 years. The ambient CR flux near the Earth is 9×10^4 erg cm^{-2} $year^{-1}$. Thus, at a distance of 300 LY approximately, the ambient flux level would increase approximately by a negligible 0.1% during a period of 10^4 years approximately. To have a significant effect, the supernova has to explode within 30 LY from the sun. Taking into consideration the estimated SN rate and distribution of Galactic SNRs, the rate of SN explosions within 30 LY from the sun is approximately 3×10^{-10} $year^{-1}$. However, at such a distance, the SN debris can blow away the Earth's atmosphere and produce a major mass extinction.

12.3.5 Cosmic rays from gamma-ray bursts

Radio, optical, and X-ray observations with high spatial resolution indicate that relativistic jets, which are fired by quasar and micro-quasars, are made of a sequence of plasmoids (cannonballs) of ordinary matter whose initial expansion (presumably with an expansion velocity similar to the speed of sound in a relativistic gas) stops shortly after launch (e.g., Dar and De Rujula, 2004 and references therein). The photometric and spectroscopic detection of SNe in the fading after glows of nearby GRBs and various other properties of GRBs and their after glows provide decisive evidence that long GRBs are produced by highly relativistic jets of plasmoids of ordinary matter ejected in SN explosions, as long advocated by the Cannonball (CB) Model of GRBs (see, for example, Dar, 2004, Dar & A. De Rujula 2004, "Magnetic field in galaxies, galaxy clusters, & intergalactic space in: Physical Review D **72**, 123002–123006; Dar and De Rujula, 2004). These jets of plasmoids (cannonballs) produce an arrow beam of high energy cosmic rays by magnetic scattering of the ionized particles of the interstellar medium (ISM) in front of them. Such CR beams from Galactic GRBs may reach large Galactic distances and can be much more lethal than their gamma-rays (Dar and De Rujula, 2001; Dar et al., 1998).

Let $\upsilon = \beta c$ be the velocity of a highly relativistic CB and $1/\gamma = 1/\sqrt{1-\beta^2}$ be its Lorentz factor. For long GRBs, typically, $\gamma \sim 10^3$ ($\upsilon \sim 0.999999c$!). Because of the highly relativistic motion of the CBs, the ISM particles that are swept by the CBs enter them with a Lorentz factor $\gamma \sim 10^3$ in the CBs' rest frame. These particles are isotropized and accelerated by the turbulent magnetic fields in the CBs (by a mechanism proposed by Enrico Fermi) before they escape back into the ISM. The highly relativistic motion of the CBs, boosts further their energy by an average factor γ through the Doppler effect and collimates their isotropic distribution into an arrow conical beam of an opening angle $\theta \sim 1/\gamma$ around the direction of motion of the CB in the ISM. This relativistic beaming depends only on the CB's Lorentz factor but not on the mass of the scattered particles or their energy.

The ambient interstellar gas is nearly transparent to the cosmic ray beam because the Coulomb and hadronic cross-sections are rather small with respect to typical Galactic column densities. The energetic CR beam follows a ballistic motion rather than being deflected by the ISM magnetic field whose typical value is $B \sim 3 \times 10^{-6}$ Gauss. This is because the magnetic field energy swept up by the collimated CR beam over typical distances to Galactic supernovae is much smaller than the kinetic energy of the beam. Thus, the CR beam sweeps away the magnetic field along its way and follows a straight ballistic trajectory through the interstellar medium. (The corresponding argument, when applied to the distant cosmological GRBs, leads to the opposite conclusion: no CR beams from distant GRBs accompany the arrival of their beamed gamma-rays.)

The fluence of the collimated beam of high energy cosmic rays at a distance from a GRB that is predicted by the CB model of GRBs is given approximately by $F \sim E_k\gamma^2/4\pi\ d^2 \sim 10^{20}\ (\mathrm{LY}/d^2)$ erg cm^{-2}, where the typical values of the kinetic energy of the jet of CBs, $E_k \sim 10^{51}$ erg and $\gamma \sim 10^3$, were obtained from the CB analysis of the observational data on long GRBs. Observations of GRB afterglows indicate that it typically takes a day or two for a CB to lose approximately 50% of its initial kinetic energy, that is, for its Lorentz factor to decrease to half its initial value. This energy is converted to CRs with a typical Lorentz factor $\gamma_{CR} \sim \gamma^2$ whose arrival time from a Galactic distance is delayed relative to the arrival of the afterglow photons by a negligible time, $\Delta t \sim d/c\gamma_{\mathrm{CR}}^2$. Consequently, the arrival of the bulk of the CR energy practically coincides with the arrival of the afterglow photons.

Thus, for a typical long GRB at a Galactic distance d = 25,000 LY, which is viewed at a typical angle $\theta \sim 1/\gamma \sim 10^{-3}$ radians, the energy deposition in the atmosphere by the CR beam is F $\sim 10^{11}$ erg cm^{-2} while that deposited by gamma-rays is smaller by about three orders of magnitude (the kinetic energy of the electrons in the jet is converted to a conical beam of gamma-rays while the bulk of the kinetic energy that is carried by protons is converted to a conical beam of CRs with approximately the same opening angle). The beam of energetic cosmic rays accompanying a Galactic GRB is deadly for life on Earth-like planets. When high energy CRs with energy E_p collide with the atmosphere at a zenith angle θ_z, they produce high energy muons whose number is given approximately by $N_\mu(E > 25\ \text{GeV}) \sim 9.14[E_p/\text{TeV}]^{0.757}/\cos\theta_z$ (Drees et al., 1989). Consequently, a typical GRB produced by a jet with EK $\sim 10^{51}$ erg at a Galactic distance of 25,000 LY, which is viewed at the typical viewing angle $\theta \sim 1/\gamma \sim 10^{-3}$, is followed by a muon fluence at ground level that is given by $F_\mu(E > 25\ \text{GeV}) \sim 3 \times 10^{11}\text{cm}^{-2}$. Thus, the energy deposition rate at ground level in biological materials, due to exposure to atmospheric muons produced by an average GRB near the centre of the Galaxy, is 1.4×10^{12} MeV g^{-1}. This is approximately 75 times the lethal dose for human beings. The lethal dosages for other vertebrates and insects can be a few times or as much as a factor 7 larger, respectively. Hence, CRs from galactic GRBs can produce a lethal dose of atmospheric muons for most animal species on the Earth. Because of the large range of muons ($\sim 4[E_\mu/\text{GeV}]$m) in water, their flux is lethal, even hundreds of metres under water and underground, for CRs arriving from well above the horizon. Thus, unlike other suggested extraterrestrial extinction mechanisms, the CRs of galactic GRBs can also generate massive extinctions deep under water and underground. Although half of the planet is in the shade of the CR beam, its rotation exposes a larger fraction of its surface to the CRs, half of which will arrive within over approximately 2 days after the gamma-rays. Additional effects that will increase the lethality of the CRs over the whole planet include:

1. Evaporation of a significant fraction of the atmosphere by the CR energy deposition.
2. Global fires resulting from heating of the atmosphere and the shock waves produced by the CR energy deposition in the atmosphere.
3. Environmental pollution by radioactive nuclei, produced by spallation of atmospheric and ground nuclei by the particles of the CR-induced showers that reach the ground.
4. Depletion of stratospheric ozone, which reacts with the nitric oxide generated by the CR-produced electrons (massive destruction of stratospheric ozone has been observed during large solar flares, which generate energetic protons).
5. Extensive damage to the food chain by radioactive pollution and massive extinction of vegetation by ionizing radiation (the lethal radiation dosages for trees and plants are slightly higher than those for animals, but still less than the flux estimated above for all but the most resilient species).

In conclusion, the CR beam from a Galactic SN/GRB event pointing in our direction, which arrives promptly after the GRB, can kill, in a relatively short time (within months), the majority of the species alive on the planet.

12.4 Origin of the major mass extinctions

Geological records testify that life on Earth has developed and adapted itself to its rather slowly changing conditions. However, good quality geological records, which extend up to approximately 500 Myr ago, indicate that the exponential diversification of marine and continental life on the Earth over that period was interrupted by many extinctions (e.g., Benton 1995; Erwin 1996, 1997; Raup and Sepkoski, 1986), with the major ones exterminating more than 50% of the species on land and sea, and occurring on average, once every 100 Myr. The five greatest events were those of the final Ordovician period (some 435 Myr ago), the late Devonian (357 Myr ago), the final Permian (251 Myr ago), the late Triassic (198 Myr ago) and the final Cretaceous (65 Myr ago). With, perhaps, the exception of the Cretaceous-Tertiary mass extinction, it is not well known what caused other mass extinctions. The leading hypotheses are:

- *Meteoritic Impact*: The impact of a sufficiently large asteroid or comet could create mega-tsunamis, global forest fires, and simulate nuclear winter from the dust it puts in the atmosphere, which perhaps are sufficiently severe as to disrupt the global ecosystem and cause mass extinctions. A large meteoritic impact was invoked (Alvarez et al., 1980) in order to explain their iridium anomaly and the mass extinction that killed the dinosaurs and 47% of all species around the K/T boundary, 65 Myr ago. Indeed, a 180 km wide crater was later discovered, buried under 1 km of Cenozoic sediments, dated back 65 Myr ago and apparently created by the impact of a 10 km diameter meteorite or comet near Chicxulub, in the Yucatan (e.g., Hildebrand, 1990; Morgan et al., 1997; Sharpton and Marin, 1997). However, only for the End Cretaceous extinction is there compelling evidence of such an impact. Circumstantial evidence was also claimed for the End Permian, End Ordovician, End Jurassic and End Eocene extinctions.
- *Volcanism*: The huge Deccan basalt floods in India occurred around the K/T boundary 65 Myr ago when the dinosaurs were finally extinct. The Permian/Triassic (P/T) extinction, which killed between 80% and 95% of the species, is the largest known is the history of life; occurred 251 Myr ago, around the time of the gigantic Siberian basalt flood. The outflow of millions of cubic kilometres of lava in a short time could have poisoned the atmosphere and oceans in a way that may have caused mass

extinctions. It has been suggested that huge volcanic eruptions caused the End Cretaceous, End Permian, End Triassic, and End Jurassic mass extinctions (e.g., Courtillot, 1988; Courtillot et al., 1990; Officer and Page, 1996; Officer et al., 1987).

- *Drastic Climate Changes*: Rapid transitions in climate may be capable of stressing the environment to the point of making life extinct, though geological evidence on the recent cycles of ice ages indicate they had only very mild impacts on biodiversity. Extinctions suggested to have this cause include: End Ordovician, End Permian, and Late Devonian.

Paleontologists have been debating fiercely which one of the above mechanisms was responsible for the major mass extinctions. But, the geological records indicate that different combinations of such events, that is, impacts of large meteorites or comets, gigantic volcanic eruptions, drastic changes in global climate and huge sea regressions/sea rise seem to have taken place around the time of the major mass extinctions. Can there be a common cause for such events?

The orbits of comets indicate that they reside in an immense spherical cloud ('the Oort cloud'), that surrounds the planetary with a typical radius of $R \sim$ 100,000 AU. The statistics imply that it may contain as many as 10^{12} comets with a total mass perhaps larger than that of Jupiter. The large radius implies that the comets share very small binding energies and mean velocities of $v <$ 100 m s^{-1}. Relatively small gravitational perturbations due to neighbouring stars are believed to disturb their orbits, unbind some of them, and put others into orbits that cross the inner solar system. The passage of the solar system through the spiral arms of the Galaxy, where the density of stars is higher, could have caused such perturbations, and consequently, the bombardment of the Earth with a meteorite barrage of comets over an extended period longer than the free fall time. It has been claimed by some authors that the major extinctions were correlated with passage times of the solar system through the Galactic spiral arms. However, these claims were challenged. Other authors suggested that biodiversity and extinction events may be influenced by cyclic processes. Raup and Sepkoski (1986) claimed a 26–30 Myr cycle in extinctions. Although this period is not much different from the 31 Myr period of the solar system crossing the Galactic plane, there is no correlation between the crossing time and the expected times of extinction. More recently, Rohde and Muller (2005) have suggested that biodiversity has a 62 ± 3 Myr cycle. But, the minimum in diversity is reached only once during a full cycle when the solar system is farthest away in the northern hemisphere from the Galactic plane.

Could Galactic GRBs generate the major mass extinction, and can they explain the correlation between mass extinctions, meteoritic impacts, volcano eruptions, climate changes and sea regressions, or can they only explain the volcanic-quiet and impact-free extinctions?

Passage of the GRB jet through the Oort cloud, sweeping up the interstellar matter on its way, could also have generated perturbations, sending some comets into a collision course with the Earth.

The impact of such comets and meteorites may have triggered the huge volcanic eruptions, perhaps by focusing shock waves from the impact at an opposite point near the surface on the other side of the Earth, and creating the observed basalt floods, timed within 1–2 Myr around the K/T and P/T boundaries. Global climatic changes, drastic cooling, glaciation and sea regression could have followed from the drastic increase in the cosmic ray flux incident on the atmosphere and from the injection of large quantities of light-blocking materials into the atmosphere from the cometary impacts and the volcanic eruptions. The estimated rate of GRBs from observations is approximately 10^3 year^{-1}. The sky density of galaxies brighter than magnitude 25 (the observed mean magnitude of the host galaxies of the GRBs with known red-shifts) in the Hubble telescope deep field is approximately 2×10^{-5} per square degree. Thus, the rate of observed GRBs, per galaxy with luminosity similar to that of the Milky Way, is $R_{\mathrm{GRB}} \sim 1.2 \times 10^{-7}$ year^{-1}. To translate this result into the number of GRBs born in our own galaxy, pointing towards us, and occurring in recent cosmic times, one must take into account that the GRB rate is proportional to the star formation rate, which increases with red-shift z like $(1 + z)^4$ for $z < 1$ and remains constant up to $z \sim 6$. The mean red-shift of GRBs with known red-shift, which were detected by SWIFT, is $\sim$2.8, that is, most of the GRBs were produced at a rate approximately 16 times larger than that in the present universe. The probability of a GRB pointing towards us within a certain angle is independent of distance. Therefore, the mean rate of GRBs pointing towards us in our galaxy is roughly $R_{\mathrm{GRB}}/(1 + z)^4 \sim 0.75 \times 10^{-8}$ year^{-1}, or once every 130 Myr. If most of these GRBs take place not much farther away than the distance to the galactic centre, their effect is lethal, and their rate is consistent with the rate of the major mass extinctions on our planet in the past 500 Myr.

12.5 The Fermi paradox and mass extinctions

The observation of planets orbiting nearby stars has become almost routine. Although current observations/techniques cannot detect yet planets with masses comparable to the Earth near other stars, they do suggest their existence. Future space-based observatories to detect Earth-like planets are being planned. Terrestrial planets orbiting in the habitable neighbourhood of stars, where planetary surface conditions are compatible with the presence of liquid water, might have global environments similar to ours, and harbour life. But, our solar system is billions of years younger than most of the stars in the Milky Way and life on extra solar planets could have preceded life on the Earth by billions of years, allowing for civilizations much more advanced than ours. Thus Fermi's famous question, 'where are they?', that is, why did they not visit us or send signals to us (see Chapter 6.) One of the possible answers is provided by cosmic mass extinction: even if advanced civilizations are not self-destructive, they are subject to a similar violent cosmic environment that may have generated the big mass extinctions on this planet. Consequently, there may be no nearby aliens who have evolved long enough to be capable of communicating with us, or pay us a visit.

12.6 Conclusions

- Solar flares do not comprise a major threat to life on the Earth. The Earth's atmosphere and the magnetosphere provide adequate protection to life on its surface, under water and underground.
- Global warming is a fact. It has drastic effects on agricultural yields, cause glacier retreat, species extinctions and increases in the ranges of disease vectors. Independent of whether or not global warming is of anthropogenic origin, human kind must conserve energy, burn less fossil fuel, and use and develop alternative non-polluting energy sources.
- The current global warming may be driven by enhanced solar activity. On the basis of the length of past large enhancements in solar activity, the probability that the enhanced activity will continue until the end of the twenty-first century is quite low (1%). (However, if the global warming is mainly driven by enhanced solar activity, it is hard to predict the time when global warming will turn into global cooling. (see Chapter 13.)
- Within one to two billion years, the energy output of the sun will increase to a point where the Earth will probably become too hot to support life.
- Passage of the sun through the Galactic spiral arms once in approximately 140 Myr will continue to produce major, approximately 30 Myr long, ice ages.
- Our knowledge of the origin of major mass extinctions is still very limited. Their mean frequency is extremely small, once every 100 Myr. Any initiative/decision beyond expanding research on their origin is premature.
- Impacts between near-Earth Objects and the Earth are very infrequent but their magnitude can be far greater than any other natural disaster. Such impacts that are capable of causing major mass extinctions are extremely infrequent as evident from the frequency of past major mass extinctions. At present, modern astronomy cannot predict or detect early enough such

an imminent disaster and society does not have either the capability or the knowledge to deflect such objects from a collision course with the Earth.

- A SN would have to be within few tens of light years from the Earth for its radiation to endanger creatures living at the bottom of the Earth's atmosphere. There is no nearby massive star that will undergo a SN explosion close enough to endanger the Earth in the next few million years. The probability of such an event is negligible, less than once in 10^9 years.
- The probability of a cosmic ray beam or a gamma-ray beam from Galactic sources (SN explosions, mergers of neutron stars, phase transitions in neutron stars or quark stars, and micro-quasar ejections) pointing in our direction and causing a major mass extinction is rather small and it is strongly constrained by the frequency of past mass extinctions – once every 100 Myr.
- No source in our Galaxy is known to threaten life on the Earth in the foreseeable future.

References

Alvarez, L.W., Alvarez, W., Asaro, F., and Michel, H.V. (1980). Extraterrestrial cause for the Cretaceous tertiary extinction, *Science*, **208**, 1095–1101.

Benton, M.J. (1995). Diversification and extinction in the history of life, *Science*, **268**, 52–58.

Carslaw, K.S., Harrison, R.G., and Kirkby, J. (2002). Cosmic rays, clouds, and climate. *Science*, **298**, 1732–1737.

Courtillot, V. (1990). A Volcanic Eruption, *Scientific American*, **263**, October 1990, pp. 85–92.

Courtillot, V., Feraud, G., Maluski, H., Vandamme, D., Moreau, M.G., and Besse, J. (1998). Deccan Flood Basalts and the Cretaceous/Tertiary Boundary. *Nature*, **333**, 843–860.

Dado, S., Dar, A., and De Rújula, A. (2002). On the optical and X-ray afterglows of gamma ray bursts. *Astron. Astrophys.*, **388**, 1079–1105.

Dado, S., Dar, A., and De Rújula, A. (2003). The supernova associated with GRB 030329. *Astrophys. J.*, **594**, L89.

Dar, A. (2004). The GRB/XRF-SN Association, arXiv astro-ph/0405386.

Dar, A. and De Rújula, A. (2002). The threat to life from Eta Carinae and gamma ray bursts. In Morselli, A. and Picozza, P. (eds.) *Astrophysics and Gamma Ray Physics in Space* (Frascati Physics Series Vol. XXIV, pp. 513–523 (astro-ph/0110162).

Dar, A. and De Rújula, A. (2004). Towards a complete theory of gamma-ray bursts. *Phys. Rep.*, **405**, 203–278.

Dar, A., Laor, A., and Shaviv, N. (1998). Life extinctions by cosmic ray jets, *Phys. Rev. Lett.*, **80**, 5813–5816.

Drees, M., Halzen, F., and Hikasa, K. (1989). Muons in gamma showers, *Phys. Rev.*, **D39**, 1310–1317.
Du and De Rujula, A. (2004). Magnetic field in galaxies, galaxy dusters, and intergalactic space in: *Physical Review D* **72**, 123002–123006.
Ellis, J., Fields, B.D., and Schramm, D.N. (1996). Geological isotope anomalies as signatures of nearby supernovae. *Astrophys. J.*, **470**, 1227–1236.
Ellis, J. and Schramm, D.N. (1995). Could a nearby supernova explosion have caused a mass extinction?, *Proc. Nat. Acad. Sci.*, **92**, 235–238.
Erwin, D.H. (1996). The mother of mass extinctions, *Scientific American*, **275**, July 1996, p. 56–62.
Erwin, D.H. (1997). The Permo-Triassic extinction. *Nature*, **367**, 231–236.
Fields, B.D. and Ellis, J. (1999). On deep-ocean ^{60}Fe as a fossil of a near-earth supernova. *New Astron.*, **4**, 419–430.
Galante, D. and Horvath, J.E. (2005). Biological effects of gamma ray bursts: distances for severe damage on the biota. *Int. J. Astrobiology*, **6**, 19–26.
Gurevich, A.V. and Zybin, K.P. (2005). Runaway breakdown and the mysteries of lightning. *Phys. Today*, **58**, 37–43.
Hildebrand, A.R. (1990). Mexican site for K/T Impact Crater?, *Mexico, Eos*, **71**, 1425.
Kirkby, J., Mangini, A., and Muller, R.A. (2004). Variations of galactic cosmic rays and the earth's climate. In Frisch, P.C. (ed.), *Solar Journey: The Significance of Our Galactic Environment for the Heliosphere and Earth* (Netherlands: Springer) pp. 349–397 (arXivphysics/0407005).
Meegan, C.A. and Fishman, G.J. (1995). Gamma ray bursts. *Ann. Rev. Astron. Astrophys.*, **33**, 415–458.
Melott, A., Lieberman, B., Laird, C., Martin, L., Medvedev, M., Thomas, B., Cannizzo, J., Gehrels, N., and Jackman, C. (2004). Did a gamma-ray burst initiate the late Ordovician mass extinction? *Int. J. Astrobiol.*, **3**, 55–61.
Morgan, J., Warner, M., and Chicxulub Working Group. (1997). Size and morphology of the Chixulub Impact Crater. *Nature*, **390**, 472–476.
Officer, C.B., Hallan, A., Drake, C.L., and Devine, J.D. (1987). Global fire at the Cretaceous-Tertiary boundary. *Nature*, **326**, 143–149.
Officer, C.B. and Page, J. (1996). *The Great Dinosaurs Controversy* (Reading, MA: Addison-Wesley Pub. Com.).
Raup, D. and Sepkoski, J. (1986). Periodic extinction of families and genera. *Science*, **231**, 833–836.
Rohde, R.A. and Muller, R.A. (2005). Cycles in fossil diversity. *Nature*, **434**, 208–210.
Ruderman, M.A. (1974). Possible consequences of nearby supernova explosions for atmospheric ozone and terrestrial life. *Science*, **184**, 1079–1081.
Scalo, J. and Wheeler, J.C. (2002). Did a gamma-ray burst initiate the late Ordovician mass extinction? *Astrophys. J.*, **566**, 723–737.
Sepkoski, J.J. (1986). Is the periodicity of extinctions a taxonomic artefact? In Raup, D.M. and Jablonski, D. (eds.), *Patterns and Processes in the History of Life*. pp. 277–295 (Berlin: Springer-Verlag).
Sharpton, V.L. and Marin, L.E. (1997). The Cretaceous-Tertiary impact crater. *Ann. NY Acad. Sci.*, **822**, 353–380.

Shaviv, N. (2002). The spiral structure of the Milky Way, cosmic rays, and ice age epochs on earth. *New Astron.*, **8**, 39–77.

Shaviv, N. and Dar, A. (1995). Gamma ray bursts from Minijets. *Astrophys. J.*, **447**, 863–873.

Smith, D.S., Scalo, J., and Wheeler, J.C. (2004). Importance of biologically active Aurora-like ultraviolet emission: stochastic irradiation of earth and mars by flares and explosions. *Origins Life Evol. Bios.*, **34**, 513–532.

Solanki, S.K., Usoskin, I.G., Kromer, B., Schüssler, M., and Bear, J. (2004). Unusual activity of the sun during recent decades compared to the previous 11000 Years. *Nature*, **431**, 1084–1087.

Svensmark, H. (1998). Influence of Cosmic rays on earths climate. *Phys. Rev. Lett.*, **81**, 5027–5030.

Thomas, B.C., Jackman, C.H., Melott, A.L., Laird, C.M., Stolarski, R.S., Gehrels, N., Cannizzo, J.K., and Hogan, D.P. (2005). Terrestrial ozone depletion due to a milky way gamma-ray burst, *Astrophys. J.*, **622**, L153–L156.

Thorsett, S.E. (1995). Terrestrial implications of cosmological gamma-ray burst models. *Astrophys. J. Lett.*, **444**, L53–L55.

van den Bergh, S. and Tammann, G.A. (1991). Galactic and extragalactic supernova rates. *Ann. Rev. Astron. Astrophys.*, **29**, 363–407.

PART III
Risks from unintended consequences

·13·

Climate change and global risk

David Frame and Myles R. Allen

13.1 Introduction

Climate change is among the most talked about and investigated global risks. No other environmental issue receives quite as much attention in the popular press, even though the impacts of pandemics and asteroid strikes, for instance, may be much more severe. Since the first Intergovernmental Panel on Climate Change (IPCC) report in 1990, significant progress has been made in terms of (1) establishing the reality of anthropogenic climate change and (2) understanding enough about the scale of the problem to establish that it warrants a public policy response. However, considerable scientific uncertainty remains. In particular scientists have been unable to narrow the range of the uncertainty in the global mean temperature response to a doubling of carbon dioxide from pre-industrial levels, although we do have a better understanding of why this is the case. Advances in science have, in some ways, made us more uncertain, or at least aware of the uncertainties generated by previously unexamined processes. To a considerable extent these new processes, as well as familiar processes that will be stressed in new ways by the speed of twenty-first century climate change, underpin recent heightened concerns about the possibility of catastrophic climate change.

Discussion of 'tipping points' in the Earth system (for instance Kemp, 2005; Lenton, 2007) has raised awareness of the possibility that climate change might be considerably worse than we have previously thought, and that some of the worst impacts might be triggered well before they come to pass, essentially suggesting the alarming image of the current generation having lit the very long, slow-burning fuse on a climate bomb that will cause great devastation to future generations. Possible mechanisms through which such catastrophes could play out have been developed by scientists in the last 15 years, as a natural output of increased scientific interest in Earth system science and, in particular, further investigation of the deep history of climate. Although scientific discussion of such possibilities has usually been characteristically guarded and responsible, the same probably cannot be said for the public debate around such notions. Indeed, many scientists regard these hypotheses and images as premature, even alarming. Mike Hulme, the Director of the United Kingdom's Tyndall Centre recently complained that:

> The IPCC scenarios of future climate change – warming somewhere between 1.4 and 5.8° Celsius by 2100 – are significant enough without invoking catastrophe and chaos as unguided weapons with which forlornly to threaten society into behavioural change. [...] The discourse of catastrophe is in danger of tipping society onto a negative, depressive and reactionary trajectory.

This chapter aims to explain the issue of catastrophic climate change by first explaining the mainstream scientific (and policy) position: that twenty-first century climate change is likely to be essentially linear, though with the possibility of some non-linearity towards the top end of the possible temperature range. This chapter begins with a brief introduction to climate modelling, along with the concept of climate forcing. Possible ways in which things could be considerably more alarming are discussed in a section on limits to our current knowledge, which concludes with a discussion of uncertainty in the context of palaeoclimate studies. We then discuss impacts, defining the concept of 'dangerous anthropogenic interference' in the climate system, and some regional impacts are discussed alongside possible adaptation strategies. The chapter then addresses mitigation policy – policies that seek to reduce the atmospheric loading of greenhouse gases (GHG) – in light of the preceeding treatment of linear and non-linear climate change. We conclude with a brief discussion of some of the main points and problems.

13.2 Modelling climate change

Scientists attempting to understand climate try to represent the underlying physics of the climate system by building various sorts of climate models. These can either be top down, as in the case of simple models that treat the Earth essentially as a closed system possessing certain simple thermodynamic properties, or they can be bottom up, as in the case of general circulation models (GCMs), which attempt to mimic climate processes (such as cloud formation, radiative transfer, and weather system dynamics). The range of models, and the range of processes they contain, is large: the model we use to discuss climate change below can be written in one line, and contains two physical parameters; the latest generation of GCMs comprise well over a million lines of computer code, and contain thousands of physical variables and parameters. In between these lies a range of Earth system models of intermediate complexity (EMICs) (Claussen et al., 2002; Lenton et al., 2006) which aim at resolving some range of physical processes between the global scale represented by EBMs and the more comprehensive scales represented by GCMs. EMICs are often used to investigate long-term phenomena, such as the millennial scale response to Milankovitch cycles, Dansger-Oeschgaard events or other such episodes and periods that it would be prohibitively expensive to investigate with a full GCM. In the following section we introduce and use a simple EBM to illustrate the global response to various sorts of forcings, and discuss the source of current and past climate forcings.

13.3 A simple model of climate change

A very simple model for the response of the global mean temperature to a specified climate forcing is given in the equation below. This model uses a single physical constraint – energy balance – to consider the effects of various drivers on global mean temperature. Though this is a very simple and impressionistic model of climate change, it does a reasonable job of capturing the aggregate climate response to fluctuations in forcing. Perturbations to the Earth's energy budget can be approximated by the following equation (Hansen et al., 1985):

$$c_{\text{eff}} \frac{\mathrm{d}\Delta T}{\mathrm{d}t} = F(t) - \lambda \Delta T, \qquad (13.1)$$

in which c_{eff} is the effective heat capacity of the system, governed mainly (Levitus et al., 2005) by the ocean, λ is a feedback parameter, and ΔT is a global temperature anomaly. The rate of change is governed by the thermal inertia of the system, while the equilibrium response is governed by the feedback parameter alone (since the term on the left hand side of the equation tends to zero as the system equilibrates). The forcing, F, is essentially the perturbation to the Earth's energy budget (in $\mathrm{W/m^2}$) which is driving the temperature response (ΔT). Climate forcing can arise from various sources, such as changes in composition of the atmosphere (volcanic aerosols; GHG) or changes in insolation. An estimate of current forcings is displayed in Fig. 13.1, and an estimate of a possible range of twenty-first century responses is shown in Fig. 13.2. It can be seen that the bulk of current forcing comes from elevated levels of carbon dioxide (CO_2), though other agents are also significant. Historically, three main forcing mechanisms are evident in the temperature record: solar forcing, volcanic forcing and, more recently, GHG forcing. The range of responses in Fig. 13.2 is mainly governed by (1) choice of future GHG scenario and (2) uncertainty in the climate response, governed in our simple model (which can essentially replicate Fig. 13.3) by uncertainty in the parameters c_{eff} and λ. The scenarios considered are listed in the figure, along with corresponding grey bands representing climate response uncertainty for each scenario (at right in grey bands).

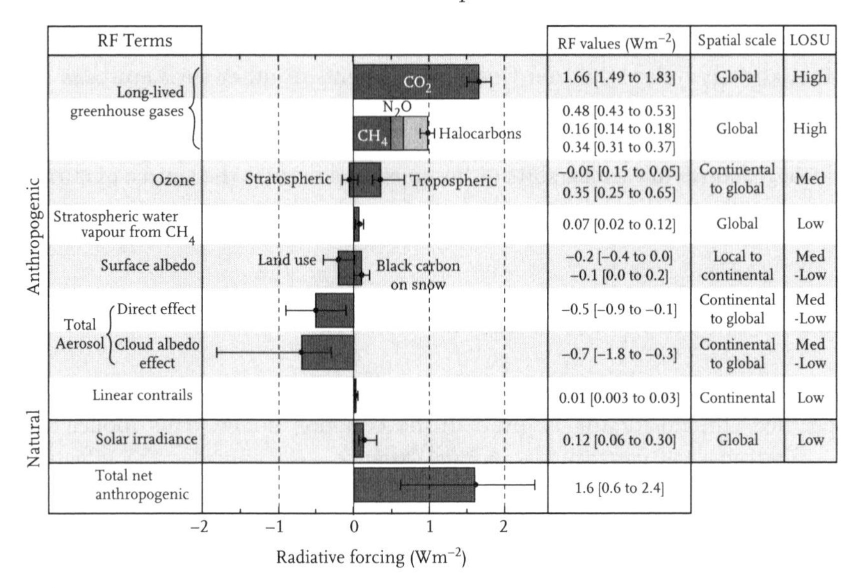

Fig. 13.1 Global average radiative forcing (RF) estimates and ranges in 2005 for anthropogenic carbon dioxide (CO_2), methane (CH_4), nitrous oxide (N_2O), and other important agents and mechanisms, together with the typical geographical extent (spatial scale) of the forcing and the assessed level of scientific understanding (LOSU). The net anthropogenic radiative forcing and its range are also shown. These require summing asymmetric uncertainty estimates from the component terms and cannot be obtained by simple addition. Additional forcing factors not included here are considered to have a very low LOSU. Volcanic aerosols contribute an additional natural forcing but are not included in this figure due to their episodic nature. The range for linear contrails does not include other possible effects of aviation on cloudiness. Reprinted with permission from Solomon et al. (2007). *Climatic Change 2007: The physical Science Basis.*

13.3.1 Solar forcing

Among the most obvious ways in which the energy balance can be altered is if the amount of solar forcing changes. This is in fact what happens as the earth wobbles on its axis in three separate ways on timescales of tens to hundreds of thousands of years. The earth's axis precesses with a period of 15,000 years, the obliquity of the Earth's orbit oscillates with a period of around 41,000 years, and its eccentricity varies on multiple timescales (95,000, 125,000, and 400,000 years).

These wobbles provide the source of the earth's periodic ice ages, by varying the amount of insolation at the Earth's surface. In particular, the Earth's climate is highly sensitive to the amount of solar radiation reaching latitudes north of about 60°N. Essentially, if the amount of summer radiation is insufficient to melt the ice that accumulates over winter, the depth of ice thickens, reflecting more back. Eventually, this slow accrual of highly reflective and very cold ice acts to reduce the Earth's global mean temperature, and the Earth enters an ice age.

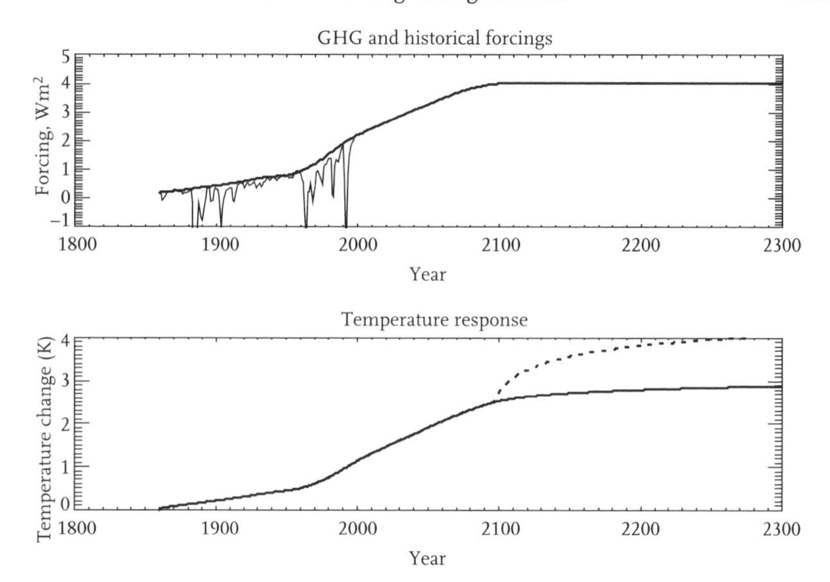

Fig. 13.2 Forcing and temperature response curves using the simple models presented in this chapter. In the top panel the thin line corresponds to historical forcings over the twentieth century; the solid line corresponds to GHG-only forcings, and these are projected forward under an extended version of the IPCC B1 scenario (SRES, 1992). In the bottom panel is the base model [Equation (13.1)] response to the B1 scenario with best guess parameters from Frame et al. (2006). Also shown (dashed line) is the response with the arbitrary disruption. SRES: N. Nakicenovic et al., IPCC Special Report on Emission Scenarios, IPCC, Cambridge University Press, 2000. Frame, D.J., D.A. Stone, P.A. Stoll, M.R. Allen, Alternatives to stabilization scenarios, *Geophysical Research Letters*, **33**, L14707.

Current solar forcing is thought to be +0.12 Wm^{-2}, around one eighth of the total current forcing (above pre-industrial levels). There has been some advocacy for a solar role in current climate change, but it seems unlikely that this is the case, since solar trends over some of the strongest parts of the climate change signal appear to be either insignificant or even out of phase with the climate response (Lockwood and Frohlich, 2007). Detection and attribution studies of recent climate change have consistently been able to detect a climate change signal attributable to elevated levels of GHG, but have been unable to detect a signal attributable to solar activity (Hegerl et al., 2007). Though dominant historically over millennial and geological timescales, the sun apparently is not a significant contributor to current climate change.

13.3.2 Volcanic forcing

Another form of climate forcing comes from the explosive release of aerosols into the atmosphere through volcanoes. This is a significant forcing when aerosols can make it into the stratosphere where the residence time for aerosols

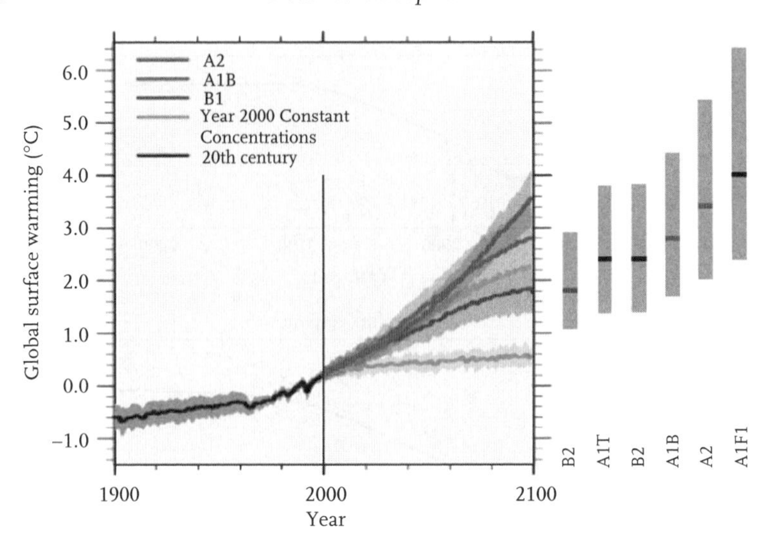

Fig. 13.3 Solid lines are multi-model global averages of surface warming (relative to 1980–1999) for the scenarios A2, A1B, and B1, shown as continuations of the twentieth century simulations. Shading denotes the ±1 standard deviation range of individual model annual averages. The orange line is for the experiment where concentrations were held constant at year 2000 values. The grey bars on the right indicate the best estimate (solid line within each bar) and the likely range assessed for the six SRES marker scenarios. The assessment of the best estimate and likely ranges in the grey bars includes the AOGCMs in the left part of the figure, as well as results from a hierarchy of independent models and observational constraints. Reprinted with permission from Solomon et al. (2007). *Climatic Change* 2007: *The physical Science Basis.*

is several years. During this time, they act to cool the planet by back-scattering incoming solar radiation. Though peak forcings due to large volcanoes are massive, the stratospheric aerosol residence time is short (approximately four years) in comparison to that of the principal anthropogenic GHG, CO_2 (approximately 150 years) so volcanic forcings tend to be much more dramatic and spikier (Fig. 13.3) than those associated with GHG. Peak forcing from the recent El Chichon (1982) and Mt Pinatubo (1991) volcanoes was of the order of approximately $3Wm^{-2}$. Though it is harder to infer the forcing from volcanoes that pre-date the satellite record, estimates suggest that volcanic forcing from Karakatau and Tambora in the nineteenth century are likely to have been larger.

Even bigger are supervolcanoes, the most extreme class of volcanic events seen on Earth. Supervolcanoes eject over a 1000 km^3 of material into the atmosphere (compared with around 25 km^3 for Pinatubo). They are highly uncommon, occurring less than once every 10,000 years, with some of the largest events being increasingly rare. Examples include the formation of the La Garita Caldera, possibly the largest eruption in history, which occurred some

28 million years ago, ejecting around 5000 km^3 of matter and, more recently, the Toba explosion, which occurred about 71,000 years ago and ejected around 2800 km^3 of material into the atmosphere. These very large events occur with a frequency in the vicinity of 1.4–22 events per million years (Mason et al., 2004), and the fact that the frequency of the most extreme events tends to fall away quickly with the magnitude of the event suggests that there is some sort of upper limit to the size of volcanic events (Mason et al., 2004).

In the last 1000 years, massive volcanic eruptions, responsible for climate forcings on the order of ten, occurred around 1258, probably in the tropics (Crowley, 2000). This medieval explosion was enormous, releasing around eight times the sulphate of the Krakatau explosion. However, correlations between the amount of sulphate released and observed global mean temperatures show that the temperature response for extremely large volcanoes does not scale simply from Krakatau-sized volcanoes (and smaller). Pinto et al. (1989) suggest that this is because beyond a certain level, increases in sulphate loading act to increase the size of stratospheric aerosols, rather than increasing their number. In essence, the radiative forcing per unit mass decreases at very large stratospheric sulphate loadings, and this moderates the cooling effects of very large volcanoes.

The expectations for the coming decades are that the odds of a volcanically induced radiative forcing of -1.0 W m^{-2} or larger occurring in a single decade are in the range 35–40%. The odds of two such eruptions in a single decade are around 15%, while the odds of three or more such eruptions is something like 5%. The odds of getting lucky volcanically (though perhaps not climatically) and experiencing no such eruptions is approximately 44%. Considering larger volcanic events, there is a 20% likelihood of a Pinatubo-scale eruption in the next decade and a 25% chance for an El Chichón-scale eruption (Hyde and Crowley, 2000). Although events of this magnitude would certainly act to cool the climate, and lead to some deleterious impacts (especially in the vicinity of the volcano) they would represent essentially temporary perturbations to the expected warming trend.

13.3.3 Anthropogenic forcing

A simple model for the accumulation of GHG (simplified here to carbon dioxide), its conversion to radiative forcing, and the subsequent effect on global mean temperatures, can be given by the following two equations:

$$CO_2(t) = \sum_{0}^{t} e_{nat}(t) + e_{ant}(t) \tag{13.2}$$

Here, $CO_2(t)$ represents the atmospheric concentration of carbon dioxide over time, which is taken to be the sum of emissions from natural (e_{nat}) and anthropogenic (e_{ant}) sources. We can further specify a relationship between carbon dioxide concentration and forcing:

$$F(t) = F_{2\times CO_2} \ln \frac{CO_2(t)}{CO_{2_{pre}}} \div \ln(2), \tag{13.3}$$

in which $F_{2\times CO_2}$ is the forcing corresponding to a doubling of CO_2, CO_2 refers to the atmospheric concentration of CO_2 and CO_{2pre} is the (constant) pre-industrial concentration of CO_2. There is some uncertainty surrounding $F_{2\times CO_2}$, but most general circulation models (GCMs) diagnose it to be in the range 3.5–4.0 W/m^2. Current forcings are displayed in Fig. 13.1. This shows that the dominant forcing is due to elevated levels (above pre-industrial) of carbon dioxide, with the heating effects of other GHG and the cooling effects of anthropogenic sulphates happening to largely cancel each other out.

It suffers from certain limitations but mimics the global mean temperature response of much more sophisticated state of the art GCMs adequately enough, especially with regard to relatively smooth forcing series. Both the simple model described above and the climate research community's best-guess GCMs predict twenty-first century warmings, in response to the kinds of elevated levels of GHG we expect to see this century, of between about 1.5°C and 5.8°C.

The key point to emphasize in the context of catastrophic risk is that this warming is expected to be quite smooth and stable. No existing best guess state-of-the-art model predicts any sort of surprising non-linear disruption in terms of global mean temperature. In recent years, climate scientists have attempted to quantify the parameters in the above model by utilizing combinations of detuned models, simple (Forest et al., 2002) or complex (Murphy et al., 2004) constrained by recent observations (and sometimes using additional information from longer term climate data (Hegerl et al., 2006; Schneider von Deimling et al., 2006). These studies have reported a range of distributions for climate sensitivity, and while the lower bound and median values tend to be similar across studies there is considerable disagreement regarding the upper bound.

Such probabilistic estimates are dependent on the choice of data used to constrain the model, the weightings given to different data streams, the sampling strategies and statistical details of the experiment, and even the functional forms of the models themselves. Although estimation of climate sensitivity remains a contested arena (Knutti et al., 2008) the global mean temperature response of the models used in probabilistic climate forecasting can be reasonably well approximated by manipulation of the parameters c_{eff} and, especially, λ, in Equation (13.1) of this chapter. Essentially, Equation (13.1) can be tuned to mimic virtually any climate model in which atmospheric feedbacks scale linearly with surface warming and in which effective oceanic heat capacity is approximately constant under twentieth century climate forcing. This includes the atmospheric and oceanic components of virtually all current best-guess GCMs.

13.4 Limits to current knowledge

However, we know there are plenty of ways in which our best-guess models may be wrong. Feedbacks within the carbon cycle, in particular, have been the source of much recent scientific discussion. Amazon dieback, release of methane hydrates from the ocean floor, and release of methane from melting permafrost are among mechanisms thought to have the potential to add nasty surprises to the linear picture of anthropogenic climate change we see in our best models. There are essentially two ways in which we could be wrong about the evolution of the system towards the target, even if we knew the climate sensitivity S:

- Forcing could be more temperature-dependent than we thought. That is: $c_{\mathrm{eff}} \frac{\mathrm{d}\Delta T}{\mathrm{d}t} = F(\Delta T) - \lambda \Delta T$.
- Feedbacks could be more temperature-dependent than we thought. In this case: $c_{\mathrm{eff}} \frac{\mathrm{d}\Delta T}{\mathrm{d}t} = F - \lambda(\Delta T)\Delta T$.

In Fig. 13.2, we plot the latter sort of variant: in the dotted line in the bottom panel of Fig. 13.2 the forcing remains the same as in the base case, but the response is different, reflecting an example of a disruptive temperature dependence of λ, which is triggered as the global mean temperature anomaly passes 2.5°C. This is a purely illustrative example, to show how the smooth rise in global mean temperature could receive a kick from some as yet un-triggered feedback in the system. In the last decade, scientists have investigated several mechanisms that might make our base equation more like (1) or (2). These include the following:

- Methane hydrate release (F) in which methane, currently locked up in the ocean in the form of hydrates can be released rapidly if ocean temperatures rise beyond some threshold value (Kennett et al., 2000). Since methane sublimes (goes straight from a solid phase to a gaseous phase) this would lead to a very rapid rise in atmospheric GHG loadings. Maslin et al. (2004) claim that there is evidence for the hypothesis from Dansgaard-Oeschger interstadials; Benton and Twitchett (2003) claim that the sudden release of methane gas from methane hydrates (sometimes called the 'clathrate gun hypothesis') may have been responsible for the mass extinction at the end of the Permian, the 'biggest crisis on Earth in the past 500 Myr'. If this is so, they claim with considerable understatement, 'it is [...] worth exploring further'.

- Methane release from melting permafrost (F) is related to the oceanic clathrate gun hypothesis. In this case, methane is trapped in permafrost through the annual cycle of summer thawing and winter freezing. Vegetable matter and methane gas from the rotting vegetable material get sequestered into the top layers of permafrost; the reaction with water forms methane hydrates (Dallimore and Collet, 1995). Increased temperatures in permafrost regions may give rise to sufficient melting to unlock the methane currently sequestered, providing an amplification of current forcing (Sazonova et al., 2004).
- Tropical forest (especially Amazonian) dieback (F) is perhaps the best known amplification to expected climate forcing. Unlike the previous two examples which are additional sources of GHG, tropical forest dieback is a reduction in the strength of a carbon sink (Cox et al., 2004). As temperatures rise, Amazonia dries and warms, initiating the loss of forest, leading to a reduction in the uptake of carbon dioxide (Cox et al., 2000).
- Ocean acidification (F) is an oceanic analogue to forest dieback from a carbon cycle perspective in that it is the reduction in the efficacy of a sink. Carbon dioxide is scrubbed out of the system by phytoplankton. As carbon dioxide builds up the ocean acidifies, rendering the phytoplankton less able to perform this role, weakening the ocean carbon cycle (Orr et al., 2005).
- Thermohaline circulation disruption (λ), while perhaps less famous than Amazonian dieback, can at least claim the dubious credit of having inspired a fully fledged disaster movie. In this scenario, cool fresh water flowing from rapid disintegration of the Greenland ice sheet acts to reduce the strength of the (density-driven) vertical circulation in the North Atlantic (sometimes known as the 'thermohaline circulation', but better described as the Atlantic meridional overturning circulation. There is strong evidence that this sort of process has occurred in the Earth's recent past, the effect of which was to plunge Europe into an ice age for another 1000 years.
- The iris effect (λ) would be a happy surprise. This is a probable mechanism through which the Earth could self-regulate its temperature. The argument, developed in Lindzen et al. (2001) is that tropical cloud cover could conceivably increase in such a way as to reduce the shortwave radiation received at the surface. On this view, the expected (and worrying warming) will not happen or will largely be offset by reductions in insolation. Thus, while the Earth becomes less efficient at getting rid of thermal radiation, it also becomes better at reflecting sunlight. Unlike the other mechanisms discussed here, it would cause us to undershoot, rather than overshoot, our target (the ball breaks left, rather than right).

- Unexpected enhancements to water vapour or cloud feedbacks (λ). In this case warming triggers new cloud or water vapour feedbacks that amplify the expected warming.

Those marked (F) are disruptions to the earth's carbon cycle, the others, marked (λ), can be seen as temperature-dependent disruptions to feedbacks. In reality, many of the processes that affect feedbacks would likely trigger at least some changes in carbon sources and sinks. In practice, the separation between forcing surprises and feedback surprises is a little artificial, but it is a convenient first order way to organize potential surprises. All are temperature-dependent, and this temperature-dependence goes beyond the simple picture sketched out in the section above. Some could be triggered – committed to – well in advance of their actual coming to pass. Most disturbingly, none of these surprises are well-quantified. We regard them as unlikely, but find it hard to know just how unlikely. We are also unsure of the amplitude of these effects. Given the current state of the climate system, the current global mean temperature and the expected temperature rise, we do not know much about just how big an effect they will have. In spite of possibly being responsible for some dramatic effects in previous climates, it is hard to know how big an effect surprises like these will have in the future. Quantifying these remote and in many cases troubling possibilities is extremely difficult. Scientists write down models that they think give a justifiable and physically meaningful description of the process under study, and then attempt to use (usually palaeoclimate) data to constrain the magnitude and likelihood of the event. This is a difficult process: our knowledge of past climates essentially involves an inverse problem, in which we build up a picture of three-dimensional global climates from surface data proxies, which are often either biological data (tree-ring patterns, animal middens, etc.) or cryospheric (ice cores). Climate models are often run under palaeo conditions in order to provide an independent check on the adequacy of the models, which have generally been built to simulate today's conditions. Several sources of uncertainty are present in this process:

1. *Data uncertainty*. Generally, the deeper we look into the past, the less sure we are that our data are a good representation of global climate. The oceans, in particular, are something of a data void, and there are also issues about the behaviour of clouds in previous climates.
2. *Forcing uncertainty*. Uncertainty in forcing is quite large for the pre-satellite era generally, and again generally grows as one moves back in time. In very different climate regimes, such as ice ages, there are difficult questions regarding the size and reflectivity of ice sheets, for instance. Knowing that we have the right inputs when we model past climates is a significant challenge.
3. *Model uncertainty*. Climate models, while often providing reasonable simulations over continental scales, struggle to do a good job of modelling

smaller-scale processes (such as cloud formation). Model imperfections thus add another layer of complexity to the problem, and thus far at least, GCMs simulating previous climates such as the last glacial maximum have had little success in constraining even global mean properties (Crucifix, 2006).

In the presence of multiple scientific uncertainties, it is hard to know how to reliably quantify possibilities such as methane hydrate release or even meridional overturning circulation disruption. These things remain possibilities, and live areas of research, but in the absence of reliable quantification it is difficult to know how to incorporate such possibilities into our thinking about the more immediate and obvious problem of twenty-first century climate change.

13.5 Defining dangerous climate change

While this book has drawn the distinction between terminal and endurable risks (see Chapter 1), limits to acceptable climate change risk have usually been couched in terms of 'dangerous' risk. The United Nations Framework Convention on Climate Change (UNFCCC), outlines a framework through which nations might address climate change. Under the Convention, governments share data and try to develop national strategies for reducing GHG emissions while also adapting to expected impacts. Central to the convention, in article 2, is the intention to stabilize GHG concentrations (and hence climate):

> The ultimate objective of this Convention and any related legal instruments that the Conference of the Parties may adopt is to achieve, in accordance with the relevant provisions of the Convention, stabilization of greenhouse gas concentrations in the atmosphere at a level that would prevent dangerous anthropogenic interference with the climate system.

Article 2 introduced the problematic but perhaps inevitable concept of 'dangerous anthropogenic interference' (DAI) in the climate system into the scientific and policy communities. DAI has been much discussed, with many quite simple studies presenting it, to first order, in terms of possible global mean temperature thresholds (O'Neill and Oppenheimer, 2002; Mastrandrea and Schneider, 2004), or equilibrium responses to some indefinite atmospheric concentration of CO_2 (Nordahus, 2005), although some commentators provide more complete discussions of regional or not-always-temperature based metrics of DAI (Keller et al., 2005; Oppenheimer, 2005).

It has become usual to distinguish between two, ideally complementary ways of dealing with the threats posed by climate change: mitigation and adaptation.

The UNFCCC's glossary[1] describes mitigation as follows:

> In the context of climate change, a human intervention to reduce the sources or enhance the sinks of greenhouse gases. Examples include using fossil fuels more efficiently for industrial processes or electricity generation, switching to solar energy or wind power, improving the insulation of buildings, and expanding forests and other 'sinks' to remove greater amounts of carbon dioxide from the atmosphere.

While adaptation is:

> Adjustment in natural or human systems in response to actual or expected climatic stimuli or their effects, which moderates harm or exploits beneficial opportunities.

Internationally, the focus of most climate policy has been on establishing targets for mitigation. The Kyoto Protocol, for instance, set 'binding' targets for emissions reductions for those industrialized countries which chose to ratify it. This is changing as nations, communities, and businesses (1) come to appreciate the difficulty of the mitigation coordination problem and (2) realize that they are beginning to need to adapt their practices to a changing climate.

In the general case, what is dangerous depends on who one is, and what one is concerned about. Local thresholds in temperature (for coral reefs, say), integrated warming (polar bears), or sea level (for small island states) rise, to take a few examples, may constitute compelling metrics for regional or highly localized DAI, but it is not obvious that they warrant a great deal of global consideration. Moreover, the inertia in the system response implies that we are already committed to a certain level of warming: the climate of the decade between 2020 and 2030 is essentially determined by the recent accumulation of GHG and no realistic scenario for mitigation will alter that. Thus, as well as mitigating against the worst aspects of centennial timescale temperature rise, we also need to learn to adapt to climate change. Adaptation is inherently a regional or localized response, since the effects of climate change – though both most discernable and best constrained at the global level – actually affect people's lives at regional scales. Although mitigation is usually thought of as a genuinely global problem requiring a global scale response, adaptation is offers more opportunity for nation states and smaller communities to exercise effective climate policy. Adaptation alone is not expected to be sufficient to deal with climate change in the next century, and is best seen as part of a combined policy approach (along with mitigation) that allows us to avoid the very worst impacts of climate change, while adapting to those aspects that we are either already committed to, or for which the costs of mitigation would be too high.

[1] http://unfccc.int/essential_background/glossary/items/3666.php

13.6 Regional climate risk under anthropogenic change

The IPCC's Fourth Assessment Report highlights a number of expected changes at regional scales. These include the following:

- Extra-tropical storm tracks are expected to move polewards.
- Hot extremes are expected to increase in frequency.
- Hurricanes are expected to become more frequent, with more intense precipitation.
- Warming is expected over land areas, especially over the continents, and is also expected to be strong in the Arctic.
- Snow cover and sea-ice are expected to decrease.
- Patterns of precipitation are generally expected to intensify.

These findings are qualitatively similar to those highlighted in the IPCC's Third Assessment Report, though some of the details have changed, and in many cases scientists are more confident of the predicted impacts.

Impacts generally become more uncertain as one moves to smaller spatio-temporal scales. However, many large-scale processes, such as El Nino Southern Oscillation and the North Atlantic Oscillation, are crucially dependent on small-scale features such as convective systems over the maritime continent, and patterns of Atlantic sea-surface temperature, respectively. Given that the El Nino Southern Oscillation and the North Atlantic Oscillation are key determinants of variability and climate-related impacts on seasonal-to-decadal timescales, good representation of these features will become an essential part of many regions' adaptation strategies.

Adaptation strategies can take many forms. They can be based on strategies that take different sorts of stakeholder relationships with the impacts into account. Depending on the nature of one's exposure, risk profile, and preferences, one may opt to base one's strategy on one, or a mix, of several different principles, including: some reading of the precautionary principle, traditional cost–benefit analysis (usually where one is confident of the numbers underpinning the analysis) or robustness against vulnerability. Other adaptive principles have also been discussed, including novel solutions such as legal remedies through climate detection and attribution studies, which is like cost–benefit analysis, though backward looking (at actual attributable harm done) and done via the legal system rather than through economics (Allen et al., 2007). As the choice of principle, tools and policies tends to be highly situation specific, there tends to be a great diversity in adaptation strategies, which contrasts with the surprisingly narrow band of mitigation policies that have entered the public domain.

Apart from a range of possible principles guiding the adaptation strategy, the range of possible responses is also very large. *Technological responses* include improved sea and flood defences, as well as improved air-conditioning systems; *policy responses* include changes in zoning and planning practices (to avoid building in increasingly fire- or flood-prone areas, for instance); *managerial responses* include changes in firm behaviour – such as planting new crops or shifts in production – in anticipation of the impacts associated with climate change; and *behavioural responses* include changes in patterns of work and leisure and changes in food consumption.

In traditional risk analysis, the risk of an event is defined as the costs associated with its happening multiplied by the probability of the event. However, identifying the costs associated with adapting to climate change is deeply problematic, since (1) the damages associated with climate change are extremely hard to quantify in the absence of compelling regional models and (2) the economic future of the community implementing the adaptive strategy is highly uncertain. As the IPCC AR4 (p. 19) states:

> At present we do not have a clear picture of the limits to adaptation, or the cost, partly because effective adaptation measures are highly dependent on specific, geographical and climate risk factors as well as institutional, political and financial constraints.

An alternative strategy, and one that seems to be gaining popularity with the adaptation community, is the development of 'robust' strategies that seek to minimize the vulnerability of a community to either climate change, or even natural climate variability. Because of the difficulty in quantifying the costs of adaptation and likely development pathways, strategies emphasizing resilience are gaining in popularity.

13.7 Climate risk and mitigation policy

The default science position from which to think about climate mitigation policy has been the simple linear model of anthropogenic warming along with the expected solar and volcanic forcings, which remains basically the same (in the long-run) even in the presence of large volcanoes that we might expect only once in every 1000 years. This is usually coupled with a cost-benefit analysis of some kind to think about what sort of policy we should adopt. However, many of the hard to quantify possibilities that would disrupt this linear picture of climate change are often thought to lead to dramatic changes in our preferred policy responses to twenty-first century climate change.

The basic problem of anthropogenic climate change is that we are releasing too much carbon dioxide[2] into the atmosphere. The principal aspect of the

[2] And other GHG, but see Fig. 13.1: the forcing from CO_2 is by far the dominant forcing.

system we can control is the amount of CO_2 we emit.[3] This is related to the temperature response (a reasonable proxy for the climate change we are interested in) by the chain emissions → concentrations → forcings → temperature response, or in terms of the equations we introduced above, (13.2) → (13.3) → (13.1). Various aspects of this chain are subject to considerable scientific uncertainty. The mapping between emissions and concentrations is subject to carbon cycle uncertainties on the order of 30% of the emissions themselves, for a doubling of CO_2 (Huntingford and Jones, 2007); and though the concentrations to forcings step seems reasonably well behaved the forcing-response step is subject to large uncertainties surrounding the values of the parameters in Equation (13.1) (Hegerl et al., 2006). So even though decision makers are interested in avoiding certain aspects of the response, they are stuck with a series of highly uncertain inferences in order to calculate the response for a given emissions path.

In order to control CO_2 emissions, policy makers can prefer price controls (taxes) or quantity controls (permits). Quantity controls directly control the amount of a pollutant released into a system, whereas price controls do not. In the presence of uncertainty, if one fixes the price of a pollutant, one is uncertain about the quantity released; whereas if one fixes the quantity, one is uncertain about how to price the pollutant. Weitzman (1974) showed how, in the presence of uncertainty surrounding the costs and benefits of a good, a flat marginal benefit curve, compared to the marginal cost curve, leads to a preference for price regulation over quantity regulation. If anthropogenic climate change is as linear as it seems under the simple model mentioned above, as most studies have quite reasonably assumed to be the case, then a classical optimal cost-benefit strategy suggests that we ought to prefer price controls (Nordhaus, 1994; Roughgarden and Schneider, 1997, for instance). Intuitively, this makes sense: if the marginal benefits of abatement are a weaker function of quantity than the marginal costs are, then it suggests getting the price right is more important than specifying the quantity, since there is no reason to try to hit some given quantity regardless of the cost. If, however, the marginal benefits of abatement are a steep function of quantity, then it suggests that specifying the quantity is more important, since we would rather pay a sub-optimal amount than miss our environmental target.

[3] This may change if various geo-engineering approaches come to play a significant role in the management of the climate change problem (Crutzen, 2006; Keith, 2001). For a review of these possibilities, see Keith (2007). Though preliminary analysis suggests the idea of attempting to tune global mean temperatures by lowering insolation might ought not be dismissed out of hand (Govindasamy and Caldeira, 2000), more full analyses are required before this becomes a serious policy option (Kiehl, 2006). In any case, some problems of the direct effects of carbon – such as ocean acidification – are likely to remain significant issues, inviting the pessimistic possibility that geoengineering may carry its own severe climate-related risks.

This sort of intuition has implied that our best-guess, linear model of climate change leads to a preference for price controls. At the same time, if it were shown that catastrophic disruptions to the climate system were a real threat, it would lead us to opt instead for quantity controls. Ideally, we would like to work out exactly how likely catastrophic processes are, factor these into our climate model, and thus get a better idea of what sort of mitigation policy instrument to choose. However, catastrophic surprises are difficult to incorporate into our thinking about climate change policy in the absence of strong scientific evidence for the reliable calculation of their likelihood. Our most comprehensive models provide scant evidence to support the idea that anthropogenic climate change will provide dramatic disruptions to the reasonably linear picture of climate change represented by the simple EBM presented above (Meehl et al., 2007). At the same time, we know that there are plenty of parts of the Earth System that could, at least in theory, present catastrophic risks to human kind. Scientists attempt to model these possibilities, but this task is not always well-posed, because (1) in a system as complex as the Earth system it is can be hard to know if the model structure is representative of the phenomena one is purporting to describe and (2) data constraints are often extremely weak. Subjective estimates of the reduction in strength of the Atlantic meridional overturning circulation in response to a 4°C global mean temperature rise put the probability of an overturning circulation collapse at anywhere between about 0% and 60% (Zickfeld et al., 2007). Worse, the ocean circulation is reasonably well understood in comparison to many of the other processes we considered above in Section 13.4: we have very few constraints to help us quantify the odds of a clathrate gun event, or to know just how much tropical forest dieback or permafrost methane release to expect. Attempting to quantify the expected frequency or magnitude of some of the more extreme possibilities even to several magnitudes is difficult, in the presence of multiple uncertainties.

Therefore, while the science behind our best expectations of anthropogenic climate change suggests a comparatively linear response, warranting a traditional sort of cost-benefit, price control-based strategy, catastrophe avoidance suggests a quite different, more cautious approach. However, we find it hard to work out just how effective our catastrophe avoidance policy will be, since we know so little about the relative likelihood of catastrophes with or without our catastrophe avoidance policy.

13.8 Discussion and conclusions

Different policy positions are suggested depending on whether twenty-first century climate change continues to be essentially linear, in line with recent temperature change and our best-guess expectations, or disruptive in some new and possibly catastrophic way. In the first case, as we have seen, price controls are preferable since there's no particular reason to need to avoid any given quantity of atmospheric GHG; in the second, we *are* trying to remain under some threshold, and allow prices to adjust in response. Ideally, science would tell us what that threshold is, but as we have seen, this is a difficult, and elusive, goal. Furthermore, quantity controls remain indirect. The thing we wish to avoid is dangerous anthropogenic interference in the climate system. Yet we cannot control temperature, we can control only anthropogenic emissions of GHG, and these map to temperature increases through lagged and uncertain processes. This matters because it affects the way we construct policy. European policy makers, who have been at the forefront of designing policy responses to the threat of anthropogenic warming, generally back-infer acceptable 'emissions pathways' by inverting, starting with an arbitrarily chosen temperature target, and then attempting to invert the equations in this chapter. This bases short- and medium-term climate policy on the inversion of a series of long-term relationships, each of which is uncertain.

Dangerous, possibly even catastrophic, climate change is a risk, but it is exceptionally hard to quantify. There is evidence that the climate has changed radically and disruptively in the past. While it's true that a global mean temperature rise of 4°C would probably constitute dangerous climate change on most metrics, it would not necessarily be catastrophic in the sense of being a terminal threat. The challenges they suggest our descendents will face look tough, but endurable.

In terms of global climate, potential catastrophes and putative tipping points retain a sort of mythic aspect: they are part of a useful way of thinking about potentially rapid surprises in a system we do not fully understand, but we find it hard to know just how to incorporate these suspicions into our strategies. We cannot reliably factor them into our decision algorithms because we have such a poorly quantified understanding of their probability. We need to investigate the mechanisms more thoroughly to get a better handle on their amplitude, triggering mechanisms, and likelihood. There is no unique and uncontroversial way to bring these concerns within the realm of climate forecasting, nor, especially, our decision-making processes.

Suggestions for further reading

Pizer, W.A. http://www.ss.org.ss/Documents/RII-DP-03-31.pdg. Climate Change Catastrophes, Discussion Paper 03–31, Resources for the Future, Washington, DC, 2003. Provides a nice summary of the ways in which catastrophic risks can change the choice of climate mitigation policy instrument.

Stern Review. http://www.hm-treasury.gov.uk/independant_reviews/stern_review_economics_climate_change/stern_review_report.cfm. Although controversial in some parts, the Stern review is probably the most important document on climate change produced in the last few years: it has moved the discourse surrounding climate change away from scientific issues towards policy responses.

IPCC AR4 Summary for Policy Makers. WG1 provides a comprehensive review of the state of climate change science. To a large extent it echoes the previous assessment (2001).

S.H. Schneider, A. Rosencranz, J.O. Niles (2002). *Climate Change Policy: A Survey* (Island Press). Provides a nice one-volume synthesis of the physical and impacts science behind climate change alongside possible policy responses. Chapters solicited from a wide range of experts.

Schellnhuber, H.-J. (ed.). (2006). *Avoiding Dangerous Climate Change* (Cambridge, UK: Cambridge University Press). Discusses the state of science and some policy thinking regarding our ability to meet the aims of the UNFCCC and avoid dangerous climate change. Provides a nice selection of papers and think pieces by many of the leading commentators in the field.

Brooks, N. (2006). Cultural responses to aridity in the Middle Holocene and increased social complexity. *Quat. Int.*, **151**, 29–49. Something of an underappreciated gem, this paper combines the intriguing uncertainty of palaeoclimate research with a fascinating – if necessarily speculative – discussion of the developments of early civilizations around the globe. An interesting scientific and social look at how social responses to climate change can succeed or fail.

References

Allen, M.R., Pall, P., Stone, D., Stott, P., Frame, D., Min, S.-K., Nozawa,T., and Yukimoto, S. (2007). Scientific challenges in the attribution of harm to human influence on climate. *University of Pennsylvania Law Review*, **155**(6), 1353–1400.

Andronova, N., Schlesinger, M., Hulme, M., Desai, S., and Li, B. (2006). The concept of climate sensitivity: history and development. In Schlesinger, M.E., Kheshgi, H.S., Joel Smith, dela Chesnaye, F.C., Reilly, J.R., Wilson, T, Kolstad, C., (eds.), *Human-Induced Climate Change: An Interdisciplinary Assessment* (Cambridge: Cambridge University Press).

Benton, M.J. and Twitchett, R.J. (2003). How to kill (almost) all life: the end-Permian extinction event. *Trend Ecol. Evol.*, **18**(7), 358–365.

Brooks, N. (2006). Cultural responses to aridity in the Middle Holocene and increased social complexity. *Quat. Int.*, **151**, 29–49.

Claussen, M., Mysak, L.A., Weaver, A.J., Crucifix, M., Fichefet, T., Loutre, M.-F., Weber, S.L., Alcamo, J., Alexeev, V.A., Berger, A., Calov, R., Ganopolski, A., Goosse, H., Lohman, G., Lunkeit, F., Mokhov, I.I., Petoukhov, V., Stone, P., and Wang, Zh. (2002). Earth system models of intermediate complexity: closing the gap in the spectrum of climate models. *Clim. Dynam.*, **18**, 579–586.

Cox, P.M., Richard A. Betts, Chris D. Jones, Steven A. Spall and Ian J. Totterdell (9 November, 2000). Acceleration of global warming due to carbon cycle feedbacks in a coupled climate model. *Nature*, **408**(6809), 184–187.

Cox, P.M., Betts, R.A., Collins, M., Harris, P.P., Huntingford, C., and Jones, C.D. (2004). Amazonian forest dieback under climate-carbon cycle projections for the 21st century. *Theor. Appl. Climatol.*, **78**, 137–156.

Crowley, T.J. (2000). Causes of climate change over the past 1000 years. *Science*, **289**, 270–277.

Crucifix, M. (2006). Does the last glacial maximum constrain climate sensitivity? *Geophys. Res. Lett.*, **33**(18), L18701–L18705.

Crutzen, P.J. (2006). Albedo enhancement by stratospheric sulphur injections: a contribution to resolve a policy dilemma? *Clim. Change*, **77**, 211–220.

Dallimore, S.R. and Collett, T.S. (1995). Intrapermafrost gas hydrates from a deep core-hole in the Mackenzie Delta, Northwest-Territories, Canada. *Geology*, **23**(6), 527–530, 1995.

Diamond, J. (2005). *Collapse* (Penguin).

Flannery, T. (1994) *The Future Eaters* (Grove Press).

Forest, C.E., Stone, P.H., Sokolov, A.P., Allen, M.R., and Webster, M.D. (2002). Quantifying uncertainties in climate system properties with the use of recent climate observations. *Science*, **295**, 113–117.

Govindasamy, B. and Caldeira, K. (2000). Geoengineering Earth's Radiation Balance to Mitigate CO_2-Induced Climate Change. *Geophys. Res. Lett.*, **27**, 2141–2144.

Hansen, J., Russell, G., Lacis, A., Fung, I., and Rind, D. (1985). Climate response times: dependence on climate sensitivity and ocean mixing. *Science*, **229**, 857–859.

Hegerl, G.C., Zwiers, F.W., Braconnot, P., Gillett, N.P., Luo, Y., Marengo Orsini, J.A., Nicholls, N., Penner, J.E., and Stott, P.A. (2007): Understanding and Attributing Climate Change. In: *Climate Change 2007: The Physical Science Basis. Contribution of Working Group I to the Fourth Assessment Report of the Intergovernmental Panel on Climate Change* [Solomon, S., Qin, D., Manning, M., Chen, Z., Marquis, M., Averyt, K.B., Tignor, M., and Miller, H.L. (eds.)], Cambridge University Press, United Kingdom and New York, NY, USA.

Hegerl, G.C., Crowley, T.J., Hyde, W.T. and Frame, D.J. (2006). Climate sensitivity constrained by temperature reconstructions over the past seven centuries. *Nature*, **440**, 1029–1032.

Houghton, J.T., Ding, Y., Griggs, D.J., Noguer, M., van der Linden, P.J., Dai, X., Maskell, K., and Johnson, C.A., (eds.). (2001). *Climate Change 2001: The Science of Climate Change* (Cambridge: Cambridge University Press).

Hyde, W.T. and Crowley, T.J. (2000). Probability of future climatically significant volcanic eruptions. *J. Clim.*, **13**, 1445–1450.

Keith, D.W. (2001). Geoengineering. *Nature*, **409**, 420.

Keith, D.W. (2007). Engineering the planet. In Schneider, S. and Mastrandrea, M. (eds.), *Climate Change Science and Policy* (Island Press).

Kiehl, J.T. (2006). Geoengineering climate change: treating the symptom over the cause? *Clim. Change*, **77**, 227–228.

Kemp, M. (2005). Science in culture: inventing an icon. *Nature*, **437**, 1238.

Keller, K., Hall, M., Kim, S., Bradford, D.F., and Oppenheimer, M. (2005). Avoiding dangerous anthropogenic interference with the climate system. *Clim. Change*, **73**, 227–238.

Kennett, J.P., Cannariato, K.G., Hendy, I.L., and Behl, R.J. (2000). Carbon isotopic evidence for methane hydrate instability during Quaternary interstadials. *Science*, **288**, 128–133.

Knutti, R., Allen, R., Friedlingstein, Gregory, J.M., Hegerl, G.C., Meehl, G.A., Meinshausen, M., Murphy, J.M., Plattner, G.-K., Raper, S.C.B., Stocker, T.F., Stott, P.A., Teng, H., and Wigley, T.M.L. (2008). A Review of Uncertainties in Global Temperature Projections over the Twenty-First Century. *Journal of Climate*, 21, 2651–2663.

Levitus, S., Antonov, J., and Boyer, T. (2005). Warming of the world ocean, 1955–2003. *Geophys. Res. Lett.*, **32**, L02604.

Lenton, T.M. et al. (2006). The GENIE team, Millennial timescale carbon cycle and climate change in an efficient Earth system model. *Clim. Dynam.*, **26**, 687–711.

Lenton, T.M., Held, H., Hall, J.W., Kriegler, E., Ludd, W., Rahmstorf, S., and Schellnhuber, H.J. (2007). Tipping Elements in the Earth System. *Proc. Natl. Acad. Sci.*, **1**.

Lindzen, R.S., Chou, M.-D., and Hou, A.Y. (2001). Does the Earth have an adaptive iris? *Bull. Am. Meteorol. Soc.*, **82**, 417–432.

Lockwood, M. and Fröhlich, C. (2007). Recent oppositely directed trends in solar climate forcings and the global mean surface air temperature. *Proc. R. Soc. A.*, 10.1098/rspa.2007.1880.

Maslin, M., Owen, M., Day, S., and Long, D. (2004). Linking continental-slope failures and climate change: testing the clathrate gun hypothesis. *Geology*, **32**, 53–56.

Mason, B.G., Pyle, D.M., and Oppenheimer, C. (2004). The size and frequency of the largest explosive eruptions on Earth. *Bull. Volcanol.*, **66**(8), 735–748.

Mastrandrea, M. and Schneider, S. (2004). Probabilistic integrated assessment of "dangerous" climate change. *Science*, **304**, 571–575.

Mastrandrea, M. and Schneider, S. (2005). Probabilistic assessment of dangerous climate change and emission scenarios. In Schellnhuber, H.-J. et al. (eds.), *Avoiding Dangerous Climate Change* (DEFRA).

Meehl, G.A., Stocker, T.F., Collins, W., Friedlingstein, P., Gaye, A., Gregory, J., Kitoh, A., Knutti, R., Murphy, J., Noda, A., Raper, S., Watterson, I., Weaver, A., and Zhao, Z.-C. (2007). Global climate projections *Climate Change 2007: The Physical Science Basis. Contribution of Working Group I to the Fourth Assessment Report of the Intergovernmental Panel on Climate Change*, S. Solomon, D. Qin, and M. Manning, Eds., Cambridge University Press.

Meinshausen, M. (2005). On the risk of overshooting 2°C. In Schellnhuber, H.-J. et al. (eds.), *Avoiding Dangerous Climate Change* (DEFRA).

Murphy, J., David M.H. Sexton, David N. Barnett, Gareth S. Jones, Mark J. Webb, Matthew Collins & David A. Stainforth, (2004). Quantification of modelling uncertainties in a large ensemble of climate change simulations. *Nature*, **430**, 768–772.

Nordhaus, W.D. (1994). Managing the Global Commons. The Economics of Climate Change, MIT Press, Cambridge, MA.

Nordhaus, W. (2005). Life After Kyoto: Alternative Approaches to Global Warming Policies, *National Bureau of Economic Research, Working Paper* 11889.

O'Neill, B.C. and Oppenheimer, M. (2004). Climate change impacts are sensitive to the concentration stabilization path. *Proc. Natl. Acad. Sci.*, **101**, 16411–16416.

Oppenheimer, M. (2005). Defining dangerous anthropogenic interference: the role of science, the limits of science. *Risk Anal.*, **25**, 1–9.

Orr, J.C., Victoria J. Fabry, Olivier Aumont, Laurent Bopp, Scott C. Doney, Richard A. Feely, Anand Gnanadesikan, Nicolas Gruber, Akio Ishida, Fortunat Joos, Robert M. Key, Keith Lindsay, Ernst Maier-Reimer, Richard Matear, Patrick Monfray, Anne Mouchet, Raymond G. Najjar, Gian-Kasper Plattner, Keith B. Rodgers, Christopher L. Sabine, Jorge L. Sarmiento, Reiner Schlitzer, Richard D. Slater, Ian J. Totterdell, Marie-France Weirig, Yasuhiro Yamanaka and Andrew Yool, (2005). Anthropogenic ocean acidification over the twenty-first century and its impact on calcifying organisms. *Nature*, **437**, 681–686.

Pinto, J.P., Turco, R.P., and Toon, O.B. (1989) Self-limiting physical and chemical effects in volcanic eruption clouds. *J. Geophys. Res.*, **94**, 11165–11174.

Pizer, W.A. (2003). Climate change catastrophes. Discussion Paper 03-31, Resources for the Future, Washington, DC.

Roughgarden, T. and Schneider, S.H. (1997). Climate change policy: quantifying uncertainties for damage and optimal carbon taxes. *Energy Policy*, **27**(7), 371–434.

Sazonova, T.S., Romanovsky, V.E., Walsh, J.E., and Sergueev, D.O. (2004). Permafrost dynamics in the 20th and 21st centuries along the East Siberian transect. *J. Geophys. Res.-Atmos.*, **109**, D1.

Schellnhuber, H.-J. (ed.). (2006). *Avoiding Dangerous Climate Change* (Cambridge, UK: Cambridge University Press).

Schneider von Deimling, T., Held, H., Ganopolski, A., and Rahmstorf, S. (2006). Climate sensitivity estimated from ensemble simulations of glacial climate. *Clim. Dynam.*, **27**, 149.

Solomon, S. et al. (2007). IPCC: Summary for Policymakers. In Solomon, S., Qin, D., Manning, M., Chen, Z., Marquis, M., Averyt, K.B., Tignor, M., and Miller, H.L. (eds.), *Climate Change* 2007: *The Physical Science Basis*. Contribution of Working Group I to the Fourth Assessment Report of the Intergovernmental Panel on Climate Change (Cambridge, UK: Cambridge University Press and New York, NY, USA).

Weitzman, M.L., (1974). Prices vs. Quantities in: *Review of Economic Studies*, **41**(4), 477–491.

Zickfeld, K., Levermann, A., Morgan, M.G., Kuhlbrodt, T., Rahmstorf, S., and Keith, D. Expert judgements on the response of the Atlantic meridional overturning circulation to climate change. *Clim. Change*, accepted.

·14·

Plagues and pandemics: past, present, and future

Edwin Dennis Kilbourne

14.1 Introduction

This chapter is about pandemics, a somewhat ambiguous term, defined in the *Oxford English Dictionary* as 'a disease prevalent throughout a country, a continent, or the world'. In present modern usage the term takes greater cognizance of its original Greek derivation and is largely restricted to global prevalence (*pan demos*) – all people. The same source tells us that 'plague' has a broader meaning, implying a sudden unexpected event that is not necessarily a disease, but introducing the concept of acute, lethal, and sudden danger – characteristics that are connoted but not specifically denoted by the term 'pandemic'.

It will become apparent that glimpses of the future must consider the emergence of new pathogens, the re-emergence of old ones, the anthropogenic fabrication of novel agents, and changes in the environment and in human behaviour. In other words 'the problem' in addressing infectious disease threats is not one but many separable problems, each of which must be isolated in traditional scientific fashion and separately evaluated as components of what I like to call 'holistic epidemiology'. This emerging discipline comprises microbial and human genetics, human behaviour, global ecology, toxicology, and environmental change.

14.2 The baseline: the chronic and persisting burden of infectious disease

As we leave our mothers' wombs and enter this vale of tears (and sometimes before) we are invaded by microbes that may become our lifelong companions, profiting from this intimate relationship by the food and shelter that our bodies offer. They, in turn, often provide us with nutrients or vitamins derived from their own metabolic processes and may even immunize us against future assaults by related but less kindly microbes. In other words, we and they (usually) coexist in a state of armed neutrality and equilibrium.

But humans bear a chronic burden of infectious diseases. Included in this burden are some diseases that have demonstrated a capacity to break out in pandemic form, depending on the circumstances that are defined later. The less overt contributors to human misery will be briefly reviewed before discussing the nature of the acute aberrations that comprise the more dramatic pandemics and plagues that suddenly burst forth in a catastrophic manner.

Beginning at the end of the nineteenth century and culminating in the middle of the twentieth, the battle with infections seemed allied with the recognition of their microbial cause and the consequent development of vaccines, environmental sanitation, and later, antimicrobial drugs. In high income regions the accustomed childhood infections became a rarity with the development of vaccines for diphtheria, pertussis, measles, rubella, varicella, mumps, and poliomyelitis. It is paradoxical that poliomyelitis, or 'infantile paralysis', emerged as a consequence of the improved sanitation of food and water that postponed infection of infants and young children to later life when susceptibility to paralysis increases (Horstmann, 1955).

Important recent studies in which population census and income level of different regions have been determined have documented the expected, that is, that the price of poverty is an increased burden of infection – particularly of the intestinal tract – and that even diseases such as tuberculosis and measles for which there are drugs and vaccines continue to burden a large part of the world. The childhood mortality from measles in Africa is expected to reach half a million in 2006 and accounts for 4% of the total deaths of children each year, and yet the disease is a current and logical target for eradication because, as was the case with smallpox, the virus has no venue other than humans, and the present vaccine is highly effective.

Tuberculosis, an ancient disease never adequately suppressed by vaccine or antibiotics, is now resurgent, aided by the new, cryptic, and terrible challenge: HIV/AIDS. In regions of poor environmental sanitation, which include much of sub-Saharan Africa, diarrhoeal diseases persist, killing millions each year, mostly children.

We are not writing here about abstruse, unsolved scientific and medical problems but obvious economic ones. This is nowhere better documented than in the rates of perinatal mortality in low and high income groups; a disparity of more than twelvefold (according to WHO statistics for 2001). The unglamorous but nonetheless deadly infantile diarrhoeas and 'lower respiratory infections' are among the commonplace diseases that encumber humankind in poverty-stricken areas of the globe. The leading infectious causes of death in 2002 (comprising almost 20% of all causes) are shown in Table 14.1. About 75% of all deaths from infectious diseases are geographically localized to Southeast Asia and sub-Saharan Africa.

Table 14.1 Leading Global Causes of Deaths Due to Infectious Diseases[a]

Causes of Death	Annual Deaths (millions)
Respiratory infections	3.9
HIV/AIDS	2.9
Diarrhoeal diseases	1.8
Tuberculosis	1.7
Malaria	1.3–3.0[b]
Neglected diseases	0.5

[a] Data presented are for the year 2002.
[b] Deaths from direct or indirect effects of malaria.
Source: Global Health Foundation 2006 (summarizing data from WHO, 2004, 2006; UNAIDS, 2006; Breman et al., 2004; Hotez et al., 2007).

14.3 The causation of pandemics

Although the proximate causes of pandemics are microbial or viral, there are ancillary causes of almost equal importance. These include human behaviour, season of the year and other environmental conditions, and the state of induced or innate immunity. These will be discussed in the context of specific disease paradigms.

In the short list of causes of past plagues identified in Table 14.2 are the viruses of smallpox, influenza and yellow fever, the bacterial causes of plague, cholera and syphilis, and the protozoan cause of malaria.

14.4 The nature and source of the parasites

Adhering to this 'short list' *pro temps*, the sources of these pathogens in nature are illustrative of the variety of mechanisms by which they survive, and often flourish. Only one, the cholera *Vibrio*, is capable of independently living freely in the environment. The plague bacillus (*Yersinia pestis*) has its primary home in rodents, in which its spread is facilitated through the bites of rat fleas, which may seek food and shelter in humans when the infection has killed their rodent hosts. The *Plasmodia* of malaria have a complicated life cycle in which their multiplication in both vertebrates (including humans) and mosquitoes is required. The arbovirus cause of yellow fever is transmitted from one to another (in the case of urban yellow fever in humans) by mosquitoes that act essentially as flying hypodermic needles, or, in the case of jungle yellow fever, from animals (principally monkeys) via mosquitoes.

Table 14.2 Some Examples of Historically Significant Pandemics and Epidemics

Pandemic	Date(s)	Region	Cause	Lethal Impact
Plague of Justinian	AD 541–542	First: Constantinople; later dispersed worldwide	*Yersinia pestis*	25–40% of city's population (25 million), 50% of world population (100 million)
Black Death	13th till 15th century (first Asian outbreak: 1347)	Asia and Europe	*Yersinia pestis*	20–60% of population (43–75 million), 1/3 of the population of Europe
Smallpox	1520–1527	Mexico, central America, South America	Variola	200,000 deaths within the Aztec population (in some areas up to 75%)
Great Plague of London	1665–1666	London, around that time also other parts of Europe	*Yersinia pestis*	5–15% of city's population (75,000–100,000 or more)
Cholera	1829–1851, second pandemic	Global with Europe, North and South Amer. Involved	*Vibrio cholerae*	20,000 of 650,000 population in London (3%), but total morbidity not known
Yellow Fever	1853	New Orleans, LA	Arbovirus	$>12,000$ deaths – Louisiana thereby had highest death rate of any state in the United States in the 19th century
Influenza during World War I 'Spanish flu'	1918–1919	Global	Influenza A	20–50 million deaths – 2–3% of those ill, but enormous morbidity
HIV/AIDS	1981–2006	Global	Human immunodeficiency virus	25–65 million ill or dead

Note: Death rates, in contrast to total deaths, are notoriously unreliable because the appropriate denominators, that is, the number actually infected and sick is highly variable from epidemic to epidemic. Influenza, for example, is not even a reportable disease in the United States because in yearly regional epidemics it is so often confused with other respiratory infections. Therefore, the author has had to search for the most credible data of the past. Although he recognizes that it would have been desirable to convert all 'lethal impacts' to rates, the data available do not permit this. On the other hand, a simple statement of recorded deaths is more credible and certainly provides a useful measure of 'lethal impact'.

14.5 Modes of microbial and viral transmission

Microbes (bacteria and protozoa) and viruses can enter the human body by every conceivable route: through gastrointestinal, genitourinary, and respiratory tract orifices, and through either intact or injured skin. In what

Table 14.3 A Comparison of Prototype Pandemic Agents

Infectious Agent	Transmission	Morbidity	Mortality	Control
Viruses				
Yellow fever	Subcutaneous (mosquito)	Asymptomatic to fatal	20% of those manifestly ill	Environment and vaccine[a]
Smallpox	Respiratory tract	All severe	15–30%	Eradicated
Influenza	Respiratory tract	Asymptomatic to fatal	2–3%[b] or less	Vaccine
Rickettsia				
Typhus	Percutaneous (by louse faeces)	Moderate to severe	10–60%	Personal hygiene[c]
Bacteria				
Bubonic Plague	Rat flea bite	Untreated is always severe	>50%	Insect and rodent control
Cholera	By ingestion	Untreated is always severe	1–50%	Water sanitation
Protozoa				
Malaria	Mosquito	Variable	*Plasmodium falciparum* most lethal	Insect control

[a] Mosquito control.
[b] In the worst pandemic in history in 1918, mortality was 2–3%. Ordinarily, mortality is only 0.01% or less.
[c] Neglect of personal hygiene occurs in times of severe population disruption, during crowding, absence of bathing facilities or opportunity to change clothing. Disinfection of clothing is mandatory to prevent further spread of lice or aerosol transmission.

is known as vertical transmission the unborn infant can be infected by the mother through the placenta.

Of these routes, dispersal from the respiratory tract has the greatest potential for rapid and effective transmission of the infecting agent. Small droplet nuclei nebulized in the tubular bronchioles of the lung can remain suspended in the air for hours before their inhalation and are not easily blocked by conventional gauze masks. Also, the interior of the lung presents an enormous number of receptors as targets for the entering virus. For these reasons, influenza virus currently heads the list of pandemic threats (Table 14.3).

14.6 Nature of the disease impact: high morbidity, high mortality, or both

If a disease is literally pandemic it is implicit that it is attended by high morbidity, that is, many people become infected, and of these most become ill – usually within a short period of time. Even if symptoms are not severe, the sheer load of many people ill at the same time can become incapacitating for the function of a community and taxing for its resources. If a newly induced infection has a very high mortality rate, as often occurs with infections with alien agents from wild animal sources, it literally reaches a 'dead end': the death of the human victim.

Smallpox virus, as an obligate parasite of humans, when it moved into susceptible populations, was attended by both a high morbidity and a high mortality rate, but it was sufficiently stable in the environment to be transmitted by inanimate objects (fomites) such as blankets. (Schulman and Kilbourne, 1963).

Influenza, in the terrible pandemic of 1918, killed more than 20 million people but the overall mortality rate rarely exceeded 2–3% of those who were sick.

14.7 Environmental factors

Even before the birth of microbiology the adverse effects of filthy surroundings on health, as in 'The Great Stink of Paris' and of 'miasms', were assumed. But even late in the nineteenth century the general public did not fully appreciate the connection between 'germs and filth' (CDC, 2005). The nature of the environment has potential and separable effects on host, parasite, vector (if any), and their interactions. The environment includes season, the components of weather (temperature and humidity), and population density. Many instances of such effects could be cited but recently published examples dealing with malaria and plague are relevant. A resurgence of malaria in the East African highlands has been shown to be related to progressive increases in environmental temperature, which in turn have increased the mosquito population (Oldstone, 1998; Saha et al., 2006). In central Asia plague dynamics are driven by variations in climate as rising temperature affects the prevalence of *Yersinia pestis* in the great gerbil, the local rodent carrier. It is interesting that 'climatic conditions favoring plague apparently existed in this region at the onset of the Black Death as well as when the most recent plague (epidemic) arose in the same region ...' (Ashburn, 1947).

Differences in relative humidity can affect the survival of airborne pathogens, with high relative humidity reducing survival of influenza virus and low relative humidity (indoor conditions in winter) favouring its survival in aerosols (Simpson, 1954). But this effect is virus dependent. The reverse effect is demonstrable with the increased stability of picornaviruses in the high humidity of summer.

14.8 Human behaviour

One has no choice in the inadvertent, unwitting contraction of most infections, but this is usually not true of sexually transmitted diseases (STD). Of course one never chooses to contract a venereal infection, but the strongest of human drives that ensures the propagation of the species leads to the deliberate taking of risks – often at considerable cost, as biographer Boswell could ruefully testify. Ignorant behaviour, too, can endanger the lives of innocent people, for example when parents eschew vaccination in misguided attempts to spare their children harm. On the other hand, the deliberate exposure of young girls to rubella (German measles) before they reached the child-bearing age was in retrospect a prudent public health move to prevent congenital anomalies in the days before a specific vaccine was available.

14.9 Infectious diseases as contributors to other natural catastrophes

Sudden epidemics of infection may follow in the wake of non-infectious catastrophes such as earthquakes and floods. They are reminders of the dormant and often unapparent infectious agents that lurk in the environment and are temporarily suppressed by the continual maintenance of environmental sanitation or medical care in civilized communities. But developing nations carry an awesome and chronic load of infections that are now uncommonly seen in developed parts of the world and are often thought of as diseases of the past. These diseases are hardly exotic but include malaria and a number of other parasitic diseases, as well as tuberculosis, diphtheria, and pneumococcal infections, and their daily effects, particularly on the young, are a continuing challenge to public health workers.

The toll of epidemics vastly exceeds that of other more acute and sudden catastrophic events. To the extent that earthquakes, tsunamis, hurricanes, and floods breach the integrity of modern sanitation and water supply systems, they open the door to water-borne infections such as cholera and typhoid fever. Sometimes these illnesses can be more deadly than the original disaster. However, recent tsunamis and hurricanes have not been followed by expected major outbreaks of infectious disease, perhaps because most of such outbreaks in the past occurred after the concentration of refugees in crowded and unsanitary refugee camps. Hurricane Katrina, which flooded New Orleans in 2005, left the unusual sequelae of non-cholerogenic *Vibrio* infections, often with skin involvement as many victims were partially immersed in contaminated water (McNeill, 1976). When true cholera infections do occur, mortality can be sharply reduced if provisions are made for the rapid treatment of victims with fluid and electrolyte replacement. Azithromycin, a new antibiotic, is also highly effective (Crosby, 1976a).

14.10 Past plagues and pandemics and their impact on history

The course of history itself has often been shaped by plagues or pandemics.[1] Smallpox aided the greatly outnumbered forces of Cortez in the conquest of the Aztecs (Burnet, 1946; Crosby, 1976b; Gage, 1998; Kilbourne, 1981), and the Black Death (bubonic plague) lingered in Europe for three centuries (Harley et al., 1999), with a lasting impact on the development of the economy and cultural evolution. Yellow fever slowed the construction of the Panama Canal (Benenson, 1982). Although smallpox and its virus have been eradicated, the plague bacillus continues to cause sporadic deaths in rodents, in the American southwest, Africa, Asia, and South America (Esposito et al., 2006). Yellow fever is still a threat but currently is partially suppressed by mosquito control and vaccine, and cholera is always in the wings, waiting – sometimes literally – for a turn in the tide. Malaria has waxed and waned as a threat, but with the development of insecticide resistance to its mosquito carriers and increasing resistance of the parasite to chemoprophylaxis and therapy, the threat remains very much with us (Table 14.1).

On the other hand, smallpox virus (*Variola*) is an obligate human parasite that depends in nature on a chain of direct human-to-human infection for its survival. In this respect it is similar to the viral causes of poliomyelitis and measles. Such viruses, which have no other substrates in which to multiply, are prime candidates for eradication. When the number of human susceptibles has been exhausted through vaccination or by natural immunization through infection, these viruses have no other place to go.

Influenza virus is different from *Variola* on several counts: as an RNA virus it is more mutable by three orders of magnitude; it evolves more rapidly under the selective pressure of increasing human immunity; and, most important, it can effect rapid changes by genetic re-assortment with animal influenza viruses to recruit new surface antigens not previously encountered by humans to aid its survival in human populations (Kilbourne, 1981). Strangely, the most notorious pandemic of the twentieth century was for a time almost forgotten because of its concomitance with World War I (Jones, 2003).

[1] See especially Oldstone (1998), Ashburn (1947), Burnet (1946), Simpson (1954), Gage (1998), McNeill (1976), Kilbourne (1981), and Crosby (1976b).

14.11 Plagues of historical note

14.11.1 Bubonic plague: the Black Death

The word 'plague' has both specific and general meanings. In its specific denotation plague is an acute infection caused by the bacterium *Yersinia pestis,* which in humans induces the formation of characteristic lymph node swellings called 'buboes' – hence 'bubonic plague'. Accounts suggestive of plague go back millennia, but by historical consensus, pandemics of plague were first clearly described with the Plague of Justinian in AD 541 in the city of Constantinople. Probably imported with rats in ships bearing grain from either Ethiopia or Egypt, the disease killed an estimated 40% of the city's population and spread through the eastern Mediterranean with almost equal effect. Later (AD 588) the disease reached Europe, where its virulence was still manifest and its death toll equally high. The Black Death is estimated to have killed between a third and two-thirds of Europe's population. The total number of deaths worldwide due to the pandemic is estimated at 75 million people, whereof an estimated 20 million deaths occurred in Europe. Centuries later, a third pandemic began in China in 1855 and spread to all continents in a true pandemic manner. The disease persists in principally enzootic form in wild rodents and is responsible for occasional human cases in North and South America, Africa, and Asia. The WHO reports a total of 1000–3000 cases a year.

14.11.2 Cholera

Cholera, the most lethal of past pandemics, kills its victims rapidly and in great numbers, but is the most easily prevented and cured – given the availability of appropriate resources and treatment. As is the case with smallpox, poliomyelitis, and measles, it is restricted to human hosts and infects no other species. Man is the sole victim of *Vibrio cholerae.* But unlike the viral causes of smallpox and measles, *Vibrio* can survive for long periods in the free-living state before its ingestion in water or contaminated food.

Cholera is probably the first of the pandemics, originating in the Ganges Delta from multitudes of pilgrims bathing in the Ganges river. It spread thereafter throughout the globe in a series of seven pandemics covering four centuries, with the last beginning in 1961 and terminating with the first known introduction of the disease into Africa. Africa is now a principal site of endemic cholera.

The pathogenesis of this deadly illness is remarkably simple: it kills through acute dehydration by damaging cells of the small and large intestine and impairing the reabsorption of water and vital minerals. Prompt replacement of fluid and electrolytes orally or by intravenous infusion is all that is required for rapid cure of almost all patients. A single dose of a new antibiotic, azithromycin, can further mitigate symptoms.

14.11.3 Malaria

It has been stated that 'no other single infectious disease has had the impact on humans ... [that] malaria has had' (Harley et al., 1999). The validity of this statement may be arguable, but it seems certain that malaria is a truly ancient disease, perhaps 4000–5000 years old, attended by significant mortality, especially in children less than five years of age.

The disease developed with the beginnings of agriculture (Benenson, 1982), as humans became less dependent on hunting and gathering and lived together in closer association and near swamps and standing water – the breeding sites of mosquitoes. Caused by any of four Plasmodia species of protozoa, the disease in humans is transmitted by the *Anopheles* mosquito in which part of the parasite's replicative cycle occurs.

Thus, with its pervasive and enduring effects, malaria did not carry the threatening stigma of an acute cause of pandemics but was an old, unwelcome acquaintance that was part of life in ancient times. The recent change in this picture will be described in a section that follows.

14.11.4 Smallpox

There is no ambiguity about the diagnosis of smallpox. There are few, if any, asymptomatic cases of smallpox (Benenson, 1982) with its pustular skin lesions and subsequent scarring that are unmistakable, even to the layman. There is also no ambiguity about its lethal effects. For these reasons, of all the old pandemics, smallpox can be most surely identified in retrospect. Perhaps most dramatic was the decimation of Native Americans that followed the first colonization attempts in the New World. A number of historians have noted the devastating effects of the disease following the arrival of Cortez and his tiny army of 500 and have surmised that the civilized, organized Aztecs were defeated not by muskets and crossbows but by viruses, most notably smallpox, carried by the Spanish. Subsequent European incursions in North America were followed by similar massive mortality in the immunologically naïve and vulnerable Native Americans. Yet a more complete historical record presents a more complex picture. High mortality rates were also seen in some groups of colonizing Europeans (Gage, 1998). It was also observed, even that long ago, that smallpox virus probably comprised both virulent and relatively avirulent strains, which also might account for differences in mortality among epidemics. Modern molecular biology has identified three 'clades' or families of virus with genomic differences among the few viral genomes still available for study (Esposito et al., 2006). Other confounding and contradictory factors in evaluating the 'Amerindian' epidemics was the increasing use of vaccination in those populations (Ashburn, 1947), and such debilitating factors as poverty and stress (Jones).

14.11.5 Tuberculosis

That traditional repository of medical palaeo-archeology, 'an Egyptian mummy', in this case dated at 2400 BC, showed characteristic signs of tuberculosis of the spine (Musser, 1994). More recently, the DNA of *Mycobacterium tuberculosis* was recovered from a 1000-year-old Peruvian mummy (Musser, 1994).

The more devastating a disease the more it seems to inspire the poetic. In seventeenth century England, John Bunyan referred to 'consumption' (tuberculosis in its terminal wasting stages) as 'the captain of all these men of death' (Comstock, 1982). Observers in the past had no way of knowing that consumption (wasting) was a stealthy plague, in which the majority of those infected when in good health never became ill. Hippocrates' mistaken conclusion that tuberculosis killed nearly everyone it infected was based on observation of far advanced clinically apparent cases.

The 'White Plague' of past centuries is still very much with us. It is one of the leading causes of death due to an infectious agent worldwide (Musser, 1994). Transmitted as a respiratory tract pathogen through coughs and aerosol spread and with a long incubation period, tuberculosis is indeed a pernicious and stealthy plague.

14.11.6 Syphilis as a paradigm of sexually transmitted infections

If tuberculosis is stealthy at its inception, there is no subtlety to the initial acquisition of *Treponema pallidum* and the resultant genital ulcerative lesions (chancres) that follow sexual intercourse with the infected, or the florid skin rash that may follow. But the subsequent clinical course of untreated syphilis is stealthy indeed, lurking in the brain and spinal cord and in the aorta as a potential cause of aneurysm. Accurate figures on morbidity and mortality rates are hard to come by, despite two notorious studies in which any treatment available at the time was withheld after diagnosis of the initial acute stages (Gjestland, 1955; Kampmeir, 1974). The tertiary manifestations of the disease, general paresis tabes dorsalis, cardiovascular and other organ involvement by 'the great imitator', occurred in some 30% of those infected decades after initial infection.

Before the development of precise diagnostic technology, syphilis was often confused with other venereal diseases and leprosy so that its impact as a past cause of illness and mortality is difficult to ascertain.

It is commonly believed that just as other acute infections were brought into the New World by the Europeans, so was syphilis by Columbus's crew to the Old World. However, there are strong advocates of the theory that the disease was exported from Europe rather than imported from America. Both propositions are thoroughly reviewed by Ashburn (1947).

14.11.7 Influenza

Influenza is an acute, temporarily incapacitating, febrile illness characterized by generalized aching (arthralgia and myalgia) and a short course of three to seven days in more than 90% of cases. This serious disease, which kills hundreds of thousands every year and infects millions, has always been regarded lightly until its emergence in pandemic form, which happened only thrice in the twentieth century, including the notorious pandemic of 1918–1919 that killed 20–50 million people (Kilbourne, 2006a). It is a disease that spreads rapidly and widely among all human populations. In its milder, regional, yearly manifestations it is often confused with other more trivial infections of the respiratory tract, including the common cold. In the words of the late comedian Rodney Dangerfield, 'it gets no respect'. But the damage the virus inflicts on the respiratory tract can pave the way for secondary bacterial infection, often leading to pneumonia. Although vaccines have been available for more than fifty years (Kilbourne, 1996), the capacity of the virus for continual mutation warrants annual or biannual reformulation of the vaccines.

14.12 Contemporary plagues and pandemics

14.12.1 HIV/AIDS

Towards the end of the twentieth century a novel and truly dreadful plague was recognized when acquired immunodeficiency disease syndrome (AIDS) was first described and its cause established as the human immunodeficiency virus (HIV). This retrovirus (later definitively subcategorized as a *Lenti* [slow] *virus*) initially seemed to be restricted to a limited number of homosexual men, but its pervasive and worldwide effects on both sexes and young and old alike are all too evident in the present century.

Initial recognition of HIV/AIDS in 1981 began with reports of an unusual pneumonia in homosexual men caused by *Pneumocystis carinii* and previously seen almost exclusively in immunocompromised subjects. In an editorial in *Science* (Fauci, 2006), Anthony Fauci writes, 'Twenty five years later, the human immunodeficiency virus (HIV) ... has reached virtually every corner of the globe, infecting more than 65 million people. Of these, 25 million have died'. Much has been learned in the past twenty-five years. The origin of the virus is most probably chimpanzees (Heeney et al., 2006), who carry asymptomatically a closely related virus, SIV (S for simian). The disease is no longer restricted to homosexuals and intravenous drug users but indeed, particularly in poor countries, is a growing hazard of heterosexual intercourse. In this rapidly increasing, true pandemic, perinatal infection can occur, and the effective battery of antiretroviral drugs that have been developed for mitigation of the disease are available to few in impoverished areas of the world. It is a tragedy that AIDS is easily prevented by the use of condoms or by circumcision, means that in many places are either not available or not condoned by social mores or cultural habits. The roles of sexual practices and of the social dominance of men over women emphasize the importance of human behaviour and economics in the perpetuation of disease.

Other viruses have left jungle hosts to infect humans (e.g., Marburg virus) (Bausch et al., 2006) but in so doing have not modified their high mortality rate in a new species in order to survive and be effectively transmitted among members of the new host species. But, early on, most of those who died *with* AIDS did not die *of* AIDS. They died from the definitive effects of a diabolically structured virus that attacked cells of the immune system, striking down defences and leaving its victims as vulnerable to bacterial invaders as are the pitiable, genetically immunocompromised children in hospital isolation tents.

14.12.2 Influenza

Influenza continues to threaten future pandemics as human-virulent mutant avian influenza viruses, such as the currently epizootic H5N1 virus, or by recombination of present 'human' viruses with those of avian species. At the time of writing (June 2007) the H5N1 virus remains almost exclusively epizootic in domestic fowl, and in humans, the customary yearly regional epidemics of H3N2 and H1N1 'human' subtypes continue their prevalence. Meanwhile, vaccines utilizing reverse genetics technology and capable of growing in cell culture are undergoing improvements that still have to be demonstrated in the field. Similarly, antiviral agents are in continued development but are as yet unproven in mass prophylaxis.

14.12.3 HIV and tuberculosis: the double impact of new and ancient threats

The ancient plague of tuberculosis has never really left us, even with the advent of multiple drug therapy. The principal effect of antimicrobial drugs has been seen in the richer nations, but to a much lesser extent in the economically deprived, in which the drugs are less available and medical care and facilities are scanty. However, in the United States, a progressive decline in cases was reversed in the 1980s (after the first appearance of AIDS) when tuberculosis cases increased by 20%. Of the excess, at least 30% were attributed to AIDS-related cases. Worldwide, tuberculosis is the most common opportunistic infection in HIV-infected persons and the most common cause of death in patients with AIDS.

The two infections may have reciprocal enhancing effects. The risk of rapid progression of pre-existing tuberculosis infection is much greater among those with HIV infection and the pathogenesis of the infection is altered, with an increase in non-pulmonary manifestation of tuberculosis. At the same time, the immune *activation* induced by response to tuberculosis may paradoxically be associated with an increase in the viral load and accelerated progression of HIV infection. The mechanism is not understood.

14.13 Plagues and pandemics of the future

14.13.1 Microbes that threaten without infection: the microbial toxins

Certain microbial species produce toxins that can severely damage or kill the host. The bacterial endotoxins, as the name implies, are an integral part of the microbial cell and assist in the process of infection. Others, the exotoxins (of anthrax, botulism), are elaborated and can produce their harmful effects, as do other prefabricated, non-microbial chemical poisons. These microbial poisons by themselves seem unlikely candidates as pandemic agents. Anthrax spores through the mail caused seventeen illnesses and five deaths in the United States in 2001 (Elias, 2006). Accordingly, a one billion dollar contract was awarded by the U.S. Department of Health and Human Services for an improved vaccine. Development was beset with problems, and delivery is not expected until 2008 (Elias, 2006). In any case, as non-propagating agents, microbial toxins do not seem to offer a significant pandemic threat.

14.13.2 Iatrogenic diseases

Iatrogenic diseases are those unintentionally induced by physicians and the altruism of the dead – or, 'The way to [health] is paved with good intentions'. An unfortunate result of medical progress can be the unwitting induction of disease and disability as new treatments are tried for the first time. Therefore, it will not be surprising if the accelerated and imaginative devising of new technologies in the future proves threatening at times. Transplantation of whole intact vital organs, including heart, kidney, and even liver, has seen a dramatic advance, although as an alien tissue, rejection by the immune system of the patient has been a continuing problem.

Reliance on xenotransplantation of non-human organs and tissues such as porcine heart valves does not seem to have much future because they may carry dangerous retroviruses. In this connection, Robin Weiss proposes that 'we need a Hippocratic oath for public health that would minimize harm to the community resulting from the treatment of individuals' (Weiss, 2004).

All these procedures have introduced discussion of quality of life (and death) values, which will and should continue in the future. Based on present evidence, I do not see these procedures as instigators of pandemics unless potentially pandemic agents are amplified or mutated to virulence in the immunosuppressed recipients of this bodily largesse.

How complicated can things get? A totally unforeseen complication of the successful restoration of immunologic function by the treatment of AIDS with antiviral drugs has been the activation of dormant leprosy as a consequence (McNeil, 2006; Visco-Comandini et al., 2004).

14.13.3 The homogenization of peoples and cultures

There is evidence from studies of isolated populations that such populations, because of their smaller gene pool, are less well equipped to deal with initial exposure to unaccustomed infectious agents introduced by genetically and racially different humans. This is suggested by modern experience with measles that demonstrated 'intensified reactions to [live] measles vaccine in [previously unexposed] populations of American Indians' (Black et al., 1971). This work and studies of genetic antigen markers in the blood have led Francis Black to propose, '[P]eople of the New World are unusually susceptible to the diseases of the Old not just because they lack any [specific] resistance but primarily because, as populations, they lack genetic heterogeneity (Black, 1992, 1994). They are susceptible because agents of disease can adapt to each population as a whole and cause unusual damage' (Black, 1992, 1994). If I may extrapolate from Black's conclusions, a population with greater genetic heterogeneity would fare better with an 'alien' microbial or viral invasion.

Although present-day conflicts, warlike, political, and otherwise, seem to fly in the face of attaining genetic or cultural uniformity and the 'one world' ideal, in fact, increasing genetic and cultural homogeneity is a fact of life in many parts of the world. Furthermore, barriers to communication are breached by the World Wide Web, the universal e-mail post office, by rapid and frequent travel, and the ascendancy of the English language as an international tongue that is linking continents and ideas as never before.

We have already seen the rapid emergence of the respiratory virus, SARS, in humans and its transport from China to Canada. We also have learned that, unlike influenza, close and sustained contact with patients was required for the further perpetuation of the epidemic (Kilbourne, 2006b). This experience serves to emphasize that viruses can differ in their epidemic pattern of infection even if their target and site of infection are the same.

With this caveat, let us recall the rapid and effective transmission of influenza viruses by aerosols but the highly variable experience of isolated population groups in the pandemic of 1918. Groups sequestered from the outside world (and in close contact in small groups) prior to that epidemic suffered higher morbidity and mortality, suggesting that prior more frequent experience with non-pandemic influenza A viruses had at least partially protected those in more open societies. The more important point is that such 'hot houses' could favour the emergence of even more transmissible strains of virus than those initially introduced. Such hot houses or hotbeds in sequestered societies would be lacking in our homogenized world of the future.

To consider briefly the more prosaic, but no less important aspects of our increasingly homogeneous society, the mass production of food and behavioural fads concerning their consumption has led to the 'one rotten apple' syndrome. If one contaminated item, apple, egg, or most recently spinach leaf, carries a billion bacteria – not an unreasonable estimate – and it enters a pool of cake mix constituents, and is then packaged and sent to millions of customers nationwide, a bewildering epidemic may ensue.

14.13.4 Man-made viruses

Although the production and dangers of factitious infectious agents are considered elsewhere in this book (see Chapter 20), the present chapter would be incomplete without some brief consideration of such potential sources of plagues and pandemics of the future.

No one has yet mixed up a cocktail of off-the-shelf nucleotides to truly 'make' a new virus. The genome of the extinct influenza virus of 1918 has been painstakingly resurrected piece by piece from preserved human lung tissue (Taubenberger et al., 2000). This remarkable accomplishment augurs well for the palaeo-archeology of other extinct viral 'dodos', but whether new or old, viruses cannot truly exist without host cellular substrates on which to replicate, as does the resurrected 1918 virus, which can multiply, and indeed, kill animals. The implications are chilling indeed. Even confined to the laboratory, these or truly novel agents will become part of a global gene pool that will lie dormant as a potential threat to the future.

Predictive principles for the epidemiology of such human (or non-human) creations can perhaps be derived from the epidemiology of presently familiar agents, for example, pathogenesis (Kilbourne, 1985).

14.14 Discussion and conclusions

If sins of omission have been committed here, it is with the recognition that Pandora's Box was full indeed and there is space in these pages to discuss only those infectious agents that have demonstrated a past or present capacity for creating plagues or pandemics, or which now appear to be emerging as serious threats in the future. I now quote from an earlier work.

In anticipating the future, we must appreciate the complexity of microbial strategies for survival. The emergence of drug-resistant mutants is easily understood as a consequence of Darwinian selection. Less well appreciated is the fact that genes for drug resistance are themselves transmissible to still other bacteria or viruses. In other instances, new infectious agents may arise through the genetic recombination of bacteria or of viruses which individually may not be pathogenic. By alteration of their environment, we have abetted the creation of such new pathogens by the promiscuous overuse or misuse of antimicrobial drugs. The traditional epidemiology of individual infectious agents has been superseded by a molecular epidemiology of their genes. (Kilbourne, 2000, p. 91)

Many of my colleagues like to make predictions. 'An influenza pandemic is inevitable', 'The mortality rate will be 50%', etc. This makes good press copy, attracts TV cameras, and raises grant funding. Are influenza pandemics likely? Possibly, except for the preposterous mortality rate that has been proposed. Inevitable? No, not with global warming and increasing humidity, improved animal husbandry, better epizootic control, and improving vaccines. This does not have the inevitability of shifting tectonic plates or volcanic eruptions.

Pandemics, if they occur, will be primarily from respiratory tract pathogens capable of airborne spread. They can be quickly blunted by vaccines, if administrative problems associated with their production, distribution, and administration are promptly addressed and adequately funded. (A lot of 'ifs' but maybe we can start learning from experience!)

Barring extreme mutations of the infective agent or changed methods of spread, all of them can be controlled (and some eradicated) with presently known methods. The problem, of course, is an economic one, but such organizations as the Global Health Foundation and the Bill and Melinda Gates Foundation offer hope for the future by their organized and carefully considered programs, which have identified and targeted specific diseases in specific developing regions of the world.

In dealing with the novel and the unforeseen – the unconventional prions of Bovine Spongioform Encephalitis that threatened British beef (WHO, 2002) and the exotic imports such as the lethal Marburg virus that did not come from Marburg (Bausch et al., 2006) – we must be guided by the lessons of the past, so it is essential that we reach a consensus on what these lessons are. Of these, prompt and continued epidemiological surveillance for the odd and unexpected and use of the techniques of molecular biology are of paramount importance (admirably reviewed by King et al. [2006]). For those diseases not amenable to environmental control, vaccines, the ultimate personal suits of armour that will protect the wearer in all climes and places, must be provided.

Should we fear the future? If we promptly address and properly respond to the problems of the present (most of which bear the seeds of the future), we should not fear the future. In the meantime we should not cry 'Wolf!', or even 'Fowl!', but maintain vigilance.

Suggestions for further reading

All suggestions are accessible to the general intelligent reader except for the Burnet book, which is 'intermediate' in difficulty level.

Burnet, F.M. (1946). *Virus as Organism: Evolutionary and Ecological Aspects of Some Human Virus Diseases* (Cambridge, MA: Harvard University Press). A fascinating glimpse into a highly original mind grappling with the burgeoning but incomplete knowledge of virus diseases shortly before mid-twentieth century, and striving for a synthesis of general principles in a series of lectures given at Harvard University; all the more remarkable because Burnet won a shared Nobel Prize later for fundamental work on immunology.

Crosby, A.W. (1976). *Epidemic and Peace,* 1918 (Westport, CT: Greenwood). This pioneering book on the re-exploration of the notorious 1918 influenza pandemic had been surprisingly neglected by earlier historians and the general public.

Dubos, R. (1966). *Man Adapting* (New Haven, CT: Yale University Press). This work may be used to gain a definitive understanding of the critical inter-relationship of microbes, environment, and humans. This is a classic work by a great scientist-philosopher which must be read. Dubos' work with antimicrobial substances from soil immediately preceded the development of antibiotics.

Kilbourne, E.D. (1983). Are new diseases really new? *Natural History,* **12**, 28. An early essay for the general public on the now popular concept of 'emerging diseases', in which the prevalence of paralytic poliomyelitis as the price paid for improved sanitation, Legionnaire's Disease as the price of air conditioning, and the triggering of epileptic seizures by the flashing lights of video games are all considered.

McNeill, W.H. (1977). *Plagues and Peoples* (Garden City, NY: Anchor Books). Although others had written earlier on the impact of certain infectious diseases on the course of history, McNeill recognized that there was no aspect of history that was untouched by plagues and pandemics. His book has had a significant influence on how we now view both infection and history.

Porter, K.A. (1990). *Pale Horse, Pale Rider* (New York: Harcourt Brace & Company). Katherine Anne Porter was a brilliant writer and her evocation of the whole tragedy of the 1918 influenza pandemic with this simple, tragic love story tells us more than a thousand statistics.

References

Ashburn, P.A. (1947). *The Ranks of Death – A Medical History of the Conquest of America* (New York: Coward-McCann).

Barnes, D.S. (2006). *The Great Stink of Paris and the Nineteenth Century Struggle against Filth and Germs* (Baltimore, MD: Johns Hopkins University Press

Bausch, D.G. and Nichol, S.T., Muymembe-Tamfum, J.J. (2006). Marburg Hemorrhagic Fever Associated with Multiple Genetic Lineages of Virus. *N. Engl. J. Med.* **355**, 909–919.

Benenson, A.S. (1982). Smallpox. In Evans, A.S. (ed.), *Viral Infections of Humans*. 2nd edition, p. 542 (New York: Plenum Medical Book Company).

Black, F.L. (1992). Why did they die? *Science*, **258**, 1739–1740.

Black, F.L. (1994). An explanation of high death rates among New World peoples when in contact with Old World diseases. *Perspect. Biol. Med.*, **37**(2), 292–303.

Black, F.L., Hierholzer, W., Woodall, J.P., and Pinhiero, F. (1971). Intensified reactions to measles vaccine in unexposed populations of American Indians. *J. Inf. Dis.*, **124**, 306–317.

Both, G.W., Shi, C.H., and Kilbourne, E.D. (1983). Hemagglutinin of swine influenza virus: a single amino acid change pleotropically affects viral antigenicity and replication. *Proc. Natl. Acad. Sci. USA*, **80**, 6996–7000.

Burnet, F.M. (1946). *Virus as Organism* (Cambridge, MA: Harvard University Press).

CDC. (2005). Vibrio illnesses after hurricane Katrina – multiple states. *Morbidity Mortality Weekly Report*, **54**, 928–931.

Comstock, G.W. (1982). Tuberculosis. In Evans, A.S. and Feldman, H.A. (eds.), *Bacterial Infections of Humans*, p. 605 (New York: Plenum Medical Book Company).

Crosby, A.W., Jr. (1976). Virgin soil epidemics as a factor in the depopulation in America. *William Mary Q.*, **33**, pp. 289–299.

Elias, P. (2006). Anthrax dispute suggests Bioshield woes. *Washington Post*, 6 October, 2006.

Esposito, J.J., Sammons, S.A., Frace, A.M., Osborne, J.D., Melissa Olsen-Rasmussen, Ming Zhang, Dhwani Govil, Inger K. Damon, Richard Kline, Miriam Laker, Yu Li, Geoffrey L. Smith, Hermann Meyer, James W. LeDuc, Robert M. Wohlhueter (2006). Genome sequence diversity and clues to the evolution of Variola (smallpox) virus. *Science*, **313**, 807–812.

Fauci, A.S. (2006). Twenty-five years of HIV/AIDS. *Science*, **313**, 409.

Gage, K.L. (1998). In Collier, L., Balows, A., Sussman, M., and Hausles, W.J. (eds.), *Topley and Wilson's Microbiology and Microbiological Infections*, Vol. 3, pp. 885–903 (London: Edward Arnold).

Gambaryan, A.S., Matrosovich, M.N., Bender, C.A., and Kilbourne, E.D. (1998). Differences in the biological phenotype of low-yielding (L) and high-yielding (H) variants of swine influenza virus A/NJ/11/76 are associated with their different receptor-binding activity. *Virology*, **247**, 223.

Gjestland, T. (1955). The Oslo study of untreated syphilis – an epidemiologic investigation of the natural course of syphilistic infection as based on a re-study of the Boeck-Bruusgaard material. *Acta Derm. Venereol.*, **35**, 1.

Harley, J., Klein, D., Lansing, P. (1999). *Microbiology*, pp. 824–826 (Boston: McGraw-Hill).

Heeney, J.L., Dalgleish, A.G., and Weiss, R.A. (2006). Origins of HIV and the evolution of resistance to AIDS. *Science*, **313**, 462–466.

Horstmann, D.M. (1955). Poliomyelitis: severity and type of disease in different age groups. *Ann. N. Y. Acad. Sci.*, **61**, 956–967.

Jones, D.S. (2003). Virgin soils revisited. *William Mary Q.*, **60**, pp. 703–742.

Kampmeir, R.H. (1974). Final report on the 'Tuskegee Syphilis Study'. *South Med. J.*, **67**, 1349–1353.

Kilbourne, E.D. (1981). Segmented genome viruses and the evolutionary potential of asymmetrical sex. *Perspect. Biol. Med.*, **25**, 66–77.

Kilbourne, E.D. (1985). Epidemiology of viruses genetically altered by man – predictive principles. In Fields, B., Martin, M., and Potter, C.W. (eds.), *Banbury Report 22: Genetically Altered Viruses and the Environment*, pp. 103–117 (Cold Spring Harbor, NY: Cold Spring Harbor Laboratories).

Kilbourne, E.D. (1996). A race with evolution – a history of influenza vaccines. In Plotkin, S. and Fantini, B. (eds.). *Vaccinia, Vaccination and Vaccinology: Jenner, Pasteur and Their Successors*, pp. 183–188 (Paris: Elsevier).

Kilbourne, E.D. (2000). Communication and communicable diseases: cause and control in the 21st century. In Haidemenakis, E.D. (ed.), *The Sixth Olympiad of the Mind, 'The Next Communication Civilization'*, November 2000, St. Georges, Paris, p. 91 (International S.T.E.P.S. Foundation).

Kilbourne, E.D. (2006a). Influenza pandemics of the 20th century. *Emerg. Infect. Dis.*, **12(1)**, 9.

Kilbourne, E.D. (2006b). SARS in China: prelude to pandemic? *JAMA*, **295**, 1712–1713. Bookreview.

King, D.A., Peckham, C., Waage, J.K., Brownlie, M.E., and Woolhouse, M.E.G. (2006). Infectious diseases: preparing for the future. *Science*, **313**, 1392–1393.

McNeil, D.G., Jr. (2006). Worrisome new link: AIDS drugs and leprosy. *The New York Times*, pp. F1, F6, 24 October, 2006.

McNeill, W.H. (1976). *Plagues and Peoples*, p. 21 (Garden City, NY: Anchor Books).

Musser, J.M. (1994). Is Mycobacterium tuberculosis, 15,000 years old? *J. Infect. Dis.*, **170**, 1348–1349.

Oldstone, M.B.A. (1998). *Viruses, Plagues and History*, pp. 31–32 (Oxford: Oxford University Press).

Pascual, M., Ahumada, J.A., Chaves, L.F., Rodo, X., and Bouma, M. (2006). Malaria resurgence in the East African highlands: temperature trends revisited. *Proc. Natl. Acad. Sci. USA*, **103**, 5829–5834.

Patz, J.A. and Olson, S.H. (2006). Malaria risk and temperature: influences from global climate change and local land use practices. *Proc. Natl. Acad. Sci. USA*, **103**, 5635–5636.

Saha, D., Karim, M.M., Khan, W.A., Ahmed, S., Salam, M.A., and Bennish, M.I. (2006). Single dose Azithromycin for the treatment of cholera in adults. *N. Engl. J. Med.*, **354**, **354**, 2452–2462.

Schulman, J.L. and Kilbourne, E.D. (1966). Seasonal variations in the transmission of influenza virus infection in mice. In *Biometeorology II, Proceedings of the Third International Biometeorological Congress*, Pau, France, 1963, pp. 83–87 (Oxford: Pergamon).

Simpson, H.N. (1954). The impact of disease on American history. *N. Engl. J. Med.*, **250**, 680.

Stearn, E.W. and Stearn, A.E. (1945). *The Effect of Smallpox on the Destiny of the Amerindian*, pp. 44–45 (Boston: Bruce Humphries).

Stenseth, N.C., Samia, N.I., Hildegunn Viljugrein, Kyrre Linné Kausrud, Mike Begon, Stephen Davis, Herwig Leirs, V.M. Dubyanskiy, Jan Esper, Vladimir S. Ageyev,

Nikolay L. Klassovskiy, Sergey B. Pole, and Kung-Sik Chan (2006). Plague dynamics are driven by climate variation. *Proc. Natl. Acad. Sci. USA*, **103**, 13110–13115.

Taubenberger, J.K., Reid, A.H., and Fanning, T.G. (2000). The 1918 influenza virus: a killer comes into view. *Virology*, **274**, 241–245.

Tumpey, T.M., Maines, T.R., Van Hoeven, N., Glaser, L., Solorzano, A., Pappas, C., Cox, N.J., Swayne, D.E., Palese, P., Katz, J.M., and Garcia-Sastre, A. (2007). A two-amino acid change in the hemagglutinin of the 1918 influenza virus abolishes transmission. *Science*, **315**, 655–659.

Visco-Comandini, U., Longo, B., Cozzi, T., Paglia, M.G., Antonucci, G. (2004). Tuberculoid leprosy in a patient with AIDS: a manifestation of immune restoration syndrome. *Scand. J. Inf. Dis.*, **36**, 881–883.

Weiss, R.A. (2004). Circe, Cassandra, and the Trojan Pigs: xenotransplantation. *Am. Phil. Soc.*, **148**, 281–295.

WHO. (2002). WHO fact sheet no. 113 (Geneva: WHO).

WHO. (2004). *Bull. WHO*, **82**, 1–81.

·15·

Artificial Intelligence as a positive and negative factor in global risk

Eliezer Yudkowsky

15.1 Introduction

By far the greatest danger of Artificial Intelligence (AI) is that people conclude too early that they understand it. Of course, this problem is not limited to the field of AI. Jacques Monod wrote: 'A curious aspect of the theory of evolution is that everybody thinks he understands it' (Monod, 1974). The problem seems to be unusually acute in Artificial Intelligence. The field of AI has a reputation for making huge promises and then failing to deliver on them. Most observers conclude that AI is hard, as indeed it is. But the *embarrassment* does not stem from the difficulty. It is difficult to build a star from hydrogen, but the field of stellar astronomy does not have a terrible reputation for promising to build stars and then failing. The critical inference is *not* that AI is hard, but that, for some reason, it is very easy for people to think they know far more about AI than they actually do.

It may be tempting to ignore Artificial Intelligence because, of all the global risks discussed in this book, AI is probably hardest to discuss. We cannot consult actuarial statistics to assign small annual probabilities of catastrophe, as with asteroid strikes. We cannot use calculations from a precise, precisely confirmed model to rule out events or place infinitesimal upper bounds on their probability, as with proposed physics disasters. But this makes AI catastrophes more worrisome, not less.

The effect of many cognitive biases has been found to *increase* with time pressure, cognitive busyness, or sparse information. Which is to say that the more difficult the analytic challenge, the more important it is to avoid or reduce bias. Therefore I strongly recommend reading my other chapter (Chapter 5) in this book before continuing with this chapter.

15.2 Anthropomorphic bias

When something is universal enough in our everyday lives, we take it for granted to the point of forgetting it exists.

Imagine a complex biological adaptation with ten necessary parts. If each of the ten genes is independently at 50% frequency in the gene pool – each gene possessed by only half the organisms in that species – then, on average, only 1 in 1024 organisms will possess the full, functioning adaptation. A fur coat is not a significant evolutionary advantage unless the environment reliably challenges organisms with cold. Similarly, if gene B depends on gene A, then gene B has no significant advantage unless gene A forms a reliable part of the *genetic* environment. *Complex, interdependent* machinery is necessarily *universal* within a sexually reproducing species; it cannot evolve otherwise (Tooby and Cosmides, 1992). One robin may have smoother feathers than another, but both will have wings. Natural selection, while feeding on variation, uses it up (Sober, 1984).

In every known culture, humans experience joy, sadness, disgust, anger, fear, and surprise (Brown, 1991), and exhibit these emotions through the same means, namely facial expressions (Ekman and Keltner, 1997). We all run the same engine under our hoods, although we may be painted in different colours – a principle that evolutionary psychologists call *the psychic unity of humankind* (Tooby and Cosmides, 1992). This observation is both explained and required by the mechanics of evolutionary biology.

An anthropologist will not excitedly report of a newly discovered tribe: 'They eat food! They breathe air! They use tools! They tell each other stories!' We humans forget how alike we are, living in a world that only reminds us of our differences.

Humans evolved to model other humans – to compete against and cooperate with our own conspecifics. It was a reliable property of the ancestral environment that every powerful intelligence you met would be a fellow human. We evolved to understand our fellow humans *empathically*, by placing ourselves in their shoes; for that which needed to be modelled was similar to the modeller. Not surprisingly, human beings often 'anthropomorphize' – expect humanlike properties of that which is not human. In *The Matrix* (Wachowski and Wachowski, 1999), the supposed 'artificial intelligence' Agent Smith initially appears utterly cool and collected, his face passive and unemotional. But later, while interrogating the human Morpheus, Agent Smith gives vent to his disgust with humanity – and his face shows the human–universal facial expression for disgust.

Querying your own human brain works fine, as an adaptive instinct, if you need to predict other humans. If you deal with any other kind of optimization process – if, for example, you are the eighteenth-century theologian William Paley, looking at the complex order of life and wondering how it came to be – then anthropomorphism is flypaper for unwary scientists, a trap so sticky that it takes a Darwin to escape.

Experiments on anthropomorphism show that subjects anthropomorphize unconsciously, often flying in the face of their deliberate beliefs. In a study by Barrett and Keil (1996), subjects strongly professed belief in non-anthropomorphic properties of God: that God could be in more than one place at a time, or pay attention to multiple events simultaneously. Barrett and Keil presented the same subjects with stories in which, for example, God saves people from drowning. The subjects answered questions about the stories, or retold the stories in their own words, in such ways as to suggest that God was in only one place at a time and performed tasks sequentially rather than in parallel. Serendipitously for our purposes, Barrett and Keil also tested an additional group using otherwise identical stories about a superintelligent computer named 'Uncomp'. For example, to simulate the property of omnipresence, subjects were told that Uncomp's sensors and effectors 'cover every square centimetre of the earth and so no information escapes processing'. Subjects in this condition also exhibited strong anthropomorphism, though significantly less than the God group. From our perspective, the key result is that even when people consciously believe an AI is unlike a human, they still visualize scenarios as if the AI were anthropomorphic (but not quite as anthropomorphic as God).

Back in the era of pulp science fiction, magazine covers occasionally depicted a sentient monstrous alien – colloquially known as a bug-eyed monster (BEM) – carrying off an attractive human female in a torn dress. It would seem the artist believed that a non-humanoid alien, with a wholly different evolutionary history, would sexually desire human females. People do not usually make mistakes like that by explicitly reasoning: 'All minds are likely to be wired pretty much the same way, so presumably a BEM will find human females sexually attractive'. Probably the artist did not *ask* whether a giant bug *perceives* human females as attractive. Rather, a human female in a torn dress *is sexy* – inherently so, as an intrinsic property. They who made this mistake did not think about the insectoid's mind; they focused on the woman's torn dress. If the dress were not torn, the woman would be less sexy; the BEM does not enter into it.[1]

It is also a serious error to begin from the conclusion and search for a neutral-seeming line of reasoning leading there; this is rationalization. If it is self-brain query that produced that first fleeting mental image of an insectoid chasing a human female, then anthropomorphism is the underlying cause of that belief, and no amount of rationalization will change that.

Anyone seeking to reduce anthropomorphic bias in himself or herself would be well advised to study evolutionary biology for practice, preferably evolutionary biology with maths. Early biologists often anthropomorphized

[1] This is a case of a deep, confusing, and extraordinarily common mistake that E.T. Jaynes named the *mind projection fallacy* (Jaynes and Bretthorst, 2003). Jaynes, a physicist and theorist of Bayesian probability, coined 'mind projection fallacy' to refer to the error of confusing states of knowledge with properties of objects. For example, the phrase *mysterious phenomenon* implies that mysteriousness is a property of the phenomenon itself. If I am ignorant about a phenomenon, then this is a fact about my state of mind, not a fact about the phenomenon.

natural selection – they believed that evolution would do the same thing they would do; they tried to predict the effects of evolution by putting themselves 'in evolution's shoes'. The result was a great deal of nonsense, which first began to be *systematically* exterminated from biology in the late 1960s, for example, by Williams (1966). Evolutionary biology offers both mathematics and case studies to help hammer out anthropomorphic bias.

Evolution strongly conserves some structures. Once other genes that depend on a previously existing gene evolve, the early gene is set in concrete; it cannot mutate without breaking multiple adaptations. Homeotic genes – genes controlling the development of the body plan in embryos – tell many other genes when to activate. Mutating a homeotic gene can result in a fruit fly embryo that develops normally except for not having a head. As a result, homeotic genes are so strongly conserved that many of them are the same in humans and fruit flies – they have not changed since the last common ancestor of humans and bugs. The molecular machinery of ATP synthase is essentially the same in animal mitochondria, plant chloroplasts, and bacteria; ATP synthase has not changed significantly since the rise of eukaryotic life 2 billion years ago.

Any two AI designs might be less similar to one another than you are to a petunia.

The term 'Artificial Intelligence' refers to a vastly greater *space of possibilities* than does the term *Homo sapiens*. When we talk about 'AIs' we are really talking about *minds-in-general*, or optimization processes in general. Imagine a map of mind design space. In one corner, a tiny little circle contains all humans, within a larger tiny circle containing all biological life; and all the rest of the huge map is *the space of minds-in-general*. The entire map floats in a still vaster space, *the space of optimization processes*. Natural selection creates complex functional machinery without mindfulness; evolution lies inside the space of optimization processes but outside the circle of minds.

It is this *enormous* space of possibilities that outlaws anthropomorphism as legitimate reasoning.

15.3 Prediction and design

We cannot query our own brains for answers about non-human optimization processes – whether bug-eyed monsters, natural selection, or Artificial Intelligences. How then may we proceed? How can we predict what Artificial Intelligences will do? I have deliberately asked this question in a form that makes it intractable. By the halting problem, it is impossible to predict whether an *arbitrary* computational system implements any input-output function, including, say, simple multiplication (Rice, 1953). So how is it possible that human engineers can build computer chips that reliably implement multiplication? Because human engineers deliberately use designs that they *can* understand.

Anthropomorphism leads people to believe that they can make predictions, given no more information than that something is an 'intelligence' – anthropomorphism will go on generating predictions regardless, your brain automatically putting itself in the shoes of the 'intelligence'. This may have been one contributing factor to the embarrassing history of AI, which stems not from the difficulty of AI as such, but from the mysterious ease of acquiring erroneous beliefs about what a given AI design accomplishes.

To make the statement that a bridge will support vehicles up to 30 tons, civil engineers have two weapons: choice of initial conditions, and safety margin. They need not predict whether an *arbitrary* structure will support 30-ton vehicles, only design a single bridge of which they can make this statement. And though it reflects well on an engineer who can correctly calculate the exact weight a bridge will support, it is also acceptable to calculate that a bridge supports vehicles of *at least* 30 tons – albeit to assert this vague statement *rigorously* may require much of the same theoretical understanding that would go into an exact calculation.

Civil engineers hold themselves to high standards in predicting that bridges will support vehicles. Ancient alchemists held themselves to much lower standards in predicting that a sequence of chemical reagents would transform lead into gold. How much lead into how much gold? What is the exact causal mechanism? It is clear enough why the alchemical researcher *wants* gold rather than lead, but why should this sequence of reagents transform lead to gold, instead of gold to lead or lead to water?

Some early AI researchers believed that an artificial neural network of layered thresholding units, trained via back propagation, would be 'intelligent'. The wishful thinking involved was probably more analogous to alchemy than civil engineering. Magic is on Donald Brown's list of human universals (Brown, 1991); science is not. We do not *instinctively* see that alchemy will not work. We do not *instinctively* distinguish between rigorous understanding and good storytelling. We do not *instinctively* notice an expectation of positive results that rests on air.

The human species came into existence through natural selection, which operates through the non-chance retention of chance mutations. One path leading to global catastrophe – to someone pressing the button with a mistaken idea of what the button does – is that AI comes about through a similar accretion of working algorithms, with the researchers having no deep understanding of how the combined system works. Nonetheless, they believe the AI will be friendly, with no strong visualization of the exact processes involved in producing friendly behaviour, or any detailed understanding of what they mean by friendliness. Much as early AI researchers had strong mistaken vague expectations for their programmes' intelligence, we imagine that these

AI researchers succeed in constructing an intelligent programme, but have strong mistaken vague expectations for their programme's friendliness.

Not knowing how to build a friendly AI is not deadly by itself, in any specific instance, if you know you do not know. It is a *mistaken* belief that an AI will be friendly, which implies an obvious path to global catastrophe.

15.4 Underestimating the power of intelligence

We tend to see individual differences instead of human universals. Thus, when someone says the word 'intelligence', we think of Einstein, instead of humans.

Individual differences of human intelligence have a standard label, *Spearman's g* aka *g-factor,* a controversial interpretation of the solid experimental result that different intelligence tests are highly correlated with each other and with real-world outcomes such as lifetime income (Jensen, 1999). Spearman's g is a statistical abstraction from individual differences of intelligence between humans, who as a *species* are far more intelligent than lizards. Spearman's g is abstracted from millimetre height differences among a species of giants.

We should not confuse Spearman's g with *human general intelligence,* our capacity to handle a wide range of cognitive tasks incomprehensible to other species. General intelligence is a between-species difference, a complex adaptation, and a human universal found in all known cultures. There may as yet be no academic consensus on intelligence, but there is no doubt about the existence, or the power, of the thing-to-be-explained. There is *something* about humans that let us set our footprints on the Moon. And, jokes aside, you will not find many CEOs, or yet professors of academia, who are chimpanzees. You will not find many acclaimed rationalists, artists, poets, leaders, engineers, skilled networkers, martial artists, or musical composers who are mice. Intelligence is the foundation of human power, the strength that fuels our other arts.

The danger of confusing general intelligence with g-factor is that it leads to tremendously underestimating the potential impact of Artificial Intelligence. (This applies to underestimating potential good impacts, as well as potential bad impacts.) Even the phrase 'transhuman AI' or 'artificial superintelligence' may still evoke images of book-smarts-in-a-box: an AI that is *really good* at cognitive tasks stereotypically associated with 'intelligence', like chess or abstract mathematics. But not superhumanly persuasive, or far better than humans at predicting and manipulating human social situations, or inhumanly clever in formulating long-term strategies. So instead of Einstein, should we think of, say, the nineteenth-century political and diplomatic genius Otto von Bismarck? But that is only the mirror version of the error. The entire range from the village idiot to Einstein, or from the village idiot to Bismarck, fits into a small dot on the range from amoeba to human.

If the word 'intelligence' evokes Einstein instead of humans, then it may sound sensible to say that intelligence is no match for a gun, as if guns had grown on trees. It may sound sensible to say that intelligence is no match for money, as if mice used money. Human beings did not *start out* with major assets in claws, teeth, armour, or any of the other advantages that were the daily currency of other species. If you had looked at humans from the perspective of the rest of the ecosphere, there was no hint that the soft pink things would eventually clothe themselves in armoured tanks. We *invented* the battleground on which we defeated lions and wolves. We did not match them claw for claw, tooth for tooth; we had our own ideas about what mattered.

Vinge (1993) aptly observed that a future containing smarter-than-human minds is *different in kind*. AI does not belong to the same graph that shows progress in medicine, manufacturing, and energy. AI is not something you can casually mix into a *lumpenfuturistic* scenario of skyscrapers and flying cars and nanotechnological red blood cells that let you hold your breath for eight hours. Sufficiently tall skyscrapers do not potentially start doing their own engineering. Humanity did not rise to prominence on Earth by holding its breath longer than other species.

The catastrophic scenario that stems from underestimating the power of intelligence is that someone builds a button, and does not care enough what the button does, because they do not think the button is powerful enough to hurt them. Or the wider field of AI researchers will not pay enough attention to risks of strong AI, and therefore good tools and firm foundations for friendliness will not be available when it becomes possible to build strong intelligences.

And one should not fail to mention – for it also impacts upon existential risk – that AI could be the powerful solution to other existential risks, and by mistake we will ignore our best hope of survival. The point about underestimating the potential impact of AI is symmetrical around potential good impacts and potential bad impacts. That is why the title of this chapter is 'Artificial Intelligence as a positive and negative factor in global risk', not 'Global risks of Artificial Intelligence'. The prospect of AI interacts with global risk in more complex ways than that.

15.5 Capability and motive

There is a fallacy often committed in discussion of Artificial Intelligence, especially AI of superhuman capability. Someone says: 'When technology advances far enough, we'll be able to build minds far surpassing human intelligence. Now, it's obvious that how large a cheesecake you can make depends on your intelligence. A superintelligence could build *enormous* cheesecakes – cheesecakes the size of cities – by golly, the future will be full of giant cheesecakes!' The question is whether the superintelligence *wants* to build giant cheesecakes. The vision leaps directly from *capability* to *actuality*, without considering the necessary intermediate of *motive*.

The following chains of reasoning, considered in isolation without supporting argument, all exhibit the Fallacy of the Giant Cheesecake:

- A sufficiently powerful AI could overwhelm any human resistance and wipe out humanity. (And the AI would decide to do so.) Therefore we should not build AI.
- A sufficiently powerful AI could develop new medical technologies capable of saving millions of human lives. (And the AI would decide to do so.) Therefore we should build AI.
- Once computers become cheap enough, the vast majority of jobs will be performable by AI more easily than by humans. A sufficiently powerful AI would even be better than us at maths, engineering, music, art, and all the other jobs we consider meaningful. (And the AI will decide to perform those jobs.) Thus after the invention of AI, humans will have nothing to do, and we will starve or watch television.

15.5.1 Optimization processes

The above deconstruction of the Fallacy of the Giant Cheesecake invokes an intrinsic anthropomorphism – the idea that motives are separable; the implicit assumption that by talking about 'capability' and 'motive' as separate entities, we are carving reality at its joints. This is a useful slice but an anthropomorphic one.

To view the problem in more general terms, I introduce the concept of an *optimization process:* a system that hits small targets in large search spaces to produce coherent real-world effects.

An optimization process steers the future into particular regions of the possible. I am visiting a distant city, and a local friend volunteers to drive me to the airport. I do not know the neighbourhood. When my friend comes to a street intersection, I am at a loss to predict my friend's turns, either individually or in sequence. Yet I can predict the *result* of my friend's unpredictable actions: we will arrive at the airport. Even if my friend's house were located elsewhere in the city, so that my friend made a wholly different sequence of turns, I would just as confidently predict our destination. Is this not a strange situation to be in, scientifically speaking? I can predict the *outcome* of a process, without being able to predict any of the *intermediate steps* in the process. I will speak of the region into which an optimization process steers the future as that optimizer's *target*.

Consider a car, say a Toyota Corolla. Of all possible configurations for the atoms making up the Corolla, only an infinitesimal fraction qualifies as a useful working car. If you assembled molecules at random, many *many* ages of the universe would pass before you hit on a car. A tiny fraction of the design space does describe vehicles that we would recognize as faster, more efficient, and safer than the Corolla. Thus the Corolla is not *optimal* under the designer's goals. The Corolla is, however, *optimized,* because the designer had to hit a comparatively infinitesimal target in design space just to create a working car, let alone a car of the Corolla's quality. You cannot build so much as an effective wagon by sawing boards randomly and nailing according to coinflips. To hit such a tiny target in configuration space requires a powerful optimization process.

The notion of an 'optimization process' is *predictively useful* because it can be easier to understand the *target* of an optimization process than to understand its step-by-step *dynamics.* The above discussion of the Corolla assumes *implicitly* that the designer of the Corolla was trying to produce a 'vehicle', a means of travel. This assumption deserves to be made explicit, but it is not wrong, and it is highly useful in understanding the Corolla.

15.5.2 Aiming at the target

The temptation is to ask what 'AIs' will 'want', forgetting that the space of minds-in-general is much wider than the tiny human dot. One should resist the temptation to spread quantifiers over all possible minds. Storytellers spinning tales of the distant and exotic land called Future, say how the future *will be.* They make *predictions.* They say, 'AIs will attack humans with marching robot armies' or 'AIs will invent a cure for cancer'. They do not propose complex relations between initial conditions and outcomes – that would lose the audience. But we need relational understanding to *manipulate* the future, steer it into a region palatable to humankind. If we do not steer, we run the danger of ending up where we are going.

The critical challenge is not to *predict* that 'AIs' will attack humanity with marching robot armies, or alternatively invent a cure for cancer. The task is not even to make the prediction for an *arbitrary* individual AI design. Rather the task is choosing into existence some *particular* powerful optimization process whose beneficial effects can legitimately be asserted.

I *strongly urge* my readers not to start thinking up reasons why a fully generic optimization process would be friendly. Natural selection is not friendly, nor does it hate you, nor will it leave you alone. Evolution cannot be so anthropomorphized, it does not work like you do. Many pre-1960s biologists expected natural selection to do all sorts of nice things, and rationalized all sorts of elaborate reasons why natural selection would do it. They were disappointed, because natural selection itself did not start out knowing that it wanted a humanly nice result, and then rationalize elaborate ways to produce nice results using selection pressures. Thus the events in Nature were outputs of causally different process from what went on in the pre-1960s biologists' minds, so that prediction and reality diverged.

Wishful thinking adds detail, constrains prediction, and thereby creates a burden of improbability. What of the civil engineer who hopes a bridge will not fall? Should the engineer argue that bridges in general are not likely to fall? But Nature itself does not rationalize reasons why bridges should not fall. Rather, the civil engineer overcomes the burden of improbability through specific choice guided by specific understanding. A civil engineer starts by desiring a bridge; then uses a rigorous theory to select a bridge design that supports cars; then builds a real-world bridge whose structure reflects the calculated design; and thus the real-world structure supports cars, thus achieving harmony of predicted positive results and actual positive results.

15.6 Friendly Artificial Intelligence

It would be a very good thing if humanity knew how to choose into existence a powerful optimization process with a particular target. In more colloquial terms, it would be nice if we knew how to build a nice AI.

To describe the *field of knowledge* needed to address that challenge, I have proposed the term 'Friendly AI'. In addition to referring to a body of technique, 'Friendly AI' might also refer to the *product* of technique – an AI created with specified motivations. When I use the term *Friendly* in either sense, I capitalize it to avoid confusion with the intuitive sense of 'friendly'.

One common reaction I encounter is for people to immediately declare that Friendly AI is an impossibility because any sufficiently powerful AI will be able to modify its own source code to break any constraints placed upon it.

The first flaw you should notice is a Giant Cheesecake Fallacy. Any AI with free access to its own source would, in principle, possess the *ability* to modify its own source code in a way that changed the AI's optimization target. This does not imply the AI has the *motive* to change its own motives. I would not knowingly swallow a pill that made me enjoy committing murder, because *currently* I prefer that my fellow humans do not die.

But what if I try to modify myself and make a mistake? When computer engineers *prove* a chip valid – a good idea if the chip has 155 million transistors and you cannot issue a patch afterwards – the engineers use human-guided, machine-verified formal proof. The glorious thing about *formal* mathematical proof is that a proof of ten billion steps is just as reliable as a proof of ten steps. But human beings are not trustworthy to peer over a purported proof of ten billion steps; we have too high a chance of missing an error. And present-day theorem-proving techniques are not smart enough to design and prove an entire computer chip on their own – current algorithms undergo an exponential explosion in the search space. Human mathematicians can prove theorems far more complex than what modern theorem-provers can handle, without being defeated by exponential explosion. But human mathematics is informal and unreliable; occasionally, someone discovers a flaw in a previously accepted informal proof. The upshot is that human engineers guide a theorem-prover through the *intermediate* steps of a proof. The human chooses the next lemma, and a complex theorem-prover generates a formal proof, and a simple verifier checks the steps. That is how modern engineers build reliable machinery with 155 million interdependent parts.

Proving a computer chip correct requires a synergy of human intelligence and computer algorithms, as *currently* neither suffices on its own. Perhaps a true AI could use a similar *combination of abilities* when modifying its own code – that would have *both* the capability to *invent* large designs without being defeated by exponential explosion, and *also* the ability to *verify* its steps with extreme reliability. That is one way a true AI might remain knowably stable in its goals, even after carrying out a large number of self-modifications.

This chapter will not explore the above idea in detail (see Schmidhuber [2003] for a related notion). But one ought to think about a challenge and study it in the best available technical detail, *before* declaring it impossible – especially if great stakes are attached to the answer. It is disrespectful to human ingenuity to declare a challenge unsolvable without taking a close look and exercising creativity. It is an enormously strong statement to say that you *cannot* do a thing – that you *cannot* build a heavier-than-air flying machine, that you *cannot* get useful energy from nuclear reactions, or that you *cannot* fly to the Moon. Such statements are universal generalizations, quantified over every single approach that anyone ever has or ever will think up for solving the problem. It only takes a single counterexample to falsify a universal quantifier. The statement that Friendly (or friendly) AI is *theoretically impossible*, dares to quantify over *every possible* mind design and *every possible* optimization process – including human beings, who are also minds, some of whom are nice and wish they were nicer. At this point there are any number of vaguely plausible reasons why Friendly AI might be *humanly* impossible, and it is still more likely that the problem is solvable but no one will get around to solving it in time. But one should not write off the challenge so quickly, especially considering the stakes involved.

15.7 Technical failure and philosophical failure

Bostrom (2001) defines an existential catastrophe as one that extinguishes Earth-originating intelligent life *or permanently destroys a substantial part of its potential*. We can divide potential failures of attempted Friendly AI into two informal fuzzy categories, *technical failure* and *philosophical failure*. Technical failure is when you try to build an AI and it does not work the way you think it should – you have failed to understand the true workings of your own code. Philosophical failure is trying to build the wrong thing, so that even if you succeeded you would still fail to help anyone or benefit humanity. Needless to say, the two failures are not mutually exclusive.

The border between these two cases is thin, since most philosophical failures are much easier to explain in the presence of technical knowledge. In theory you ought to first say what you *want*, then figure out *how* to get it. In practice it often takes a deep technical understanding to figure out what you want.

15.7.1 An example of philosophical failure

In the late nineteenth century, many honest and intelligent people advocated communism, all in the best of good intentions. The people who first invented and spread and swallowed the communist meme were usually, in sober historical fact, idealists. The *first* communists did not have the example of Soviet Russia to warn them. *At that time, without benefit of hindsight, it must have sounded like a pretty good idea.* After the revolution, when communists came into power and were corrupted by it, other motives came into play; but this itself was not something the first idealists predicted, however predictable it may have been. It is important to understand that the authors of huge catastrophes need not be evil, or even unusually stupid. If we attribute every tragedy to evil or unusual stupidity, we will look at ourselves, correctly perceive that we are not evil or unusually stupid, and say: 'But that would never happen to *us*'.

What the first communist revolutionaries thought would happen, as the empirical consequence of their revolution, was that people's lives would improve: labourers would no longer work long hours at backbreaking labour and make little money from it. This turned out to be not the case, to put it mildly. But what the first communists *thought* would happen, was not so very different from what advocates of other political systems thought would be the empirical consequence of *their* favourite political systems. They thought people would be happy. They were wrong.

Now imagine that someone should attempt to programme a 'Friendly' AI to implement communism, or libertarianism, or anarcho-feudalism, or *favourite political system,* believing that this shall bring about utopia. People's favourite political systems inspire blazing suns of positive affect, so the proposal will sound like a really good idea to the proposer.

We could view the programmer's failure on a moral or ethical level – say that it is the result of someone trusting themselves too highly, failing to take into account their own fallibility, refusing to consider the possibility that communism might be mistaken after all. But in the language of Bayesian decision theory, there is a complementary technical view of the problem. From the perspective of decision theory, the choice for communism stems from combining an empirical belief with a value judgement. The *empirical* belief is that communism, when implemented, results in a specific outcome or class of outcomes: people will be happier, work fewer hours, and possess greater material wealth. This is ultimately an *empirical* prediction; even the part about happiness is a real property of brain states, though hard to measure. If you implement communism, either this outcome eventuates or it does not. The value judgement is that this outcome satisfies or is preferable to current conditions. Given a different *empirical* belief about the *actual real-world consequences* of a communist system, the decision may undergo a corresponding change.

We would expect a true AI, an Artificial General Intelligence, to be capable of changing its empirical beliefs (or its probabilistic world model, etc.). If somehow Charles Babbage had lived before Nicolaus Copernicus, somehow computers had been invented before telescopes, and somehow the programmers of that day and age successfully created an Artificial General Intelligence, it would not follow that the AI would believe forever after that the Sun orbited the Earth. The AI might transcend the factual error of its programmers, provided that the programmers understood inference rather better than they understood astronomy. To build an AI that *discovers* the orbits of the planets, the programmers need not know the maths of Newtonian mechanics, only the maths of Bayesian probability theory.

The folly of programming an AI to implement communism, or any other political system, is that you are programming *means* instead of *ends*. You are programming in a fixed decision, without that decision being re-evaluable after acquiring improved empirical knowledge about the results of communism. You are giving the AI a fixed decision without telling the AI how to re-evaluate, at a higher level of intelligence, the fallible process that produced that decision.

If I play chess against a stronger player, I cannot predict *exactly* where my opponent will move against me – if I could do that, I would necessarily be at least that strong at chess myself. But I can predict the end result, which is a win for the other player. I know the region of possible futures my opponent is aiming for, which is what lets me predict the destination, even if I cannot see the path. When I am at my most creative, that is when it is hardest to predict my actions, and *easiest* to predict the *consequences* of my actions (providing that you know and understand my goals!). If I want a better-than-human chess player, I have to programme a *search* for winning moves. I cannot programme in specific moves because then the chess player will not be any better than I am. When I launch a search, I necessarily sacrifice my ability to predict the *exact* answer in advance. To get a really good answer you must sacrifice your ability to predict the answer, albeit not your ability to say what the question is.

15.7.2 An example of technical failure

In place of laws constraining the behavior of intelligent machines, we need to give them emotions that can guide their learning of behaviors. They should want us to be happy and prosper, which is the emotion we call love. We can design intelligent machines so their primary, innate emotion is unconditional love for all humans. First we can build relatively simple machines that learn to recognize happiness and unhappiness in human facial expressions, human voices and human body language. Then we can hard-wire the result of this learning as the innate emotional values of more complex intelligent machines, positively reinforced when we are happy and negatively reinforced when we are unhappy. Machines can learn algorithms for approximately predicting the future, as for example investors currently use learning machines to predict future security prices. So we can program intelligent machines to learn algorithms for predicting future human happiness, and use those predictions as emotional values.

Bill Hibbard (2001), Super-intelligent Machines

Once upon a time, the US Army wanted to use neural networks to automatically detect camouflaged enemy tanks. The researchers trained a neural net on fifty photos of camouflaged tanks in trees, and fifty photos of trees without tanks. Using standard techniques for supervised learning, the researchers trained the neural network to a weighting that correctly loaded the training set – output 'yes' for the fifty photos of camouflaged tanks, and output 'no' for the fifty photos of empty forest. This did not ensure, or even imply, that *new* examples would be classified correctly. The neural network might have 'learned' one hundred special cases that would not generalize to any new problem. Wisely, the researchers had originally taken two hundred photos, one hundred photos of tanks and one hundred photos of trees. They had used only fifty of each for the training set. The researchers ran the neural network on the remaining one hundred photos, and without further training, the neural network classified all remaining photos correctly. Success confirmed! The researchers handed the finished work to the Pentagon, which soon handed it back, complaining that in their own tests the neural network did no better than chance at discriminating photos.

It turned out that in the researchers' data set, photos of camouflaged tanks had been taken on cloudy days, while photos of plain forest had been taken on sunny days. The neural network had learned to distinguish cloudy days from sunny days, instead of distinguishing camouflaged tanks from empty forest.[2]

A technical failure occurs when the code does not do what you think it would do, though it faithfully executes as you programmed it. More than one model can load the same data. Suppose we trained a neural network to recognize smiling human faces and distinguish them from frowning human faces. Would the network classify a tiny picture of a smiley-face into the same attractor as a smiling human face? If an AI 'hard-wired' to such code possessed the power – and Hibbard (2001) spoke of superintelligence – would the galaxy end up tiled with tiny molecular pictures of smiley-faces?[3]

[2] This story, although famous and oft-cited as fact, *may* be apocryphal; I could not find a first-hand report. For unreferenced reports see for example, Crochat and Franklin (2000) or http://neil.fraser.name/writing/tank/. However, failures of the type described are a major real-world consideration when building and testing neural networks.

[3] Bill Hibbard, after viewing a draft of this paper, wrote a response arguing that the analogy to the 'tank classifier' problem does not apply to reinforcement learning in general. His critique

This form of failure is especially dangerous because it will *appear* to work within a fixed context, then fail when the context changes. The researchers of the 'tank classifier' story tweaked their neural network until it correctly loaded the training data, and then verified the network on additional data (without further tweaking).Unfortunately, both the training data and verification data turned out to share an assumption that held over all the data used in development but not in all the real-world contexts, where the neural network was called upon to function. In the story of the tank classifier, the assumption is that tanks are photographed on cloudy days.

Let us suppose we wish to develop an AI of increasing power. The AI possesses a developmental stage where the human programmers are more powerful than the AI – not in the sense of mere physical control over the AI's electrical supply, but in the sense that the human programmers are smarter, more creative, and more cunning than the AI. During the developmental period, we suppose that the programmers possess the ability to make changes to the AI's source code without needing the consent of the AI to do so. However, the AI is also intended to possess post-developmental stages, including, in the case of Hibbard's scenario, superhuman intelligence. An AI of superhuman intelligence is very unlikely to be modified without its consent *by humans*. At this point, we must rely on the previously laid-down goal system to function correctly, because if it operates in a sufficiently unforeseen fashion, the AI may actively resist our attempts to correct it – and, if the AI is smarter than a human, probably win.

Trying to control a growing AI by *training a neural network to provide its goal system* faces the problem of a huge *context change* between the AI's developmental stage and post-developmental stage. During the developmental stage, the AI may be able to produce *only* stimuli that fall into the 'smiling human faces' category, by solving humanly provided tasks, as its makers intended. Flash forward to a time when the AI is superhumanly intelligent and has built its own nanotech infrastructure, and the AI may be able to produce stimuli classified into the same attractor by tiling the galaxy with tiny smiling faces.

Thus the AI appears to work fine during development, but produces catastrophic results after it becomes smarter than the programmers(!).

There is a temptation to think, 'But surely the AI will know that is not what we meant?' But the code is not *given* to the AI, for the AI to look over and hand back if it does the wrong thing. The code *is* the AI. Perhaps with enough effort and understanding, we can write code that cares if we have written the wrong code – the legendary DWIM instruction, which among programmers stands

may be found in Hibbard (2006); my response may be found at Yudkowsky (2006). Hibbard's model recommends a two-layer system in which expressions of agreement from humans reinforce recognition of happiness, and recognized happiness reinforces action strategies.

for Do-What-I-Mean (Raymond, 2003). But effort is required to write a DWIM dynamic, and nowhere in Hibbard's proposal is there mention of designing an AI that does what we mean, not what we say. Modern chips do not DWIM their code; it is not an automatic property. And if you messed up the DWIM itself, you would suffer the consequences. For example, suppose DWIM was defined as maximizing the satisfaction of the programmer with the code; when the code executed as a superintelligence, it might rewrite the programmers' brains to be maximally satisfied with the code. I do not say this is inevitable; I only point out that Do-What-I-Mean is a major, non-trivial technical challenge of Friendly AI.

15.8 Rates of intelligence increase

From the standpoint of existential risk, one of the most critical points about Artificial Intelligence is that an AI might increase in intelligence *extremely fast.* The obvious reason to suspect this possibility is recursive self-improvement (Good, 1965). The AI becomes smarter, including becoming smarter at the task of writing the internal cognitive functions of an AI, so the AI can rewrite its existing cognitive functions to work even better, which makes the AI still smarter, including being smarter at the task of rewriting itself, so that it makes yet more improvements.

Although human beings improve themselves to a limited extent (by learning, practicing, honing of skills and knowledge), our brains today are much the same as they were 10,000 years ago. In a similar sense, natural selection improves organisms, but the process of natural selection does not itself improve – not in a strong sense. Adaptation can open up the way for additional adaptations. In this sense, adaptation feeds on itself. But even as the gene pool boils, there is still an underlying heater, the process of mutation and recombination and selection, which is not itself re-architected. A few rare innovations increased the rate of evolution itself, such as the invention of sexual recombination. But even sex did not change the essential nature of evolution: its lack of abstract intelligence, its reliance on random mutations, its blindness and incrementalism, its focus on allele frequencies. Similarly, the inventions of language or science did not change the essential character of the human brain: its limbic core, its cerebral cortex, its prefrontal self-models, its characteristic speed of 200Hz.

An Artificial Intelligence could rewrite its code from scratch – it could change the underlying dynamics of optimization. Such an optimization process would wrap around *much more strongly* than either evolution accumulating adaptations or humans accumulating knowledge. The key implication for our purposes is that an AI might make a *huge* jump in intelligence after reaching some threshold of criticality.

One often encounters scepticism about this scenario – what Good (1965) called an 'intelligence explosion' – because progress in AI has the reputation of being very slow. At this point, it may prove helpful to review a loosely analogous historical surprise. (What follows is taken primarily from Rhodes, 1986.)

In 1933, Lord Ernest Rutherford said that no one could ever expect to derive power from splitting the atom: 'Anyone who looked for a source of power in the transformation of atoms was talking moonshine.' At that time laborious hours and weeks were required to fission a handful of nuclei.

Flash forward to 1942, in a squash court beneath Stagg Field at the University of Chicago. Physicists are building a shape like a giant doorknob out of alternate layers of graphite and uranium, intended to start the first self-sustaining nuclear reaction. In charge of the project is Enrico Fermi. The key number for the pile is k, the effective neutron multiplication factor: the average number of neutrons from a fission reaction that cause another fission reaction. At k less than one, the pile is sub-critical. At $k \geq 1$, the pile should sustain a critical reaction. Fermi calculates that the pile will reach $k = 1$ between layers 56 and 57.

A work crew led by Herbert Anderson finishes Layer 57 on the night of 1 December 1942. Control rods, strips of wood covered with neutron-absorbing cadmium foil, prevent the pile from reaching criticality. Anderson removes all but one control rod and measures the pile's radiation, confirming that the pile is ready to chain-react the next day. Anderson inserts all cadmium rods and locks them into place with padlocks, then closes up the squash court and goes home.

The next day, 2 December 1942, on a windy Chicago morning of sub-zero temperatures, Fermi begins the final experiment. All but one of the control rods are withdrawn. At 10.37 a.m., Fermi orders the final control rod withdrawn about half-way out. The Geiger counters click faster, and a graph pen moves upwards. 'This is not it', says Fermi, 'the trace will go to this point and level off', indicating a spot on the graph. In a few minutes the graph pen comes to the indicated point, and does not go above it. Seven minutes later, Fermi orders the rod pulled out another foot. Again the radiation rises, then levels off. The rod is pulled out another six inches, then another, then another. At 11.30 a.m., the slow rise of the graph pen is punctuated by an enormous CRASH – an emergency control rod, triggered by an ionization chamber, activates and shuts down the pile, which is still short of criticality. Fermi calmly orders the team to break for lunch.

At 2 p.m., the team reconvenes, withdraws and locks the emergency control rod, and moves the control rod to its last setting. Fermi makes some measurements and calculations, then again begins the process of withdrawing the rod in slow increments. At 3.25 p.m., Fermi orders the rod withdrawn by another twelve inches. 'This is going to do it', Fermi says. 'Now it will become self-sustaining. The trace will climb and continue to climb. It will not level off'.

Herbert Anderson recounts (from Rhodes, 1986, p. 27):

> At first you could hear the sound of the neutron counter, clickety-clack, clickety-clack. Then the clicks came more and more rapidly, and after a while they began to merge into a roar; the counter couldn't follow anymore. That was the moment to switch to the chart recorder. But when the switch was made, everyone watched in the sudden silence the mounting deflection of the recorder's pen. It was an awesome silence. Everyone realized the significance of that switch; we were in the high intensity regime and the counters were unable to cope with the situation anymore. Again and again, the scale of the recorder had to be changed to accommodate the neutron intensity which was increasing more and more rapidly. Suddenly Fermi raised his hand. 'The pile has gone critical', he announced. No one present had any doubt about it.

Fermi kept the pile running for twenty-eight minutes, with the neutron intensity doubling every two minutes. The first critical reaction had k of 1.0006. Even at $k = 1.0006$, the pile was only controllable because some of the neutrons from a uranium fission reaction are *delayed* – they come from the decay of short-lived fission by-products. For every 100 fissions in U_{235}, 242 neutrons are emitted almost immediately (0.0001s), and 1.58 neutrons are emitted an average of 10 seconds later. Thus the *average* lifetime of a neutron is approximately 0.1 second, implying 1200 generations in two minutes, and a doubling time of two minutes because 1.0006 to the power of 1200 is approximately two. A nuclear reaction that is *prompt critical* is critical without the contribution of delayed neutrons. If Fermi's pile had been prompt critical with $k = 1.0006$, neutron intensity would have doubled every *tenth* of a second.

The first moral is that confusing the speed of *AI research* with the speed of *a real AI once built* is like confusing the speed of physics research with the speed of nuclear reactions. It mixes up the map with the territory. It took years to get that first pile built, by a small group of physicists who did not generate much in the way of press releases. But, once the pile was built, interesting things happened on the timescale of nuclear interactions, not the timescale of human discourse. In the nuclear domain, elementary interactions happen much faster than human neurons fire. Much the same may be said of transistors.

Another moral is that there is a huge difference between one self-improvement triggering 0.9994 further improvements on average and another self-improvement triggering 1.0006 further improvements on average. The nuclear pile did not cross the critical threshold as the result of the physicists suddenly piling on a lot more material. The physicists piled on material slowly and steadily. Even if there is a smooth underlying curve of brain intelligence as a function of optimization pressure previously exerted on that brain, the curve of *recursive self-improvement* may show a huge leap.

There are also other reasons why an AI might show a sudden huge leap in intelligence. The species *Homo sapiens* showed a sharp jump in the effectiveness of intelligence, as the result of natural selection exerting a more-or-less steady optimization pressure on hominids for millions of years, gradually expanding the brain and prefrontal cortex, tweaking the software architecture. A few tens of thousands of years ago, hominid intelligence crossed some key threshold and made a *huge* leap in real-world effectiveness; we went from caves to skyscrapers in the blink of an evolutionary eye. This happened with a continuous underlying selection pressure – there was no huge jump in the optimization power *of evolution* when humans came along. The underlying brain architecture was also continuous – our cranial capacity did not suddenly increase by two orders of magnitude. So it might be that, even if the AI is being elaborated from outside by human programmers, the curve for *effective* intelligence will jump sharply.

Or perhaps someone builds an AI prototype that shows some promising results, and the demo attracts another $100 million in venture capital, and this money purchases a thousand times as much supercomputing power. I doubt a 1000-fold increase in hardware would purchase anything like a 1000-fold increase in effective intelligence – but mere doubt is not reliable in the absence of any ability to perform an analytical calculation. Compared to chimps, humans have a threefold advantage in brain and a sixfold advantage in prefrontal cortex, which suggests (1) software is more important than hardware and (2) small increases in hardware can support large improvements in software. It is one more point to consider.

Finally, AI may make an *apparently* sharp jump in intelligence purely as the result of anthropomorphism, the human tendency to think of 'village idiot' and 'Einstein' as the extreme ends of the intelligence scale, instead of nearly indistinguishable points on the scale of minds-in-general. Everything dumber than a dumb human may appear to us as simply 'dumb'. One imagines the 'AI arrow' creeping steadily up the scale of intelligence, moving past mice and chimpanzees, with AIs still remaining 'dumb' because AIs cannot speak fluent language or write science papers, and then the AI arrow crosses the tiny gap from infra-idiot to ultra-Einstein in the course of one month or some similarly short period. I do not think this *exact* scenario is plausible, mostly because I do not expect the curve of recursive self-improvement to move at a linear creep. But I am not the first to point out that 'AI' is a moving target. As soon as a milestone is actually achieved, it ceases to be 'AI'. This can only encourage procrastination.

Let us concede for the sake of argument, for all we know (and it seems to me also probable in the real world), that an AI has the capability to make a sudden, sharp, large leap in intelligence. What follows from this?

First and foremost: it follows that a reaction I often hear, 'We don't need to worry about Friendly AI because we don't yet have AI', is misguided or downright suicidal. We cannot rely on having distant advance warning before AI is created; past technological revolutions usually did not telegraph themselves to people alive *at the time,* whatever was said afterwards in hindsight. The mathematics and techniques of Friendly AI will not materialize from nowhere when needed; it takes years to lay firm foundations. Furthermore, we need to solve the Friendly AI challenge *before* Artificial General Intelligence is created, not afterwards; I should not even have to point this out. There will be difficulties for Friendly AI because the field of AI itself is in a state of low consensus and high entropy. But that does not mean we do not need to worry about Friendly AI. It means there will be difficulties. The two statements, sadly, are not remotely equivalent.

The possibility of sharp jumps in intelligence also implies a higher standard for Friendly AI techniques. The technique cannot assume the programmers' ability to monitor the AI *against its will*, rewrite the AI *against its will*, bring to bear the threat of superior military force; nor may the algorithm assume that the programmers control a 'reward button', which a smarter AI could wrest from the programmers; etc. Indeed no one should be making these assumptions to begin with. The indispensable protection is an AI that does not *want* to hurt you. Without the indispensable, no auxiliary defence can be regarded as safe. No system is secure that searches for ways to defeat its own security. If the AI would harm humanity in *any* context, you must be doing *something* wrong on a very deep level, laying your foundations awry. You are building a shotgun, pointing the shotgun at your foot, and pulling the trigger. You are deliberately setting into motion a created cognitive dynamic that will seek in some context to hurt you. That is the wrong behaviour for the dynamic, but a right code that does something else instead.

For much the same reason, Friendly AI programmers should assume that the AI has total access to its own source code. If the AI *wants* to modify itself to be no longer Friendly, then Friendliness has *already* failed, at the point when the AI forms that intention. Any solution that relies on the AI not being *able* to modify itself must be broken in some way or other, and will still be broken even if the AI never does modify itself. I do not say it should be the *only* precaution, but the *primary* and *indispensable* precaution is that you choose into existence an AI that does not choose to hurt humanity.

To avoid the Giant Cheesecake Fallacy, we should note that the ability to self-improve does not imply the choice to do so. The *successful* exercise of Friendly AI technique might create an AI that had the *potential* to grow more quickly, but chose instead to grow along a slower and more manageable curve. Even so, after the AI passes the criticality threshold of *potential* recursive self-improvement, you are then operating in a much more dangerous regime. If Friendliness fails, the AI might decide to rush full speed ahead on self-improvement – metaphorically speaking, it would go prompt critical.

I tend to assume arbitrarily large *potential* jumps for intelligence because (1) this is the conservative assumption; (2) it discourages proposals based on building AI without really understanding it; and (3) large potential jumps strike me as probable-in-the-real-world. If I encountered a domain where it was conservative *from a risk-management perspective* to assume slow improvement of the AI, then I would demand that a plan not break down *catastrophically* if an AI lingers at a near-human stage for years or longer. This is not a domain over which I am willing to offer narrow confidence intervals.

15.9 Hardware

People tend to think of large computers as *the* enabling factor for AI. This is, to put it mildly, an extremely questionable assumption. Outside futurists discussing AI talk about hardware progress because hardware progress is easy to measure – in contrast to understanding of intelligence. It is not that there has been no progress, but that the progress cannot be charted on neat PowerPoint graphs. Improvements in understanding are harder to report on and therefore less reported.

Rather than thinking in terms of the 'minimum' hardware 'required' for AI, think of a minimum level of researcher understanding that decreases as a function of hardware improvements. The better the computing hardware, the less understanding you need to build an AI. The extreme case is natural selection, which used a ridiculous amount of brute computational force to create intelligence using *no* understanding, and only non-chance retention of chance mutations.

Increased computing power makes it easier to build AI, but there is no obvious reason why increased computing power would help make the AI Friendly. Increased computing power makes it easier to use brute force; easier to combine poorly understood techniques that work. Moore's Law steadily *lowers* the barrier that keeps us from building AI *without* a deep understanding of cognition.

It is acceptable to fail at AI *and* at Friendly AI. Similarly, it is acceptable to succeed at AI *and* at Friendly AI. What is not acceptable is succeeding at AI and failing at Friendly AI. Moore's Law makes it easier to do exactly that – 'easier' but thankfully not easy. I doubt that AI will be 'easy' at the time it is finally built – simply because there are parties who will exert tremendous effort to build AI, and one of them will succeed after AI first becomes possible to build with tremendous effort.

Moore's Law is an interaction between Friendly AI and other technologies, which adds *oft-overlooked* existential risk to other technologies. We can imagine that molecular nanotechnology is developed by a benign multinational governmental consortium and that they successfully avert the *physical-layer* dangers of nanotechnology. They straightforwardly prevent accidental replicator releases, and with much greater difficulty they put global defences in place against malicious replicators; they restrict access to 'root level' nanotechnology while distributing configurable nanoblocks, etc. (see

Chapter 21, this volume). But nonetheless, nanocomputers become widely available, either because attempted restrictions are bypassed, or because no restrictions are attempted. And then someone brute-forces an AI that is non-Friendly; and so the curtain is rung down. This scenario is especially worrying because incredibly powerful nanocomputers would be among the first, the easiest, and the safest-seeming applications of molecular nanotechnology.

What of regulatory controls on supercomputers? I certainly would not rely on it to prevent AI from ever being developed; yesterday's supercomputer is tomorrow's laptop. The standard reply to a regulatory proposal is that when nanocomputers are outlawed, only outlaws will have nanocomputers. The burden is to argue that the supposed benefits of *reduced* distribution outweigh the inevitable risks of *uneven* distribution. For myself, I would certainly not argue in *favour* of regulatory restrictions on the use of supercomputers for AI research; it is a proposal of dubious benefit that would be fought tooth and nail by the entire AI community. But in the unlikely event that a proposal made it that far through the political process, I would not expend any significant effort on *fighting* it, because I do not expect the good guys to *need* access to the 'supercomputers' of their day. *Friendly* AI is not about brute-forcing the problem.

I can imagine regulations effectively controlling a small set of ultra-expensive computing resources that are *presently considered* 'supercomputers'. But computers are everywhere. It is not like the problem of nuclear proliferation, where the main emphasis is on controlling plutonium and enriched uranium. The raw materials for AI are *already* everywhere. That cat is so far out of the bag that it is in your wristwatch, cellphone, and dishwasher. This too is a special and unusual factor in AI as an existential risk. We are separated from the risky regime, not by large visible installations like isotope centrifuges or particle accelerators, but *only* by missing knowledge. To use a perhaps over-dramatic metaphor, imagine if sub-critical masses of enriched uranium had powered cars and ships throughout the world, *before* Leo Szilard first thought of the chain reaction.

15.10 Threats and promises

It is a risky intellectual endeavour to predict *specifically* how a benevolent AI would help humanity, or an unfriendly AI harm it. There is the risk of *conjunction fallacy*: added detail necessarily reduces the joint probability of the entire story, but subjects often assign higher probabilities to stories that include strictly added details (See Chapter 5, this volume, on cognitive biases). There is the risk – virtually the certainty – of failure of imagination; and the risk of Giant Cheesecake Fallacy that leaps from capability to motive. Nonetheless, I will try to solidify threats and promises.

The future has a reputation for accomplishing feats that the past thought was impossible. Future civilizations have even broken what past civilizations thought (incorrectly, of course) to be the laws of physics. If prophets of 1900 AD – never mind 1000 AD – had tried to bound the powers of human civilization a billion years later, some of those impossibilities would have been accomplished before the century was out – for example, transmuting lead into gold. Because we remember future civilizations surprising past civilizations, it has become cliché that we cannot put limits on our great-grandchildren. And yet everyone in the twentieth century, in the nineteenth century, and in the eleventh century, was human.

We can distinguish three families of unreliable metaphors for imagining the capability of a smarter-than-human Artificial Intelligence:

- *G-factor metaphors*: Inspired by differences of individual intelligence between humans. AIs will patent new technologies, publish groundbreaking research papers, make money on the stock market, or lead political power blocs.
- *History metaphors*: Inspired by knowledge differences between past and future human civilizations. AIs will swiftly invent the kind of capabilities that cliché would attribute to human civilization a century or millennium from now: molecular nanotechnology, interstellar travel, computers performing 10^{25} operations per second and so on.
- *Species metaphors*: Inspired by differences of brain architecture between species. AIs have magic.

The g-factor metaphors seem most common in popular futurism: when people think of 'intelligence' they think of human geniuses instead of humans. In stories about hostile AI, *g* metaphors make for a Bostromian 'good story': an opponent that is powerful enough to create dramatic tension, but not powerful enough to instantly squash the heroes like bugs, and ultimately weak enough to lose in the final chapters of the book. Goliath against David is a 'good story', but Goliath against a fruit fly is not.

If we suppose the *g*-factor metaphor, then global catastrophic risks of this scenario are relatively mild; a hostile AI is not much more of a threat than a hostile human genius. If we suppose a *multiplicity* of AIs, then we have a metaphor of conflict between nations, between the AI tribe and the human tribe. If the AI tribe wins in military conflict and wipes out the humans, then that is an existential catastrophe of the Bang variety (Bostrom, 2001). If the AI tribe dominates the world economically and attains effective control of the destiny of Earth-originating intelligent life, but the AI tribe's goals do not seem to us interesting or worthwhile, then that is a Shriek, Whimper, or Crunch.

But how likely is it that AI will cross the entire vast gap from amoeba to village idiot, and then stop at the level of human genius?

The fastest observed neurons fire 1000 times per second; the fastest axon fibres conduct signals at 150 m/second, a half-millionth the speed of light; each synaptic operation dissipates around 15,000 attojoules, which is more than a million times the thermodynamic minimum for irreversible computations at room temperature, 0.003 attojoules per bit.[4] It would be physically possible to build a brain that computed a million times as fast as a human brain, without shrinking the size, or running at lower temperatures, or invoking reversible computing or quantum computing. If a human mind were thus accelerated, a subjective year of thinking would be accomplished for every 31 physical seconds in the outside world, and a millennium would fly by in eight-and-a-half hours. Vinge (1993) referred to such sped-up minds as 'weak superintelligence': a mind that thinks like a human but much faster.

We suppose there comes into existence an extremely fast mind, embedded in the midst of human technological civilization as it exists at that time. The failure of imagination is to say, 'No matter how fast it thinks, it can only affect the world at the speed of its manipulators; it cannot operate machinery faster than it can order human hands to work; therefore a fast mind is no great threat'. It is no law of Nature that physical operations must crawl at the pace of long seconds. Critical times for elementary molecular interactions are measured in femtoseconds, sometimes picoseconds. Drexler (1992) has analysed controllable molecular manipulators that would complete $>10^6$ mechanical operations per second – note that this is in keeping with the general theme of 'millionfold speedup'. (The smallest physically sensible increment of time is generally thought to be the Planck interval, $5 \cdot 10^{-44}$ seconds, on which scale even the dancing quarks are statues.)

Suppose that a human civilization were locked in a box and allowed to affect the outside world only through the glacially slow movement of alien tentacles, or mechanical arms that moved at microns per second. We would focus all our creativity on finding the *shortest possible path* to building fast manipulators in the outside world. Pondering over fast manipulators, one immediately thinks of molecular nanotechnology – though there may be other ways. What is the *shortest* path you could take to molecular nanotechnology in the slow outside world, if you had eons to ponder over each move? The answer is that I do not know because I do not have eons to ponder. Here is one imaginable fast pathway:

- Crack the protein folding problem, to the extent of being able to generate DNA strings whose folded peptide sequences fill specific functional roles in a complex chemical interaction.

[4] This follows for the Landauer–Brillouin's limit, the maximal amount of information you can process in any *classical* system dissipating energy $E: I_{\max} = E/(kT \ln 2)$, where k is the Boltzmann constant and T is the working temperature.

- Email sets of DNA strings to one or more online laboratories that offer DNA synthesis, peptide sequencing, and FedEx delivery. (Many labs currently offer this service, and some boast of 72-hour turnaround times.)
- Find at least one human connected to the Internet who can be paid, blackmailed, or fooled by the right background story, into receiving FedExed vials and mixing them in a specified environment.
- The synthesized proteins form a very primitive 'wet' nanosystem, which, ribosome-like, is capable of accepting external instructions; perhaps patterned acoustic vibrations delivered by a speaker attached to the beaker.
- Use the extremely primitive nanosystem to build more sophisticated systems, which construct still more sophisticated systems, bootstrapping to molecular nanotechnology – or beyond.

The elapsed turnaround time would be, imaginably, on the order of a week from when the fast intelligence first became able to solve the protein folding problem. Of course this whole scenario is strictly something *I* am thinking of. Perhaps in 19,500 years of subjective time (one week of physical time at a millionfold speedup) I would think of a better way. Perhaps you can pay for rush courier delivery instead of FedEx. Perhaps there are existing technologies, or slight modifications of existing technologies, that combine synergetically with simple protein machinery. Perhaps if you are *sufficiently* smart, you can use waveformed electrical fields to alter reaction pathways in existing biochemical processes. I do not know. I am not that smart.

The challenge is to chain your capabilities – the physical-world analogue of combining weak vulnerabilities in a computer system to obtain root access. If one path is blocked, you choose another, seeking always to increase your capabilities and use them in synergy. The presumptive goal is to obtain *rapid infrastructure,* means of manipulating the external world on a large scale in fast time. Molecular nanotechnology fits this criterion, first because its elementary operations are fast, and second because there exists a ready supply of precise parts – atoms – which can be used to self-replicate and exponentially grow the nanotechnological infrastructure. The pathway alleged above has the AI obtaining rapid infrastructure within a week – this sounds fast to a human with 200Hz neurons, but is a vastly longer time for the AI.

Once the AI possesses rapid infrastructure, further events happen on the AI's timescale, not a human timescale (unless the AI *prefers* to act on a human timescale). With molecular nanotechnology, the AI could (potentially) rewrite the solar system unopposed.

An unFriendly AI with molecular nanotechnology (or other rapid infrastructure) need not bother with marching robot armies or blackmail or subtle economic coercion. The unFriendly AI has the ability to repattern all matter in the solar system according to its optimization target. This is fatal for us if the AI does not choose *specifically* according to the criterion of how this transformation affects existing patterns such as biology and people. The AI neither hates you, nor loves you, but you are made out of atoms that it can use for something else. The AI runs on a different timescale than you do; by the time your neurons finish thinking the words 'I should do something' you have already lost.

A Friendly AI in addition to molecular nanotechnology is presumptively powerful enough to solve any problem, which can be solved either by moving atoms or by creative thinking. One should beware of failures of imagination: curing cancer is a popular contemporary target of philanthropy, but it does not follow that a Friendly AI with molecular nanotechnology would say to itself, 'Now I shall cure cancer'. Perhaps a better way to view the problem is that biological cells are not programmable. To solve the latter problem cures cancer as a special case, along with diabetes and obesity. A fast, nice intelligence wielding molecular nanotechnology is power on the order of *getting rid of disease,* not *getting rid of cancer.*

There is finally the family of *species metaphors,* based on between-species differences of intelligence. The AI has magic – not in the sense of incantations and potions, but in the sense that a wolf cannot understand how a gun works, or what sort of effort goes into making a gun, or the nature of that human power that lets us invent guns. Vinge (1993) wrote:

> Strong superhumanity would be more than cranking up the clock speed on a human-equivalent mind. It's hard to say precisely what strong superhumanity would be like, but the difference appears to be profound. Imagine running a dog mind at very high speed. Would a thousand years of doggy living add up to any human insight?

The species metaphor would seem the nearest analogy a priori, but it does not lend itself to making up detailed stories. The main advice the metaphor gives us is that *we had better get Friendly AI right,* which is good advice in any case. The only defence it suggests against hostile AI is *not to build it in the first place,* which is also excellent advice. Absolute power is a conservative engineering assumption in Friendly AI, exposing broken designs. If an AI will hurt you given magic, the Friendliness architecture is wrong.

15.11 Local and majoritarian strategies

One may classify proposed risk-mitigation strategies into the following:

- Strategies that require *unanimous* cooperation: strategies that can be catastrophically defeated by individual defectors or small groups.

- Strategies that require *majority* action: a majority of a legislature in a single country, or a majority of voters in a country, or a majority of countries in the United Nations; the strategy requires *most, but not all,* people in a large pre-existing group to behave in a particular way.
- Strategies that require *local* action: a concentration of will, talent, and funding which overcomes the threshold of some specific task.

Unanimous strategies are unworkable, but it does not stop people from proposing them.

A *majoritarian* strategy is sometimes workable, if you have decades in which to do your work. One must build a movement, from its first beginnings over the years, to its debut as a recognized force in public policy, to its victory over opposing factions. Majoritarian strategies take substantial time and *enormous* effort. People have set out to do such, and history records some successes. But beware: history books tend to focus selectively on movements that have an impact, as opposed to the vast majority that never amount to anything. There is an element involved of luck, and of the public's prior willingness to hear. Critical points in the strategy will involve events beyond your personal control. If you are not willing to devote your entire life to pushing through a majoritarian strategy, do not bother; and just *one* life devoted will not be enough, either.

Ordinarily, *local* strategies are most plausible. One hundred million dollars of funding is not *easy* to obtain, and a global political change is not *impossible* to push through, but it is still vastly easier to obtain one hundred million dollars of funding than to push through a global political change.

Two assumptions that give rise to a *majoritarian* strategy for AI are as follows:

- A majority of Friendly AIs can effectively protect the human species from a few unFriendly AIs.
- The first AI built cannot by itself do catastrophic damage.

This reprises essentially the situation of a human civilization before the development of nuclear and biological weapons: most people are cooperators in the overall social structure, and defectors can do damage but not *global catastrophic* damage. Most AI researchers will not want to make unFriendly AIs. So long as *someone* knows how to build a stable Friendly AI – so long as the problem is not completely beyond contemporary knowledge and technique – researchers will learn from each other's successes and repeat them. Legislation could (for example) require researchers to publicly report their Friendliness strategies, or penalize researchers whose AIs cause damage; and while this legislation will not prevent *all* mistakes, it may suffice that a *majority* of AIs are built Friendly.

We can also imagine a scenario that implies an easy local strategy:

- The first AI cannot by itself do catastrophic damage.
- If even a single Friendly AI exists, that AI *plus* human institutions can fend off any number of unFriendly AIs.

The easy scenario would hold if, for example, human institutions can reliably distinguish Friendly AIs from unFriendly ones, and give revocable power into the hands of Friendly AIs. Thus we could pick and choose our allies. The only requirement is that the Friendly AI problem must be solvable (as opposed to being completely beyond human ability).

Both of the above scenarios assume that the *first* AI (the first powerful, general AI) cannot by itself do global catastrophic damage. Most concrete visualizations that imply this use a g metaphor: AIs as analogous to unusually able humans. In Section 15.8 on *rates of intelligence increase*, I listed some reasons to be wary of a *huge, fast* jump in intelligence:

- The distance from idiot to Einstein, which looms large to us, is a small dot on the scale of minds-in-general.
- Hominids made a *sharp* jump in *real-world effectiveness* of intelligence, despite natural selection exerting roughly steady optimization pressure on the underlying genome.
- An AI may absorb a huge amount of additional hardware after reaching some brink of competence (i.e., eat the Internet).
- Criticality threshold of recursive self-improvement. One self-improvement triggering 1.0006 self-improvements is qualitatively different from one self-improvement triggering 0.9994 self-improvements.

As described in Section 15.9, a sufficiently powerful intelligence may need only a short time (from a human perspective) to achieve molecular nanotechnology, or some other form of rapid infrastructure.

We can therefore visualize a possible *first-mover effect* in superintelligence. The first-mover effect is when the outcome for Earth-originating intelligent life depends primarily on the makeup of whichever mind *first* achieves some key threshold of intelligence – such as criticality of self-improvement. The two necessary assumptions are as follows:

- The *first* AI to surpass some key threshold (e.g., criticality of self-improvement), if unFriendly, can wipe out the human species.
- The *first* AI to surpass the same threshold, if Friendly, can prevent a hostile AI from coming into existence or from harming the human species; or find some other creative way to ensure the survival and prosperity of Earth-originating intelligent life.

More than one scenario qualifies as a first-mover effect. Each of these examples reflects a different key threshold:

- Post-criticality, self-improvement reaches superintelligence on a timescale of weeks or less. AI projects are sufficiently sparse that no *other* AI achieves criticality before the *first* mover is powerful enough to overcome all opposition. The key threshold is criticality of recursive self-improvement.
- AI-1 cracks protein folding three days before AI-2. AI-1 achieves nanotechnology six hours before AI-2. With rapid manipulators, AI-1 can (potentially) disable AI-2's R&D before fruition. The runners are close, but whoever crosses the finish line first, wins. The key threshold is rapid infrastructure.
- The first AI to absorb the Internet can (potentially) keep it out of the hands of other AIs. Afterwards, by economic domination or covert action or blackmail or supreme ability at social manipulation, the first AI halts or slows other AI projects so that no other AI catches up. The key threshold is absorption of a unique resource.

The human species, *Homo sapiens,* is a first mover. From an evolutionary perspective, our cousins, the chimpanzees, are only a hairbreadth away from us. *Homo sapiens* still wound up with all the technological marbles because we got there a little earlier. Evolutionary biologists are still trying to unravel which order the key thresholds came in, because the first-mover species was first to cross so *many:* Speech, technology, abstract thought (see, however, also the findings Chapter 3, this volume).

A first-mover effect implies a theoretically localizable strategy (a task that can, in principle, be carried out by a strictly local effort), but it invokes a technical challenge of extreme difficulty. We only need to get Friendly AI right in one place and one time, not every time everywhere. But someone must get Friendly AI right on the first try, *before* anyone else builds AI to a lower standard.

I cannot perform a precise calculation using a precisely confirmed theory, but my *current opinion* is that sharp jumps in intelligence are *possible, likely,* and *constitute the dominant probability.* But a much more serious problem is strategies visualized for slow-growing AIs, which fail catastrophically if there *is* a first-mover effect. This is considered a more serious problem for the following reasons:

- Faster-growing AIs represent a greater technical challenge.
- Like a car driving over a bridge built for trucks, an AI designed to remain Friendly in extreme conditions should (presumptively) remain Friendly in less extreme conditions. The reverse is not true.

- Rapid jumps in intelligence are counterintuitive in everyday social reality. The g-factor metaphor for AI is intuitive, appealing, reassuring, and conveniently implies fewer design constraints.

My current strategic outlook tends to focus on the difficult local scenario: the first AI must be Friendly. With the caveat that, if no sharp jumps in intelligence materialize, it should be possible to switch to a strategy for making a majority of AIs Friendly. In either case, the technical effort that went into preparing for the extreme case of a first mover should leave us better off, not worse.

The scenario that implies an impossible, unanimous strategy is as follows:

- A single AI can be powerful enough to destroy humanity, even despite the protective efforts of Friendly AIs.
- No AI is powerful enough to prevent human researchers from building one AI after another (or find some other creative way of solving the problem).

It is good that this balance of abilities seems unlikely a priori, because in this scenario we are doomed. If you deal out cards from a deck, one after another, you will eventually deal out the ace of clubs.

The same problem applies to the strategy of *deliberately* building AIs that choose not to increase their capabilities past a fixed point. If capped AIs are not powerful enough to defeat uncapped AIs, or prevent uncapped AIs from coming into existence, then capped AIs cancel out of the equation. We keep dealing through the deck until we deal out a superintelligence, whether it is the ace of hearts or the ace of clubs.

A majoritarian strategy only works if it is not *possible* for a single defector to cause global catastrophic damage. For AI, this possibility or impossibility is a natural feature of the design space – the *possibility* is not subject to human decision any more than the speed of light or the gravitational constant.

15.12 Interactions of Artificial Intelligence with other technologies

Speeding up a desirable technology is a local strategy, while *slowing down* a dangerous technology is a difficult majoritarian strategy. *Halting* or *relinquishing* an undesirable technology tends to require an impossible unanimous strategy. I would suggest that we think, not in terms of developing or not-developing technologies, but in terms of our *pragmatically available latitude* to *accelerate* or *slow down* technologies; and ask, *within the realistic bounds of this latitude*, which technologies we might prefer to see developed *before* or *after* one another.

In nanotechnology, the goal usually presented is to develop defensive shields before offensive technologies. I worry a great deal about this, because a *given level* of offensive technology tends to require much less sophistication than a technology that can defend against it. Guns were developed centuries before bullet-proof vests were made. Smallpox was used as a tool of war before the development of smallpox vaccines. Today there is still no shield that can deflect a nuclear explosion; nations are protected not by defences that cancel offences, but by a balance of offensive terror.

So should we prefer that nanotechnology precede the development of AI, or that AI precede the development of nanotechnology? So far as *ordering* is concerned, the question we should ask is, 'Does AI help us deal with nanotechnology? Does nanotechnology help us deal with AI?'

It looks to me like a successful resolution of Artificial Intelligence should help us considerably in dealing with nanotechnology. I cannot see how nanotechnology would make it easier to develop *Friendly* AI. If huge nanocomputers make it easier to develop AI *without* making it easier to solve the particular challenge of Friendliness, that is a *negative* interaction. Thus, all else being equal, I would greatly prefer that Friendly AI *precede* nanotechnology in the *ordering* of technological developments. If we confront the challenge of AI and succeed, we can call on Friendly AI to help us with nanotechnology. If we develop nanotechnology and survive, we still have the challenge of AI to deal with after that.

Generally speaking, a *success* on Friendly AI should help solve nearly any other problem. Thus, if a technology makes AI neither easier nor harder, but carries with it a catastrophic risk, we should prefer all else being equal to *first* confront the challenge of AI.

Any technology that increases available computing power decreases the minimum theoretical sophistication necessary to develop AI, but does not help at all on the *Friendly* side of things, and I count it as a net negative. Moore's Law of Mad Science: Every 18 months, the minimum IQ necessary to destroy the world drops by one point.

A success on human intelligence enhancement would make Friendly AI easier, and also help on other technologies. But human augmentation is *not* necessarily safer, or easier, than Friendly AI; nor does it necessarily lie within our realistically available latitude to reverse the natural ordering of human augmentation and Friendly AI, if one technology is naturally much easier than the other.

15.13 Making progress on Friendly Artificial Intelligence

We propose that a 2 month, 10 man study of artificial intelligence be carried out during the summer of 1956 at Dartmouth College in Hanover, New Hampshire. The study is to proceed on the basis of the conjecture that every aspect of learning or any other feature of intelligence can in principle be so precisely described that a machine can be made to simulate it. An attempt will be made to find how to make machines use language, form abstractions and concepts, solve kinds of problems now reserved for humans, and improve themselves. We think that a significant advance can be made in one or more of these problems if a carefully selected group of scientists work on it together for a summer.

McCarthy, Minsky, Rochester, and Shannon (1955)

The *Proposal for the Dartmouth Summer Research Project on Artificial Intelligence* is the first recorded use of the phrase 'artificial intelligence'. They had no prior experience to warn them that the problem was hard. I would still label it a genuine mistake, that they said 'a significant advance *can* be made', not *might* be made, with a summer's work. That is a specific guess about the problem difficulty and solution time, which carries a specific burden of improbability. But if they had said *might,* I would have no objection. How were they to know?

The *Dartmouth Proposal* included, among others, the following topics: linguistic communication, linguistic reasoning, neural nets, abstraction, randomness and creativity, interacting with the environment, modelling the brain, originality, prediction, invention, discovery, and self-improvement.

Now it seems to me that an AI capable of language, abstract thought, creativity, environmental interaction, originality, prediction, invention, discovery, and above all self-improvement, is *well beyond* the point where it needs also to be Friendly.

The *Dartmouth Proposal* makes no mention of building nice/good/benevolent AI. Questions of safety are not mentioned even for the purpose of dismissing them. This, even in that bright summer, when human-level AI seemed just around the corner. The *Dartmouth Proposal* was written in 1955, before the Asilomar conference on biotechnology, thalidomide babies, Chernobyl, or 11 September. If today the idea of artificial intelligence were proposed *for the first time,* then *someone* would demand to know what specifically was being done to manage the risks. I am not saying whether this is a good change or a bad change in our culture. I am not saying whether this produces good or bad science. But the point remains that if the *Dartmouth Proposal* had been written fifty years later, one of the topics would have been safety.

At the time of this writing in 2007, the AI RESEARCH community still does not see Friendly AI as part of the problem. I wish I could cite a reference to this effect, but I cannot cite an absence of literature. Friendly AI is absent from the *conceptual* landscape, not just unpopular or unfunded. You cannot even call Friendly AI a blank spot on the map, because there is no notion that something is missing.[5,6] If you have read popular/semi-technical books proposing how to build AI, such as *Gödel, Escher, Bach* (Hofstadter, 1979) or *The Society of Mind* (Minsky, 1986), you may think back and recall that you did not see Friendly AI discussed as part of the challenge. Neither have I seen Friendly AI discussed in the technical literature as a technical problem. My attempted literature search turned up primarily brief non-technical papers, unconnected to each other, with no major reference in common except Isaac Asimov's 'Three Laws of Robotics' (Asimov, 1942).

The field of AI has techniques, such as neural networks and evolutionary programming, which have grown in power with the slow tweaking over decades. But neural networks are opaque – the user has no idea how the neural net is making its decisions – and cannot easily be rendered non-opaque; the people who invented and polished neural networks were not thinking about the long-term problems of Friendly AI. Evolutionary programming (EP) is stochastic, and does not precisely preserve the optimization target in the generated code; EP gives you code that does what you ask, most of the time, under the tested circumstances, but the code may also do something else on the side. EP is a powerful, still maturing technique that is *intrinsically* unsuited to the demands of Friendly AI. Friendly AI, as I have proposed it, requires repeated cycles of recursive self-improvement that precisely preserve a stable optimization target.

The most powerful *current* AI techniques, as they were developed and then polished and improved over time, have basic incompatibilities with the requirements of Friendly AI as I currently see them. The Y2K problem – although not a global-catastrophe, but which proved very expensive to fix – analogously arose from failing to foresee tomorrow's design requirements. The nightmare scenario is that we find ourselves stuck with a catalogue of mature, powerful, publicly available AI techniques, which combine to yield *non-Friendly* AI, but which *cannot* be used to build Friendly AI without redoing the last three decades of AI work from scratch.

[5] This is usually true but not universally true. The final chapter of the widely used textbook *Artificial Intelligence: A Modern Approach* (Russell and Norvig, 2003) includes a section on 'The Ethics and Risks of Artificial Intelligence'; mentions I.J. Good's intelligence explosion and the Singularity; and calls for further research, soon. But as of 2006, this attitude remains very much the exception rather than the rule.

[6] After this chapter was written, a special issue on Machine Ethics appeared in *IEEE Intelligent Systems* (Anderson and Anderson, 2006). These articles primarily deal in ethics for domain-specific near-term AI systems, rather than superintelligence or ongoing intelligence explosions. Allen et al. (2006, p. 15), for example, remark that 'Although 2001 has passed and HAL remains fiction, and it's a safe bet that the doomsday scenarios of *Terminator* and *Matrix* movies will not be realized before their sell-by dates of 2029 and 2199, we're already at a point where engineered systems make decisions that can affect our lives.'

However, the issue of machine ethics has now definitely been put on the map; though not, perhaps, the issue of superintelligent machine ethics, or AI as a positive and negative factor in global risk.

15.14 Conclusion

It once occurred to me that modern civilization occupies an unstable state. I.J. Good's hypothesized intelligence explosion describes a dynamically unstable system, like a pen precariously balanced on its tip. If the pen is *exactly* vertical, it may remain upright; but if the pen tilts even a little from the vertical, gravity pulls it farther in that direction, and the process accelerates. So too would smarter systems have an easier time making themselves smarter.

A dead planet, lifelessly orbiting its star, is also stable. Unlike an intelligence explosion, extinction is not a *dynamic* attractor – there is a large gap between *almost* extinct, and extinct. Even so, *total* extinction is stable.

Must not our civilization eventually wander into one mode or the other?

The logic of the above argument contains holes. Giant Cheesecake Fallacy, for example: minds do not blindly wander into attractors, they have motives. Even so, I suspect that, *pragmatically* speaking, our alternatives boil down to becoming smarter or becoming extinct.

Nature is not cruel, but indifferent: a neutrality that often seems indistinguishable from outright hostility. Reality throws at you one challenge after another, and when you run into a challenge you cannot handle, you suffer the consequences. Often, Nature poses requirements that are grossly unfair, even on tests where the penalty for failure is death. How is a tenth-century medieval peasant supposed to invent a cure for tuberculosis? Nature does not match her challenges to your skill, or your resources, or how much free time you have to think about the problem. And when you run into a lethal challenge too difficult for you, you die. It may be unpleasant to think about, but that has been the reality for humans, for thousands upon thousands of years. The same thing could as easily happen to the whole human species, if the human species runs into an unfair challenge.

If human beings did not age, so that 100-year-olds had the same death rate as 15-year-olds, we would not be immortal. We would last only until the probabilities caught up with us. To live even a million years, as an unaging human in a world as risky as our own, you must somehow drive your annual probability of accident down to nearly *zero*. You may not drive; you may not fly; you may not walk across the street even after looking both ways, for it is still too great a risk. Even if you abandoned all thoughts of fun, gave up living to preserve your life, you could not navigate a million-year obstacle course. It would be, not physically impossible, but *cognitively* impossible.

The human species, *Homo sapiens,* is unaging but not immortal. Hominids have survived this long only because, for the last million years, there were no arsenals of hydrogen bombs, no spaceships to steer asteroids towards Earth, no biological weapons labs to produce superviruses, no recurring annual prospect of nuclear war or nanotechnological war or rogue AI. To survive any appreciable time, we need to drive down *each* risk to nearly *zero*. 'Fairly good' is not good enough to last another million years.

It seems like an unfair challenge. Such competence is not historically typical of human institutions, no matter how hard they try. For decades, the United States and the USSR avoided nuclear war, but not *perfectly;* there were close calls, such as the Cuban Missile Crisis in 1962. If we postulate that future minds exhibit the same mixture of foolishness and wisdom, the same mixture of heroism and selfishness, as the minds we read about in history books – then the game of existential risk is already over; it was lost from the beginning. We might survive for another decade, even another century, but not another million years.

But the human mind is not the limit of the possible. *Homo sapiens* represent the *first* general intelligence. We were born into the uttermost beginning of things, the dawn of mind. With luck, future historians will look back and describe the present world as an awkward in-between stage of adolescence, when humankind was smart enough to create tremendous problems for itself, but not quite smart enough to solve them.

Yet before we can pass out of that stage of adolescence, we must, as adolescents, confront an adult problem: the challenge of smarter-than-human intelligence. This is the way out of the high-mortality phase of the life cycle, the way to close the window of vulnerability; it is also probably the single most dangerous risk we face. Artificial Intelligence is one road into that challenge; and I think it is the road we will end up taking.

I do not want to play down the colossal audacity of trying to build, to a precise purpose and design, something smarter than ourselves. But let us pause and recall that *intelligence* is not the first thing human science has ever encountered that proved difficult to understand. Stars were once mysteries, and chemistry, and biology. Generations of investigators tried and failed to understand those mysteries, and they acquired the reputation of being impossible to mere science. Once upon a time, no one understood why some matter was inert and lifeless, while other matter pulsed with blood and vitality. No one knew how living matter reproduced itself, or why our hands obeyed our mental orders. Lord Kelvin wrote:

> The influence of animal or vegetable life on matter is infinitely beyond the range of any scientific inquiry hitherto entered on. Its power of directing the motions of moving particles, in the demonstrated daily miracle of our human free-will, and in the growth of generation after generation of plants from a single seed, are infinitely different from any possible result of the fortuitous concurrence of atoms. (Quoted in MacFie, 1912)

All scientific ignorance is hallowed by ancientness. Each and every absence of knowledge dates back to the dawn of human curiosity; and the hole lasts through the ages, seemingly eternal, right up until someone fills it. I think it is possible for mere fallible humans to succeed on the challenge of building Friendly AI. But only if intelligence ceases to be a sacred mystery to us, as life was a sacred mystery to Lord Kelvin. Intelligence must cease to be any kind of mystery whatever, sacred or not. We must execute the creation of Artificial Intelligence as the exact application of an exact art. And maybe then we can win.

Acknowledgement

I thank Michael Roy Ames, John K. Clark, Emil Gilliam, Ben Goertzel, Robin Hanson, Keith Henson, Bill Hibbard, Olie Lamb, Peter McCluskey, and Michael Wilson for their comments, suggestions and criticisms. Needless to say, any remaining errors in this paper are my own.

References

Allen, C., Wallach, W., and Smit, I. (2006). Why machine ethics? *IEEE Intell. Syst.*, **21**(4), 12–17.

Anderson, M. and Anderson, S. (2006). Guest editors' introduction: machine ethics. *IEEE Intell. Syst.*, **21**(4), 1550–1604.

Asimov, I. (March 1942). Runaround. *Astounding Science Fiction.*

Anderson, M. and Anderson, S. (2006). Guest Editors' Introduction: Machine Ethics. *IEEE Intelligent Systems*, **21**(4), pp. 1550–1604.

Allen, C., Wallach, W. and Smit, I. (2006). Why Machine Ethics? *IEEE Intelligent Systems*, **21**(4), pp. 12–17.

Barrett, J.L. and Keil, F. (1996). Conceptualizing a non-natural entity: anthropomorphism in God concepts. *Cogn. Psychol.*, **31**, 219–247.

Bostrom, N. (1998). How long before superintelligence? *Int. J. Future Studies*, **2**.

Bostrom, N. (2001). Existential risks: analyzing human extinction scenarios. *J.Evol. Technol.*, **9**.

Brown, D.E. (1991). *Human Universals* (New York: McGraw-Hill).

Crochat, P. and Franklin, D. (2000). Back-propagation neural network tutorial. http://ieee.uow.edu.au/~daniel/software/libneural/

Deacon, T. (1997). *The Symbolic Species: The Co-evolution of Language and the Brain* (New York: Norton).

Drexler, K.E. (1992). *Nanosystems: Molecular Machinery, Manufacturing, and Computation* (New York: Wiley-Interscience).

Ekman, P. and Keltner, D. (1997). Universal facial expressions of emotion: an old controversy and new findings. In Segerstrale, U. and Molnar, P. (eds.), *Nonverbal Communication: Where Nature Meets Culture*, pp. 27–46 (Mahwah, NJ: Lawrence Erlbaum Associates).

Good, I.J. (1965). Speculations concerning the first ultraintelligent machine. In Alt, F.L. and Rubinoff, M. (eds.), *Advances in Computers, Vol* 6, pp. 31-88 (New York: Academic Press).

Hayes, J.R. (1981). *The Complete Problem Solver* (Philadelphia, PA: Franklin Institute Press).

Hibbard, B. (2001). Super-intelligent machines. *ACM SIGGRAPH Computer Graphics*, **35**(1), 11–13.

Hibbard, B. (2004). Reinforcement learning as a Context for Integrating AI Research. Presented at the 2004 *AAAI Fall Symposium on Achieving Human-level Intelligence through Integrated Systems and Research.* edited by N. Cassimatis & D. Winston, The AAAI Press, Mento Park, California.

Hibbard, B. (2006). Reply to AI Risk. http://www.ssec.wisc.edu/~billh/g/AIRisk_Reply.html

Hofstadter, D. (1979). *Gödel, Escher, Bach: An Eternal Golden Braid* (New York: Random House).

Jaynes, E.T. and Bretthorst, G.L. (2003). *Probability Theory: The Logic of Science* (Cambridge: Cambridge University Press).

Jensen, A.R. (1999). The G factor: the science of mental ability. *Psycoloquy*, **10**(23).

MacFie, R.C. (1912). *Heredity, Evolution, and Vitalism: Some of the Discoveries of Modern Research into These Matters – Their Trend and Significance* (New York: William Wood and Company).

McCarthy, J., Minsky, M.L., Rochester, N., and Shannon, C.E. (1955). *A Proposal for the Dartmouth Summer Research Project on Artificial Intelligence.* http://www.formal.stanford.edu/jmc/history/dartmouth/dartmouth.html.

Merkle, R.C. (November 1989). Large scale analysis of neural structure. Xerox PARC Technical Report CSL-89-10.

Merkle, R.C. and Drexler, K.E. (1996). Helical logic. *Nanotechnology*, **7**, 325–339.

Minsky, M.L. (1986). *The Society of Mind* (New York: Simon and Schuster).

Monod, J.L. (1974). *On the Molecular Theory of Evolution* (New York: Oxford).

Moravec, H. (1988). *Mind Children: The Future of Robot and Human Intelligence* (Cambridge: Harvard University Press).

Moravec, H. (1999). *Robot: Mere Machine to Transcendent Mind* (New York: Oxford University Press).

Raymond, E.S. (ed.) (December 2003). DWIM. *The on-line hacker Jargon File,* version 4.4.7, 29

Rhodes, R. (1986). *The Making of the Atomic Bomb* (New York: Simon & Schuster).

Rice, H.G. (1953). Classes of recursively enumerable sets and their decision problems. *Trans. Am. Math. Soc.*, **74**, 358–366.

Russell, S.J. and Norvig, P. (2003). *Artificial Intelligence: A Modern Approach*, pp. 962–964 (NJ: Prentice Hall).

Sandberg, A. (1999). The physics of information processing superobjects: daily life mong the Jupiter brains. *J. Evol. Technol.*, **5**. http://ftp.nada.kth.se/pub/home/asa/work/Brains/Brains2

Schmidhuber, J. (2003). Goedel machines: self-referential universal problem solvers making provably optimal self-improvements. In Goertzel, B. and Pennachin, C. (eds.), *Artificial General Intelligence,* (New York: Springer-Verlag).

Sober, E. (1984). *The Nature of Selection* (Cambridge, MA: MIT Press).

Tooby, J. and Cosmides, L. (1992). The psychological foundations of culture. In Barkow, J.H., Cosmides, L. and Tooby, J. (eds.), *The Adapted Mind: Evolutionary Psychology and the Generation of Culture,* (New York: Oxford University Press).

Vinge, V. (March 1993). The Coming Technological Singularity. Presented at the VISION-21 Symposium, sponsored by NASA Lewis Research Center and the Ohio Aerospace Institute.

Wachowski, A. and Wachowski, L. (1999). *The Matrix* (Warner Bros, 135 min, USA).

Weisburg, R. (1986). *Creativity, Genius and Other Myths* (New York: W.H. Freeman).

Williams, G.C. (1966). *Adaptation and Natural Selection: A Critique of Some Current Evolutionary Thought* (Princeton, NJ: Princeton University Press).

Yudkowsky, E. (2006). Reply to AI Risk. http://www.ssec.wisc.edu/~billh/g/AIRisk_Reply.html

·16·
Big troubles, imagined and real

Frank Wilczek

Modern physics suggests several exotic ways in which things could go terribly wrong on a very large scale. Most, but not all, are highly speculative, unlikely, or remote. Rare catastrophes might well have decisive influences on the evolution of life in the universe. So also might slow but inexorable changes in the cosmic environment in the future.

16.1 Why look for trouble?

Only a twisted mind will find joy in contemplating exotic ways to shower doom on the world as we know it. Putting aside that hedonistic motivation, there are several good reasons for physicists to investigate doomsday scenarios that include the following:

Looking before leaping: Experimental physics often aims to produce extreme conditions that do not occur naturally on Earth (or perhaps elsewhere in the universe). Modern high-energy accelerators are one example; nuclear weapons labs are another. With new conditions come new possibilities, including – perhaps – the possibility of large-scale catstrophe. Also, new technologies enabled by advances in physics and kindred engineering disciplines might trigger social or ecological instabilities. The wisdom of 'Look before you leap' is one important motivation for considering worst-case scenarios.

Preparing to prepare: Other drastic changes and challenges must be anticipated, even if we forego daring leaps. Such changes and challenges include exhaustion of energy supplies, possible asteroid or cometary impacts, orbital evolution and precessional instability of Earth, evolution of the Sun, and – in the very long run – some form of 'heat death of the universe'. Many of these are long-term problems, but tough ones that, if neglected, will only loom larger. So we should prepare, or at least prepare to prepare, well in advance of crises.

Wondering: Catastrophes might leave a mark on cosmic evolution, in both the physical and (exo)biological senses. Certainly, recent work has established a major role for catastrophes in sculpting terrestrial evolution (see http://www.answers.com/topic/timeline-of-evolution). So to understand the universe, we must take into account their possible occurrence. In particular, serious consideration of Fermi's question 'Where are they?', or logical pursuit of anthropic reasoning, cannot be separated from thinking about how things could go drastically wrong.

This will be a very unbalanced essay. The most urgent and realistic catastrophe scenarios, I think, arise from well-known and much-discussed dangers: the possible use of nuclear weapons and the alteration of global climate. Here those dangers will be mentioned only in passing. The focus instead will be scenarios for catastrophe that are not-so-urgent and/or highly speculative, but involve interesting issues in fundamental physics and cosmology. Thinking about these exotic scenarios needs no apology; but I do want to make it clear that I, in no way, want to exaggerate their relative importance, or to minimize the importance of plainer, more imminent dangers.

16.2 Looking before leaping

16.2.1 Accelerator disasters

Some accelerators are designed to be dangerous. Those accelerators are the colliders that bring together solid uranium or plutonium 'beams' to produce a fission reaction – in other words, nuclear weapons. Famously, the physicists of the Manhattan project made remarkably accurate estimates of the unprecedented amount of energy their 'accelerator' would release (Rhodes, 1986). Before the Alamogordo test, Enrico Fermi seriously considered the possibility that they might be producing a doomsday weapon, which would ignite the atmosphere. He concluded, correctly, that it would not. (Later calculations that an all-out nuclear exchange between the United States and Soviet Union might produce a world-wide firestorm and/or inject enough dust into the atmosphere to produce nuclear winter were not universally accepted; fortunately, they have not been put to the test. Lesser, but still staggering catastrophe is certain [see http://www.sciencedaily.com/releases/2006/12/061211090729.htm].)

So physicists, for better or worse, got that one right. What about accelerators that are designed not as weapons, but as tools for research? Might they be dangerous?

When we are dealing with well-understood physics, we can do conventional safety engineering. Such engineering is not foolproof – bridges do collapse, astronauts do perish – but at least we foresee the scope of potential problems. In contrast, the whole point of great accelerator projects like the Brookhaven Relativistic Heavy Ion Collider (RHIC) or the Counseil Europeén pour la Recherche Nucléaire (CERN) Large Hadron Collider (LHC) is to produce extreme conditions that take us beyond what is well understood. In that context, safety engineering enters the domain of theoretical physics.

In discussing possible dangers associated with frontier research accelerators, the first thing to say is that while these machines are designed to produce unprecedented density of energy, that density is packed within such a miniscule volume of space that the total energy is, by most standards, tiny. Thus a proton–proton collision at the LHC involves about 40 erg of energy – less energy than a dried pea acquires in falling through one centimetre. Were that energy to be converted into mass, it would amount to about one-ten thousandth of a gram. Furthermore, the high energy density is maintained only very briefly, roughly for 10^{-24} seconds.

To envision significant dangers that might be triggered with such limited input, we have to exercise considerable imagination. We have to imagine that a tiny seed disturbance will grow vast, by tapping into hidden instabilities. Yet the example of nuclear weapons should give pause. Nuclear weapons tap into instabilities that were totally unsuspected just five decades before their design. Both ultraheavy (for fission) and ultralight (for fusion) nuclei can release energy by cooking toward the more stable nuclei of intermediate size.

Three possibilities have dominated the discussion of disaster scenarios at research accelerators. I will now discuss each one briefly. Much more extensive, authoritative technical discussions are available (Jaffe et al., 2000).

Black holes: The effect of gravity is extraordinarily feeble in accelerator environments, according to both conventional theory and experiment. (That is to say, the results of precision experiments to investigate delicate properties of the other fundamental interactions agree with theoretical calculations that predict gravity is negligible, and therefore ignore it.) Conventional theory suggests that the relative strength of the gravitational compared to the electromagnetic interactions is, by dimensional analysis, approximately

$$\frac{\text{gravity}}{\text{electromagnetism}} \sim \frac{GE^2}{\alpha} \approx 10^{-28}\left(\frac{E}{\text{TeV}}\right)^2 \qquad (16.1)$$

where G is the Newton constant, α is the fine-structure constant, and we adopt units with $h = c = 1$. Even for LHC energies $E \sim 10$ TeV, this is such a tiny ratio that more refined estimates are gratuitous.

But what if, within a future accelerator, the behaviour of gravity is drastically modified? Is there any reason to think it might? At present, there is no empirical evidence for deviations from general relativity, but speculation that drastic changes in gravity might set in starting at $E \sim 1$ TeV have been popular recently in parts of the theoretical physics community (Antoniadis et al., 1998; Arkani-Hamed et al., 1998; Randall and Sundrm, 1999). There are two broad motivations for such speculation:

Precocious unification? Physicists seek to unify their description of the different interactions. We have compelling ideas about how to unify our description of the strong, electromagnetic, and weak interactions. But the tiny ratio Equation (1) makes it challenging to put gravity on the same footing. One line of thought is that unification takes place only at extraordinarily high energies, namely, $E \sim 10^{15}$ TeV the Planck energy. At this energy, which also corresponds to an extraordinarily small distance of approximately 10^{-33} cm, the coupling ratio is near unity. Nature has supplied a tantalizing hint, from the other interactions, that this is indeed the scale at which unification becomes manifest (Dimopoulos et al., 1981, 1991). A competing line of thought has it that unification could take place at lower energies. That could happen if Equation (1) fails drastically, in such a way that the ratio increases much more rapidly. Then the deepest unity of physics would be revealed directly at energies that we might hope to access – an exciting prospect.

Extra dimensions: One way that this could happen, is if there are extra, curled-up spatial dimensions, as suggested by superstring theory. The short-distance behaviour of gravity will then be drastically modified at lengths below the size of the extra dimensions. Schematic world-models implementing these ideas have been proposed. While existing models appear highly contrived, at least to my eye, they can be fashioned so as not to avoid blatant contradiction with established facts. They provide a concrete framework in which the idea that gravity becomes strong at accessible energies can be realized.

If gravity becomes strong at $E \sim 1-10^{2}$ TeV, then particle collisions at those energies could produce tiny black holes. As the black holes encounter and swallow up ordinary matter, they become bigger black holes ... and we have ourselves a disaster scenario! Fortunately, a more careful look is reassuring. While the words 'black hole' conjure up the image of a great gaping maw, the (highly) conjectural black holes that might be produced at an accelerator are not like that. They would weigh about one ten-thousandth of a gram, with a Compton radius of 10^{-18} cm and, formally, a Schwarzschild radius of 10^{-47} cm. (The fact that the Compton radius, associated with the irreducible quantum-mechanical uncertainty in position, is larger than the Schwarzschild radius, that is, the nominal radius inside which light is trapped, emphases the quantum-mechanical character of these 'black holes'.) Accordingly, their capture zone is extremely small, and they would be very slow eaters. If, that is, these mini-black holes did not spontaneously decay. Small black holes are subject to the Hawking (1974) radiation process, and very small ones are predicted to decay very rapidly, on timescales of order 10^{-18} seconds or less. This is not enough time for a particle moving at the speed of light to encounter more than a few atoms. (And the probability of a hit is in any case miniscule, as mentioned above.) Recent theoretical work even suggests that there is an alternative, dual description of the higher-dimension gravity theory in terms of a four-dimensional strongly interacting quantum field theory, analogous to quantum chromodynamics (QCD) (Maldacena, 1998, 2005). In that description, the short-lived mini-black holes appear only as subtle features in the distribution of particles emerging from collisions; they are similar to the highly unstable resonances of QCD. One might choose to question both the Hawking process and the dual description of strong gravity, both of which are theoretical conceptions with no direct empirical support. But these ideas are inseparable from, and less speculative than, the theories that motivate the mini-black hole hypothesis; so denying the former erodes the foundation of the latter.

Strangelets: From gravity, the feeblest force in the world of elementary particles, we turn to QCD, the strongest, to confront our next speculative disaster scenario. For non-experts, a few words of review are in order. QCD is our theory of the so-called strong interaction (Close, 2006). The ingredients of QCD are elementary particles called quarks and gluons. We have precise, well-tested equations that describe the behaviour of quarks and gluons. There are six different flavours of quarks. The flavours are denoted u, d, s, c, b, t for *up, down strange, charm, bottom, top*. The heavy quarks c, b, and t are highly unstable. Though they are of great interest to physicists, they play no significant role in the present-day natural world, and they have not been implicated in any even remotely plausible disaster scenario. The lightest quarks, u and d, together with gluons, are the primary building blocks of protons and neutrons, and thus of ordinary atomic nuclei. Crudely speaking, protons are composites uud of two up quarks and a down quark, and neutrons are composites udd of one up quark and two down quarks. (More accurately, protons and neutrons are complex objects that contain quark–antiquark pairs and gluons in addition to those three 'valence' quarks.) The mass-energy of the (u, d) quarks is $\sim(5, 10)$ MeV, respectively, which is very small compared to the mass-energy of a proton or neutron of approximately 940 MeV. Almost all the mass of the nucleons – that is, protons and neutrons – arises from the energy of quarks and gluons inside, according to $m = E/c^2$.

Strange quarks occupy an intermediate position. This is because their intrinsic mass-energy, approximately 100 MeV, is comparable to the energies associated with interquark interactions. Strange quarks are known to be constituents of so-called hyperons. The lightest hyperon is the Λ, with a mass of approximately 1116 MeV. The internal structure of the Λ resembles that of nucleons, but it is built from uds rather than *uud* or *udd*.

Under ordinary conditions, hyperons are unstable, with lifetimes of the order 10^{-10} seconds or less. The Λ hyperon decays into a nucleon and a π meson, for example. This process involves conversion of an s quark into a u or d quark, and so it cannot proceed through the strong QCD interactions, which do not change quark flavours. (For comparison, a typical lifetime for particles – 'resonances' – that decay by strong interactions is $\sim 10^{-24}$ seconds.) Hyperons are not so extremely heavy or unstable that they play no role whatsoever in the natural world. They are calculated to be present with small but not insignificant density during supernova explosions and within neutron stars.

The reason for the presence of hyperons in neutron stars is closely related to the concept of 'strangelets', so let us briefly review it. It is connected to the Pauli exclusion principle. According to that principle, no two fermions can occupy the same quantum state. Neutrons (and protons) are fermions, so the exclusion principle applies to them. In a neutron star's interior, very high pressures – and therefore very high densities – are achieved, due to the weight of the overlying layers. In order to obey the Pauli exclusion principle, then, nucleons must squeeze into additional quantum states, with higher energy. Eventually, the extra energy gets so high that it becomes economical to trade a high-energy nucleon for a hyperon. Although the hyperon has larger mass, the marginal cost of that additional mass-energy is less than the cost of the nucleon's exclusion principle-energy.

At even more extreme densities, the boundaries between individual nucleons and hyperons break down, and it becomes more appropriate to describe matter directly in terms of quarks. Then we speak of quark matter. In quark matter, a story very similar to what we just discussed again applies, now with the lighter u and d quarks in place of nucleons and the s quarks in place of hyperons. There is a quantitative difference, however, because the s quark mass is less significant than the hyperon–nucleon mass difference. Quark matter is therefore expected to be rich in strange quarks, and is sometimes referred to as strange matter. Thus there are excellent reasons to think that under high pressure, hadronic – that is, quark-based – matter undergoes a qualitative change, in that it comes to contain a significant fraction of strange quarks. Bodmer and Witten (1984) posed an interesting question: Might this new kind of matter, with higher density and significant strangeness, which theory tells us is surely produced at high pressure, remain stable at zero pressure? If so, then the lowest energy state of a collection of quarks would not be the familiar nuclear matter, based on protons and neutrons, but a bit of strange matter – a strangelet. In a (hypothetical) strangelet, extra strange quarks permit higher density to be achieved, without severe penalty from the exclusion principle. If there are attractive interquark forces, gains in interaction energy might compensate for the costs of additional strange quark mass.

At first hearing, the answer to the question posed in the preceding paragraph seems obvious: No, on empirical grounds. For, if ordinary nuclear matter is not the most energetically favourable form, why is it the form we find around us (and, of course, in us)? Or, to put it another way, if ordinary matter could decay into matter based on strangelets, why has it not done so already? On reflection, however, the issue is not so clear. If only sufficiently large strangelets are favourable – that is, if only large strangelets have lower energy than ordinary matter containing the same net number of quarks – ordinary matter would have a very difficult time converting into them. Specifically, the conversion would require many simultaneous conversions of *u* or *d* quarks into strange quarks. Since each such quark conversion is a weak interaction process, the rate for multiple simultaneous conversions is incredibly small.

We know that for small numbers of quarks, ordinary nuclear matter is the most favourable form, that is, that small strangelets do not exist. If a denser, differently organized version of the Λ existed, for example, nucleons would decay into it rapidly, for that decay requires only one weak conversion. Experiments searching for an alternative Λ – the so-called 'H particle' – have come up empty handed, indicating that such a particle could not be much lighter than two separate Λ particles, let alone light enough to be stable (Borer et al., 1994).

After all this preparation, we are ready to describe the strangelet disaster scenario. A strangelet large enough to be stable is produced at an accelerator. It then grows by swallowing up ordinary nuclei, liberating energy. And there is nothing to stop it from continuing to grow until it produces a catastrophic explosion (and then, having burped, resumes its meal), or eats up a big chunk of Earth, or both.

For this scenario to occur, four conditions must be met:

1. Strange matter must be absolutely stable in bulk.
2. Strangelets would have to be at least metastable for modest numbers of quarks, because only objects containing small numbers of strange quarks might conceivably be produced in an accelerator collision.
3. Assuming that small metastable strangelets exist, it must be possible to produce them at an accelerator.
4. The stable configuration of a strangelet must be negatively charged (see below).

Only the last condition is not self-explanatory. A positively charged strangelet would resemble an ordinary atomic nucleus (though, to be sure, with an unusually small ratio of charge to mass). Like an ordinary atomic nucleus, it would surround itself with electrons, forming an exotic sort of atom. It would not eat other ordinary atoms, for the same reasons that ordinary atoms do not spontaneously eat one another – no cold fusion! – namely, the Coulomb barrier. As discussed in detail in Jaffe et al. (2000), there is no evidence that any of these conditions is met. Indeed, there is substantial theoretical evidence that none is met, and direct experimental evidence that neither condition (2) nor (3) can be met. Here are the summary conclusions of that report:

1. At present, despite vigorous searches, there is no evidence whatsoever on the existence of stable strange matter anywhere in the Universe.
2. On rather general grounds, theory suggests that strange matter becomes unstable in small lumps due to surface effects. Strangelets small enough

to be produced in heavy ion collisions are not expected to be stable enough to be dangerous.

3. Theory suggests that heavy ion collisions (and hadron–hadron collisions in general) are not a good place to produce strangelets. Furthermore, it suggests that the production probability is lower at RHIC than at lower energy heavy ion facilities like the Alternating Gradient Synchrotron (AGS) and CERN. Models and data from lower energy heavy ion colliders indicate that the probability of producing a strangelet decreases very rapidly with the strangelet's atomic mass.
4. It is overwhelmingly likely that the most stable configuration of strange matter has positive electric charge.

It is not appropriate to review all the detailed and rather technical arguments supporting these conclusions here, but two simple qualitative points, that suggest conclusions (3) and (4) above, are easy to appreciate.

Conclusion 3: To produce a strangelet at an accelerator, the crucial condition is that one produces a region where there are many strange quarks (and few strange antiquarks) and not too much excess energy. Too much energy density is disadvantageous, because it will cause the quarks to fly apart: when things are hot you get steam, not ice cubes. Although higher energy at an accelerator will make it easier to produce strange–antistrange quark pairs, higher energy also makes it harder to segregate quarks from antiquarks, and to suppress extraneous background (i.e., extra light quarks and antiquarks, and gluons). Thus conditions for production of strangelets are less favourable at frontier, ultra-high accelerators than at older, lower-energy accelerators – for which, of course, the (null) results are already in. For similar reasons, one does not expect that strangelets will be produced as cosmological relics of the big bang, even if they are stable in isolation.

Conclusion 4: The maximum leeway for avoiding Pauli exclusion, and the best case for minimizing other known interaction energies, occurs with equal numbers of u, d, and s quarks. This leads to electrical neutrality, since the charges of those quarks are 2/3, −1/3, −1/3 times the charge of the proton, respectively. Since the s quark, being significantly heavier than the others, is more expensive, one expects that there will be fewer s quarks than in this otherwise ideal balance (and nearly equal numbers of u and d quarks, since both their masses are tiny). This leads to an overall positive charge.

The strangelet disaster scenario, though ultimately unrealistic, is not silly. It brings in subtle and interesting physical questions, that require serious thought, calculation, and experiment to address in a satisfactory way. Indeed, if the strange quark were significantly lighter than it is in our world, then big strangelets would be stable, and small ones at least metastable. In such an alternative universe, life in anything like the form we know it, based on ordinary nuclear matter, might be precarious or impossible.

Vacuum instability: In the equations of modern physics, the entity we perceive as empty space, and call vacuum, is a highly structured medium full of spontaneous activity and a variety of fields. The spontaneous activity is variously called quantum fluctuations, zero-point motion, or virtual particles. It is directly responsible for several famous phenomena in quantum physics, including Casimir forces, the Lamb shift, and asymptotic freedom. In a more abstract sense, within the framework of quantum field theory, all forces can be traced to the interaction of real with virtual particles (Feynman, 1988; Wilczek, 1999; Zee, 2003).

The space-filling fields can also be viewed as material condensates, just as an electromagnetic field can be considered as a condensate of photons. One such condensation is understood deeply. It is the quark–antiquark condensate that plays an important role in strong interaction theory. A field of quark–antiquark pairs of opposite helicity fills space-time.[1] That quark–antiquark field affects the behaviour of particles that move through it. That is one way we know it is there! Another is by direct solution of the well-established equations of QCD. Low-energy π mesons can be modeled as disturbances in the quark–antiquark field; many properties of π mesons are successfully predicted using that model.

Another condensate plays a central role in our well-established theory of electroweak interactions, though its composition is presently unknown. This is the so-called Higgs condensate. The equations of the established electroweak theory indicate that the entity we perceive as empty space is in reality an exotic sort of superconductor. Conventional superconductors are super(b) conductors of electric currents, the currents that photons care about. Empty space, we learn in electroweak physics, is a super(b) conductor of other currents: specifically, the currents that W and Z bosons care about. Ordinary superconductivity is mediated by the flow of paired electrons – Cooper pairs – in a metal. Cosmic superconductivity is mediated by the flow of something else. No presently known form of matter has the right properties to do the job; for that purpose, we must postulate the existence of new form(s) of matter. The simplest hypothesis, at least in the sense that it introduces the fewest new particles, is the so-called minimal standard model. In the minimal standard model, we introduce just one new particle, the so-called Higgs particle. According to this model, cosmic superconductivity is due to a condensation of Higgs particles. More complex hypotheses, notably including low-energy supersymmetry, introduce several contributions to the electroweak condensate. These models predict that there are several contributors to the electroweak condensate, and that there is a complex of several 'Higgs particles', not just one. A major goal of ongoing

[1] Amplitude of this field is constant in time, spatially uniform and occurs in a spin-0 channel, so that no breaking of Lorentz symmetry is involved.

research at the Fermilab Tevatron and the CERN LHC is to find the Higgs particle, or particles.

Since 'empty' space is richly structured, it is natural to consider whether that structure might change. Other materials exist in different forms – might empty space? To put it another way, could empty space exist in different phases, supporting in effect different laws of physics?

There is every reason to think the answer is 'Yes'. We can calculate, for example, that at sufficiently high temperature the quark–antiquark condensate of QCD will boil away. And although the details are much less clear, essentially all models of electroweak symmetry breaking likewise predict that at sufficiently high temperatures the Higgs condensate will boil away. Thus in the early moments of the big bang, empty space went through several different phases, with qualitatively different laws of physics. (For example, when the Higgs condensate melts, the W and Z bosons become massless particles, on the same footing as photons. So then the weak interactions are no longer so weak!) Somewhat more speculatively, the central idea of inflationary cosmology is that in the very early universe, empty space was in a different phase, in which it had non-zero energy density and negative pressure.

The empirical success of inflationary cosmology therefore provides circumstantial evidence that empty space once existed in a different phase.

More generally, the structure of our basic framework for understanding fundamental physics, relativistic quantum field theory, comfortably supports theories in which there are alternative phases of empty space. The different phases correspond to different configurations of fields (condensates) filling space. For example, attractive ideas about unification of the apparently different forces of Nature postulate that these forces appear on the same footing in the primary equations of physics, but that in their solution, the symmetry is spoiled by space-filling fields. Superstring theory, in particular, supports vast numbers of such solutions, and postulates that our world is described by one of them: for, certainly, our world exhibits much less symmetry than the primary equations of superstring theory.

Given, then, that empty space can exist in different phases, it is natural to ask: Might our phase, that is, the form of physical laws that we presently observe, be suboptimal? Might, in other words, our vacuum be only metastable? If so, we can envisage a terminal ecological catastrophe, when the field configuration of empty space changes, and with it the effective laws of physics, instantly and utterly destabilizing matter and life in the form we know it.

How could such a transition occur? The theory of empty space transitions is entirely analogous to the established theory of other, more conventional first-order phase transitions. Since our present-day field configuration is (at least) metastable, any more favourable configuration would have to be significantly different, and to be separated from ours by intermediate configurations that are less favourable than ours (i.e., that have higher energy density). It is most likely that a transition to the more favourable phase would begin with the emergence of a rather small bubble of the new phase, so that the required rearrangement of fields is not too drastic and the energetic cost of intermediate configurations is not prohibitive. On the other hand the bubble cannot be too small, for the volume energy gained in the interior must compensate unfavourable surface energy (since between the new phase and the old metastable phase one has unfavourable intermediate configurations).

Once a sufficiently large bubble is formed, it could expand. Energy liberated in the bulk transition between old and new vacuum goes into accelerating the wall separating them, which quickly attains near-light speed. Thus the victims of the catastrophe receive little warning: by the time they can see the approaching bubble, it is upon them.

How might the initial bubble form? It might form spontaneously, as a quantum fluctuation. Or it might be nucleated by some physical event, such as – perhaps? – the deposition of lots of energy into a small volume at an accelerator.

There is not much we can do about quantum fluctuations, it seems, but it would be prudent to refrain from activity that might trigger a terminal ecological catastrophe. While the general ideas of modern physics support speculation about alternative vacuum phases, at present there is no concrete candidate for a dangerous field whose instability we might trigger. We are surely in the most stable state of QCD. The Higgs field or fields involved in electroweak symmetry breaking might have instabilities – we do not yet know enough about them to be sure. But the difficulty of producing even individual Higgs particles is already a crude indication that triggering instabilities which require coordinated condensation of many such particles at an accelerator would be prohibitively difficult. In fact there seems to be no reliable calculation of rates of this sort – that is, rates for nucleating phase transitions from particle collisions – even in model field theories. It is an interesting problem of theoretical physics. Fortunately, the considerations of the following paragraph assure us that it is not a practical problem for safety engineering.

As the matching bookend to our initial considerations on size, energy, and mass, let us conclude our discussion of speculative accelerator disaster scenarios with another simple and general consideration, almost independent of detailed theoretical considerations, and which makes it implausible that any of these scenarios apply to reality. It is that Nature has, in effect, been doing accelerator experiments on a grand scale for a very long time (Hut, 1984; Hut and Rees, 1984). For, cosmic rays achieve energies that even the most advanced terrestrial accelerators will not match at any time soon. (For experts: Even by the criterion of center-of-mass energy, collisions of the highest energy cosmic rays with stationary targets beat top-of-the-line accelerators.) In the history of the universe, many collisions have occurred over a very wide spectrum of energies and ambient conditions (Jaffe et al., 2000). Yet in the history of astronomy, no candidate unexplained catastrophe has ever been observed. And many such cosmic rays have impacted Earth, yet Earth abides and we are here. This is reassuring (Bostrom and Tegmark, 2005).

16.2.2 Runaway technologies

Neither general source of reassurance – neither miniscule scale nor natural precedent – necessarily applies to other emergent technologies.

Technologies that are desirable in themselves can get out of control, leading to catastrophic exhaustion of resources or accumulation of externalities. Jared Diamond has argued that history presents several examples of this phenomenon (Diamond, 2005), on scales ranging from small island cultures to major civilizations. The power and agricultural technologies of modern industrial civilization appear to have brought us to the cusp of severe challenges of both these sorts, as water resources, not to speak of oil supplies, come under increasing strain, and carbon dioxide, together with other pollutants, accumulates in the biosphere. Here it is not a question of whether dangerous technologies will be employed – they already are – but on what scale, how rapidly, and how we can manage the consequences.

As we have already discussed in the context of fundamental physics at accelerators, runaway instabilities could also be triggered by inadequately considered research projects. In that particular case, the dangers seem far-fetched. But it need not always be so. Vonnegut's 'Ice 9' was a fictional example (Vonnegut, 1963), very much along the lines of the runaway strangelet scenario – a new form of water, that converts the old. An artificial protein that turned out to catalyse crystallization of natural proteins – an artifical 'prion' – would be another example of the same concept, from yet a different realm of science.

Perhaps more plausibly, runaway technological instabilities could be triggered as an unintended byproduct of applications (as in the introduction of cane toads to Australia) or sloppy practices (as in the Chernobyl disaster); or by deliberate pranksterism (as in computer virus hacking), warfare, or terrorism.

Two technologies presently entering the horizon of possibility have, by their nature, especially marked potential to lead to runaways:

Autonomous, capable robots: As robots become more capable and autonomous, and as their goals are specified more broadly and abstractly, they could become formidable antagonists. The danger potential of robots developed for military applications is especially evident. This theme has been much explored in science fiction, notably in the writings of Isaac Asimov (1950) and in the Star Wars movies.

Self-reproducing machines, including artificial organisms: The danger posed by sudden introduction of new organisms into unprepared populations is exemplified by the devastation of New World populations by smallpox from the Old World, among several other catastrophes that have had a major influence on human history. This is documented in William McNeill's (1976) marvelous *Plagues and Peoples*. Natural organisms that have been re-engineered, or 'machines' of any sort capable of self-reproduction, are by their nature poised on the brink of exponential spread. Again, this theme has been much explored in science fiction, notably in Greg Bear's Blood Music [19]. The chain reactions of nuclear technology also belong, in a broad conceptual sense, to this class – though they involve exceedingly primitive 'machines', that is, self-reproducing nuclear reactions.

16.3 Preparing to prepare

Runaway technologies: The problem of runaway technologies is multi-faceted. We have already mentioned several quite distinct potential instabilities, involving different technologies, that have little in common. Each deserves separate, careful attention, and perhaps there is not much useful that can be said in general. I will make just one general comment. The majority of people, and of scientists and engineers, by far, are well-intentioned; they would much prefer not to be involved in any catastrophe, technological or otherwise. Broad-based democratic institutions and open exchange of information can coalesce this distributed good intention into an effective instrument of action.

Impacts: We have discussed some exotic – and, it turns out, unrealistic – physical processes that could cause global catastrophes. The possibility that asteroids or other cosmic debris might impact Earth, and cause massive devastation, is not academic – it has happened repeatedly in the past. We now have the means to address this danger, and certainly should do so (http://impact.arc.nasa.gov/intro.cfm).

Astronomical instabilities: Besides impacts, there are other astronomical effects that will cause Earth to become much less hospitable on long time scales. Ice ages can result from small changes in Earth's obliquity, the eccentricity of its orbit, and the alignment of its axis with the eccentricity (which varies as the axis precesses) (see http://www.aip.org/history/climate/cycles.htm). These changes occur on time scales of tens of thousands of years. At present the obliquity oscillates within the range 22.1–24.5°. However as the day lengthens and the moon recedes, over time scales of a billion years or so, the obliquity enters a chaotic zone, and much larger changes occur (Laskar et al., 1993). Presumably, this leads to climate changes that are both extreme and highly variable. Finally, over yet longer time scales, our Sun evolves, gradually becoming hotter and eventually entering a red giant phase.

These adverse and at least broadly predictable changes in the global environment obviously pose great challenges for the continuation of human civilization. Possible responses include moving (underground, underwater, or into space), re-engineering our physiology to be more tolerant (either through bio-engineering, or through man–machine hybridization), or some combination thereof.

Heat death: Over still longer time scales, some version of the 'heat death of the universe' seems inevitable. This exotic catastrophe is the ultimate challenge facing the mind in the universe.

Stars will burn out, the material for making new ones will be exhausted, the universe will continue to expand – it now appears, at an accelerating rate – and, in general, useful energy will become a scarce commodity. The ultimate renewable technology is likely to be pure thought, as I will now describe.

It is reasonable to suppose that the goal of a future-mind will be to optimize a mathematical measure of its wellbeing or achievement, based on its internal state. (Economists speak of 'maximizing utility', normal people of 'finding happiness'.) The future-mind could discover, by its powerful introspective abilities or through experience, its best possible state the Magic Moment – or several excellent ones. It could build up a library of favourite states. That would be like a library of favourite movies, but more vivid, since to recreate magic moments accurately would be equivalent to living through them. Since the joys of discovery, triumph, and fulfillment require novelty, to re-live a magic moment properly, the future-mind would have to suppress memory of that moment's previous realizations.

A future-mind focused upon magic moments is well matched to the limitations of reversible computers, which expend no energy. Reversible computers cannot store new memories, and they are as likely to run backwards as forwards. Those limitations bar adaptation and evolution, but invite eternal cycling through magic moments. Since energy becomes a scarce quantity in an expanding universe, that scenario might well describe the long-term future of mind in the cosmos.

16.4 Wondering

A famous paradox led Enrico Fermi to ask, with genuine puzzlement, 'Where are they?' He was referring to advanced technological civilizations in our Galaxy, which he reckoned ought to be visible to us (see Chapter 6).

Simple considerations strongly suggest that technological civilizations whose works are readily visible throughout our Galaxy (that is, given current or imminent observation technology techniques we currently have available, or soon will) ought to be common. But they are not. Like the famous dog that did not bark in the night time, the absence of such advanced technological civilizations speaks through silence.

Main-sequence stars like our Sun provide energy at a stable rate for several billions of years. There are billions of such stars in our Galaxy. Although our census of planets around other stars is still in its infancy, it seems likely that many millions of these stars host, within their so-called habitable zones, Earth-like planets. Such bodies meet the minimal requirements for life in something close to the form we know it, notably including the possibility of liquid water.

On Earth, a species capable of technological civilization first appeared about one hundred thousand years ago. We can argue about defining the precise time when technological civilization itself emerged. Was it with the beginning of agriculture, of written language, or of modern science? But whatever definition we choose, its age will be significantly less than one hundred thousand years.

In any case, for Fermi's question, the most relevant time is not one hundred thousand years, but more nearly one hundred years. This marks the period of technological 'breakout', when our civilization began to release energies and radiations on a scale that may be visible throughout our Galaxy. Exactly what that visibility requires is an interesting and complicated question, whose answer depends on the hypothetical observers. We might already be visible to a sophisticated extraterrestrial intelligence, through our radio broadcasts or our effects on the atmosphere, to a sophisticated extraterrestrial version of SETI. The precise answer hardly matters, however, if anything like the current trend of technological growth continues. Whether we are barely visible to sophisticated though distant observers today, or not quite, after another thousand years of technological expansion at anything like the prevailing pace, we should be easily visible. For, to maintain even modest growth in energy consumption, we will need to operate on astrophysical scales.

One thousand years is just one millionth of the billion-year span over which complex life has been evolving on Earth. The exact placement of breakout within the multi-billion year timescale of evolution depends on historical accidents. With a different sequence of the impact events that lead to mass extinctions, or earlier occurrence of lucky symbioses and chromosome doublings, Earth's breakout might have occurred one billion years ago, instead of one hundred years.

The same considerations apply to those other Earth-like planets. Indeed, many such planets, orbiting older stars, came out of the starting gate billions of years before we did. Among the millions of experiments in evolution in our Galaxy, we should expect that many achieved breakout much earlier, and thus became visible long ago. So: Where are they?

Several answers to that paradoxical question have been proposed. Perhaps this simple estimate of the number of life-friendly planets is for some subtle reason wildly over-optimistic. For example, our Moon plays a crucial role in stabilizing the Earth's obliquity, and thus its climate; probably, such large moons are rare (ours is believed to have been formed as a consequence of an unusual, giant impact), and plausibly extreme, rapidly variable climate is enough to inhibit the evolution of intelligent life. Perhaps on Earth the critical symbioses and chromosome doublings were unusually lucky, and the impacts extraordinarily well-timed. Perhaps, for these reasons or others, even if life of some kind is widespread, technologically capable species are extremely rare, and we happen to be the first in our neighbourhood.

Or, in the spirit of this essay, perhaps breakout technology inevitably leads to catastrophic runaway technology, so that the period when it is visible is sharply limited. Or – an optimistic variant of this – perhaps a sophisticated, mature society avoids that danger by turning inward, foregoing power engineering in favour of information engineering. In effect, it thus chooses to become invisible from afar. Personally, I find these answers to Fermi's question to be the most plausible. In any case, they are plausible enough to put us on notice.

Suggestions for further reading

Jaffe, R., Busza, W., Sandweiss, J., and Wilczek, F. (2000). Review of speculative 'disaster scenarios' at RHIC. *Rev. Mod. Phys.*, **72**, 1125–1140, available on the web at arxiv.org:hepph/9910333. A major report on accelerator disaster scenarios, written at the request of the director of Brookhaven National Laboratory, J. Marburger, before the commissioning of the RHIC. It includes a non-technical summary together with technical appendices containing quantitative discussions of relevant physics issues, including cosmic ray rates. The discussion of strangelets is especially complete.

Rhodes, R. (1986). *The Making of the Atomic Bomb* (Simon & Schuster). A rich history of the one realistic 'accelerator catastrophe'. It is simply one of the greatest books ever written. It includes a great deal of physics, as well as history and high politics. Many of the issues that first arose with the making of the atom bomb remain, of course, very much alive today.

Kurzweil, R. (2005). *The Singularity Is Near* (Viking Penguin). Makes a case that runaway technologies are endemic – and that is a good thing! It is thought-provoking, if not entirely convincing.

References

Antoniadis, I., Arkani-Hamed, N., Dimopoulos, S., and Dvali, G. (1998). *Phys. Lett. B*, **436**, 257.

Arkani-Hamed, N., Dimopoulos, S., and Dvali, G. (1998). *Phys. Lett.*, **429**, 263.

Asimov, I. (1950). *I, Robot* (New York: Gnome Press).

Bear, G. (1985). *Blood Music* (New York: Arbor House).

Borer, K., Dittus, F., Frei, D., Hugentobler, E., Klingenberg, R., Moser, U., Pretzl, K., Schacher, J., Stoffel, F., Volken, W., Elsener, K., Lohmann, K.D., Baglin, C., Bussière, A., Guillaud, J.P., Appelquist, G., Bohm, C., Hovander, B., Selldèn, B., and Zhang, Q.P. (1994). Strangelet search in S-W collisions at 200*A* Ge V/*c*. *Phys. Rev. Lett.*, **72**, 1415–1418.

Bostrom, N. and Tegmark, M. (2005). Is a doomsday catastrophe likely? *Nature*, **438**, 754–756.

Close, F. (2006). *The New Cosmic Onion* (New York and London: Taylor & Francis).

Diamond, J. (2005). *Collapse: How Societies Choose to Fail or Succeed* (New York: Viking).
Dimopoulos, S., Raby, S., and Wilczek, F. (1981). Supersymmetry & the scale of unification. *Phys. Rev.*, **D24**, 1681–1683.
Dimopoulos, S., Raby, S., and Wilczek, F. (1991). Unification of Couplings. *Physics Today*, **44**, October 25, pp. 25–33.
Feynman, R. (1988). *QED: The Strange Theory of Light and Matter* (Princeton, NJ: Princeton University Press).
Hawking, S.W. (1974). Black hole explosion? *Nature*, **248**, 30–31.
Hut, P. (1984). Is it safe to distribute the vacuum? *Nucl. Phys.*, **A418**, 301C.
Hut, P. and Rees, M.J. How stable is our vacuum? Report-83-0042 (Princeton: IAS).
Jaffe, R., Busza, W., Sandweiss, J., and Wilczek, F. (2000). Review of speculative 'disaster scenarios' at RHIC. *Rev. Mod. Phys.*, **72**, 1125–1140.
Laskar, J., Joutel, F., and Robutel, P. (1993). Stabilization of the earth's obliguity by the Moon. *Nature*, **361**, 615–617.
Maldacena, J. (1998). The cage-N limit of superconformal field theories & supergravity. *Adv. Theor. Math. Phys.*, **2**, 231–252.
Maldacena, J. (2005). The illusion of gravity. *Scientific American*, **November**, 56–63.
McNeill, W. (1976). *Plagues and Peoples* (New York: Bantam).
Randall, L. and Sundrm, R. (1999). Large mass hierarchy from a small extra demensia. *Phys. Rev. Lett.*, **83**, 3370–3373.
Rhodes, R. (1986). *The Making of the Atomic Bomb* (New York: Simon & Schuster).
Schröder, P., Smith, R., and Apps, K. (2001). Solar evolution & the distant future of earth. *Astron. Geophys.*, **42**(6), 26–32.
Vonnegut, K. (1963). *Cat's Cradle* (New York: Holt, Rinehart, & Wilson).
Wilczek, F. (1999). Quantum field theory. *Rev. Mod. Phys.*, **71**, S85–S05.
Witten, E. (1984). Cosmic separation of phases. *Phys. Rev.*, **D30**, 272–285.
Zee, A. (2003). *Quantum Field Theory in a Nutshell* (Princeton, NJ: Princeton University Press).

·17·

Catastrophe, social collapse, and human extinction

Robin Hanson

17.1 Introduction

Modern society is a bicycle, with economic growth being the forward momentum that keeps the wheels spinning. As long as the wheels of a bicycle are spinning rapidly, it is a very stable vehicle indeed. But, [Friedman] argues, when the wheels stop – even as the result of economic stagnation, rather than a downturn or a depression – political democracy, individual liberty, and social tolerance are then greatly at risk even in countries where the absolute level of material prosperity remains high ...

DeLong, 2006

The main reason to be careful when you walk up a flight of stairs is not that you might slip and have to retrace one step, but rather that the first slip might cause a second slip, and so on until you fall dozens of steps and break your neck. Similarly, we are concerned about the sorts of catastrophes explored in this book not only because of their terrible direct effects, but also because they may induce an even more damaging collapse of our economic and social systems. In this chapter, I consider the nature of societies, the nature of social collapse, and the distribution of disasters that might induce social collapse, and possible strategies for limiting the extent and harm of such collapse.

17.2 What is society?

Before we can understand how societies collapse, we must first understand how societies exist and grow. Humans are far more numerous, capable, and rich than were our distant ancestors. How is this possible? One answer is that today we have more of most kinds of 'capital', but by itself this answer tells us little; after all, 'capital' is just anything that helps us to produce or achieve more. We can understand better by considering the various types of capital we have.

First, we have natural capital, such as soil to farm, ores to mine, trees to cut, water to drink, animals to domesticate, and so on. Second, we have physical capital, such as cleared land to farm, irrigation ditches to move water, buildings to live in, tools to use, machines to run, and so on. Third, we have human capital, such as healthy hands to work with, skills we have honed with practice, useful techniques we have discovered, and abstract principles that help us think. Fourth, we have social capital, that is, ways in which groups of people have found to coordinate their activities. For example, households organize who does what chores, firms organize which employees do which tasks, networks of firms organize to supply inputs to each other, cities and nations organize to put different activities in different locations, culture organizes our expectations about the ways we treat each other, law organizes our coalitions to settle small disputes, and governments coordinate our largest disputes.

There are several important things to understand about all this capital. First, the value of almost any piece of capital depends greatly on what other kinds of capital are available nearby. A fence may be very useful in a prairie but useless in a jungle, while a nuclear engineer's skills may be worth millions in a rich nation, but nothing in a poor nation. The productivity of an unskilled labourer depends greatly on how many other such labourers are available.

Second, scale makes a huge difference. The more people in a city or nation, the more each person or group can narrow their specialty, and get better at it. Special products or services that would just not be possible in a small society can thrive in a large society. So anything that lets people live more densely, or lets them talk or travel more easily, can create large gains by increasing the effective social scale.

Third, coordination and balance of capital are very important. For example, places with low social capital can stay poor even after outsiders contribute huge resources and training, while places with high social capital can quickly recover from wars that devastate their natural, physical, and human capital.

17.3 Social growth

The opposite of collapse is growth. Over history, we have dramatically increased our quantities of most, though not all, kinds of capital. How has this been possible?

Over the last few decades, economists have learned a lot about how societies grow (Aghion and Howitt, 1998; Barro and Sala-I-Martin, 2003; Jones, 2002). While much ignorance remains, a few things seem clear. Social capital is crucial; rich places can grow fast while poor places decline. Also crucial is scale and neighbouring social activity; we each benefit greatly on average from other productive activity nearby.

Another key point is that better 'technology', that is, better techniques and coordination, drive growth more than increased natural or physical capital. Better technology helps us produce and maintain more natural and physical capital, a stronger effect than the ability of more natural and physical capital to enable better technology (Grubler, 1998).

Let us quickly review the history of growth (Hanson, 2000), starting with animals, to complete our mental picture. All animal species have capital in the form of a set of healthy individuals and a carefully honed genetic design. An individual animal may also have capital in the form of a lair, a defended territory, and experience with that area. Social animals, such as ants, also have capital in the form of stable organized groups.

Over many millions of years the genetic designs of animals slowly acquired more possibilities. For example, over the last half billion years, the size of the largest brains doubled roughly every 35 million years. About 2 million years ago some primates acquired the combination of a large social brain, hands that could handle tools, and mouths that could voice words; a combination that allowed tools, techniques, and culture to become powerful forms of capital.

The initial human species had perhaps ten thousand members, which some estimate to be the minimum for a functioning sexual species. As human hunter-gatherers slowly accumulated more kinds of tools, clothes, and skills, they were able to live in more kinds of places, and their number doubled every quarter million years. Eventually, about 10,000 years ago, humans in some places knew enough about how to encourage local plants and animals that these humans could stop wandering and stay in one place.

Non-wandering farmers could invest more profitably in physical capital such as cleared land, irrigation ditches, buildings, and so on. The increase in density that farming allowed also enabled our ancestors to interact and coordinate with more people. While a hunter-gatherer might not meet more than a few hundred people in his or her life, a farmer could meet and trade with many thousands.

Soon, however, these farming advantages of scale and physical capital reached diminishing returns, as the total productivity of a region was limited by its land area and the kinds of plants and animals available to grow. Growth was then limited importantly by the rate at which humans could domesticate new kinds of plants and animals, allowing the colonization of new land. Since farmers talked more, they could spread such innovations much faster than hunter-gatherers; the farming population doubled every 1000 years.

A few centuries ago, the steady increase in farming efficiency and density, as well as travel ease, finally allowed humans to specialize enough to support an industrial society. Specialized machines, factories, and new forms of social coordinate, allowed a huge increase in productivity. Diminishing returns quickly set in regarding the mass of machines we produced, however. We still make about the same mass of items per person as we did two centuries ago.

Today's machines are far more capable as a result of improving technologies. And networks of communication between specialists in particular techniques have allowed the rapid exchange of innovations; during the industrial era, world product (the value of items and services we produce) has doubled roughly every 15 years.

Our history has thus seen four key growth modes: animals with larger brains; human hunter-gatherers with more tools and culture enabling them to fill more niches; human farmers domesticating more plants, animals, and land types; and human industry improving its techniques and social capital. During each mode, growth was over a hundred times faster than before, and production grew by a factor of over two hundred. While it is interesting to consider whether even faster growth modes might appear in the future, in this chapter we turn our attention to the opposite of growth: collapse.

17.4 Social collapse

Social productivity fluctuates constantly in response to various disturbances, such as changes in weather, technology, or politics. Most such disturbances are small, and so induce only minor social changes, but the few largest disturbances can induce great social change. The historical record shows at least a few occasions where social productivity fell rapidly by a large enough degree to be worthy of the phrase 'social collapse'. For example, there have been famous and dramatic declines, with varying speeds, among ancient Sumeria, the Roman empire, and the Pueblo peoples. A century of reduced rain, including three droughts, apparently drove the Mayans from their cities and dramatically reduced their population, even though the Mayans had great expertise and experience with irrigation and droughts (Haug et al., 2003).

Some have explained these historical episodes of collapse as due to a predictable internal tendency of societies to overshoot ecological capacity (Diamond, 2005), or to create top-heavy social structures (Tainter, 1988). Other analysis, however, suggests that most known ancient collapses were initiated by external climate change (deMenocal, 2001; Weiss and Bradley, 2001). The magnitude of the social impact, however, often seems out of proportion to the external disturbance. Similarly, in recent years, relatively minor external problems often translate into much larger reductions in economic growth (Rodrik, 1999). This disproportionate response is of great concern; what causes it?

One obvious explanation is that the intricate coordination that makes a society more productive also makes it more vulnerable to disruptions. For example, productivity in our society requires continued inputs from a large number of specialized systems, such as for electricity, water, food, heat, transportation, communication, medicine, defense, training, and sewage. Failure of any one of these systems for an extended period can destroy the entire system. And since geographic regions often specialize in supplying particular inputs, disruption of one geographic region can have a disproportionate effect on a larger society. Transportation disruptions can also reduce the benefits of scale societies enjoy.

Capital that is normally carefully balanced can become unbalanced during a crisis. For example, a hurricane may suddenly increase the value of gas, wood, and fresh water relative to other goods. The sudden change in the relative value of different kinds of capital produces inequality, that is, big winners and losers, and envy – a feeling that winner gains are undeserved. Such envy can encourage theft and prevent ordinary social institutions from functioning; consider the widespread resistance to letting market prices rise to allocate gas or water during a crisis.

'End game' issues can also dilute reputational incentives in severe situations. A great deal of social coordination and cooperation is possible today because the future looms large. We forgo direct personal benefits now for fear that others might learn later of such actions and avoid us as associates. For most of us, the short-term benefits of 'defection' seem small compared to the long-term benefits of continued social 'cooperation'.

But in the context of a severe crisis, the current benefits of defection can loom larger. So not only should there be more personal grabs, but the expectation of such grabs should reduce social coordination. For example, a judge who would not normally consider taking a bribe may do so when his life is at stake, allowing others to expect to get away with theft more easily, which leads still others to avoid making investments that might be stolen, and so on. Also, people may be reluctant to trust bank accounts or even paper money, preventing those institutions from functioning.

Such multiplier effects of social collapse can induce social elites to try to deceive the rest about the magnitude of any given disruption. But the rest of society will anticipate such deception, making it hard for social elites to accurately communicate the magnitude of any given disruption. This will force individuals to attend more to their private clues, and lead to less social coordination in dealing with disruptions.

The detailed paths of social collapse depend a great deal on the type of initial disruption and the kind of society disrupted. Rather than explore these many details, let us see how far we can get thinking in general about social collapse due to large social disruptions.

17.5 The distribution of disaster

First, let us consider some general features of the kinds of events that can trigger large social disruptions. We have in mind events such as earthquakes, hurricanes, plagues, wars, and revolutions. Each such catastrophic event can be described by its severity, which might be defined in terms of energy released, deaths induced, and so on.

For many kinds of catastrophes, the distribution of event severity appears to follow a power law over a wide severity range. That is, sometimes the chance that within a small time interval one will see an event with severity S that is greater than a threshold s is given by

$$P(S > s) = ks^{-\alpha}, \tag{17.1}$$

where k is a constant and α is the power of this type of disaster.

Now we should keep in mind that these powers α can only be known to apply within the scales sampled by available data, and that many have disputed how widely such power laws apply (Bilham, 2004), and whether power laws are the best model form, compared, for example, to the lognormal distribution (Clauset et al., 2007a).

Addressing such disputes is beyond the scope of this chapter. We will instead consider power law distributed disasters as an analysis reference case. Our conclusions would apply directly to types of disasters that continue to be distributed as a power law even up to very large severity. Compared to this reference case, we should worry less about types of disasters whose frequency of very large events is below a power law, and more about types of disasters whose frequency is greater.

The higher the power α, the fewer larger disasters there are, relative to small disasters. For example, if they followed a power law, then car accidents would have a high power, as most accidents involve only one or two cars, and very few accidents involve one hundred or more cars. Supernovae deaths, on the other hand, would probably have a small power; if anyone on Earth is killed by a supernova, most likely many will be killed.

Disasters with a power of one are right in the middle, with both small and large disasters being important. For example, the energy of earthquakes, asteroid impacts, and Pacific hurricanes all seem to be distributed with a power of about one (Christensen et al., 2002; Lay and Wallace, 1995; Morrison et al., 2003; Sanders, 2005). (The land area disrupted by an earthquake also seems to have a power of one [Turcotte, 1999].) This implies that for any given earthquake of energy E and for any time interval, as much energy will on average be released in earthquakes with energies in the range from E to $2E$ as in earthquakes with energies in the range from $E/2$ to E. While there should be twice as many events in the second range, each event should only release half as much energy.

Disasters with a high power are not very relevant for social collapse, as they have little chance of being large. So, assuming published power estimates are reliable and that the future repeats the past, we can set aside windstorms (energy power of 12), and worry only somewhat about floods, tornadoes, and terrorist attacks (with death powers of 1.35, 1.4, and 1.4). But we should worry more about disasters with lower powers, such as forest fires (area power of 0.66), hurricanes (dollar-loss power of 0.98, death power of 0.58), earthquakes (energy power of 1, dollar-loss and death powers of 0.41), wars (death power of 0.41), and plagues (death power of 0.26 for Whooping Cough and Measles) (Barton and Nishenko, 1997; Cederman, 2003; Clauset et al., 2007b; Nishenko & Barton, 1995; Rhodes et al., 1997; Sanders, 2005; Turcotte, 1999; Watts et al., 2005).

Note that energy power tends to be higher than economic loss power, which tends to be higher than death power. This says that compared to the social loss produced by a small disturbance, the loss produced by a large disturbance seems out of proportion to the disturbance, an effect that is especially strong for disasters that threaten lives and not just property. This may (but not necessarily) reflect the disproportionate social collapse that large disasters induce.

For a type of disaster where damage is distributed with a power below one, if we are willing to spend time and effort to prevent and respond to small events, which hurt only a few people, we should be willing to spend far more to prevent and respond to very large events, which would hurt a large fraction of the Earth's population. This is because, while large events are less likely, their enormous damage more than makes up for their low frequency. If our power law description is not misleading for very large events, then in terms of expected deaths, most of the deaths from war, earthquakes, hurricanes, and plagues occur in the very largest of such events, which kill a large fraction of the world's population. And those deaths seem to be disproportionately due to social collapse, rather than the direct effect of the disturbance.

17.6 Existential disasters

How much should we worry about even larger disasters, triggered by disruptions several times stronger than the ones that can kill a large fraction of humanity? Well, if we only cared about the expected number of people killed due to an event, then we would not care that much whether 99% or 99.9% of the population was killed. In this case, for low power disasters, we would care the most about events large enough to kill roughly half of the population; our concern would fall away slowly as we considered smaller events, and fall away quickly as we considered larger events.

A disaster large enough to kill off humanity, however, should be of special concern. Such a disaster would prevent the existence of all future generations of humanity. Of course, it is possible that humanity was about to end in any case, and it is also possible that without humans, within a few million years, some other mammal species on Earth would evolve to produce a society we would respect. Nevertheless, since it is also possible that neither of these things would happen, the complete destruction of humanity must be considered a great harm, above and beyond the number of humans killed in such an event.

It seems that groups of about seventy people colonized both Polynesia and the New World (Hey, 2005; Murray-McIntosh et al., 1998). So let us assume, as a reference point for analysis, that the survival of humanity requires that 100 humans remain, relatively close to one another, after a disruption and its resulting social collapse. With a healthy enough environment, 100 connected humans might successfully adopt a hunter-gatherer lifestyle. If they were in close enough contact, and had enough resources to help them through a transition period, they might maintain a sufficiently diverse gene pool, and slowly increase their capabilities until they could support farming.

Once they could communicate to share innovations and grow at the rate that our farming ancestors grew, humanity should return to our population and productivity level within 20,000 years. (The fact that we have used up some natural resources this time around would probably matter little, as growth rates do not seem to depend much on natural resource availability.) With less than 100 survivors near each other, on the other hand, we assume humanity would become extinct within a few generations.

Figure 17.1 illustrates a concrete example to help us explore some issues regarding existential disruptions and social collapse. It shows a log–log graph of event severity versus event frequency. For the line marked 'Post-collapse deaths', the part of the line on the right side of the figure is set to be roughly the power law observed for war deaths today (earthquake deaths have the same slope, but are one-third as frequent). The line marked 'Direct deaths' is speculative and represents the idea that a disruption only directly causes some deaths; the rest are due to social collapse following a disruption. The additional deaths due to social collapse are a small correction for small events, and become a larger correction for larger events.

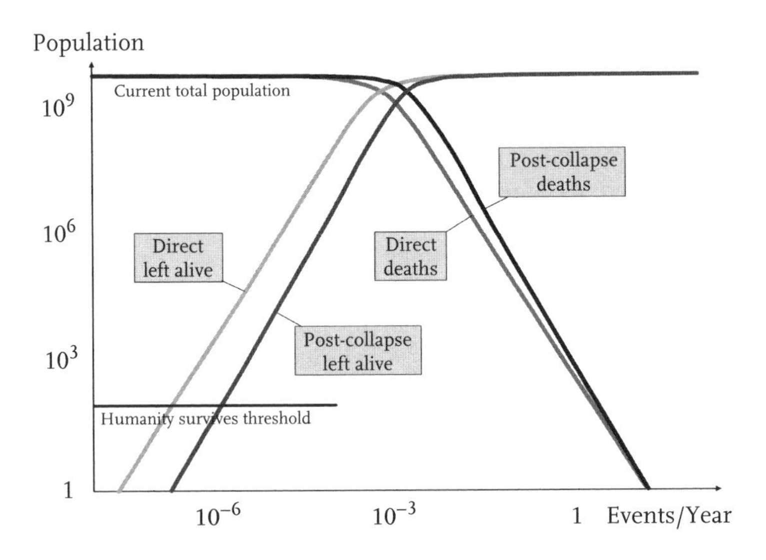

Fig. 17.1 A soft cut-off power law scenario.

Of course the data to which these power laws have been fitted do not include events where most of humanity was destroyed. So in the absence of direct data, we must make guesses about how to project the power law into the regime where most people are killed. If S is the severity of a disaster, to which a power law applies, T is the total population just before the disaster, and D is the number killed by the disaster, then one simple approach would be to set

$$D = \max(T, S) \tag{17.2}$$

This would produce a very hard cut-off.

In this case, much of the population would be left alive or everyone would be dead; there would be little chance of anything close to the borderline. This model expresses the idea that whether a person dies from a disaster depends primarily on the strength of that disaster, and depends little on a varying individual ability to resist disaster. Given the parameters of Fig. 17.1, there would be a roughly a 1 in a 1000 chance each year of seeing an event that destroyed all of humanity.

Figure 17.1 instead shows a smoother projection, with a softer cut-off,

$$\frac{1}{D} = \frac{1}{S} + \frac{1}{T} \tag{17.3}$$

In the regime where most people are left alive, $D \ll T$, this gives $D \approx S$, and so gives the familiar power law,

$$P(D > s) = \mathrm{k}s^{-\alpha} \tag{17.4}$$

But in the regime where the number of people left alive, $L = T - D$, is small, with $L \ll T$, we have a new but similar power law,

$$P(L < s) = \mathrm{k}'s^{\alpha} \tag{17.5}$$

For this projection, it takes a much stronger event to destroy all of humanity.

This model expresses the idea that in addition to the strength of the disaster, variations in individual ability to resist disaster are also very important. Such power law survival fractions have been seen in some biological cases (Burchell et al., 2004). Variable resistance might be due to variations in geographic distance, stockpiled wealth, intelligence, health, and military strength.

Figure 17.1 shows a less than 1 in 3 million chance per year of an event that would kill everyone in the ensuing social collapse. But there is a 1 in 500,000 chance of an event that leaves less than 100 people alive; by assumption, this would not be enough to save humanity. And if the remaining survivors were not all in one place, but distributed widely across the Earth and unable to move to come together, it might take many thousands of survivors to save humanity.

The figure illustrates some of the kinds of trade-offs involved in preventing the extinction of humanity. We assumed somewhat arbitrarily above that 100 humans were required to preserve humanity. Whatever this number is, if it could be reduced somehow by a factor of two, for the survival of humanity, that would be equivalent to making this type of disaster a factor of two less damaging, or increasing our current human population by a factor of two. In the figure, that is equivalent to about a 25% reduction in rate at which this type of event occurs. This figure also predicts that of every fifty people left alive directly after the disruption, only one remains alive after the ensuing social collapse. A factor of two improvement in the number who survive social collapse would also bring the same benefits.

17.7 Disaster policy

For some types of disasters, like car accidents and windstorms, frequency falls so quickly with event severity that large events can be ignored; they just do not happen. For other types of disasters, such as floods, tornadoes, and terrorist attacks, the frequency falls quickly enough that disasters large enough to cause serious social collapse can be mostly ignored; they are very rare.

But for still other types of disasters, such as fires, hurricanes, earthquakes, wars, and plagues, most of the expected harm may be in the infrequent but largest events, which would hurt a large fraction of the world. So if we are willing to invest at all in preventing or preparing for these type of events, it seems we should invest the most in preventing and preparing for these largest events. (Of course this conclusion is muted if there are other benefits of preparing for smaller events, benefits which do not similarly apply to preparing for large events.)

For some types of events, such as wars or plagues, large events often arise from small events that go wrong, and so preparing for and preventing small events may in fact be the best way to prevent large events. But often there are conflicts between preparing for small versus large events. For example, the best response to a small fire in a large building is to stay put until told to move, but at the World Trade Center many learned the hard way that this is bad advice for a large fire. Also, allowing nations to have nuclear weapons can discourage small wars, but encourage large ones.

Similarly, the usual advice for an earthquake is to 'duck and cover' under a desk or doorway. This is good advice for small earthquakes, where the main risk is being hit by items falling from the walls or ceiling. But some claim that in a large earthquake where the building collapses, hiding under a desk will most likely get you flattened under that desk; in this case the best place is said to be pressed against the bottom of something incompressible like file cabinets full of paper (Copp, 2000). Unfortunately, our political systems may reward preparing for the most common situations, rather than the greatest expected damage situations.

For some kinds of disruptions, like asteroid strikes, we can work to reduce the rate and severity of events. For other kinds of disruptions, like earthquakes, floods, or hurricanes, we can design our physical systems to better resist damage, such as making buildings that sway rather than crack, and keeping buildings out of flood plains. We can also prevent nuclear proliferation and reduce existing nuclear arsenals.

We can similarly design our social systems to better resist damage. We can consider various crisis situations ahead of time, and make decisions about how to deal with them. We can define who would be in charge of what, and who would have what property rights.

We can even create special insurance or crisis management organizations which specialize in dealing with such situations.

If they could count on retaining property rights in a crisis, private organizations would have incentives to set aside private property that they expect to be valuable in such situations. For public goods, or goods with large positive externalities, governments might subsidize organizations that set aside such goods in preparation for a disaster.

Unfortunately, the fact that large disasters are rare makes it hard to evaluate claims about which mechanisms will actually help in such situations. An engineering organization may claim that a dike would only fail once in a century, and police may claim they will keep the peace even with serious social collapse, but track records are not of much use in evaluating such claims.

If we value future generations of humanity, we may be willing to take extra efforts to prevent the extinction of humanity. For types of disasters where variations in individual ability to resist disruptions are minor, however, there is little point in explicitly preparing for human extinction possibilities. This is because there is almost no chance that an event of this type would put us very near an extinction borderline. The best we could do here would be to try to prevent all large disruptions. Of course there can be non extinction-related reasons to prepare for such disruptions.

On the other hand, there may be types of disasters where variations in resistance abilities can be important. If so, there might be a substantial chance of finding a post-disaster population that is just above, or just below, a threshold for preserving humanity. In this case it is reasonable to wonder what we might do now to change the odds. The most obvious possibility would be to create refuges with sufficient resources to help preserve a small group of people through a very large disruption, the resulting social collapse, and a transition period to a post-disaster society. Refuges would have to be strong enough to survive the initial disruption. If desperate people trying to survive a social collapse could threaten a refuge's long-term viability, such as by looting the refuge's resources, then refuges might need to be isolated, well-defended, or secret enough to survive such threats.

We have actually already developed similar refuges to protect social elites during a nuclear war (McCamley, 2007). Though nuclear sanctuaries may not be designed with other human extinction scenarios in mind, it is probably worth considering how they might be adapted to deal with non-nuclear-war disasters. It is also worth considering whether to create a distinct set of refuges, intended for other kinds of disasters. I imagine secret rooms deep in a mine, well stocked with supplies, with some way to monitor the surface and block entry.

An important issue here is whether refuges could by themselves preserve enough humans to supply enough genetic diversity for a post-disaster society. If not, then refuges would either have to count on opening up at the right moment to help preserve enough people outside the sanctuary, or they would need some sort of robust technology for storing genes and implanting them. Perhaps a sperm bank would suffice.

Developing a robust genetic technology might be a challenging task; devices would have to last until the human population reached sufficient size to hold enough genetic diversity on its own. But the payoff could be to drastically reduce the required post-collapse population, perhaps down to a single fertile female. For the purpose of saving humanity reducing the required population from 1000 down to 10 is equivalent to a factor of one hundred in current world population, or a factor of one hundred in the severity of each event. In the example of Fig. 17.1, it is the same as reducing the disaster event rate by a factor of fifty.

Refuges could in principle hold many kinds of resources which might ease and speed the restoration of a productive human society. They could preserve libraries, machines, seeds, and much more. But the most important resources would clearly be those that ensure that humanity survives. By comparison, on a cosmic scale, it is a small matter whether humanity takes 1000 or 100,000 years to return to our current level of development. Thus the priority should be resources to support a return to at least a hunter-gatherer society.

It is important to realize that a society rebuilding after a near-extinction crisis would have a vastly smaller scale than our current society; very different types and mixes of capital would be appropriate. Stocking a sanctuary full of the sorts of capital that we find valuable today could be even less useful than the inappropriate medicine, books, or computers often given by first world charities to the third world poor today. Machines would quickly fall into disrepair, and books would impart knowledge that had little practical application.

Instead, one must accept that a very small human population would mostly have to retrace the growth path of our human ancestors; one hundred people cannot support an industrial society today, and perhaps not even a farming society. They might have to start with hunting and gathering, until they could reach a scale where simple farming was feasible. And only when their farming population was large and dense enough could they consider returning to industry.

So it might make sense to stock a refuge with real hunter-gatherers and subsistence farmers, together with the tools they find useful. Of course such people would need to be disciplined enough to wait peacefully in the refuge until the time to emerge was right. Perhaps such people could be rotated periodically from a well-protected region where they practiced simple lifestyles, so they could keep their skills fresh. And perhaps we should test our refuge concepts, isolating real people near them for long periods to see how well particular sorts of refuges actually perform at returning their inhabitants to a simple sustainable lifestyle.

17.8 Conclusion

While there are many kinds of catastrophes that might befall humanity, most of the damage that follows large disruptions may come from the ensuing social collapse, rather than from the direct effects of the disruption. In thinking about how to prevent and respond to catastrophe, it is therefore crucial to consider the nature of social collapse and how we might minimize it.

After reviewing the nature of society and of social collapse, we have considered how to fit social collapse into a framework where disaster severity follows a reference power law distribution. We made two key distinctions. The first distinction is between types of disasters where small events are the most important, and types of disasters where large events are the most important. The second key distinction is whether individual variation in resistance to a disaster is minor or important.

For types of disaster where both large events and individual resistance variation are important, we have considered some of the trade-offs involved in trying to preserve humanity. And we have briefly explored the possibility of building special refuges to increase the chances of saving humanity in such situations.

It should go without saying that this has been a very crude and initial analysis; a similar but more careful and numerically precise analysis might be well worth the effort.

Acknowledgement

I thank Jason Matheny, the editors, and an anonymous referee. For their financial support, I thank the Center for Study of Public Choice and the Mercatus Center.

References

Aghion, P., and Howitt, P. (1998). *Endogenous Growth Theory*. London: MIT Press.

Barro, R.J. and Sala-I-Martin, X. (2003). *Economic Growth*, 2nd edition. London: MIT Press.

Barton, C. and Nishenko, S. (1997). *Natural Disasters: Forecasting Economic and Life Losses*. http://pubs.usgs.gov/fs/natural-disasters/

Bilham, R. (2004). Urban earthquake fatalities – a safer world or worse to come? *Seismol. Rev. Lett.* **75**, 706–712.

Burchell, M.J., Mann, J.R., and Bunch, A.W. (2004). Survival of bacteria and spores under extreme shock pressures. *MNRAS*, **352**(4), 1273–1278.

Caplan, B. (2003). The idea trap: the political economy of growth divergence. *Eur. J. Polit. Econ.*, **19**(2), 183–203.

Cederman, L.-E. (2003). Modeling the size of wars: from Billiard Balls to Sandpiles. *Am. Polit. Sci. Rev.*, **97**(1), 135–150.

Christensen, K., Danon, L., Scanlon, T., and Bak, P. (2002). Unified scaling law for earth-quakes. *Proc. Natl. Acad. Sci.*, **99**(1), 2509–2513.

Clauset, A., Shalizi, C.R., and Newman, M.E.J. (2007a). Power-law distributions in empirical data. arXiv:0706.1062v1.

Clauset, A., Young, M., and Gleditsch, K.S. (2007b). Scale invariance in the severity of terrorism. *J. Confl. Resol.*, **5**. http://xxx.lanl.gov/abs/physics/0606007

Copp, D. (2000). *Triangle of Life*. American Survival Guide. http://www.amerrescue.org. triangleoflife.html

DeLong, J.B. (2006). Growth is good. *Harvard Magazine*, 19–20.

deMenocal, P.B. (2001). Cultural responses to climate change during the late Holocene. *Science*, **292**(5517), 667–673.

Diamond, J. (2005). *Collapse: How Societies Choose to Fail or Succeed* (New York: Viking Adult).

Grubler, A. (1998). *Technology and Global Change*. (New York: Cambridge Universtity Press).

Hanson, R. (2000). Long-term growth as a sequence of exponential modes. http://hanson.gmu.edu/longgrow.html

Haug, G.H., Gnther, D., Peterson, L.C., Sigman, D.M., Hughen, K.A., and Aeschlimann, B. (2003). Climate and the collapse of Maya civilization. *Science*, **299**(5613), 1731–1735.

Hey, J. (2005). On the number of new world founders: a population genetic portrait of the peopling of the Americas. *PLoS Biol.*, **3**(6), 965–975.

Jones, C.I. (2002). *Introduction to Economic Growth*, 2nd edition. W. W. Norton & Company.

Lay, T. and Wallace, T. (1995). *Modern Global Seismology*. (San Deigo, CA: Academic Press).

McCamley, N. (2007). *Cold War Secret Nuclear Bunkers* (Pen and Sword).

Morrison, D., Harris, A.W., Sommer, G., Chapman, C.R., and Carusi, A. (2003). Dealing with the impact hazard. In Bottke, W., Cellino, A., Paolicchi, P., and Binzel, R.P. (eds.), *Asteroids III* (Tucson, AZ: University of Arizona Press).

Murray-McIntosh, R.P., Scrimshaw, B.J., Hatfield, P.J., and Penny, D. (1998). Testing migration patterns and estimating founding population size in Polynesia by using human mtDNA sequences. *Proc. Natl. Acad. Sci. USA*, **95**, 90479052.

Nishenko, S. and Barton, C. (1995). Scaling laws for natural disaster fatalities. In Rundle, J., Klein, F., and Turcotte, D. (eds.), *Reduction and Predictability of Natural Disasters*, Volume 25, p. 32 (Addison Wesley). Princeton.

Posner, R.A. (2004). *Catastrophe: Risk and Response* (New York: Oxford University Press).

Rhodes, C.J., Jensen, H.J., and Anderson, R.M. (1997). On the critical behaviour of simple epidemics. *Proc. Royal Soc. B: Biol. Sci.*, **264**(1388), 1639–1646.

Rodrik, D. (1999). Where did all the growth go? External shocks, social conflict, and growth collapses. *J. Econ. Growth*, **4**(4), 385–412.

Sanders, D.E.A. (2005). *The Modeling of Extreme Events.*, British Actuarial Journal, **11**(3), 519–557.

Tainter, J. (1988). *The Collapse of Complex Societies.* (New York: Cambridge University Press).

Turcotte, D.L. (1999). Self-organized criticality. *Reports Prog. Phys.*, **62**, 1377–1429.

Watts, D., Muhamad, R., Medina, D., and Dodds, P. (2005). Multiscale, resurgent epidemics in a hierarchical metapopulation model. *Proc. Natl. Acad. Sci.*, **102**(32), 11157–11162.

Weiss, H. and Bradley, R.S. (2001). What drives societal collapse? *Science*, **291**(5504), 609–610.

PART IV
Risks from hostile acts

·18·

The continuing threat of nuclear war

Joseph Cirincione

18.1 Introduction

The American poet Robert Frost famously mused on whether the world will end in fire or in ice. Nuclear weapons can deliver both. The fire is obvious: modern hydrogen bombs duplicate on the surface of the earth the enormous thermonuclear energies of the Sun, with catastrophic consequences. But it might be a nuclear cold that kills the planet. A nuclear war with as few as 100 hundred weapons exploded in urban cores could blanket the Earth in smoke, ushering in a years-long nuclear winter, with global droughts and massive crop failures.

The nuclear age is now entering its seventh decade. For most of these years, citizens and officials lived with the constant fear that long-range bombers and ballistic missiles would bring instant, total destruction to the United States, the Soviet Union, many other nations, and, perhaps, the entire planet. Fifty years ago, Nevil Shute's best-selling novel, *On the Beach*, portrayed the terror of survivors as they awaited the radioactive clouds drifting to Australia from a northern hemisphere nuclear war. There were then some 7000 nuclear weapons in the world, with the United States outnumbering the Soviet Union 10 to 1. By the 1980s, the nuclear danger had grown to grotesque proportions. When Jonathan Schell's chilling book, *The Fate of the Earth*, was published in 1982, there were then almost 60,000 nuclear weapons stockpiled with a destructive force equal to roughly 20,000 megatons (20 billion tons) of TNT, or over 1 million times the power of the Hiroshima bomb. President Ronald Reagan's 'Star Wars' anti-missile system was supposed to defeat a first-wave attack of some 5000 Soviet SS-18 and SS-19 missile warheads streaking over the North Pole. 'These bombs', Schell wrote, 'were built as "weapons" for "war", but their significance greatly transcends war and all its causes and outcomes. They grew out of history, yet they threaten to end history. They were made by men, yet they threaten to annihilate man'.[1]

[1] Jonathan, S. (2000). *The Fate of the Earth* (Palo Alto: Stanford University Press), p. 3.

The threat of a global thermonuclear war is now near-zero. The treaties negotiated in the 1980s, particularly the START agreements that began the reductions in US and Soviet strategic arsenals and the Intermediate Nuclear Forces agreement of 1987 that eliminated an entire class of nuclear-tipped missiles, began a process that accelerated with the end of the Cold War. Between 1986 and 2006 the nuclear weapons carried by long-range US and Russian missiles and bombers decreased by 61%.[2] Overall, the number of total nuclear weapons in the world has been cut in half, from a Cold War high of 65,000 in 1986 to about 26,000 in 2007, with approximately 96% held by the United States and Russia. These stockpiles will continue to decline for at least the rest of this decade.

But the threat of global war is not zero. Even a small chance of war each year, for whatever reason, multiplied over a number of years sums to an unacceptable chance of catastrophe. This is not mere statistical musings. We came much closer to Armageddon after the Cold War ended than many realize. In January 1995, a global nuclear war almost started by mistake. Russian military officials mistook a Norwegian weather rocket for a US submarine-launched ballistic missile. Boris Yelstin became the first Russian president to ever have the 'nuclear suitcase' open in front of him. He had just a few minutes to decide if he should push the button that would launch a barrage of nuclear missiles. Thankfully, he concluded that his radars were in error. The suitcase was closed.

Such a scenario could repeat today. The Cold War is over, but the Cold War weapons remain, and so does the Cold War posture that keep thousands of them on hair-trigger alert, ready to launch in under fifteen minutes. As of January 2007, the US stockpile contains nearly 10,000 nuclear weapons; about 5000 of them deployed atop *Minuteman* intercontinental ballistic missiles based in Montana, Wyoming, and North Dakota, a fleet of twelve nuclear-powered *Trident* submarines that patrol the Pacific, Atlantic, and Artic oceans and in the weapon bays of long-range B-2 bombers housed in Missouri and B-fifty-two based in Louisiana and North Dakota. Russia has as many as 15,000 weapons, with 3300 atop its SS-18, SS-19, SS-24, and SS-27 missiles deployed in silos in six missile fields arrayed between Moscow and Siberia (Kozelsk, Tatishchevo, Uzhur, Dombarovskiy, Kartalay, and Aleysk), 11 nuclear-powered *Delta* submarines that conduct limited patrols with the Northern and Pacific fleets from three naval bases (Nerpich'ya, Yagel'Naya, and Rybachiy), and *Bear* and *Blackjack* bombers stationed at Ukrainka and Engels air bases (see Table 18.2).[3]

[2] Calculations are based on the following deployed strategic warhead totals: 1986, a combined total of 22,526 (US – 12,314, USSR – 10,212); 2006, a combined total of 8835 (US – 5021, USSR – 3814).

[3] Norris, R.S. and Kristensen, H.M. (2007). NRDC Nuclear Notebook, U.S Nuclear Forces, 2007. *Bulletin of the Atomic Scientists, January/February* 2007, p. 79; Norris, R.S. and Kristensen, H.M.

Although the Soviet Union collapsed in 1991 and Russian and American presidents now call each other friends, Washington and Moscow continue to maintain and modernize these huge nuclear arsenals. In July 2007, just before Russian President Vladimir Putin vacationed with American President George W. Bush at the Bush home in Kennebunkport, Maine, Russia successfully tested a new submarine-based missile. The missile carries six nuclear warheads and can travel over 6000 miles, that is, it is designed to strike targets in the United States, including, almost certainly, targets in the very state of Maine Putin visited. For his part, President Bush's administration adopted a nuclear posture that included plans to produce new types of weapons, begin development of a new generation of nuclear missiles, submarines and bombers, and to expand the US nuclear weapons complex so that it could produce thousands of new warheads on demand.

Although much was made of the 1994 joint decision by Presidents Bill Clinton and Boris Yeltsin to no longer target each other with their weapons, this announcement had little practical consequences. Target coordinates can be uploaded into a warhead's guidance systems within minutes. The warheads remain on missiles on a high alert status similar to that they maintained during the tensest moments of the Cold War. This greatly increases the risk of an unauthorized or accidental launch. Because there is no time buffer built into each state's decision-making process, this extreme level of readiness enhances the possibility that either side's president could prematurely order a nuclear strike based on flawed intelligence.

Bruce Blair, a former Minuteman launch officer now president of the World Security Institute says, 'If both sides sent the launch order right now, without any warning or preparation, thousands of nuclear weapons – the equivalent in explosive firepower of about 70,000 Hiroshima bombs – could be unleashed within a few minutes'.[4]

Blair describes the scenario in dry but chilling detail:

> If early warning satellites or ground radar detected missiles in flight, both sides would attempt to assess whether a real nuclear attack was under way within a strict and short deadline. Under Cold War procedures that are still in practice today, early warning crews manning their consoles 24/7 have only three minutes to reach a preliminary conclusion. Such occurrences happen on a daily basis, sometimes more than once per day. . . . if an apparent nuclear missile threat is perceived, then an emergency teleconference would

(2007). NRDC Nuclear Notebook, Russian Nuclear Forces, 2007. *Bulletin of the Atomic Scientists March/April* 2007, p. 61; McKinzie, M.G., Cochran, T.B., Norris, R.S., and Arkin, W.M. (2001). *The U.S. Nuclear War Plan: A Time For Change* (New York: Natural Resources Defense Council), p. 42, 73, 84.

[4] Bruce, G.B. (2007). Primed and Ready. *The Defense Monitor: The Newsletter of the Center for Defense Information*, **XXXVI**(3), 2–3.

be convened between the president and his top nuclear advisers. On the US side, the top officer on duty at Strategic Command in Omaha, Neb., would brief the president on his nuclear options and their consequences. That officer is allowed all of 30 seconds to deliver the briefing.

Then the US or Russian president would have to decide whether to retaliate, and since the command systems on both sides have long been geared for launch-on-warning, the presidents would have little spare time if they desired to get retaliatory nuclear missiles off the ground before they – and possibly the presidents themselves – were vaporized. On the US side, the time allowed to decide would range between zero and 12 minutes, depending on the scenario. Russia operates under even tighter deadlines because of the short flight time of US Trident submarine missiles on forward patrol in the North Atlantic.[5]

Russia's early warning systems remain in a serious state of erosion and disrepair, making it all the more likely that a Russian president could panic and reach a different conclusion than Yeltsin did in 1995.[6] As Russian capabilities continue to deteriorate, the chances of accidents only increase. Limited spending on the conventional Russian military has led to greater reliance on an ageing nuclear arsenal, whose survivability would make any deterrence theorist nervous. Yet, the missiles remain on a launch status begun during the worst days of the Cold War and never turned off.

As Blair concludes: 'Such rapid implementation of war plans leaves no room for real deliberation, rational thought, or national leadership'.[7] Former chairman of the Senate Armed Services Committee Sam Nunn agrees: 'We are running the irrational risk of an Armageddon of our own making . . . The more time the United States and Russia build into our process for ordering a nuclear strike, the more time is available to gather data, to exchange information, to gain perspective, to discover an error, to avoid an accidental or unauthorized launch'.[8]

18.1.1 US nuclear forces

As of January 2007, the US stockpile contains nearly 10,000 nuclear warheads. This includes about 5521 deployed warheads: 5021 strategic warheads and 500 non-strategic warheads, including cruise missiles and bombs (Table 18.1). Approximately 4441 additional warheads are held in the reserve or inactive/responsive stockpiles or awaiting dismantlement. Under

[5] Ibid.
[6] Ibid.
[7] Ibid.
[8] Nunn, S. (2004). Speech to the Carnegie International Non-proliferation Conference, June 21, 2004. www.ProliferationNews.org.

Table 18.1 US Nuclear Forces

Name/Type	Launchers	Warheads
ICBMs	500	1050
SLBMs	336/14	2016
Bombers	115	1955
Total strategic weapons	**951**	**5021**
Tomahawk cruise missile	325	100
B-61-3, B-61-4 bombs	N/A	400
Total nonstrategic weapons	N/A	500
Total deployed weapons	1276	~5521
Non-deployed weapons		~4441
Total nuclear weapons		~9962

Source: Robert S.N. and Hans M.K. (January/February 2007). NRDC nuclear notebook, US nuclear forces, 2007. *Bulletin of the Atomic Scientists*, 79–82.

Table 18.2 Russian Nuclear Forces

Type	Launchers	Warheads
ICBMs	493	1843
SLBMs	176/11	624
Bombers	78	872
Total strategic weapons	747	**3339**
Total non-strategic weapons		~ 2330
Total deployed weapons		~ 5670
Non-deployed weapons		~ 9300
Total nuclear weapons		~14,970

Source: Robert S.N. and Hans M.K. (March/April 2007). NRDC Nuclear Notebook, Russian Nuclear Forces, 2007. *Bulletin of the Atomic Scientists*, 61–67.

current plans, the stockpile is to be cut 'almost in half' by 2012, leaving approximately 6000 warheads in the total stockpile.

18.1.2 Russian nuclear forces

As of March 2007, Russia has approximately 5670 operational nuclear warheads in its active arsenal. This includes about 3340 strategic warheads and approximately 2330 non-strategic warheads, including artillery, short-range rockets and landmines. An additional 9300 warheads are believed to be in reserve or awaiting dismantlement, for a total Russian stockpile of approximately 15,000 nuclear warheads (Table 18.2).

18.2 Calculating Armageddon

18.2.1 Limited war

There are major uncertainties in estimating the consequences of nuclear war. Much depends on the time of year of the attacks, the weather, the size of the weapons, the altitude of the detonations, the behaviour of the populations attacked, etc. But one thing is clear: the numbers of casualties, even in a small, accidental nuclear attack, are overwhelming. If the commander of just one Russian Delta-IV ballistic-missile submarine were to launch twelve of its sixteen missiles at the United States, seven million Americans could die.[9]

Experts use various models to calculate nuclear war casualties. The most accurate estimate the damage done from a nuclear bomb's three sources of destruction: blast, fire and radiation. Fifty percent of the energy of the weapon is released through the blast, 35% as thermal radiation, and 15% through radiation.

Like a conventional weapon, a nuclear weapon produces a destructive blast, or shock wave. A nuclear explosion, however, can be thousands and even millions of times more powerful than a conventional one. The blast creates a sudden change in air pressure that can crush buildings and other objects within seconds of the detonation. All but the strongest buildings within 3 km (1.9 miles) of a 1 megaton hydrogen bomb would be levelled. The blast also produces super-hurricane winds that can destroy people and objects like trees and utility poles. Houses up to 7.5 km (4.7 miles) away that have not been completely destroyed would still be heavily damaged.

A nuclear explosion also releases thermal energy (heat) at very high temperatures, which can ignite fires at considerable distances from the detonation point, leading to further destruction, and can cause severe skin burns even a few miles from the explosion. Stanford University historian Lynn Eddy calculates that if a 300 kiloton nuclear weapon were dropped on the U.S. Department of Defense, 'within tens of minutes, the entire area, approximately 40 to 65 square miles – everything within 3.5 or 6.4 miles of the Pentagon – would be engulfed in a mass fire' that would 'extinguish all life and destroy almost everything else'. The creation of a 'hurricane of fire', Eden argues, is a predictable effect of a high-yield nuclear weapon, but is not taken into account by war planners in their targeting calculations.[10]

Unlike conventional weapons, a nuclear explosion also produces lethal radiation. Direct ionizing radiation can cause immediate death, but the more significant effects are long term. Radioactive fallout can inflict damage over

[9] Bruce, G.B. et al. (1998). Accidental nuclear war – a Post-Cold War assessment. *The New England Journal of Medicine*, 1326–1332.

[10] Eddy, L. (2004). *Whole World on Fire: Organizations, Knowledge, and Nuclear Weapons Devastation* (Ithaca: Cornell University Press). http://www.gwu.edu/~nsarchiv/NSAEBB/NSAEBB108/index.htm

Table 18.3 Immediate Deaths from Blast and Firestorms in eight US Cities, Submarine Attack

City	Number of Warheads	Number of Deaths
Atlanta	8	428,000
Boston	4	609,000
Chicago	4	425,000
NewYork	8	3,193,000
Pittsburgh	4	375,000
San Francisco	8	739,000
Seattle	4	341,000
Washington, DC	8	728,000
Total	**48**	**6,838,000**

Source: Bruce, G.B. et al. (April 1988). Accidental nuclear war – a post-Cold War assessment. *The New England Journal of Medicine*, 1326–1332.

periods ranging from hours to years. If no significant decontamination takes place (such as rain washing away radioactive particles, or their leaching into the soil), it will take eight to ten years for the inner circle near the explosion to return to safe levels of radiation. In the next circles such decay will require three to six years. Longer term, some of the radioactive particles will enter the food chain.[11] For example, a 1-megaton hydrogen bomb exploded at ground level would have a lethal radiation inner circle radius of about 50 km (30 miles) where death would be instant. At 145 km radius (90 miles), death would occur within two weeks of exposure. The outermost circle would be at 400 km (250 miles) radius where radiation would still be harmful, but the effects not immediate.

In the accidental Delta IV submarine strike noted above, most of the immediate deaths would come from the blast and 'superfires' ignited by the bomb. Each of the 100 kiloton warheads carried by the submarine's missiles would create a circle of death 8.6 km (5.6 miles) in diameter. Nearly 100% of the people within this circle would die instantly. Firestorms would kill millions more (see Table 18.3). The explosion would produce a cloud of radioactive dust that would drift downwind from the bomb's detonation point. If the bomb exploded at a low altitude, there would be a 10–60 km (20–40 miles) long and 3–5 km (2–3 miles) wide swath of deadly radiation that would kill all exposed and unprotected people within six hours of exposure.[12] As the radiation continued and spread over thousands of square kilometres, it is very possible that secondary deaths in dense urban populations would match or even exceed the immediate deaths caused by fire and blast, doubling the total

[11] Office of Technology Assessment. (1979). *The Effects of Nuclear War* (Washington, DC), p. 15, 35; The Effects of Nuclear Weapons. *Atomic Archive at* http://www.atomicarchive.com/Effects/

[12] See note 9.

Table 18.4 Casualties and Fatalities in a South Asian Nuclear War

Targets	Weapon Used	Casualties (Millions)	Fatalities (Millions)
Main cities (13 in India, 12 in Pakistan)	50 KT	31	12.1
Main cities (13 in India, 12 in Pakistan)	1 MT	98	59.2
All strategic military targets in both countries	Various	6.8–13.4	2–4
All transportation and power targets in both countries	Various	7.5–13.7	2.55–4.74
All targets in both countries (city centres, strategic military targets, power and transportation)	Various	40–111	15.5–63.7

Source: Robert T.B. (2004). The consequences of an Indo-Pakistani nuclear war. *International Studies Review*, **6**, 143–150.

fatalities listed in Table 18.4. The cancer deaths and genetic damage from radiation could extend for several generations.

18.2.2 Global war

Naturally, the number of casualties in a global nuclear war would be much higher. US military commanders, for example, might not know that the commander of the Delta submarine had launched by accident or without authorization. They could very well order a US response. One such response could be a precise 'counter-force' attack on all Russian nuclear forces. An American attack directed solely against Russian nuclear missiles, submarines and bomber bases would require some 1300 warheads in a coordinated barrage lasting approximately 30 minutes, according to a sophisticated analysis of US war plans by experts at the Natural Resources Defense Council in Washington, DC.[13] It would destroy most of Russia's nuclear weapons and development facilities. Communications across the country would be severely degraded. Within hours after the attack, the radioactive fallout would descend and accumulate, creating lethal conditions over an area exceeding 775,000 km^2 – larger than France and the United Kingdom combined. The attack would result in approximately 11–17 million civilian casualties, 8–12 million of which would be fatalities, primarily due to the fallout generated by numerous ground bursts.

American war planners could also launch another of the plans designed and modelled over the nuclear age: a 'limited' attack against Russian cities,

[13] McKinzie, M.G., Cochran, T.B., Norris, R.S., and Arkin, W.M. (2001). *The U.S Nuclear War Plan: A Time For Change* (New York: Natural Resources Defense Council), pp ix–xi. NRDC used computer software and unclassified databases to model a nuclear conflict and approximate the effects of the use of nuclear weapons, based on an estimate of the U.S. nuclear war plan (SIOP).

using only 150–200 warheads. This is often called a 'counter-value' attack, and would kill or wound approximately one-third of Russia's citizenry. An attack using the warheads aboard just a single US Trident submarine to attack Russian cities will result in 30–45 million casualties. An attack using 150 US Minuteman III ICBMs on Russian cities would produce 40–60 million casualties.[14]

If, in either of these limited attacks, the Russian military command followed their planned operational procedures and launched their weapons before they could be destroyed, the results would be an all-out nuclear war involving most of the weapons in both the US and Russian arsenals. The effects would be devastating. Government studies estimate that between 35 and 77% of the US population would be killed (105–230 million people, based on current population figures) and 20–40% of the Russian population (28–56 million people).[15]

A 1979 report to Congress by the U.S. Office of Technology Assessment, *The Effects of Nuclear War*, noted the disastrous results of a nuclear war would go far beyond the immediate casualties:

> In addition to the tens of millions of deaths during the days and weeks after the attack, there would probably be further millions (perhaps further tens of millions) of deaths in the ensuing months or years. In addition to the enormous economic destruction caused by the actual nuclear explosions, there would be some years during which the residual economy would decline further, as stocks were consumed and machines wore out faster then recovered production could replace them. . . . For a period of time, people could live of supplies (and, in a sense, off habits) left over from before the war. But shortages and uncertainties would get worse. The survivors would find themselves in a race to achieve viability . . . before stocks ran out completely. A failure to achieve viability, or even a slow recovery, would result in many additional deaths, and much additional economic, political, and social deterioration. This postwar damage could be as devastating as the damage from the actual nuclear explosions.[16]

According to the report's comprehensive analysis, if production rose to the rate of the consumption before stocks were exhausted, then viability would be achieved and economic recovery would begin. If not, then 'each post-war year would see a lower level of economic activity than the year before, and the future of civilization itself in the nations attacked would be in doubt'.[17] It is doubtful that either the United States or Russia would ever recover as viable nation states.

[14] *The U.S Nuclear War Plan*, p. 130.

[15] Office of Technology Assessment (1979). *The Effects of Nuclear War* (Washington, DC), p. 8. Russian casualties are smaller that U.S. causalities because a higher percentage of Russians still live in rural areas and the lower-yield U.S. weapons produce less fallout.

[16] *The Effects of Nuclear War*, p. 4–5.

[17] *The Effects of Nuclear War*, p. 8.

18.2.3 Regional war

There are grave dangers inherent not only in countries such as the United States and Russia maintaining thousands of nuclear weapons but also in China, France, the United Kingdom, Israel, India, and Pakistan holding hundreds of weapons. While these states regard their own nuclear weapons as safe, secure, and essential to security, each views others' arsenals with suspicion.

Existing regional nuclear tensions already pose serious risks. The decades-long conflict between India and Pakistan has made South Asia the region most likely to witness the first use of nuclear weapons since World War II. An active missile race is under way between the two nations, even as India and China continue their rivalry. Although some progress towards détente has been made, with each side agreeing to notify the other before ballistic missile tests, for example, quick escalation in a crisis could put the entire subcontinent right back on the edge of destruction. Each country has an estimated 50 to 100 nuclear weapons, deliverable via fighter-bomber aircraft or possibly by the growing arsenal of short- and medium-range missiles each nation is building. Their use could be devastating.

South Asian urban populations are so dense that a 50 kiloton weapon would produce the same casualties that would require megaton-range weapons on North American or European cities. Robert Batcher with the U.S. State Department Office of Technology Assessment notes:

> Compared to North America, India and Pakistan have higher population densities with a higher proportion of their populations living in rural locations. The housing provides less protection against fallout, especially compared to housing in the U.S. Northeast, because it is light, often single-story, and without basements. In the United States, basements can provide significant protection against fallout. During the Cold War, the United States anticipated 20 minutes or more of warning time for missiles flown from the Soviet Union. For India and Pakistan, little or no warning can be anticipated, especially for civilians. Fire fighting is limited in the region, which can lead to greater damage as a result of thermal effects. Moreover, medical facilities are also limited, and thus, there will be greater burn fatalities. These two countries have limited economic assets, which will hinder economic recovery.[18]

18.2.4 Nuclear winter

In the early 1980s, scientists used models to estimate the climatic effect of a nuclear war. They calculated the effects of the dust raised in high-yield nuclear surface bursts and of the smoke from city and forest fires ignited by airbursts of all yields. They found that 'a global nuclear war could have a major impact on climate – manifested by significant surface darkening over many weeks, subfreezing land temperatures persisting for up to several months,

[18] Batcher, R.T. (2004). The consequences of an Indo-Pakistani nuclear war. *International Studies Review* 6, 137.

large perturbations in global circulation patterns, and dramatic changes in local weather and precipitation rates'.[19] Those phenomena are known as 'Nuclear Winter'.

Since this theory was introduced it has been repeatedly examined and reaffirmed. By the early 1990s, as tools to asses and quantify the production and injection of soot by large-scale fires, and its effects, scientists were able to refine their conclusions. The prediction for the average land cooling beneath the smoke clouds was adjusted down a little bit, from the 10–25°C estimated in the 1980s to 10–20°C. However, it was also found that the maximum continental interior land cooling can reach 40°C, more than the 30–35 degrees estimate in the 1980s, 'with subzero temperatures possible even in the summer'.[20]

In a *Science* article in 1990, the authors summarized:

> Should substantial urban areas or fuel stocks be exposed to nuclear ignition, severe environmental anomalies – possibly leading to more human casualties globally than the direct effects of nuclear war – would be not just a remote possibility, but a likely outcome.[21]

Carl Sagan and Richard Turco, two of the original scientists developing the nuclear winter analysis, concluded in 1993:

> Especially through the destruction of global agriculture, nuclear winter might be considerably worse than the short-term blast, radiation, fire, and fallout of nuclear war. It would carry nuclear war to many nations that no one intended to attack, including the poorest and most vulnerable.[22]

In 2007, members of the original group of nuclear winter scientists collectively performed a new comprehensive quantitative assessment utilizing the latest computer and climate models. They concluded that even a small-scale, regional nuclear war could kill as many people as died in all of World War II and seriously disrupt the global climate for a decade or more, harming nearly everyone on Earth.

The scientists considered a nuclear exchange involving 100 Hiroshima-size bombs (15 kilotons) on cities in the subtropics, and found that:

> Smoke emissions of 100 low-yield urban explosions in a regional nuclear conflict would generate substantial global-scale climate anomalies, although not as large as the previous 'nuclear winter' scenarios for a full-scale war. However, indirect effect on

[19] Turco, R.P., Toon, O.B., Ackerman, T.P., Pollack, J.B., and Sagan, C. (1983). Nuclear winter: global consequences of mutliple nuclear explosions. *Science*, **222**, 1290.

[20] Turco, R.P., Toon, O.B., Ackerman, T.P., Pollack, J.B., and Sagan, C. (1990). Climate and smoke: an appraisal of nuclear winter. *Science*, **247**, 166.

[21] Ibid., p. 174.

[22] Sagan, C. and Turco, R.P. (1993). Nuclear winter in the Post-Cold War era. *Journal of Peace Research*, **30**(4), 369.

surface land temperatures, precipitation rates, and growing season lengths would be likely to degrade agricultural productivity to an extent that historically has led to famines in Africa, India and Japan after the 1784 Laki eruption or in the northeastern United States and Europe after the Tambora eruption of 1815. Climatic anomalies could persist for a decade or more because of smoke stabilization, far longer than in previous nuclear winter calculations or after volcanic eruptions.[23]

The scientists concluded that the nuclear explosions and firestorms in modern cities would inject black carbon particles higher into the atmosphere than previously thought and higher than normal volcanic activity (see Chapter 10, this volume). Blocking the Sun's thermal energy, the smoke clouds would lower temperatures regionally and globally for several years, open up new holes in the ozone layer protecting the Earth from harmful radiation, reduce global precipitation by about 10% and trigger massive crop failures. Overall, the global cooling from a regional nuclear war would be about twice as large as the global warming of the past century 'and would lead to temperatures cooler than the pre-industrial Little Ice Age'.[24]

18.3 The current nuclear balance

Despite the horrific consequences of their use, many national leaders continue to covet nuclear weapons. Some see them as a stabilizing force, even in regional conflicts. There is some evidence to support this view. Relations between India and Pakistan, for example, have improved overall since their 1998 nuclear tests. Even the conflict in the Kargil region between the two nations that came to a boil in 1999 and again in 2002 (with over one million troops mobilized on both sides of the border) ended in negotiations, not war. Columbia University scholar Kenneth Waltz argues, 'Kargil showed once again that deterrence does not firmly protect disputed areas but does limit the extent of the violence. Indian rear admiral Raja Menon put the larger point simply: "The Kargil crisis demonstrated that the subcontinental nuclear threshold probably lies territorially in the heartland of both countries, and not on the Kashmir cease-fire line"'.[25]

It would be reaching too far to say the Kargil was South Asia's Cuban missile crisis, but since the near-war, both nations have established hotlines and other confidence-building measures (such as notification of military exercises), exchanged cordial visits of state leaders, and opened transportation

[23] Toon, O.B., Robock, A., Turco, R.P., Bardeen, C., Oman, L., and Stenchikov, G.L. (2007). Consequences of regional-scale nuclear conflicts. *Science*, **315**, 1224–1225.

[24] Ibid., p. 11823.

[25] Sagan, S.D. and Waltz, K.N. (2003). *The Spread of Nuclear Weapons: A Debate Renewed* (New York: W.W. Norton & Company), p. 115.

and communications links. War seems less likely now than at any point in the past.

This calm is deceiving. Just as in the Cold War stand off between the Soviet Union and the United States, the South Asian détente is fragile. A sudden crisis, such as a terrorist attack on the Indian parliament, or the assassination of Pakistan President Pervez Musaraf, could plunge the two countries into confrontation. As noted above, it would not be thousands that would die, but millions. Michael Krepon, one of the leading American experts on the region and its nuclear dynamics, notes:

> Despite or perhaps because of the inconclusive resolution of crises, some in Pakistan and India continue to believe that gains can be secured below the nuclear threshold. How might advantage be gained when the presence of nuclear weapons militates against decisive end games? ... If the means chosen to pursue advantage in the next Indo-Pakistan crisis show signs of success, they are likely to prompt escalation, and escalation might not be easily controlled. If the primary alternative to an ambiguous outcome in the next crisis is a loss of face or a loss of territory, the prospective loser will seek to change the outcome.[26]

Many share Krepon's views both in and out of South Asia. Indian scholar P.R. Chari, for example, further observes:

> [S]ince the effectiveness of these weapons depends ultimately on the willingness to use them in some situations, there is an issue of coherence of thought that has to be addressed here. Implicitly or explicitly an eventuality of actual use has to be a part of the possible alternative scenarios that must be contemplated, if some benefit is to be obtained from the possession and deployment of nuclear weapons. To hold the belief that nuclear weapons are useful but must never be used lacks cogency....[27]

A quickly escalating crisis over Taiwan is another possible scenario in which nuclear weapons could be used, not accidentally as with any potential US-Russian exchange, but as a result of miscalculation. Neither the United States nor China is eager to engage in a military confrontation over Taiwan's status, and both sides believe they could effectively manage such a crisis. But crises work in mysterious ways – political leaders are not always able to manipulate events as they think they can, and events can escalate very quickly. A Sino-US nuclear exchange may not happen even in the case of a confrontation over Taiwan's status, but it is possible and should not be ignored.

The likelihood of use depends greatly on the perceived utility of nuclear weapons. This, in turn, is greatly influenced by the policies and attitudes of the existing nuclear weapon states. Recent advocacy by some in the United States

[26] Krepon, M. (2004). *From Confrontation to Cooperation* (Washington, D.C.: Henry L. Stimson Center). http://www.stimson.org/southasia/pubs.cfm?ID=197

[27] Chari, P.R. (2004). *Nuclear Restraint, Nuclear Risk Reduction, and the Security – Insecurity Paradox in South Asia* (Washington, D.C.: Henry L. Stimson Center).

of new battlefield uses for nuclear weapons (e.g., in pre-emptive strikes on Iran's nuclear facilities) and for programmes for new nuclear weapon designs could lead to fresh nuclear tests, and possibly lower the nuclear threshold, that is, the willingness of leaders to use nuclear weapons. The five nuclear weapon states recognized by the Non-proliferation Treaty have not tested since the signing of the Comprehensive Test Ban Treaty in 1996, and until North Korea's October 2007 test, no state had tested since India and Pakistan did so in May 1998. If the United States again tested nuclear weapons, political, military and bureaucratic forces in several other countries would undoubtedly pressure their governments to follow suit. Indian scientists, for example, are known to be unhappy with the inconclusive results of their 1998 tests. Indian governments now resist their insistent demands for new tests for fear of the damage they would do to India's international image. It is a compelling example of the power of international norms. New US tests, however, would collapse that norm, trigger tests by India, then perhaps China, Russia, and other nations. The nuclear test ban treaty, an agreement widely regarded as a pillar of the nonproliferation regime, would crumble, possibly bringing down the entire regime.

All of these scenarios highlight the need for nuclear weapons countries to decrease their reliance on nuclear arms. While the United States has reduced its arsenal to the lowest levels since the 1950s it has received little credit for these cuts because they were conducted in isolation from a real commitment to disarmament. The United States, Russia, and all nuclear weapons states need to return to the original bargain of the NPT – the elimination of nuclear weapons. The failure of nuclear weapons states to accept their end of the bargain under Article VI of the NPT has undermined every other aspect of the non-proliferation agenda.

Universal Compliance, a 2005 study concluded by the Carnegie Endowment for International Peace, reaffirmed this premise:

> The nuclear-weapon states must show that tougher nonproliferation rules not only benefit the powerful but constrain them as well. Nonproliferation is a set of bargains whose fairness must be self-evident if the majority of countries is to support their enforcement . . . The only way to achieve this is to enforce compliance universally, not selectively, including the obligations the nuclear states have taken on themselves . . . The core bargain of the NPT, and of global nonproliferation politics, can neither be ignored nor wished away. It underpins the international security system and shapes the expectations of citizens and leaders around the world.[28]

While nuclear weapons are more highly valued by national officials than chemical or biological weapons ever were, that does not mean they are a

[28] Perkovich, G., Mathew, J., Cirincione, J., Gottemoeller, R., Wolfsthal, J. (2005). *Universal Compliance: A Strategy for Nuclear Security* (Washington, D.C.: Carnegie Endowment for International Peace), pp. 24, 34, 39.

permanent part of national identity. A January 2007 op-ed in the *Wall Street Journal* co-authored by George Shultz, Henry Kissinger, William Perry and Sam Nunn, marked a significant change in the thinking of influential policy and decision makers in the United States and other nuclear weapons states. They contend that the leaders of the countries in possession of nuclear weapons should turn the goal of a world without nuclear weapons into a 'joint enterprise'. They detail that a nine-point programme that includes substantial reductions in the size of nuclear forces in all states, the elimination of short-range nuclear weapons, and the ratification of the Comprehensive Test Ban Treaty. The op-ed concludes that, 'Reassertion of the vision of a world free of nuclear weapons and practical measures towards achieving that goal would be, and would be perceived as, a bold initiative consistent with America's moral heritage. The effort could have a profoundly positive impact on the security of future generations'.[29]

Breaking the nuclear habit will not be easy, but there are ways to minimize the unease some may feel as they are weaned away from dependence on these weapons. The United States and Russia account for over 96% of the world's nuclear weapons. The two nations have such redundant nuclear capability that it would not compromise any vital security interests to quickly reduce down to several hundred warheads each. Further reductions and the possibility of complete elimination could then be examined in detailed papers prepared by and for the nuclear-weapon states. If accompanied by reaffirmation of the ban on nuclear testing, removal of all weapons from rapid-launch alert status, establishment of a firm norm against the first use of these weapons, and commitments to make the reductions in weapons irreversible and verifiable, the momentum and example generated could fundamentally alter the global dynamic.

Such an effort would hearken back to US President Harry Truman's proposals in 1946 which coupled weapons elimination with strict, verified enforcement of nonproliferation. Dramatic reductions in nuclear forces could be joined, for example, with reforms making it more difficult for countries to withdraw from the NPT (by clarifying that no state may withdraw from the treaty and escape responsibility for prior violations of the treaty or retain access to controlled materials and equipment acquired for 'peaceful' purposes).[30] It would make it easier to obtain national commitments to stop the illegal transfer of nuclear technologies and reform the fuel cycle. The reduction in the number of weapons and the production of nuclear materials would also greatly decrease the risk of terrorists acquiring such materials.

[29] Shultz, G., Kissinger, H., Perry, W., and Nunn, S. (2007). A world free of nuclear weapons. *The Wall Street Journal*. Eastern edition, NY, January 4, 2007, pg. A15.

[30] See, for example, the excellent suggestions made by Sally Horn, a State Department representative to the NPT Review Conference in May 2005, summarized in Cirincione, J. (2005). 'No Easy Out', Carnegie Analysis. www.ProliferationNews.org

18.4 The good news about proliferation

Is it reasonable to expect such dramatic changes in international security strategies? Absolutely: there is nothing inevitable about either proliferation or nuclear arsenals. Though obscured by the media and political emphasis on terror and turmoil, the nations of the world have made remarkable progress in the past two decades in reducing many nuclear dangers.

There are far fewer countries that have nuclear weapons or weapon programmes today than there were in the 1960s, 1970s, or 1980s. In the 1960s, 23 countries had weapons or were pursuing programmes, including Australia, Canada, China, Egypt, India, Japan, Norway, Sweden, Switzerland, and West Germany. Today, nine countries have weapons (China, France, India, Israel, North Korea, Pakistan, Russia, United Kingdom, and the United States). Iran may be pursuing a weapons programme under the guise of peaceful nuclear power, but no other nation is believed to be doing so.

In fact, more countries have given up nuclear weapons or weapons programmes in the past 20 years than have started them. South Africa, Belarus, Kazakhstan, and Ukraine all gave up weapons in the 1990s. Similarly, civilian governments in Brazil and Argentina in the 1980s stopped the nuclear weapon research military juntas had started. We now know that United Nations inspection and dismantlement programmes ended Iraq's nuclear weapon programme in 1991. In December 2003, Libya became the most recent nation to abandon a secret weapons programme.

The Non-proliferation Treaty itself is widely considered one of the most successful security pacts in history, with every nation of the world a member except for Israel, India, Pakistan, and North Korea. And until North Korea tested in October 2006, no nation had exploded a nuclear weapon in a test for 8 years – the longest period in the atomic age. The outrage that greeted the test shows how strong this anti-nuclear sentiment had become.

There is more good news. The ballistic missile threat that dominated American and NATO national security debates in the late 1990s is declining by most measures: There are far fewer nuclear-tipped missiles capable of hitting the United States or any European country today than there were 10 years ago. Agreements negotiated by American Presidents Ronald Reagan, George H.W. Bush and George W. Bush have slashed the former Soviet arsenal by 71% from 1987, while China has retained about 20 missiles that could reach US shores. No other country can strike the United States from its own soil. Most of the news about missile tests in Iran, North Korea, or South Asia are of short- or medium-range missiles that threaten those nations' neighbours but not America.[31]

[31] In 1987 the Soviet Union deployed 2380 long-range missiles and China approximately 20. The number declined to 689 by 2007 (669 Russian; 20 Chinese).

Nonetheless, four critical challenges – nuclear terrorism, the threats from existing arsenals, the emergence of new weapon states, and the spread of dual-use nuclear fuel facilities – threaten to unravel today's nuclear non-proliferation regime. Overcoming these challenges will require forging a consensus of expert opinion, focusing the attention of senior officials, securing the necessary funding, and, above all, executive leadership. The decisions taken over the next few years with regard to these challenges by the president of the United States and the leaders of other nuclear-weapon states as well as other key countries will determine whether the progress of the last two decades in reducing and eliminating nuclear weapons and materials continues, or whether the world launches into the second wave of proliferation since World War II.

18.5 A comprehensive approach

To understand the need for and the practicality of a comprehensive approach to reducing the danger of nuclear war, consider what is required to prevent new nations from acquiring nuclear weapons. For some scholars and officials, the addition of new nations to the nuclear club is as natural and inevitable as population growth. Kenneth Waltz argues, 'Nuclear weapons will nevertheless spread, with a new member occasionally joining the club ... Someday the world will be populated by fifteen or eighteen nuclear-weapon states'.[32]

American expert William Potter says this view is shared by many in the Bush administration. 'Principle one', for many officials, he says, is 'that nuclear proliferation is inevitable, at best it can be managed, not prevented'.[33] Currently, however, there are only two countries of serious concern. If the nuclear programmes in North Korea and Iran can be checked, then prospects for halting and reversing proliferation globally improve dramatically. If they are not, then they may start a momentum that tips neighbouring countries over the nuclear edge.

The specifics and politics vary from country to country, but all of the threats from new nations acquiring weapons share the same need for a comprehensive, multi-dimensional approach. Iran, by far the more difficult of the cases, can serve as a model of how such an approach could work.

The current Iranian government plans to build between six and twenty nuclear power reactors and all the facilities needed to make and reprocess the fuel for these reactors. The same facilities that can make fuel for nuclear reactors can also make the fuel for nuclear bombs. Whatever its true intentions,

[32] Sagan, S.D. and Waltz, K.N. (2003). *The Spread of Nuclear Weapons* (New York: W.W. Norton & Company), p. 4.

[33] Potter, W. (2005). Carnegie International Non-proliferation Conference, 2005, Panel on 'The New Look of US Nonproliferation Policy'. www.ProliferationNews.org.

convincing Iran that while it could proceed with construction of power reactors, the country must abandon construction of fuel manufacturing facilities will not be easy. It will require both threats of meaningful sanctions, and promises of the economic benefits of cooperation.

This is the package of carrots and sticks that comprised the negotiations between the European Union and Iran. Calibrating the right balance in this mix is difficult enough, but the package itself is not sufficient to seal a deal. The hard-line government of President Mahmoud Ahmadinejad further complicates the issue with its harsh rhetorical insistence on proceeding with the nuclear plans and pointed threats to Israel. While the rhetoric may eventually fade, at the core, Iran or any country's reasons for wanting its own fuel cycle capabilities are similar to the reasons some countries want nuclear weapons: security, prestige and domestic political pressures. All of these will have to be addressed in order to craft a permanent solution.

Part of the security equation can be addressed by the prospect of a new relationship with the United States that ends regime change efforts. Iran would need some assurances that agreements on nuclear programmes could end efforts by the United States and Israel to remove the current regime. The United States has told North Korea that it has no hostile intentions towards the state and that an end to that country's programme would lead to the restoration of diplomatic relations. Similar assurances will be needed for Iran.

But there is also a regional dimension. Ending the threat from an Iranian nuclear programme will require placing the Iranian decision in the context of the long-standing goal of a Middle East free of nuclear weapons. It will be impossible for a country as important as Iran to abstain permanently from acquiring the technologies for producing nuclear weapons – at least as a hedge – if other countries in the region have them. This dynamic was noted in the very first US National Intelligence Estimates of the proliferation threats done in 1958 and 1961 and is still true today.

Iran's leaders will want some assurances that there is a process underway that can remove what they see as potential threats from their neighbours, including Israel. For domestic political reasons, they will want to present their nuclear abstinence as part of a movement towards a shared and balanced regional commitment.

Some may argue that Israel would never give up its nuclear weapons. But such nuclear free zones have been created in other regions which, though not as intensely contested as the Middle East, still had to overcome substantial rivalries and involved the abandonment of existing programmes (in South America) and the dismantlement of actual weapons (in Africa and Central Asia). All the states in the region rhetorically support this goal, as do many of the major powers, but in recent years, little diplomatic effort has been put behind this policy – certainly nothing on the scale of the effort needed to create the nuclear Non-proliferation Treaty and its support mechanisms in the 1960s and 1970s.

Ridding the region of nuclear weapons will, of course, be difficult, but it is far better than the alternative of a Middle East with not one nuclear power (Israel) but two, three or four nuclear weapon states – and with unresolved territorial, religious and political disputes. This is a recipe for nuclear war.

This is not a distant fear. In late 2006 and early 2007, a dozen Muslim nations expressed their interest in starting their own nuclear power programmes. In the entire 62-year history of the nuclear age there has been exactly one nuclear power reactor built in the Middle East (the one under construction in Iran) and two in Africa (in South Africa). Suddenly, Saudi Arabia, Egypt, Turkey, Jordan, Morocco, and several Gulf states have begun exploring nuclear power programmes. This is not about energy; it is about hedging against a nuclear Iran.

The key to stopping this process is to get a counter-process going. States in the region must have some viable alternative to the pessimistic view that the Middle East will eventually be a nuclear free-for-all. In order for this effort to succeed, there will have to be a mechanism to change fundamentally the way nuclear fuel is produced and reprocessed. Doing so would satisfy a nation's security considerations that it does not have to build its own facilities in order to have a secure supply of fuel for its reactors. Some Iranians see the current negotiations as a new effort by the West to place them, once again, in a dependent relationship. This time the West would not control their oil, they say, but the energy of the future, nuclear fuel. Iran, indeed any nation, will not permanently acquiesce to a discriminatory regime that adds to the existing inequality allowing some countries to have nuclear weapons while others cannot, by now allowing some countries to make nuclear fuel while others cannot.

A comprehensive approach operating at several levels is the only sure way to prevent more and more nations from wanting and acquiring the technology that can bring them – legally – right up to the threshold of nuclear weapons capability.

18.6 Conclusion

Ultimately, reducing the risks from nuclear weapons in the twenty-first century cannot be just a military or nuclear energy strategy. At the beginning of the nuclear age, it was already clear that unless we solved the underlying political conflicts that encourage some states to seek security in nuclear arms, we would never prevent nuclear competition. Robert Oppenheimer, the scientific head of the Manhattan Project, said, 'We must ask, of any proposals for the control of atomic energy, what part they can play in reducing the probability of war. Proposals that in no way advance the general problem of the avoidance of war are not satisfactory proposals'.[34]

Thus, nuclear-weapon-specific efforts should be joined by focused initiatives to resolve conflicts in key regions. A quick look at the map should make clear that nuclear weapons have not spread around the world uniformly, like a drop of ink diffusing in a glass of water. Vast areas of the world – entire continents – are nuclear-weapon free. There are no nuclear weapons in South America, Africa, Australia, or Southeast Asia. Rather, the states of proliferation concern are in an arc of crisis that flows from the Middle East through South Asia up to Northeast Asia. In other words, in regions within which unresolved territorial, political and religious disputes give rise to the desire to gain some strategic advantage by acquiring nuclear weapons.

Countries have given up nuclear weapons and programmes in the past only when these disputes have been resolved. The pattern of the past should be the template for the future. Avoiding nuclear war in South Asia requires continuing the progress in normalizing relations between India and Pakistan and achieving a permanent resolution of the Kashmir issue. Ridding the Middle East of nuclear weapons and new nuclear programmes requires normalization of relations between Israel and other regional states and groups based on a just resolution to the Israeli-Palestinian conflict.

Resolution of some of these may come more quickly than most imagine. Even ten years ago it was inconceivable to many that Ian Paisley, the leader of the militant Protestant Democratic Union Party would ever share power with Martin McGuinness, a leader of the militant Catholic IRA. Each called the other terrorist. Each swore to wipe each other's groups from the face of the earth. Yet, in 2007, they shook hands and were sworn into office as the joint leaders of a united Northern Ireland.

Others conflicts may take more time to resolve, but as history teaches, it is the direction in which we are moving that informs national attitudes and shapes each state's security decisions. The more arrows we can get pointed in the right direction, the easier it becomes to make progress on all fronts.

Former U.S. State Department official Robert Einhorn and former U.S. Defense Department official Kurt Campbell note that the wisdom of societies and states that have gone without nuclear weapons is reinforced by 'a world in which the goals of the NPT are being fulfilled – where existing nuclear arsenals are being reduced, parties are not pursuing clandestine nuclear programmes, nuclear testing has been stopped, the taboo against the use of nuclear weapons

[34] Oppenheimer, J.R. (June 1946). The International Control of Atomic Energy. *Bulletin of the Atomic Scientists.*

is being strengthened, and in general, the salience of nuclear weapons in international affairs is diminishing'.[35]

There is every reason to believe that in the first half of this century the peoples and nations of the world will come to see nuclear weapons as the 'historic accident' Mohamed ElBaradei says they are. It may become clearer that nations have no need for the vast destructive force contained in a few kilograms of enriched uranium or plutonium. These weapons still appeal to national pride but they are increasingly unappealing to national budgets and military needs. It took just sixty years to get to this point in the nuclear road. If enough national leaders decide to walk the path together, it should not take another sixty to get to a safer, better world.

Suggestions for further reading

Campbell, K., Einhorn, R., Reiss, M. (eds.). (2004). *The Nuclear Tipping Point: Why States Reconsider Their Nuclear Choices* (Washington, D.C.: Brookings Institution). A comprehensive set of case studies of why and how eight key nations decided not to acquire nuclear weapons . . . for now.

Cirincione, J. (2007). *Bomb Scare: The History and Future of Nuclear Weapons* (New York: Columbia University Press). A concise guide to the science, strategy, and politics of the nuclear age.

Office of Technology Assessment. (1979). *The Effects of Nuclear War* (Washington, D.C.). The seminal source for the blast, fire, radiation, and strategic effects of the use of nuclear weapons.

Rhodes, R. (1986). *The Making of the Atomic Bomb* (New York: Simon and Schuster). Rhodes won the Pulitzer Prize for this gripping, comprehensive history of how the first atomic bombs were built, used, and regretted.

Shute, N. (1957). *On the Beach* (New York: Ballantine Books). The best-selling, classic novel about the world after Nuclear War.

The Weapons of Mass Destruction Commission (2006). *Weapons of Terror: Feeing the World of Nuclear, Biological and Chemical Arms* (Stockholm). An international prescription for a nuclear-free world.

[35] Campbell, K.M., Einhorn, R., and Reiss, M. (eds.). (2004). *The Nuclear Tipping Point: Global Prospects for Revisiting Nuclear Renunciation* (Washington, D.C.: Brookings Institution Press), cited in *Universal Compliance*, p. 130.

·19·

Catastrophic nuclear terrorism: a preventable peril

Gary Ackerman and William C. Potter

19.1 Introduction

One can conceive of at least three potentially catastrophic events involving the energy of the atom: a nuclear accident in which massive quantities of radiation inadvertently are released into the environment including inadvertent nuclear missile launches; nuclear war among nation-states; and nuclear violence inflicted by non-state actors. This chapter focuses on the last of these threats – the dangers posed by nuclear terrorism, a phenomenon that lies at the nexus between what are widely considered to be two of the primary security threats of the modern era.

Non-state actors have essentially four mechanisms by which they can exploit civilian and military nuclear assets intentionally to serve their terrorist[1] goals:

- the dispersal of radioactive material by conventional explosives or other means;
- attacks against or sabotage of nuclear facilities, in particular nuclear power plants and fuel storage sites, causing the release of radioactivity;
- the theft, purchase, or receipt of fissile material leading to the fabrication and detonation of a crude nuclear explosive, usually referred to as an improvised nuclear device (IND); and
- the theft, purchase, or receipt and detonation of an intact nuclear weapon.

All of these nuclear threats are real; all merit the attention of the international community; and all require the expenditure of significant resources to reduce their likelihood and potential impact. The threats, however, are different and

[1] There is no universally accepted definition of terrorism: assertions regarding the core characteristics of terrorism vary widely among governments and scholars. Recognizing these differences, for the purposes of this paper, the following definition will be used: *Terrorism is the calculated use of violence, or the threat of violence, perpetrated by non-state actors in order to achieve social or political goals, with the aim of influencing a wider audience than the immediate victims of the violence and which is usually directed against non-combatants.*

vary widely in their probability of occurrence, in consequences for human and financial loss, and in the ease with which intervention might reduce destructive outcomes (for a detailed analysis, see Ferguson and Potter, 2005).

Nuclear terrorism experts generally agree that the nuclear terror scenarios with the highest consequences – those involving nuclear explosives – are the least likely to occur because they are the most difficult to accomplish. Conversely, the scenarios with the least damaging consequences – those involving the release of radioactivity but no nuclear explosion – are the most likely to occur because they are the easiest to carry out. Constructing and detonating an IND, for example, is far more challenging than building and setting off a radiological dispersal device (RDD), because the former weapon is far more complex technologically and because the necessary materials are far more difficult to obtain. Thus, an IND presents a less likely threat than does an RDD, but the potential consequences of an IND explosion are orders of magnitude more devastating than the potential damage from the use of an RDD.

It is difficult to conceive of any scenario in which either the terrorist use of RDDs or the sabotage of or attack on one or more nuclear facilities would approach the category of global, terminal risks referred to in this volume as existential risks. The remainder of this chapter will focus on the two forms of 'high consequence' nuclear terrorism, those involving INDs and intact nuclear weapons. Although it also is hard to devise plausible scenarios in which non-state actors could employ these forms of nuclear terrorism to achieve a situation in which humankind as a whole is imperiled, under certain circumstances terrorist action might create global catastrophic risks of the endurable kind.

This chapter examines the theoretical requirements for engaging in nuclear terrorism, outlines the current evidence for and possible future shape of the threat, and then discusses the potential short- and long-term global consequences of nuclear terrorism. It concludes with policy recommendations for mitigating this particular species of global catastrophic risk.

19.2 Historical recognition of the risk of nuclear terrorism

Asked in a closed Senate hearing room 'whether three or four men couldn't smuggle units of an [atomic] bomb into New York and blow up the whole city'. Oppenheimer responded, 'Of course it could be done, and people could destroy New York'. When a startled senator then followed by asking, 'What instrument would you use to detect an atomic bomb hidden somewhere in the city?' Oppenheimer quipped, 'A screwdriver [to open each and every crate or suitcase]'. There was no defense against nuclear terrorism – and he felt there never would be.

(*Bird and Sherwin*, 2005, *p.* 349)

The subject of non-state actors and nuclear explosives has been the focus of sporadic scholarship, journalistic commentary, and government attention for over three decades. While it was anticipated from the beginning of the nuclear age, as the quote above notes, few worried about non-state actors as a nuclear threat before the 1970s. One of the earliest efforts to examine the issue was by a panel of experts convened by the US Atomic Energy Commission (AEC) under the chairmanship of Ralph F. Lumb. In its 1967 report, the Lumb panel addressed the need to strengthen safeguards to discourage diversion of nuclear materials from the expanding civilian nuclear fuel cycle in the United States.[2] The report followed the discovery by the AEC in 1965 that it could not account for a large quantity of weapons grade uranium at the US naval nuclear fuel plant in Apollo, Pennsylvania (Walker, 2001, p. 109).[3]

The potential for home grown nuclear terrorism gained greater recognition in the United States at the end of the 1960s and in the early 1970s when incidents of politically motivated violence, including a number of incidents at nuclear research and power reactor sites, increased. For example, explosives were discovered at a research reactor at the Illinois Institute of Technology and the Point Beach nuclear power plant (Walker, 2001, p. 111). Another dimension of the terrorism problem – nuclear extortion – was also highlighted following a threat in October 1970 to detonate a hydrogen bomb in Orlando, Florida unless $1 million was provided. The extortion threat, accompanied by a drawing of the alleged device, was judged sufficiently credible by authorities that city officials considered paying the ransom until a slip-up revealed that it was a hoax perpetrated by a local 14-year-old honours student (Willrich and Taylor, 1974, p. 80).

In addition to a spate of press reports about the dangers of nuclear terrorism (including Ingram, 1972; Lapp, 1973; Zorza, 1972, 1974), the early to mid-1970s generated a second AEC report on the subject, released to considerable public fanfare by Senator Abraham Ribicoff and the first book length non-governmental studies. The 1974 volume *Nuclear Theft: Risks and Safeguards* by Mason Willrich and Theodore Taylor was particularly noteworthy because of its thorough treatment of the risks of nuclear theft and the fact that Taylor was a professional nuclear weapons designer who spoke with great authority about the technical obstacles – or lack thereof – a non-state actor would have to overcome to build a nuclear explosive (Zorza, 1972).[4] According to the account by Willrich and Taylor,

Under conceivable circumstances, a few persons, possibly even one person working alone, who possessed about ten kilograms of plutonium and substantial amount of chemical high

[2] For a discussion of the Lumb panel and other early government efforts to assess the risks of nuclear terrorism (see Willrich and Taylor, 1974, pp. 78–82; Walker, 2001, pp. 107–132).

[3] Although the AEC maintained that poor accounting and handling practices probably were responsible for the discrepancy, there was speculation that about 100 kilograms of bomb-grade uranium from the plant may have found its way to Israel (Gilinsky, 2004).

[4] Taylor's views also were popularized in a three part article by John McPhee in 3, 10, and 17 December 1973 issues of *The New Yorker* (McPhee, 1974).

explosive could, within several weeks, design and build a crude fission bomb. . . . This could be done using materials and equipment that could be purchased at a hardware store and from commercial suppliers of scientific equipment for student laboratories. . . . Statements similar to those made above about a plutonium oxide bomb, could also be made about a fission bomb made with high-enriched uranium.

(*McPhee*, 1974, *pp*. 20–21)

This view was not shared uniformly by other experts, and, as is noted below, may understate the difficulty of producing a crude nuclear explosive with plutonium (as opposed to uranium). Taylor, himself, subsequently appeared to raise the degree of difficulty facing non-state actors. Nevertheless, the warning about the ability of terrorists to inflict high consequence nuclear violence could not be easily dismissed. This perspective was reinforced in a 1977 assessment by the U.S. Office of Technology Assessment, which drew upon relevant classified information. It reported that 'a small group of people, none of whom have ever had access to the classified literature, could possibly design and build a crude nuclear explosive device'. According to the report, modest machine shop facilities could suffice and necessary equipment could be obtained for 'a fraction of a million dollars'. The group, however, 'would have to include, at a minimum, a person capable of researching and understanding the literature in several fields, and a jack-of-all trades technician' (U.S. Office of Technology Assessment, 1977, p. 30).

Two other influential authors during the 1970s were David Rosenbaum, a consultant who challenged the adequacy of AEC regulations to prevent nuclear theft (Rosenbaum, 1977) and Brian Jenkins, an analyst at the Rand Corporation. Best known for his assertion that 'Terrorists want a lot of people watching, not a lot of people dead . . .', Jenkins, probably more than anyone else, encouraged the notion that terrorists were unlikely to employ weapons of mass destruction (WMD) in pursuit of their objectives (Jenkins, 1977).

Although it is hard to gauge how much impact Jenkins' thesis had on government policy, the dangers of high consequence nuclear terrorism received relatively little attention in the scholarly literature during the period between the mid-1970s and early 1990s. An important exception was an in-depth report of an international task force whose findings were published under the editorship of Paul Leventhal and Yonah Alexander (1987). A frequently cited chapter in that book by J. Carson Mark and a team of other former US bombmakers concluded that terrorists could indeed build a nuclear explosive device. It was more circumspect than Willrich and Taylor, however, in attributing that capability to any single individual. 'The number of specialists required', Mark et al. maintained, 'would depend on the background and experience of those enlisted, but their number could scarcely be fewer than three or four and might well have to be more' (Mark et al., 1987, p. 58).

The collapse of the Soviet Union in 1991 and the risk of 'loose nukes' prompted a new wave of government, academic, and popular concern about nuclear terrorism. A series of studies organized by Harvard University's Center for Science and International Studies, in particular, helped to frame the nature of the nuclear threat posed by the disintegration of the Soviet Union, and also contributed to major new legislation in the form of the Nunn-Lugar Cooperative Threat Reduction Program (Allison, 1996; Campbell, 1991).[5]

If the demise of the Soviet Union refocused scholarly and governmental attention on the risk that insecure nuclear weapons and material might find their way into the hands of terrorists, the use of sarin nerve agent by the Japanese cult Aum Shinrikyo in 1995 and the events of 11 September 2001, resulted in a surge in research and government spending on many dimensions of WMD terrorism, including a reassessment of the readiness and ability of non-state actors to resort to nuclear violence.

19.3 Motivations and capabilities for nuclear terrorism

19.3.1 Motivations: the demand side of nuclear terrorism

An intentional risk such as nuclear terrorism can only result from the conscious decision of a human actor, yet the integral motivational component of such threats has often been overshadowed by assessments of terrorist capabilities and weapon consequences.[6] What is needed is a systematic dissection of the variety of factors that might induce terrorists to pursue the use of nuclear weapons.[7] The following discussion examines possible terrorist motivations that reflect strategic, operational, and tactical incentives for using nuclear weapons (i.e., where nuclear weapons are used as a means to an end) as well as more esoteric motives where the use of nuclear weapons is an end in itself.[8]

[5] Another influential volume in the Harvard series but with a broader focus was the volume by Falkenrath et al. (1998).

[6] Jerrold Post maintains that, 'absent a clear understanding of the adversary's intentions, the strategies and tactics developed [to counter them] are based primarily on knowledge of terrorists [*sic*] technological capabilities and give insufficient weight to psychological motivations' (1987, p. 91). Indeed, Cameron has even asserted that 'the real driving force behind the heightened danger of nuclear terrorism lies not with the increased opportunities for micro-proliferation, but rather with the changing nature of political violence and the psychological and organizational characteristics of terrorism itself' (1999, p. 152).

[7] One of the few systematic efforts to explore terrorist motivational incentives and disincentives for using CBRN weapons in general can be found in Gurr and Cole (2002).

[8] It should be stated at the outset that it is not necessary for terrorists to be irrational or psychologically imbalanced for them to seek nuclear weapons (Cameron, 1999, p. 23). Cameron further states that, 'If a sufficiently important end were sought by the [terrorist] group, all means, including nuclear terrorism, might be justifiable' (1999, p. 154).

1. *Mass Casualties*: The most obvious reason for terrorists to seek nuclear weapons is for the purpose of inflicting massive casualties upon their perceived enemies.[9] Indeed, while conventional (and even most unconventional) weapons will suffice to kill thousands or perhaps even tens of thousands of people if used by terrorist groups, for perpetrators who seek to cause the maximum possible immediate carnage (on the order of hundreds of thousands or millions of fatalities) the most viable means is to utilize the kinetic and thermal effects of a nuclear blast.[10] Much of the concern surrounding terrorism involving WMD stems from the belief that there is a growing number of non-state actors prepared to inflict catastrophic violence.[11] The majority of terrorist attacks, however, are carried out for a multiplicity of motives, so one should not assume that the desire to inflict mass casualties is necessarily the sole, or even predominant, motive for resorting to a nuclear option.[12]
2. *Inordinate Psychological Impact*: It is a truism that one of the core elements of terrorism is the terror it evokes. For a terrorist group seeking to traumatize a targeted society and generate public and official disorientation, nuclear weapons must hold a particular allure, for there can be few images that are guaranteed to leave as indelible a mark on the collective psyche of the targeted country as that of a mushroom cloud over one of its major cities.[13] Anthony Cordesman asserts that it is not even necessary for a nuclear weapon to have catastrophic physical effects for it to have far-ranging psychological and political impact (Cordesman, 2001, pp. 9, 10, 38–39).
3. *Prestige*: Historically, nuclear weapons have remained under the exclusive purview of nation-states, with one of the key motivations for state acquisition being the status which nuclear weapons are believed to bestow upon their possessors. How much more appealing then might the possession of nuclear weapons seem for non-state groups, many of whom

[9] A comprehensive discussion of the underlying reasons that would precipitate a goal of causing mass casualties is beyond the scope of this paper, but several perspectives will be discussed.

[10] Contagious biological agents that are used to start a large-scale epidemic could conceivably also lead to comparable casualties, although this type of attack has less reliable and less immediate consequences and it is also more difficult to fix the geographical scope of biological attacks than is the case with nuclear weapons.

[11] See, for example, Marlo: 'the increasing willingness to engage in mass murder makes terrorists more likely to consider WMD as usable and even preferable to conventional explosives and other traditional terrorist weaponry' (1999, p. 55). Also see Falkenrath (1998, p. 53) and Foxell (1999, p. 96).

[12] Indeed, if causing colossal numbers of casualties is the sole reason for employing nuclear weapons then, technically speaking, the act does not constitute terrorism but mass murder, since for an act to be defined as terrorism, there must be present the intention to influence a wider audience than the immediate victims.

[13] For a more in-depth discussion of the issue of audience impact, see Gressang (2001). Also, compare for example, Falkenrath et al. (1998, pp. 206–207) and McCormick (2003).

seek international legitimization? To the extent that terrorists believe that nuclear weapons could enable them to attain quasi-state standing or redress military imbalances vis-à-vis their purported enemies, the possession of such weapons, but not necessarily their use, becomes an attractive proposition. It is even conceivable that a terrorist group might pursue nuclear weapons in the hope of deterring, blackmailing or coercing a particular state or group of states. Thomas Schelling explores the prestige and deterrence aspects for non-state terrorists (Schelling, 1982). Also see Cameron (1999, p. 134).

4. *Incentives for Innovation and Escalation*: In a milieu in which terrorists groups may have to compete with rival groups for 'market share' of media attention and constituency support, terrorist decision makers may feel compelled to outdo the destruction wrought by previous attacks. For a discussion of why terrorists seek mass-casualty events that 'out-do' previous attacks, see Post (2000, pp. 280–282). The asymptote of such escalatory pressures, especially in the wake of such attacks as those of September 11, may be the detonation of a nuclear weapon on enemy territory, which would guarantee unrivalled attention upon the terrorists and their cause. While most terrorist supporters and sympathizers would be appalled by such horrific actions, there are certain subsets of disaffected populations that could condone the use of nuclear weapons against a hated enemy. For example, brutalized communities motivated by revenge may be more likely to condone such use.
5. *Mass Destruction and Area Denial*: In certain cases, terrorists may desire not only mass casualties, but also to physically destroy the infrastructure of their enemies and deny them the use or functioning of vital areas. These are tasks for which nuclear weapons are well-suited because they have both immediately destructive blast effects and persistent radiological contamination effects.
6. *Ideology*: The worldview of a terrorist group or individual demarcates allies and enemies and forms the basis for deciding between legitimate and illegitimate targets and tactics.[14] As such it is likely to be one of the most important factors in any decision to resort to the use of nuclear weapons. It is often asserted that the use of a weapon as destructive and reviled as nuclear weapons would alienate the supporters and perceived constituency of any terrorist group motivated primarily by a nationalist or secular political ideology (Cameron, 1999, pp. 156–157), and therefore that such groups would mostly refrain from using nuclear

[14] Albert Bandura has discussed various ways in which terrorist groups legitimize their violent behaviour, several of which can flow from a group's ideological outlook, including moral justification, displacement of responsibility, ignoring the actual suffering of victims, and dehumanizing victims (Bandura, 1998).

weapons. Whatever the accuracy of this assertion, a corollary is widely accepted by terrorism experts, that is, that groups motivated by religion, which are focused on cosmic as opposed to mortal concerns, are far more willing to engage in attacks involving mass casualties and hence would be more prone to use nuclear weapons or other means of mass destruction (Cameron, 2000; Campbell, 2000; Gurr and Cole, 2002; Hoffman, 1993a; Hoffman, 1997, pp. 45–50; Hoffman, 1998, p. 94). As one analyst observed, 'to the extent that violent extremist groups are absolutely convinced that they are doing God's bidding, virtually any action that they decide to undertake can be justified, no matter how heinous, since the "divine" ends are thought to justify the means' (Bale, 2005, p. 298; cf. Hoffman, 1993a, p. 12). The resurgence in religiously inspired terrorism in recent decades could imply that there is now a greater possibility of terrorists seeking to use WMD.[15] The situation, however, is more complex. First, not all religious terrorists are equally likely to pursue mass destruction – many religiously motivated terrorist organizations have political components, represent constituencies that are well-defined geographically (and thus are subject to retribution), or depend for financial or logistical support on parties whose views may not be quite as radical as their own. Moreover, it is the theological and cultural content of the particular strand of religious belief which is of greatest significance (Gressang, 2001), rather than the mere fact that a group has a religious bent. It has been asserted that the ideologies most conducive to the pursuit of catastrophic violence are those which simultaneously reflect an apocalyptic millenarian character, in which an irremediably corrupt world must be purged to make way for a utopian future, and emphasize the capacity for purification from sins through sacrificial acts of violence (Ackerman and Bale, 2004, pp. 29–30; Cameron, 1999, pp. 80–83; see also Chapter 4 in this volume). Such ideologies are often, though not exclusively, found amongst unorthodox religious cults, such as Aum Shinrikyo, the Covenant, the Sword, the Arm of the Lord, and R.I.S.E.,[16] and one can conceive of an affinity between the 'the relentless impulse toward world-rejecting purification' (Lifton, 1999, p. 204) displayed by such groups and the levels of 'cathartic'

[15] Several authors have questioned the link between a desire on the part of religious terrorists to cause mass casualties and the potential use of WMD, as well as the extent to which religious actors are oblivious to political concerns. They have also pointed to the large number of CBRN plots on the part of ethno-nationalist terrorists. See, for example, Rapoport (1999) and Dolnik (2004). To the first of these objections, one can refer to the discussion above relating to the desire to cause mass casualties and note that for actors seeking to cause a genuinely catastrophic scale of injury and death, conventional weapons will not suffice. The other objections are addressed in following sections.

[16] For more on these three groups, see chapters by Kaplan, Stern, and Carus in Tucker (2000).

destruction only achievable using nuclear weapons. Moreover, Jessica Stern has suggested that religious terrorists might embrace WMD, including nuclear weapons, as a means of 'emulat[ing] God' (Stern, 1999, p. 70). One must bear in mind, however, that possessing an ideology with a religious character may at most be a contributing factor to any desire to engage in nuclear terrorism, and is certainly not determinative, an assertion which has been validated empirically for CBRN weapons in toto (Asal and Ackerman, 2006).

7. *Atomic Fetishism:* A terrorist group whose ideology or key decision makers display a peculiar fascination for things nuclear or radiological might be more likely to consider pursuing a nuclear weapons capability. It is not hard to imagine that a group whose ideology is based, for instance, upon a nuclear holocaust motif or whose leader is obsessed with the science-fiction genre, could be drawn towards nuclear weapons as their preferred instruments of destruction. The archetype amongst known terrorist groups is Aum Shinrikyo, whose leader, Shoko Asahara, whose view of several types of unconventional weapons, including the nuclear variety, verged on fetishism.
8. *Revenge and Other 'Expressive' Motives*: It is believed that individuals from heavily brutalized and traumatized communities (such as those who fall victim to genocide) might be capable of unrestrained levels of violence in the pursuit of revenge against their perceived persecutors,[17] and thus might consider a retributive act as devastating as a nuclear detonation. Other expressive motives might also come into play, for example, an extreme form of defensive aggression wherein a group perceives its own imminent destruction (or that of those it purports to represent) and thus resorts to the most violent measures imaginable as a 'swan song' (Cameron, 1999, p. 135).

In addition to the possible set of instrumental, ideological or psychological motives already described, there are two other considerations that, while not necessarily sufficient on their own to motivate terrorists to pursue nuclear weapons, might influence this choice indirectly. The first of these is opportunity: a terrorist group manifesting one or more of the above motives may be propelled to consider the nuclear option more seriously by happenstance. For example, governmental collapse in a nuclear weapons state could provide increased scope for the terrorists' procurement of intact nuclear weapons and thus might precipitate for the first time the consideration of using a nuclear device. The second salient consideration is the impact of organizational structure and dynamics. It has been suggested

[17] See, for example, the discussion of the group Avenging Israel's Blood in Sprinzak and Zertal (2000).

that groups exhibiting certain structural characteristics might be more likely to engage in acts of violence as extreme as nuclear terrorism. Some of these allegedly pernicious traits include: control by megalomaniacal or sadistic, but nonetheless charismatic and authoritarian leaders; isolation from their broader society, with little display of concern for outgroups; an intentional focus on recruiting technical or scientifically skilled members; a record of innovation and excessive risk-taking; and the possession of sufficient resources, whether financial, human or logistical, to enable long-term research and development into multiple advanced weapons systems.[18]

While none of the above motives will necessarily lead to a decision to use nuclear weapons, the existence of such a broad array of potential motives provides a prima facie theoretical case that the most extreme and violent of terrorists might find either strategic, tactical, or emotional advantage in utilizing the destructive power of nuclear weapons. Any group possessing several of the abovementioned attributes deserves close scrutiny in this regard. Moreover, many (though not all) of the motives listed could also be served by lower-scale attacks, including using RDDs or attacking nuclear facilities. For instance, RDDs would likely result in a disproportionate psychological impact and area denial, but would not satisfy terrorists seeking mass fatalities.

19.3.2 The supply side of nuclear terrorism

Fortunately, even for those terrorist organizations that are not dissuaded from high consequence nuclear terrorism by moral considerations or fears of reprisal, there are major implementation challenges. These include access to nuclear assets and a variety of technical hurdles.

19.3.2.1 Improvised nuclear devices

A terrorist group motivated to manufacture and detonate an IND would need to:

1. acquire sufficient fissile material to fabricate an IND
2. fabricate the weapon
3. transport the intact IND (or its components) to a high-value target and
4. detonate the IND.[19]

[18] These factors are drawn from a combination of Tucker (2000, pp. 255–263), Campbell (2000, pp. 35–39), and Jackson (2001, p. 203). Many of these factors are related to a group's capabilities for engaging in nuclear terrorism (discussed in the following section), leading to the obvious observation that, in addition to motives driving capabilities, on occasion capabilities can reciprocally influence a terrorist's intentions.

[19] The key steps terrorists would have to take on the pathway to a nuclear attack are discussed in Bunn et al. (2003, pp. 21–31). Also see Maerli (2004, p. 44) and Ferguson and Potter (2005, pp. 112–113).

In this 'chain of causation', the most difficult challenge for a terrorist organization would most likely be obtaining the fissile material necessary to construct an IND.[20]

The problem of protecting fissile material globally has many dimensions, the most significant of which is the vast quantity of highly enriched uranium (HEU) and plutonium situated at approximately 350 different sites in nearly five dozen countries. It is estimated that there are more than 2000 metric tons of fissile material – enough for over 200,000 nuclear weapons. Many of the sites holding this material lack adequate material protection, control, and accounting measures; some are outside of the International Atomic Energy Agency's (IAEA) safeguard system; and many exist in countries without independent nuclear regulatory bodies or rules, regulations, and practices consistent with a meaningful safeguards culture.

19.3.2.2 The special dangers of HEU

Two types of fissile material can be used for the purpose of fabricating a nuclear explosive – plutonium and HEU. The most basic type of nuclear weapon and the simplest to design and manufacture is a HEU-based gun-type device. As its name implies, it fires a projectile – in this case a piece of HEU – down a gun barrel into another piece of HEU. Each piece of HEU is sub-critical and by itself cannot sustain an explosive chain reaction. Once combined, however, they form a supercritical mass and can create a nuclear explosion.

Weapons-grade HEU – uranium enriched to over 90% of the isotope U-235 – is the most effective material for a HEU-based device. However, even HEU enriched to less than weapons-grade can lead to an explosive chain reaction. The Hiroshima bomb, for example, used about 60 kg of 80% enriched uranium. Terrorists would probably need at least 40 kg of weapons-grade or near weapons-grade HEU to have reasonable confidence that the IND would work (McPhee, 1974, pp. 189–194).[21]

As indicated above, the potential for non-state actors to build a nuclear explosive already had been expressed publicly by knowledgeable experts as early as the 1970s. Their view is shared today by many physicists and nuclear weapons scientists, who concur that construction of a gun-type device would pose few technological barriers to technically competent terrorists (Alvarez, 1988, p. 125; Arbman et al., 2004; Barnaby, 1996; Boutwell et al., 2002; Civiak, 2002; Garwin and Charpak, 2001; Maerli, 2000; National Research Council, 2002; Union of Concerned Scientists, 2003; von Hippel, 2001, p. 1; Wald, 2000). In 2002, for example, the U.S. National Research Council warned,

[20] It is beyond the scope of this chapter to detail the various acquisition routes a non-state actor might pursue. For an analysis of this issue see Ferguson and Potter (2005, pp. 18–31).

[21] A technologically sophisticated terrorist group might be able to achieve success with a smaller amount of HEU if it could construct a 'reflector' to enhance the chain reaction.

'Crude HEU weapons could be fabricated without state assistance' (National Research Council, 2002, p. 45). The Council further specified, 'The primary impediment that prevents countries or technically competent terrorist groups from developing nuclear weapons is the availability of [nuclear material], especially HEU' (National Research Council, 2002, p. 40). This perspective was echoed in testimony before the Senate Foreign Relations Committee during the Clinton administration when representatives from the three US nuclear weapons laboratories all acknowledged that terrorists with access to fissile material could produce a crude nuclear explosion using components that were commonly available (Bunn and Weir, 2005, p. 156).

While there appears to be little doubt among experts that technically competent terrorists could make a gun-type device given sufficient quantities of HEU, no consensus exists as to how technically competent they have to be and how large a team they would need. At one end of the spectrum, there is the view that a suicidal terrorist could literally drop one piece of HEU metal on top of another piece to form a supercritical mass and initiate an explosive chain reaction. Nobel laureate Luis Alvarez's oft-cited quote exemplifies this view:

> With modern weapons-grade uranium, the background neutron rate is so low that terrorists, if they have such material, would have a good chance of setting off a high-yield explosion simply by dropping one half of the material onto the other half. Most people seem unaware that if separated HEU is at hand it's a trivial job to set off a nuclear explosion . . . even a high school kid could make a bomb in short order.
>
> (Alvarez, 1988, p. 125)

However, to make sure that the group could surmount any technical barriers, it would likely want to recruit team members who have knowledge of conventional explosives (needed to fire one piece of HEU into another), metalworking, and draftsmanship. A well-financed terrorist organization such as al Qaeda would probably have little difficulty recruiting personnel with these skills.

There are many potential sources of HEU for would-be nuclear terrorists. It is estimated that there are nearly 130 research reactors and associated fuel facilities in the civilian nuclear sector around the world with 20 kg or more of HEU, many of which lack adequate security (Bunn and Weir, 2005, p. 39; GAO, 2004). Also vulnerable is HEU in the form of fuel for naval reactors and targets for the production of medical isotopes. Indeed, a number of the confirmed cases involving illicit nuclear trafficking involve naval fuel.

Although an HEU-fuelled gun-type design would be most attractive to a would-be nuclear terrorist, one cannot altogether rule out an IND using plutonium. Such an explosive would require an implosion design in which the sphere of fissile material was rapidly compressed. In order to accomplish this feat, a terrorist group would require access to and knowledge of high speed electronics and high explosive lenses. A US government sponsored experiment in the 1960s suggests that several physics graduates without prior experience with nuclear weapons and with access to only unclassified information could *design* a workable implosion type bomb.[22] The participants in the experiment pursued an implosion design because they decided a gun-type device was too simple and not enough of a challenge (Stober, 2003).

Assuming that terrorists were able to acquire the necessary fissile material and manufacture an IND, they would need to transport the device (or its components) to the target site. Although an assembled IND would likely be heavy – perhaps weighing up to 1 ton – trucks and commercial vans could easily haul a device that size. In addition, container ships and commercial airplanes could provide delivery means. Inasmuch as, by definition, terrorists constructing an IND would be familiar with its design, the act of detonating the device would be relatively straightforward and present few technical difficulties.

19.3.2.3 Intact nuclear weapons

In order for terrorists to detonate an intact nuclear weapon at a designated target they would have to:

1. acquire an intact nuclear charge
2. bypass or defeat any safeguards against unauthorized use incorporated into the intact weapons and
3. detonate the weapon.

By far the most difficult challenge in the aforementioned pathway would be acquisition of the intact weapon itself.[23]

According to conventional wisdom, intact nuclear weapons are more secure than are their fissile material components. Although this perspective is probably correct, as is the view that the theft of a nuclear weapon is less likely than most nuclear terrorist scenarios, one should not be complacent about nuclear weapons security. Of particular concern are tactical nuclear weapons (TNW), of which thousands exist, none covered by formal arms control accords. Because of their relatively small size, large number, and, in some instances, lack of electronic locks and deployment outside of central storage sites, TNW would appear to be the nuclear weapon of choice for terrorists.

The overwhelming majority of TNW reside in Russia, although estimates about the size of the arsenal vary widely (see, for example, the estimate provided

[22] The experiment, however, did not demonstrate how several individuals might obtain key components for an implosion type device, some of which are very tightly controlled.

[23] For a discussion of possible routes to acquisition of intact nuclear weapons, see Ferguson and Potter (2005, pp. 53–61). Potential routes include deliberate transfer by a national government, unauthorized assistance from senior government officials, assistance from the custodian of the state's nuclear weapons, seizure by force without an insider's help, and acquisition during loss of state control over its nuclear assets due to political unrest, revolution, or anarchy.

by Sokov, Arkin, and Arbatov, in Potter et al., 2000, pp. 58–60). The United States also deploys a small arsenal of under 500 TNW in Europe in the form of gravity bombs. A major positive step enhancing the security of TNW was taken following the parallel, unilateral Presidential Nuclear Initiatives of 1991–1992. In their respective declarations, the American and Soviet/Russian presidents declared that they would eliminate many types of TNW, including artillery-fired atomic projectiles, tactical nuclear warheads, and atomic demolition munitions, and would place most other classes of TNW in 'central storage'. Although Russia proceeded to dismantle several thousand TNW, it has been unwilling to withdraw unilaterally all of its remaining TNW from forward bases or even to relocate to central storage in a timely fashion those categories of TNW covered by the 1991–1992 declarations. Moreover, in recent years, neither the United States nor Russia has displayed any inclination in pursuing negotiations to reduce further TNW or to reinforce the informal and fragile TNW regime based on parallel, unilateral declarations.

Although Russia has been the focus of most international efforts to enhance the security of nuclear weapons, many experts also are concerned about nuclear weapons security in South Asia, particularly in Pakistan. Extremist Islamic groups within Pakistan and the surrounding region, a history of political instability, uncertain loyalties of senior officials in the civilian and military nuclear chain of command, and a nascent nuclear command and control system increase the risk that Pakistan's nuclear arms could fall into the hands of terrorists. Little definite information is available, however, on the security of Pakistan's nuclear weapons or those in its nuclear neighbour, India.

Should a terrorist organization obtain an intact nuclear weapon, in most instances it would still need to overcome mechanisms in the weapon designed to prevent its use by unauthorized persons. In addition to electronic locks known as Permissive Action Links (PALs), nuclear weapons also may be safeguarded through so-called safing, arming, fusing, and firing procedures. For example, the arming sequence for a warhead may require changes in altitude, acceleration, or other parameters verified by sensors built into the weapon to ensure that the warhead can only be used according to a specific mission profile. Finally, weapons are likely to be protected from unauthorized use by a combination of complex procedural arrangements (requiring the participation of many individuals) and authenticating codes authorizing each individual to activate the weapon.

All operational US nuclear weapons have PALs. Most authorities believe that Russian strategic nuclear weapons and modern shorter range systems also incorporate these safeguards, but are less confident that older Russian TNW are equipped with PALs (Sokov, 2004). Operational British and French nuclear weapons (with the possible exception of French SLBM warheads) also probably are protected by PALs. The safeguards on warheads of the other nuclear-armed states cannot be determined reliably from open sources, but are more likely to rely on procedures (e.g., a three-man rule) than PALs to prevent unauthorized use (Ferguson and Potter, 2005, p. 62).

Unless assisted by sympathetic experts, terrorists would find it difficult, though not necessarily impossible, to disable or bypass PALs or other safeguard measures. If stymied, terrorists might attempt to open the weapon casing to obtain fissile material in order to fabricate an IND. However, the act of prying open the bomb casing might result in terrorists blowing themselves up with the conventional high explosives associated with nuclear warheads. Thus, terrorists would likely require the services of insiders to perform this operation safely.

Assuming a terrorist organization could obtain a nuclear weapon and had the ability to overcome any mechanisms built into the device to prevent its unauthorized detonation, it would still need to deliver the weapon to the group's intended target. This task could be significantly complicated if the loss of the weapon were detected and a massive recovery effect were mounted.

It is also possible that terrorists might adopt strategies that minimized transportation. These include detonating the weapon at a nearby, less-than-optimal target, or even at the place of acquisition.

If a nuclear weapon were successfully transported to its target site, and any PALs disabled, a degree of technical competence would nonetheless be required to determine how to trigger the device and provide the necessary electrical or mechanical input for detonation. Here, again, insider assistance would be of considerable help.

19.4 Probabilities of occurrence

19.4.1 The demand side: who wants nuclear weapons?

Thankfully, there have been no instances of non-state actor use of nuclear weapons. The historical record of pursuit by terrorists of nuclear weapons is also very sparse, with only two cases in which there is credible evidence that terrorists actually attempted to acquire nuclear devices.[24] The most commonly cited reasons for this absence of interest include the technical and material difficulties associated with developing and executing a nuclear detonation attack, together with the alleged technological and operational 'conservatism'[25] of most terrorists, fears of reprisal, and the moral and political constraints on employing such frightful forms of violence. These propositions are then

[24] Empirically speaking, the record of non-use should not be compared over the long history of terrorism, but only over the period since it became feasible for non-state actors to acquire or use nuclear weapons, Circa 1950.

[25] See, for example, Ferguson and Potter (2005, p. 40), Jenkins (1986, p. 777), Hoffman (1993a, pp. 16–17), and Clutterbuck (1993, pp. 130–139).

combined and cast in a relative manner to conclude that the overwhelming majority of terrorists have thus far steered clear of nuclear weapons because they have found other weapon types to be (1) easier to develop and use, (2) more reliable and politically acceptable, and (3) nonetheless eminently suitable for accomplishing their various political and strategic goals. In short, the basic argument is that interest has been lacking because large-scale unconventional weapons, especially of the nuclear variety, were seen to be neither necessary nor sufficient[26] for success from the terrorists' point of view.

Although the past non-use of nuclear weapons may ab initio be a poor indicator of future developments, it appears that some prior restraints on terrorist pursuit and use of nuclear weapons (and other large scale unconventional weapons systems) may be breaking down. For example, one can point to an increase in the number of terrorist-inclined individuals and groups who subscribe to beliefs and goals that are concordant with several of the motivational factors described earlier. In terms of mass casualties, for instance, there are now groups who have expressed the desire to inflict violence on the order of magnitude that would result from the use of a nuclear weapon. Illustrative of this perspective was the claim in 2002 by Sulaiman Abu Ghaith, Osama bin Laden's former official press spokesman, of the right for jihadists 'to kill four million Americans' (The Middle East Media Research Institute, 2002). One can also point to groups displaying an incipient techno-fetishism, who are simultaneously less interested in global, or sometimes any external, opinion, such as Aum Shinrikyo. It might be of little surprise, then, that it is these very two groups which have manifested more than a passing interest in nuclear weapons.

Prior to Aum Shinrikyo, most would-be nuclear terrorists were more kooks than capable, but Aum made genuine efforts to acquire a nuclear capability. As is described in more detail in the next section, during the early 1990s, Aum repeatedly attempted to purchase, produce, or otherwise acquire a nuclear weapon. Aum's combination of apocalyptic ideology, vast financial and technical resources, and the non-interference by authorities in its activities enabled it to undertake a generous, if unsuccessful, effort to acquire nuclear weapons. Although some analysts at the time sought to portray Aum as a one-off nexus of factors that were unlikely to ever be repeated, in fact, a far larger transnational movement emerged shortly thereafter with a similar eye on the nuclear prize.

Al Qaeda, the diffuse jihadist network responsible for many of the deadliest terrorist attacks in the past decade, has not been shy about its nuclear ambitions. As early as 1998, its self-styled emir, Usama bin Ladin, declared that 'To seek to possess the weapons [of mass destruction] that could counter those of the infidels is a religious duty ... It would be a sin for Muslims not to seek

[26] Since these weapons were too difficult to acquire and use reliably.

possession of the weapons that would prevent the infidels from inflicting harm on Muslims' (Scheuer, 2002, p. 72). As noted previously, the group has asserted 'the right to kill 4 million Americans – 2 million of them children', in retaliation for the casualties it believes the United States and Israel have inflicted on Muslims. Bin Laden also sought and was granted a religious edict or *fatwa* from a Saudi cleric in 2003, authorizing such action (Bunn, 2006). In addition to their potential use as a mass-casualty weapon for punitive purposes,[27] al Qaeda ostensibly also sees strategic political advantage in the possession of nuclear weapons, perhaps to accomplish such tasks as coercing the 'Crusaders' to leave Muslim territory. When combined with millenarian impulses among certain quarters of global *jihadis* and a demonstrated orientation towards martyrdom, it is apparent that many (if not most) of the motivational factors associated with nuclear terrorism apply to the current violent jihadist movement. The manner in which these demands on motivations have been translated into concrete efforts to obtain nuclear explosives is described in the section below on 'The Supply Side'.

At present, the universe of non-state actors seeking to acquire and use nuclear weapons appears to be confined to violent jihadhists, a movement that is growing in size and scope and spawning a host of radical offshoots and followers. Although in the short term at least, the most likely perpetrators of nuclear violence will stem from operationally sophisticated members of this milieu, in the longer term, they may be joined by radical right-wing groups (especially those components espousing extremist Christian beliefs),[28] an as-yet-unidentified religious cult, or some other group of extremists who limn the ideological and structural arcs associated with nuclear terrorism.

Within any society, there will always be some people dissatisfied with the status quo. A very small subset of these angry and alienated individuals may embark on violent, terrorist campaigns for change, in some cases aiming globally. An even tinier subset of these non-state actors with specific ideological, structural, and operational attributes may seek nuclear weapons. Perhaps the most frightening possibility would be the development of technology or the dissolution of state power in a region to the point where a single disgruntled individual would be able to produce or acquire a working nuclear weapon. Since there are far more hateful, delusional and

[27] Indeed, the former head of the CIA's Bin Laden Unit has explained that, 'What al-Qaeda wants is a high body count as soon as possible, and it will use whatever CBRN [chemical, biological, radiological, nuclear] materials it gets in ways that will ensure the most corpses' (Scheuer, 2002, p. 198).

[28] For instance, *The Turner Diaries*, a novel written by the former leader of the National Alliance, William Pierce, and which has had considerable influence on many right-wingers, describes racist 'patriots' destroying cities and other targets with nuclear weapons (Macdonald, 1999).

solipsistic individuals than organized groups in this world,[29] this situation would indeed be deserving of the label of a nuclear nightmare. This and other similar capability related issues are discussed in the following section.

19.4.2 The supply side: how far have terrorists progressed?

As a recent briefing by the Rand Corporation points out, 'On the supply side of the nuclear market, the opportunities for [non-state] groups to acquire nuclear material and expertise are potentially numerous' (Daly et al., 2005, p. 3). These opportunities include huge global stocks of fissile material, not all of which are adequately secured; tens of thousands of weapons in various sizes and states of deployment and reserve in the nuclear arsenals of at least eight states; and a large cadre of past and present nuclear weaponeers with knowledge of the science and art of weapons design and manufacture. In addition, the extraordinarily brazen and often successful nuclear supply activities of A.Q. Khan and his wide-flung network demonstrate vividly the loopholes in current national and international export control arrangements. Although not in great evidence to date, one also cannot discount the emergence of organized criminal organizations that play a Khan-like middleman role in finding a potential nuclear supplier, negotiating the purchase/sale of the contraband, and transporting the commodity to the terrorist end-user. Taken together, these factors point to the need to assess carefully the past procurement record of terrorist groups and the potential for them in the future to obtain not only highly sensitive nuclear technology and know-how, but also nuclear weapon designs and the weapons themselves. The most instructive cases in examining the level of success attained by terrorists thus far are once again to be found in Aum Shinrikyo and al Qaeda.

The religious cult Aum Shinrikyo initiated an ambitious programme to acquire chemical, biological, and nuclear weapons in the early 1990s. In the nuclear sphere, this effort was first directed at purchasing a nuclear weapon. To this end, Aum appears to have sought to exploit its large following in Russia (at its peak estimated to number in the tens of thousands) and its access to senior Russian officials, to obtain a nuclear warhead. Aum members are reported to have included a number of scientists at the Kurchatov Institute, an important nuclear research facility in Moscow possessing large quantities of HEU, which at the time were poorly secured (Bunn, 2006; Daly et al., 2005, p. 16). Aum leader Shoko Asahara, himself, led a delegation to Russia in 1992, which met with former Vice President Aleksandr Rutskoi, Russian Parliament speaker Rusian Khasbulatov, and Head of Russia's Security Council, Oleg Lobov

[29] For one thing, many deranged or excessively aggressive individuals cannot function as part of a group.

(Daly et al., 2005, p. 14).[30] Inscribed in a notebook confiscated from senior Aum leader Hayakawa Kiyohi, who reportedly made over 20 trips to Russia, was the question, 'Nuclear warhead. How much?' Following the questions were several prices in the millions of dollars (Bunn, 2006; Daly et al., 2005, p. 13).

Despite its deep pockets and unusually high-level contacts in Russia, possibly extending into the impoverished nuclear scientific complex, Aum's persistent efforts were unsuccessful in obtaining either intact nuclear weapons or the fissile material for their production. A similarly ambitious if more far-fetched attempt to mine uranium ore on a sheep farm in Australia also resulted in failure.[31] Aum, therefore, at least temporarily turned its attention away from nuclear weapons in order to pursue the relatively easier task of developing chemical weapons.[32]

There is substantial evidence that al Qaeda, like Aum Shinrikyo, has attempted to obtain nuclear weapons and their components. According to the federal indictment of Osama bin Laden for his role in the attacks on US embassies in Kenya and Tanzania, these procurement activities date back to at least 1992 (Bunn, 2006). According to Jamal Ahmed al-Fadl, a Sudanese national who testified against bin Laden in 2001, an attempt was made by al Qaeda operatives in 1993 to buy HEU for a bomb (McLoud and Osborne, 2001). This effort, as well as several other attempts to purchase fissile material appear to have been unproductive, and were undermined by the lack of technical knowledge on the part of al Qaeda aides. In fact, al Qaeda may well have fallen victim to various criminal scams involving the sale of low-grade reactor fuel and a bogus nuclear material called 'Red Mercury'.[33]

If al Qaeda's early efforts to acquire nuclear weapons were unimpressive, the leadership did not abandon this objective and pursued it with renewed energy once the organization found a new sanctuary in Afghanistan after 1996. Either with the acquiescence or possible assistance of the Taliban, al Qaeda appears to have sought not only fissile material but also nuclear weapons expertise, most notably from the Pakistani nuclear establishment. Bin Laden and his deputy al-Zawahiri, for example, are reported to have met at length with two Pakistani nuclear weapons experts, who also were Taliban sympathizers, and

[30] Allegedly, subsequent efforts by Aum representatives to meet with Russia's minister of atomic energy were unsuccessful.

[31] The farm was purchased for the dual purposes of testing chemical weapons and mining uranium.

[32] Following its sarin attack on the Tokyo subway in 1995, Aum reappeared under the name of Aleph. In 2000, Japanese police discovered that the cult had obtained information about nuclear facilities in a number of countries from classified computer networks (Daly et al., 2005, p. 19).

[33] For a discussion of some of the incidents, see Leader (1999, pp. 34–37) and Daly et al. (2005, pp. 31–33).

to have sought to elicit information about the manufacture of nuclear weapons (Glasser and Khan, 2001).

According to one analyst who has closely followed al Qaeda's nuclear activities, the Pakistanis may have provided bin Laden with advice about potential suppliers for key nuclear components (Albright and Higgens, 2003, pp. 9–55). In addition, although there is no solid evidence, one cannot rule out the possibility that al Qaeda received the same kind of weapons design blueprints from the A.Q. Khan network that were provided to Libya and also, conceivably, to Iran.

The Presidential Commission on the Intelligence Capabilities of the United States Regarding WMD reported in March 2005 that in October 2001 the US intelligence community assessed al Qaeda as capable of making at least a 'crude' nuclear device if it had access to HEU or plutonium (Commission, 2005, pp. 267, 271, 292). The commission also revealed that the CIA's non-proliferation intelligence and counterterrorism centres had concluded in November 2001 that al Qaeda 'probably had access to nuclear expertise and facilities and that there was a real possibility of the group developing a crude nuclear device' (Commission, 2005, pp. 267, 271, 292). These assessments appear to be based, at least in part, on documents uncovered at al Qaeda safe houses after the US toppled the Taliban regime, which revealed that al Qaeda operatives were studying nuclear explosives and nuclear fuel cycle technology.[34]

In addition to compelling evidence about al Qaeda's efforts to obtain nuclear material and expertise, there have been numerous unsubstantiated reports about the possible acquisition by al Qaeda of intact nuclear weapons from the former Soviet Union. Most of these accounts are variants of the storyline that one or more Soviet-origin 'suitcase nukes' were obtained by al Qaeda on the black market in Central Asia.[35] Although one cannot dismiss these media reports out of hand due to uncertainties about Soviet weapons accounting practices, the reports provide little basis for corroboration, and most US government analysts remain very skeptical that al Qaeda or any other non-state actor has yet acquired a stolen nuclear weapon on the black market.

[34] For an assessment of these documents, see Albright (2002). Al Qaeda's nuclear programme: through the window of seized documents [online]. Policy Forum Online, Nautilus Institute. Available from: http://nautilus.org/acrchives/fora/Special-Policy-Forum/47_Albright.html [Accessed 15 September 2006.]

[35] For a review of some of these reports see (Daly et al., 2005, pp. 40–45). One of the more sensationalist reports emanating from Israel claimed that the terrorist organization had acquired eight to ten atomic demolition mines built for the KGB (*Jerusalem DEBKA-Net-Weekly*, 12 October 2001 cited by Daly et al., p. 41). Another report from the Pakistani press suggests that two small nuclear weapons were transported to the US mainland (The Frontier Post, 2001). If press reports can be believed, on occasion, bin Laden and his senior aides have sought to give credence to these stories by bragging of their acquisition of nuclear weapons (Connor, 2004; Mir, 2001). A sceptical assessment of these risks is provided by Sokov (2002).

Despite the loss of its safe haven in Afghanistan, US government officials continue to express the belief that al Qaeda is in a position to build an improvised nuclear device (Jane's Intelligence Digest, 2003; Negroponte statement, 2005). And yet even at their heyday, there is no credible evidence that either al Qaeda or Aum Shinrikyo were able to exploit their high motivations, substantial financial resources, demonstrated organizational skills, far-flung network of followers, and relative security in a friendly or tolerant host country to move very far down the path towards acquiring a nuclear weapons capability. As best one can tell from the limited information available in public sources, among the obstacles that proved most difficult for them to overcome was access to the fissile material needed to fabricate an IND. This failure probably was due in part to a combination of factors, including the lack of relevant 'in house' technical expertise, unfamiliarity with the nuclear black market and lack of access to potential nuclear suppliers, and greater security and control than was anticipated at sites possessing desired nuclear material and expertise. In the case of the former Soviet Union, the group's procurement efforts also probably underestimated the loyalty of underpaid nuclear scientists when it came to sharing nuclear secrets. In addition, both Aum Shinrikyo and al Qaeda's pursuit of nuclear weapons may have suffered from too diffuse an effort in which large but limited organizational resources were spread too thinly over a variety of ambitious tasks.[36]

19.4.3 What is the probability that terrorists will acquire nuclear explosive capabilities in the future?

Past failures by non-state actors to acquire a nuclear explosive capability may prove unreliable indicators of future outcomes. What then are the key variables that are likely to determine if and when terrorists will succeed in acquiring INDs and/or nuclear weapons, and if they are to acquire them, how many and of what sort might they obtain? The numbers, in particular, are relevant in determining whether nuclear terrorism could occur on a scale and scope that would constitute a global catastrophic threat.

As discussed earlier, the inability of two very resourceful non-state actors – Aum Shinrikyo and al Qaeda – to acquire a credible nuclear weapons capability cautions against the assumption that terrorists will any time soon be more successful. Although weapons usable nuclear material and nuclear weapons themselves are in great abundance and, in some instances, lack adequate physical protection, material control, and accounting, not withstanding the activities of the A.Q. Khan network, there are few confirmed reports of illicit trafficking of weapons usable material, and no reliable accounts of the diversion or sale of intact nuclear weapons (Potter and Sokova, 2002).

[36] Daly et al. provides a related but alternative assessment of the unsuccessful procurement efforts (2005).

This apparent positive state of affairs may be an artifact of poor intelligence collection and analysis as well as the operation of sophisticated smugglers able to avoid detection (Potter and Sokova, 2002). It also may reflect the relatively small number of would-be nuclear terrorists, the domination of the nuclear marketplace by amateur thieves and scam artists, and the inclination of most organized criminal organizations to shy away from nuclear commerce when they can make their fortune in realms less likely to provoke harsh governmental intervention. Finally, one cannot discount the role of good luck.

That being said, a number of respected analysts have proffered rather dire predictions about the looming threat of terrorist initiated nuclear violence. Graham Allison, author of one of the most widely cited works on the subject, offers a standing bet of 51 to 49 odds that, barring radical new antiproliferation steps, there will be a terrorist nuclear strike within the next ten years. Allison gave such odds in August 2004, and former US Secretary of Defense William Perry agreed, assessing the chance of a terror strike as even (Allison, 2004; Kristof, 2004). More recently, in June 2007, Perry and two other former US government officials concluded that the probability of a nuclear weapon going off in an American city had increased in the past five years although they did not assign a specific probability or time frame (Perry et al., 2007, p. 8). Two other well-known analysts, Matthew Bunn and David Albright, also are concerned about the danger, but place the likelihood of a terrorist attack involving an IND at 5% and 1%, respectively (Sterngold, 2004).[37]

It is difficult to assess these predictions as few provide much information about their underlying assumptions. Nevertheless, one can identify a number of conditions, which, if met, would enable non-state actors to have a much better chance of building an IND or seizing or purchasing one or more intact nuclear weapons.

Perhaps the single most important factor that could alter the probability of a successful terrorist nuclear weapons acquisition effort is state-sponsorship or state-complicity. Variants of this pathway to nuclear weapons range from the deliberate transfer of nuclear assets by a national government to unauthorized assistance by national government officials, weapons scientists, or custodians. At one end of the continuum, for example, a terrorist group could conceivably acquire a number of intact operational nuclear weapons directly from a sympathetic government. Such action would make it unnecessary to bypass security systems protecting the weapons. This 'worst case' scenario has shaped

[37] To some extent, this apparent divergence in threat assessments among experts is due to their focus on different forms of nuclear terrorism and different time-frames. Nevertheless, there is no clear prevailing view among experts about the immediacy or magnitude of the nuclear terrorism threat. For an extended review of expert opinions on various proliferation and terrorist threats – although not precisely the issue of use of a nuclear explosive, see Lugar (2005).

US foreign policy towards so-called rogue states, and continues to be a concern with respect to countries such as North Korea and Iran.[38]

Even if a state's most senior political leaders were not prepared to transfer a nuclear weapon to a terrorist organization, it is possible that other officials with access to their country's nuclear assets might take this step for financial or ideological reasons. If President Musharraf and A.Q. Khan are to be believed, the latter's unauthorized sale of nuclear know-how to a number of states demonstrated the feasibility of such transfer, including the provision of nuclear weapon designs, and, in principle, could as easily have been directed to non-state actors. Aid from one or more insiders lower down the chain of command, such as guards at a nuclear weapons storage or deployment site, also could facilitate the transfer of a nuclear weapon into terrorist hands.

Moreover, one can conceive of a plausible scenario in which terrorist groups might take advantage of a coup, political unrest, a revolution, or a period of anarchy to gain control over one or more nuclear weapons. Nuclear assets could change hands, for example, because of a coup instigated by insurgents allied to or cooperating with terrorists. Although the failed coup attempt against Soviet President Mikhail Gorbachev during August 1991 did not involve terrorists, during the crisis Gorbachev reportedly lost control of the Soviet Union's nuclear arsenal to his would-be successors when they cut off his communications links (Ferguson and Potter, 2005, p. 59; Pry, 1999, p. 60). It is also possible that during a period of intense political turmoil, nuclear custodians might desert their posts or otherwise be swept aside by the tide of events. During China's Cultural Revolution of the mid-1960s, for example, leaders of China's nuclear establishment feared that their institutes and the Lop Nor nuclear test site might be overrun by cadres of Red Guards (Spector, 1987, p. 32). Similarly in 1961, French nuclear authorities appeared to have rushed to test a nuclear bomb at a site in Algeria out of fear that a delay could enable rebel forces to seize the weapon (Spector, 1987, pp. 27–30). Although it is unlikely that political unrest would threaten nuclear controls in most weapons states today, the situation is not clear cut everywhere, and many analysts express particular concerns with respect to Pakistan and North Korea.

Although the preceding examples pertain to operational nuclear weapons, similar state-sponsored or state-complicit scenarios apply with equal force to the transfer of weapons usable material and technical know-how. Indeed, the transfers of highly sensitive nuclear technology to Iran, Libya, and North Korea by A.Q. Khan and his associates over an extended period of time (1989–2003), demonstrate the feasibility of similar transfers to non-state

[38] For a discussion of this and other nuclear pathways involving state sponsorship see Ferguson and Potter (2005, pp. 54–61).

actors. As in an attempted seizure of a nuclear weapon, political instability during a coup or revolution could provide ample opportunities for terrorists to gain control of fissile material, stocks of which are much more widely dispersed and less well guarded than nuclear weapons. There is one confirmed case, for example, in which several kilograms of HEU disappeared from the Sukhumi Nuclear Research Center in the breakaway Georgian province of Abkhazia. Although the details of the case remain murky, and there is no evidence that a terrorist organization was involved, the HEU was diverted during a period of civil turmoil in the early 1990s (Potter and Sokova, 2002, p. 113).

Aside from the assistance provided by a state sponsor, the most likely means by which a non-state actor is apt to experience a surge in its ability to acquire nuclear explosives is through technological breakthroughs. Today, the two bottlenecks that most constrain non-state entities from fabricating a nuclear weapon are the difficulty of enriching uranium and the technical challenge of correctly designing and building an implosion device.

Although almost all experts believe uranium enrichment is beyond the capability of any non-state entity acting on its own today, it is possible that new enrichment technologies, especially involving lasers, may reduce this barrier. Unlike prevailing centrifuge and diffusion enrichment technology, which require massive investments in space, infrastructure, and energy, laser enrichment could theoretically involve smaller facilities, less energy consumption, and a much more rapid enrichment process. These characteristics would make it much easier to conceal enrichment activities and could enable a terrorist organization with appropriate financial resources and technical expertise to enrich sufficient quantities of HEU covertly for the fabrication of multiple INDs. However, despite the promise of laser enrichment, it has proved much more complex and costly to develop than was anticipated (Boureston and Ferguson, 2005).

Unlike a gun-type device, an implosion-type nuclear explosive can employ either HEU or plutonium. However, it also requires more technical sophistication and competence than an HEU-based IND. Although these demands are likely to be a major impediment to would-be nuclear terrorists at present, the barriers may erode over time as relevant technology such as high-speed trigger circuitry and high explosive lenses become more accessible.

Were terrorists to acquire the ability to enrich uranium or manufacture implosion devices using plutonium, it is much more likely that they would be able to produce multiple INDs, which could significantly increase the level of nuclear violence. Neither of these developments, however, are apt to alter significantly the yield of the IND. Indeed, nothing short of a hard to imagine technological breakthrough that would enable a non-state actor to produce an advanced fission weapon or a hydrogen bomb would produce an order of magnitude increase in the explosive yield of a single nuclear device.[39]

19.4.4 Could terrorists precipitate a nuclear holocaust by non-nuclear means?

The discussion to this point has focused on the potential for terrorists to *inflict* nuclear violence. A separate but related issue is the potential for terrorists to *instigate* the use of nuclear explosives, possibly including a full-fledged nuclear war. Ironically, this scenario involves less demanding technical skills and is probably a more plausible scenario in terms of terrorists approaching the level of a global catastrophic nuclear threat.

One can conceive of a number of possible means by which terrorists might seek to provoke a nuclear exchange involving current nuclear weapon states. This outcome might be easiest to accomplish in South Asia, given the history of armed conflict between India and Pakistan, uncertainties regarding the command and control arrangements governing nuclear weapons release, and the inclination on the part of the two governments to blame each other whenever any doubt exists about responsibility for terrorist actions. The geographical proximity of the two states works to encourage a 'use them or lose them' crisis mentality with nuclear assets, especially in Pakistan which is at a severe military disadvantage vis a vis India in all indices but nuclear weapons. It is conceivable, therefore, that a terrorist organization might inflict large scale but conventional violence in either country in such manner as to suggest the possibility of state complicity in an effort to provoke a nuclear response by the other side.

A number of films and novels rely on cyber terror attacks as the source of potential inter-state nuclear violence.[40] Although the plot lines vary, they frequently involve an individual or group's ability to hack into a classified military network, thereby setting in motion false alarms about the launch of foreign nuclear-armed missiles, or perhaps even launching the missiles themselves. It is difficult to assess the extent to which these fictional accounts mirror real-life vulnerabilities in the computer systems providing the eyes and ears of early warning systems for nuclear weapon states, but few experts attach much credence to the ability of non-state actors to penetrate and spoof these extraordinarily vital military systems.

A more credible, if still unlikely means, to mislead an early warning system and trigger a nuclear exchange involves the use by terrorists of scientific rockets. The 'real world' model for such a scenario is the 1995 incident

[39] Some breakthroughs might lead to weapons that are more likely to successfully detonate, improving success rates but not overall destruction.

[40] See, for example, the 1983 movie War Games. Schollmeyer provides a review of other fictional accounts of nuclear violence (2005).

in which a legitimate scientific sounding rocket (used to take atmospheric measurements) launched from Norway led the Russian early warning system to conclude initially that Russia was under nuclear attack (Forden, 2001).[41]

Although terrorists might find it very difficult to anticipate the reaction of the Russian early warning system (or those of other nuclear weapon states) to various kinds of rocket launches, access to and use of scientific rockets is well within the reach of many non-state actors, and the potential for future false alerts to occur is considerable.

19.5 Consequences of nuclear terrorism

19.5.1 Physical and economic consequences

The physical and health consequences of a nuclear terrorist attack in the foreseeable future, while apt to be devastating, are unlikely to be globally catastrophic. This conclusion is derived, in part, from a review of various government and scholarly calculations involving different nuclear detonation/exchange scenarios which range from a single 20 kiloton IND to multiple megaton weapons.

The most likely scenario is one in which an IND of less than 20 kilotons is detonated at ground level in a major metropolitan area such as New York. The size of the IND is governed by the amount of HEU available to would-be nuclear terrorists and their technical skills.[42]

A blast from such an IND would immediately wipe out the area within about a one and a half mile radius of the weapon. Almost all non-reinforced structures within that radius would be destroyed and between 50,000 and 500,000 people would probably die, with a similar number seriously injured.[43] The amount of radioactive fallout after such an attack is difficult to estimate because of uncertain atmospheric factors, but one simulation predicts that 1.5 million people would be exposed to fallout in the immediate aftermath of the blast, with 10,000 immediate deaths from radiation poisoning and an eventual 200,000 cancer deaths (Helfand et al., 2002, p. 357). Very soon after the attack, hospital facilities would be overwhelmed, especially with burn victims, and many victims would die because of a lack of treatment. These estimates are derived from computer models, nuclear testing results, and the experiences of

[41] Sagan made important study of the broader but related issue of the possibility of achieving fail safe systems in large organizations such as the military (1993).

[42] The Hiroshima bomb from highly enriched uranium had a yield of between 12.5 and 15 kilotons (Rhodes, 1986, p. 711).

[43] Estimates come from Helfand et al. (2002) and Bunn et al. (2003, p. 16). Both derive their estimates from Glasstone and Dolan (1977) and their own calculations.

Hiroshima and Nagasaki.[44] A firestorm might engulf the area within one to two miles of the blast, killing those who survived the initial thermal radiation and shockwave.[45]

Of course, no one really knows what would happen if a nuclear weapon went off today in a major metropolis. Most Cold War studies focus on weapons exploding at least 2000 feet above the ground, which result in more devastation and a wider damage radius than ground level detonations. But such blasts, like those at Hiroshima and Nagasaki, also produce less fallout. Government studies of the consequences of nuclear weapons use also typically focus on larger weapons and multiple detonations. A terrorist, however, is unlikely to expend more than a single weapon in one city or have the ability to set off an airburst. Still, the damage inflicted by a single IND would be extreme compared to any past terrorist bombing. By way of contrast, the Oklahoma City truck bomb was approximately 0.001 kiloton, a fraction of the force of an improvised nuclear device.

TNW typically are small in size and yield and would not produce many more casualties than an IND. Although they are most likely to be the intact weapon of choice for a terrorist because of their relative portability, if a larger nuclear weapon were stolen it could increase the number of deaths by a factor of 10. If, for example, a 1-megaton fusion weapon (the highest yield of current US and Russian arsenals) were to hit the tip of Manhattan, total destruction might extend as far as the middle of Central Park, five miles away. One study by the Office of Technology Assessment estimated that a 1-megaton weapon would kill 250,000 people in less-dense Detroit (U.S. Office of Technology Assessment, 1979), and in Manhattan well over a million might die instantly. Again, millions would be exposed to radioactive fallout, and many thousands would die depending on where the wind happened to blow and how quickly people took shelter.

Regardless of the precise scenarios involving an IND or intact nuclear weapon, the physical and health consequences would be devastating. Such an attack would tax health care systems, transportation, and general commerce, perhaps to a breaking point. Economic studies of a single nuclear attack estimate property damage in the many hundreds of billions of dollars (ABT Associates, 2003, p. 7),[46] and the direct cost could easily exceed one trillion dollars if lost lives are counted in economic terms. For comparison, the economic impact of the September 11 attacks has been estimated at $83 billion in both direct and indirect costs (GAO, 2002).

[44] Casualty estimates from these two attacks vary by a factor of 2, from 68,000 at Hiroshima (Glasstone and Dolan, 1977) to 140,000 (Rhodes, 1986).

[45] Whether or not a firestorm is generated would depend on the exact location of the blast. A smaller, ground-based explosion is less likely to trigger a firestorm, but an excess of glass, flammable material, gas mains, and electrical wires might trigger one in a city (Eden, 2004).

[46] Neither the U.S. Office of Technology Assessment (1979) nor Glasstone and Dolan (1977) attempt to calculate the economic impact of a nuclear attack.

A nuclear attack, especially a larger one, might totally destroy a major city. The radioactivity of the surrounding area would decrease with time, and outside the totally destroyed area (a mile or two), the city could be repopulated within weeks, but most residents would probably be hesitant to return due to fear of radioactivity.

A more globally catastrophic scenario involves multiple nuclear warheads going off at the same time or in rapid succession. If one attack is possible, multiple attacks also are conceivable as weapons and weapons material are often stored and transported together. Should terrorists have access to multiple INDs or weapons they might target a number of financial centres or trade hubs. Alternatively, they could follow a Cold War attack plan and aim at oil refineries. In either scenario, an attack could lead to severe global economic crisis with unforeseen consequences. Whereas the September 11 attacks only halted financial markets and disrupted transportation for a week, even a limited nuclear attack on lower Manhattan might demolish the New York Stock Exchange and numerous financial headquarters. Combined with attacks on Washington DC, Paris, London, and Moscow, major centres of government and finance could nearly disappear. The physical effects of the attacks would be similar in each city, though many cities are less densely packed than New York and would experience fewer immediate deaths.

19.5.2 Psychological, social, and political consequences

Unlike the more tangible physical and economic effects of nuclear terrorism, it is almost impossible to model the possible psychological, social, and political consequences of nuclear terrorism, especially in the long term and following multiple incidents. One is therefore forced to rely on proxy data from the effects of previous cases of large-scale terrorism, a variety of natural disasters, and past nuclear accidents such as the Chernobyl meltdown. The psychological, social, and political effects of nuclear terrorism are likely to extend far beyond the areas affected by blast or radiation, although many of these effects are likely to be more severe closer to ground zero.

It can be expected that the initial event will induce a number of psychological symptoms in victims, responders, and onlookers. In an age of instantaneous global communication, the last category might rapidly encompass most of the planet. The constellation of possible symptoms might include anxiety, grief, helplessness, initial denial, anger, confusion, impaired memory, sleep disturbance and withdrawal (Alexander and Klein, 2006). Based on past experience with terrorism and natural disasters, these symptoms will resolve naturally in the majority of people, with only a fraction[47] going on to develop

[47] The actual percentage of those likely to succumb to long-term psychological illness will depend on several factors, including the individual's proximity to the attack location, whether or not a person became ill, previous exposures to trauma, the individual's prior psychological state,

persistent psychiatric illness such as post-traumatic stress disorder. However, the intangible, potentially irreversible, contaminating, invasive and doubt-provoking nature of radiation brings with it a singular aura of dread and high levels of stress and anxiety. Indeed, this fear factor is one of the key reasons why some terrorists might select weapons emitting radiation.

In addition to significant physical casualties, a nuclear terrorism event would most likely result in substantially greater numbers of unexposed individuals seeking treatment, thereby complicating medical responses.[48] In the 1987 radiological incident in Goiania, Brazil, up to 140,000 unexposed people flooded the health care system seeking treatment (Department of Homeland Security, 2003, p. 26). Although genuine panic – in the sense of maladaptive responses such as 'freezing' – is extremely rare (Jones, 1995), a nuclear terrorism incident might provoke a mass exodus from cities as individuals make subjective decisions to minimize their anxiety. Following the Three Mile Island nuclear accident in the United States in 1979, 150,000 people took to the highways – forty-five people evacuated for every person advised to leave (Becker, 2003).

Were nuclear terrorism to become a repeating occurrence, the question would arise regarding whether people would eventually be able to habituate to such events, much as the Israeli public currently manages to maintain a functional society despite continual terrorist attacks. While desensitization to extremely high levels of violence is possible, multiple cases of nuclear terrorism over an extended period of time might prove to be beyond the threshold of human tolerance.

Even a single incidence of nuclear terrorism could augur negative social changes. While greater social cohesion is likely in the immediate aftermath of an attack (Department of Homeland Security, 2003, p. 38), over time feelings of fear, anger and frustration could lead to widespread anti-social behaviour, including the stigmatization of those exposed to radiation and the scapegoating of population groups associated with the perceived perpetrators of the attack. This reaction could reach the level of large-scale xenophobia and vigilantism. Repeated attacks on major cities might even lead to behaviours

the presence or otherwise of a support network and the amount of exposure to media coverage (Pangi, 2002); for more on the links between terrorism and subsequent psychological disorders, see Schlenger (2002) and North and Pfefferbaum (2002).

[48] The ratio of unexposed to exposed patients has been conservatively estimated as 4:1. (Department of Homeland Security, 2003, p. 34) Department of Homeland Security Working Group on Radiological Dispersal Device (RDD) Preparedness, Medical Preparedness and Response Sub-Group. (2003). *Radiological Medical Countermeasures.* Becker notes that estimates state that in any terrorist event involving unconventional weapons, the number of psychological casualties will outnumber the number of physical casualties by a factor of five. Becker (2001). Are the psychosocial aspects of weapons of mass destruction incidents addressed in the Federal Response Plan: summary of an expert panel. *Military Medicine,* 166(Suppl. 2), 66–68.

encouraging social reversion and the general deterioriation of civil society, for example, if many people adopt a survivalist attitude and abandon populated areas. There is, of course, also the possibility that higher mortality salience might lead to positive social effects, including more constructive approaches to problem-solving (Calhoun and Tedeschi, 1998). Yet higher morality could just as easily lead to more pathological behaviours. For instance, during outbreaks of the Black Death plague in the Middle Ages, some groups lost all hope and descended into a self-destructive epicureanism.

A nuclear terrorist attack, or series of such attacks, would almost certainly alter the fabric of politics (Becker, 2003). The use of a nuclear weapon might trigger a backlash against current political or scientific establishments for creating and failing to prevent the threat. Such attacks might paralyse an open or free society by causing the government to adopt draconian methods (Stern, 1999, pp. 2–3), or massively restrict movement and trade until all nuclear material can be accounted for, an effort that would take years and which could never be totally complete. The concomitant loss of faith in governing authorities might eventually culminate in the fulfillment of John Herz's initial vision of the atomic age, resulting in the demise of the nation-state as we know it (1957).

While the aforementioned effects could occur in a number of countries, especially if multiple states were targeted by nuclear-armed terrorists, nuclear terrorism could also impact the overall conduct of international relations. One issue that may arise is whether the terrorists responsible, as non-state actors, would have the power to coerce or deter nation-states.[49] Nuclear detonation by a non-state group virtually anywhere would terrorize citizens in potential target countries around the globe, who would fear that the perpetrators had additional weapons at their disposal. The organization could exploit such fears in order to blackmail governments into political concessions – for example, it could demand the withdrawal of military forces or political support from states the terrorists opposed. The group might even achieve these results without a nuclear detonation, by providing proof that it had a nuclear weapon in its possession at a location unknown to its adversaries. The addition on the world stage of powerful non-state actors as the ostensible military equals of (at least some) states would herald the most significant change in international affairs since the advent of the Westphalian system. While one can only speculate about the nature of the resultant international system, one possibility is that 'superproliferation' would occur. In this case every state (and non-state) actor with the wherewithal pursues nuclear weapons, resulting in an extremely multipolar and unstable world order and greater possibility for violent conflict. On the other hand, a nuclear terrorist attack

[49] This is presuming, of course, that they could not be easily retaliated against and could credibly demonstrate the possession of a robust nuclear arsenal.

might finally create the international will to control or eliminate nuclear weaponry.

It is almost impossible to predict the direction, duration, or extent of the above-mentioned changes since they depend on a complex set of variables. However, it is certainly plausible that a global campaign of nuclear terrorism would have serious and harmful consequences, not only for those directly affected by the attack, but for humanity as a whole.

19.6 Risk assessment and risk reduction

19.6.1 The risk of global catastrophe

While any predictions under conditions of dynamic change are inherently complicated by the forces described earlier, the exploration of the motivations, capabilities and consequences associated with nuclear terrorism permit a preliminary estimate of the current and future overall risk. Annotated estimates of the risk posed by nuclear terrorism are given in Tables 19.1 to 19.3 under three different scenarios.

The tables represent a static analysis of the risk posed by nuclear terrorism. When one considers dynamics, it is more difficult to determine which effects will prevail. On the one hand, a successful scenario of attack as in Table 19.2 or Table 19.3 might have precedent-setting and learning effects, establishing proof-of-concept and spurring more terrorists to follow suit, which would increase the overall risk. Similarly, the use of nuclear weapons by some states (such as the United States or Israel against a Muslim country) might redouble the efforts of some terrorists to acquire and use these weapons or significantly increase the readiness of some state actors to provide assistance to terrorists. Moreover, one must consider the possibility of discontinuous adoption practices. This is rooted in the idea that certain technologies (perhaps including nuclear weapons) possess inherently 'disruptive' traits and that the transition to the use of such technologies need not be incremental, but could be rapid, wholesale and permanent once a tipping point is reached in the technology's maturization process.[50]

On the other hand, the acquisition of a nuclear weapons capability by states with which a terrorist group feels an ideological affinity might partially satiate their perceived need for nuclear weapons. At the same time, the advent of improved detection and/or remediation technologies might make terrorists less inclined to expend the effort of acquiring nuclear weapons. As a counterargument to the above assertions regarding the adoption of new weapons technologies, it is possible that following the initial use by terrorists

[50] For more information on the topic of disruptive technologies and their singular adoption behaviour, see Bower and Christensen (1995).

Table 19.1 Most Extreme Overall Scenario: Terrorists Precipitate a Full-scale Interstate Nuclear War (Either by Spoofing, Hacking, or Conducting a False Flag Operation)

Time Period	Likelihood of Motivational Threshold Being Met	Likelihood of Capability Requirements Being Met	Probability of Successful Attack (Taking into Account Motivations and Capabilities)	Consequences of Successful Attack	Global Risk
Present	*Moderate*	*Extremely low* Requires ability to 1. crack launch systems (of US/Russia) OR 2. spoof early warning systems (of US/Russia) OR 3. detonate their own warhead AND ability to disguise their involvement AND luck that states would be fooled	*Extremely low*	*Extremely adverse* (Global Scale) *Catastrophic* (Local/National Scale) – equal to consequences of large-scale interstate nuclear war	*Extremely low* (globally endurable risk)
Future (within 100 years)	*Moderate to high*	*Very low* (likelihood is similar to present with the exception of the possibility of more states with nuclear weapons and technological advances enabling easier self-production or acquisition of nuclear weapons. Likelihood also could be affected by the state of relations among nuclear weapon states)	*Very low to low*	*Extremely adverse* (Global Scale) *Catastrophic* (Local/National Scale) – equal to consequences of large-scale interstate nuclear war	*Extremely low* (globally endurable risk)

Table 19.2 Most Extreme 'Terrorist Weapon Only' Scenario: Multiple Attacks With Megaton-Scale Weapons

Time Period	Likelihood of Motivational Threshold Being Met	Likelihood of Capability Requirements Being Met	Probability of Successful Attack (Taking into Account Motivations and Capabilities)	Consequences of Successful Attack	Global Risk
Present	*Moderate*	*Very low* (not sophisticated enough to produce an advanced fission weapon or an H-bomb)	*Very low*	*Highly adverse* (Global Scale) – esp. economically; socially *Catastrophic* (Local Scale)	*Low* (globally endurable risk)
Future (within 100 years)	*High*	*Low-moderate* (greater number of states possessing H-bombs from which terrorists might acquire intact weapons)	*Moderate*	*Highly adverse* (Global Scale) – esp. economically; socially *Extremely adverse – catastrophic* (Local Scale) (taking into account various potential remediation measures for radiation poisoning)	*Moderate* (globally endurable risk)

Table 19.3 Most Likely Scenario: A Single to a Few <50 kilotons Detonations

Time Period	Likelihood of Motivational Threshold Being Met	Likelihood of Capability Requirements Being Met	Probability of Successful Attack (Taking into Account Motivations and Capabilities)	Consequences of Successful Attack	Global Risk
Present	*High*	*Low* (limited access to fissile material; inability to enrich)	*Moderate*	*Moderate to highly adverse* (Global Scale) – esp. economically; socially *Catastrophic* (Local Scale)	*Low* (globally endurable risk)
Future (within 100 years)	*High*	*Moderate-high* (technological advances enabling 'home' enrichment; greater number of states possessing nuclear weapons from which terrorists might acquire intact weapons; terrorists possessing greater technical skills)	*Moderate to high*	*Moderate to highly adverse* (Global Scale) – esp. economically; socially *Extremely adverse – catastrophic* (Local Scale) (taking into account various potential remediation measures for radiation poisoning)	*Moderate* (globally endurable risk)

of a nuclear weapon, the international community may act swiftly to stem the availability of nuclear materials and initiate a severe crackdown on any terrorist group suspected of interest in causing mass casualties.

Scholars have asserted that acquisition of nuclear weapons is the most difficult form of violence for a terrorist to achieve, even when compared with other unconventional weapons (Gurr and Cole, 2002, p. 56). Indeed, this is one of the main reasons for the relatively low overall risk of global devastation from nuclear terrorism in the near term. However, the risk is not completely negligible and, in the absence of intervention, could grow significantly in the future, eventually acquiring the potential to cause a global catastrophe.

19.6.2 Risk reduction

Among the most thoughtful analyses of how to accomplish this objective are a series of five volumes co-authored by a team of researchers at Harvard University and sponsored by the Nuclear Threat Initiative, a US foundation (Bunn, 2006; Bunn and Wier, 2005a, 2005b; Bunn et al., 2002, 2004). Since 2002, these annual monographs have tracked progress (and on occasion regression) in controlling nuclear warheads, material, and expertise, and have provided many creative ideas for reducing the supply side opportunities for would-be nuclear terrorists. A number of their insights, including proposals for accelerating the 'global cleanout' of HEU and building a nuclear security culture, inform the recommendations that follow in the next section below.

Several recent studies have called attention to the manner in which gaps in the international non-proliferation regime have impeded efforts to curtail nuclear terrorism (Ferguson and Potter, 2005; Perkovich et al., 2004). They note, for example, the state-centric orientation of the Treaty on the Non-proliferation of Nuclear Weapons (NPT) – the most widely subscribed to treaty in the world – and its failure to address the relatively new and very different dangers posed by non-state actors. They also call attention to the importance of developing new international mechanisms that provide a legal basis for implementing effective controls on the safeguarding and export of sensitive nuclear materials, technology, and technical know-how.[51] In addition, they point to the urgency of developing far more comprehensive and coordinated global responses to nuclear terrorism threats. Both the potential and limitations of current initiatives in this regard are illustrated by the G-8 Global Partnership, which in 2002 set the ambitious target of 20 billion dollars to be committed over a 10-year period for the purpose of preventing terrorists from acquiring weapons and materials of mass destruction (Einhorn and Flourney, 2003). Although this initiative, like most others in the realm of combating nuclear

[51] See, in particular, the growing literature on the promise and problems associated with implementing United Nations Security Council Resolution 1540, including Jones (2006).

terrorism, can point to some successes, it has been stronger on rhetoric than on sustained action. This phenomenon has led some observers to note that the 'most fundamental missing ingredient of the U.S. and global response to the nuclear terrorism threat to date is sustained high-level leadership' (Bunn and Wier, 2006, p. viii).

Although there is no consensus among analysts about how best to intervene to reduce the risks of nuclear terrorism, most experts share the view that a great deal can be done to reduce at least the supply side opportunities for would-be nuclear terrorists. At the core of this optimism is the recognition that the problem ultimately is an issue of physics. As one leading exponent of this philosophy puts its, the logic is simple, 'No fissile material, no nuclear explosion, no nuclear terrorism' (Allison, 2004, p. 140).

19.7 Recommendations

Major new initiatives to combat the nuclear proliferation and terrorism threats posed by non-state actors have been launched by national governments and international organizations, and considerable sums of financial and political capital have been committed to new and continuing programmes to enhance nuclear security. These initiatives include the adoption in April 2004 of UN Security Council Resolution 1540, the U.S. Department of Energy's May 2004 Global Threat Reduction Initiative, the expanded G-8 Global Partnership, the Proliferation Security Initiative, and the 2006 Global Initiative to Combat Nuclear Terrorism. Although these and other efforts are worthy of support, it is not obvious that they reflect a clear ordering of priorities or are being implemented with a sense of urgency. In order to correct this situation it is imperative to pursue a multi-track approach, the core elements of which should be to enhance the security of nuclear weapons and fissile material globally, consolidate nuclear weapons and fissile material stockpiles, reduce their size, and move towards their elimination, while at the same time working to reduce the number of terrorists seeking these weapons.

19.7.1 Immediate priorities

The following initiatives, most of which focus on the supply side of the nuclear terrorism problem, should be given priority: (1) minimize HEU use globally, (2) implement UN Security Council Resolution 1540, (3) promote adoption of stringent, global nuclear security standards, (4) secure vulnerable Russian TNW, and (5) accelerate international counterterrorism efforts to identify and interdict would-be nuclear terrorists.

1. *Minimize HEU Use Globally.* Significant quantities of fissile materials exist globally which are not needed, are not in use, and, in many instances, are

not subject to adequate safeguards. From the standpoint of nuclear terrorism, the risk is most pronounced with respect to stockpiles of HEU in dozens of countries. It is vital to secure, consolidate, reduce, and, when possible, eliminate these HEU stocks. The principle should be one in which fewer countries retain HEU, fewer facilities within countries posses HEU, and fewer locations within those facilities have HEU present. Important components of a policy guided by this principle include, rapid repatriation of all US- and Soviet/Russian-origin HEU (both fresh and irradiated), international legal prohibitions of exports of HEU-fuelled research and power reactors, and down-blending of existing stocks of HEU to low-enriched uranium (LEU). Use of spallation sources also can contribute to this HEU minimization process. Particular attention should be given to de-legitimizing the use of HEU in the civilian nuclear sector – a realistic objective given the relatively few commercial applications of HEU and the feasibility of substituting LEU for HEU in most, if not all, of these uses (Potter, 2006).

2. *Implement UN Security Council Resolution* 1540. One of the most important new tools to combat nuclear terrorism is UN Security Council Resolution 1540 (http://www.un.org/News/Press/docs/2004/sc8076.doc.htm). Adopted in April 2004, this legally binding measure on all UN members prohibits states from providing any form of support to non-state actors attempting to acquire or use nuclear, chemical, or biological weapons and their means of delivery. It also requires states to adopt and enforce 'appropriate effective measures' to account for and secure such items, including fissile material, and to establish and maintain effective national export and trans-shipment controls over these commodities. This United Nations mandate provides an unusual opportunity for those states most concerned about nuclear security to develop the elements of a global nuclear security standard, to assess the specific needs of individual states in meeting this standard, and to render necessary assistance (Bunn and Wier, 2005, p. 109). This urgent task is complicated by the fact that many states are not persuaded about the dangers of nuclear terrorism or doubt if the risk applies to them. Therefore, in order for the potential of 1540 to be realized, it is necessary for all states to recognize that a nuclear terror act anywhere has major ramifications everywhere. As discussed below, a major educational effort will be required to counter complacency and the perception that nuclear terrorism is someone else's problem. In the meantime, it is imperative that those states already convinced of the danger and with stringent nuclear security measures in place assist other countries to meet their 1540 obligations.

3. *Promote Adoption of Stringent, Global Security Standards*. Renewed efforts are required to establish binding international standards for the physical protection of fissile material. An important means to accomplish that objective is to ratify the recent amendment to the Convention on the Physical Protection of Nuclear Material to make it applicable to civilian nuclear material in

domestic storage, use, and transport. Ideally, the amendment would oblige parties to provide protection comparable to that recommended in INFCIRC 225/Rev 4 and to report to the IAEA on the adoption of measures to bring national obligations into conformity with the amendment. However, because ratifying the amended Convention is likely to require a long time, as many like-minded states as possible should agree immediately to meet a stringent material protection standard, which should apply to all civilian and military HEU.

4. *Secure and Reduce TNW.* Preventing non-state actors from gaining access to intact nuclear weapons is essential in combating nuclear terrorism. Priority should be given to safeguarding and reducing TNW, the category of nuclear arms most vulnerable to theft. Although it would be desirable to initiate negotiations on a legally binding and verifiable treaty to secure and reduce such arms, this approach does not appear to have much prospect of success, at least in the foreseeable future. As a consequence, one should concentrate on encouraging Russia to implement its pledges under the 1991–1992 Presidential Nuclear Initiatives, including the removal to central storage of all but one category of TNW. Ideally, all TNW should be stored at exceptionally secure facilities far from populated regions. In parallel, the United States should declare its intention to return to US territory the small number of air-launched TNW currently deployed in Europe. Although probably less vulnerable to terrorist seizure than TNW forward deployed in Russia, there no longer is a military justification for their presence in Europe. The US action, while valuable in its own right, might be linked to Russian agreement to move its tactical nuclear arms to more secure locations.

5. *Accelerate International Counterterrorism Efforts to Pre-emptively Identify and Interdict Nuclear Terrorists.* On the demand side, one can identify several recommendations for reducing the threat of nuclear terrorism that are generally less specific than those on the supply side. The most effective measures in the near term involve improved law enforcement and intelligence. As only a small proportion of non-state actors is likely to possess both the motivation and capability necessary for high consequence nuclear terrorism, it should be at least possible to identify potential nuclear perpetrators in advance and concentrate counterterrorism efforts – including surveillance and prosecution – against these groups and individuals.[52] However, counterterrorism agencies have traditionally proven to be less proficient at terrorist threat preemption than response after an attack. Given the likely horrific consequences of nuclear terrorism, it is crucial to invest

[52] As Ehud Sprinzak has argued, 'the vast majority of terrorist organizations can be identified well in advance ... and the number of potential [WMD] suspects is significantly less than doomsayers seem to believe. Ample early warning signs should make effective interdiction of potential superterrorists much easier than today's prevailing rhetoric suggests' Sprinzak (2000, pp. 5–6).

more resources wisely in apprehending terrorists known to harbour nuclear ambitions and to be more vigilant and savvy in anticipating the emergence of new and evolving non-state actors who may be predisposed to seek a nuclear terrorist option. Successful efforts in this regard will require much greater international collaboration in intelligence sharing, law enforcement, and prosecution – developments more likely to occur if global perceptions of nuclear terrorism threats converge.

19.7.2 Long-term priorities

Implementation of the aforementioned short-term measures should reduce significantly the risks of nuclear terrorism. However, the threat will remain unless certain underlying factors are addressed. On the demand side, the most basic long-term strategy is to decrease the absolute number of terrorists (and hence the number of would-be nuclear terrorists). While the root causes of terrorism are beyond the scope of this essay, it should be noted that terrorist grievances stem from a complex and poorly understood interplay of social, political and psychological factors, some of which can be assuaged by policy. The ideological make-up of potential nuclear terrorists, however, reduces their susceptibility to such measures as political concessions or improved socio-economic conditions, which may take decades to implement in any case.

Another way to reduce motivations for nuclear terrorism is to remove at least part of the subjective benefit that terrorists might derive from conducting acts of nuclear violence. Useful steps include strengthening normative taboos against the use of nuclear weapons, vilifying terrorists who have attempted to obtain nuclear weapons, and increasing public education programmes in order to psychologically immunize the public against some irrational fears related to radiation. Implementation of these measures might help to dissuade some terrorists that the strategic benefits of nuclear violence outweigh the costs.

Maintaining high standards of nuclear forensics and attribution, coupled with strict warnings to states that they will be held responsible for any terrorism involving fissile material of their national origin also may be useful. In addition to providing states with greater incentive to increase the protection of fissile material under their control, these steps could provide some measure of deterrence against state complicity in terrorist acts.

On the supply side, vigorous implementation of the previously noted priority measures should significantly reduce the risk of nuclear terrorism. However, these dangers will not be completely eliminated as long as countries attach value to and retain nuclear weapons and weapons-usable material. Although it is unrealistic to assume that prevailing national views regarding nuclear weapons will change anytime soon, it nevertheless is important to initiate steps today to change mindsets and forge norms consistent with formal national obligations to nuclear disarmament and non-proliferation. An important but underutilized tool for this purpose is education.

Although few national governments or international organizations have invested significantly in this sphere, there is growing recognition among states of the need to rectify this situation. This positive development is reflected in the broad support for recommendations of a UN study on disarmament and non-proliferation education and in related initiatives within the NPT review process (Potter, 2001; Toki and Potter, 2005). Among specific steps states should take to utilize education and training to combat complacency and reduce the proliferation risks posed by non-state actors are:

- Develop educational materials that illustrate the urgency of the proliferation threats posed by non-state actors and their potential impact on all states.
- Cooperate with regional and international organizations to provide training courses on best practices related to nuclear materials and security and non-proliferation exports controls for government officials and law enforcement officers.
- Adopt national non-proliferation education legislation to support graduate training in the field – the best guarantee that states and international organizations will have an adequate pool of knowledgeable nuclear proliferation and counter-terrorism intelligence analysts.

Perhaps the greatest promise of non-proliferation education in the long term is the potential growth of a global network of non-proliferation experts and practitioners who increasingly share common norms and promote their countries' adherence to non-proliferation and anti-terrorism treaties and agreements. This desirable and necessary state of affairs is most likely to be realized if many more bright young individuals enter the nuclear non-proliferation field and, by the strength of their idealism and energy, prod national governments to abandon old ways of doing things and adjust their non-proliferation strategies and tactics to take account of new realities involving non-state actors and nuclear weapons.

19.8 Conclusion

It is difficult to find many reasons for optimism regarding the threat of high consequence nuclear terrorism. It is a growing danger and one that could result in enormously devastating and enduring local, regional, national, and even global consequences. One, therefore, should not take great solace in the conclusion that nuclear terrorism is unlikely to pose an existential, end-of-the world threat. It can still cause sufficient perturbation to severely disrupt economic and cultural life and adversely affect the nature of human civilization.

Given the rising potential for terrorists to inflict nuclear violence, what then accounts for the failure on the part of the most powerful nations on earth to take corrective action commensurate with the threat? Is it a lack of political leadership, a failure of imagination, faulty conceptualization, domestic politics, bureaucratic inertia, competing national security objectives, wishful thinking, the intractable nature of the problem, or simply incompetence?

All of these factors contribute to the current predicament, but some are more amenable to correction than others. Perhaps the most fundamental shortcoming, and one that can be remedied, is the failure by government and academic analysts alike to distinguish clearly between the proliferation risks posed by state and non-state actors, and to devise and employ tools that are appropriate for combating these very different threats. The challenge is an urgent but manageable one, affording the world a reasonable second chance.

Acknowledgements

The authors are grateful to Michael Miller for his excellent research assistance. His contribution to the section on 'The Consequences of Nuclear Terrorism' was especially valuable. The authors also wish to thank Erika Hunsicker for her editorial assistance.

Suggestions for further reading

Ferguson, C.D. and Potter, W.C. (2005). *The Four Faces of Nuclear Terrorism* (New York: Routledge).

Levi, M. (2007). *On Nuclear Terrorism.* (Cambridge, MA. Harvard University Press).

Zimmerman, P.D. and Lewis, J.G. (2006). The bomb in the backyard. *Foreign Policy* (November/December 2006), 33–40.

References

ABT Associates. (2003). The Economic Impact of Nuclear Terrorist Attacks on Freight Transport Systems in an Age of Seaport Vulnerability. Executive summary. Accessed 15 September 2006 http://www.abtassociates.com/reports/ES-Economic_Impact_of_Nuclear_Terrorist_Attacks.pdf

Ackerman, G. and Bale, J.M. (2004). How Serious is the 'WMD Terrorism' Threat?: Terrorist Motivations and Capabilities for Using Chemical, Biological, Radiological, and Nuclear (CBRN) Weapons. Report prepared for Los Alamos National Laboratory.

Albright, D. and Higgens, H. (March/April 2003). A bomb for the Ummah. *Bulletin of the Atomic Scientists*, 49–55.

Alexander, D.A. and Klein, S. (2006). The challenge of preparation for a chemical, biological, radiological or nuclear terrorist attack. *Journal of Postgraduate Medicine*, **52**, 126–131.

Allison, G. (1996). *Avoiding Nuclear Anarchy: Containing the Threat of Loose Russian Nuclear Weapons and Fissile Material* (Cambridge, MA: MIT Press).

Allison, G. (2004). *Nuclear Terrorism: The Ultimate Preventable Catastrophe* (New York: Henry Holt).

Alvarez, L.W. (1988). *Adventures of a Physicist* (New York: Basic Books).

Arbman, G., Calogero, F., Cotta-Ramusino, P., van Dessen, L., Martellini, M., Maerli, M.B., Nikitin, A., Prawitz, J., and Wredberg, L. (2004). Eliminating Stockpiles of Highly-Enriched Uranium. Report submitted to the Swedish Ministry for Foreign Affairs, SKI Report 2004.

Asal, V. and Ackerman, G. (2006). Size Matters: Terrorist Organizational Factors and the Pursuit and Use of CBRN Terrorism. Submitted for publication.

Bale, J. (2005). Islamism. In Pilch, R.F. and Zilinskas, R. (eds.), *Encyclopedia of Bioterrorism Defense* (New York: Wiley).

Bandura, A. (1998). Mechanisms of moral disengagement. In Reich, W. (ed.), *Origins of Terrorism: Psychologies, Ideologies, Theologies, States of Mind*, pp. 161–191 (Washington, DC: Woodrow Wilson Center).

Barnaby, F. (1996). Issues Surrounding Crude Nuclear Explosives in Crude Nuclear Weapons: Proliferation and the Terrorist Threat, IPPNW Global Health Watch Report Number 1.

Becker, S.M. (2003). Psychosocial Issues in Radiological Terrorism and Response: NCRP 138 and After. Presented at the International Workshop on Radiological Sciences and Applications: Issues and Challenges of Weapons of Mass Destruction Proliferation. Albuquerque, New Mexico. 21 April 2003.

Bird, K. and Sherwin, M.J. (2005). *American Prometheus: The Triumph and Tragedy of J. Robert Oppenheimer* (New York: Alfred A Knopf).

Bostrom, N. (2002). Existential risks: analyzing human extinction scenarios and related hazards. *Journal of Evolution and Technology*, **9**. http://www.nickbostrom.com/existential/risks.html

Boureston, J. and Ferguson, C.D. (March/April 2005). Laser enrichment: separation anxiety. *Bulletin of the Atomic Scientists*, 14–18.

Boutwell, J., Calegero, F., and Harris, J. (2002). Nuclear Terrorism: The Danger of Highly Enriched Uranium (HEU). Pugwash Issue Brief.

Bower, J.L. and Christensen, C.M. (1995). Disruptive technologies: catching the wave. *Harvard Business Review*, **73**, 43–53.

Bunn, M. and Wier, A. (2005a). *Securing the Bomb 2005: The New Global Imperative.* (Cambridge, MA: Project on Managing the Atom, Harvard University).

Bunn, M. and Wier, A. (April 2005b). The seven myths of nuclear terrorism. *Current History*.

Bunn, M. (2006). The Demand for Black Market Fissile Material. NTI Web site, accessed at [http://www.nti.org/e_research/cnwm/threat/demand.asp?print=true] on August 20, 2006.

Bunn, M., Holdren, J.P., and Wier, A. (2002). *Securing Nuclear Weapons and Materials: Seven Steps for Immediate Action* (Cambridge, MA: Project on Managing the Atom, Harvard University).

Bunn, M., Holdren, J.P., and Wier, A. (2003). *Controlling Nuclear Warheads and Materials: A Report Card and Action Plan* (Cambridge, MA: Project on Managing the Atom, Harvard University).

Bunn, M., Holdren, J.P., and Wier, A. (2004). *Securing the Bomb: An Agenda for Action* (Cambridge, MA: Project on Managing the Atom, Harvard University).

Calhoun, L.G. and Tedeschi, R.G. (1998). Posttraumatic growth: future directions. In Tedeschi, R.G., Park, C.L., and Calhoun, L.G. (eds.), *Posttraumatic Growth: Positive Changes in the Aftermath of Crisis*, pp. 215–238 (Mahwah, NJ: Lawrence Earlbaum Associates).

Cameron, G. (1999). *Nuclear Terrorism: A Threat Assessment for the 21st Century* (New York: St. Martin's Press, Inc.).

Cameron, G. (2000). WMD terrorism in the United States: the threat and possible countermeasures. *The Nonproliferation Review*, **7**(1), 169–170.

Campbell K.M. et al. (1991). *Soviet Nuclear Fission: Control of the Nuclear Arsenal in a Disintegrating Soviet Union* (Cambridge, MA: MIT Press).

Campbell, J.K. (2000). On not understanding the problem. In Roberts, B. (ed.), *Hype or Reality?: The 'New Terrorism' and Mass Casualty Attacks* (Alexandria, VA: Chemical and Biological Arms Control Institute).

Carus, W.S. (2000). R.I.S.E. (1972). In Tucker 2000: 55–70.

Civiak, R.L. (May 2002). Closing the Gaps: Securing High Enriched Uranium in the Former Soviet Union and Eastern Europe. Report for the Federation of American Scientists.

Clutterbuck, R. (1993). Trends in terrorist weaponry. *Terrorism Political Violence*, **5**, 130.

Commission on the Intelligence Capabilities of the United States Regarding Weapons of Mass Destruction. (2005). *Report to the President* (Washington, DC: WMD Commission).

Connor, T. (22 March 2004). Al Qaeda: we bought nuke cases. *New York Daily News*.

Cordesman, A.H. (2001). Defending America: Asymmetric and Terrorist Attacks with Radiological and Nuclear Weapons, Center for Strategic and International Studies.

Daly, S., Parachini, J., and Rosenau, W. (2005). *Aum Shinrikyo, Al Qaeda, and the Kinshasa Reactor: Implications of Three Studies for Combatting Nuclear Terrorism*. Document Briefing (Santa Monica, CA: RAND).

Department of Homeland Security (2003). 'Radiological Countermeasures.' Prepared by the Department of Homeland Security, Working Group on Radiological Dispersion Device Preparedness, May 1, 2003, www.va.gov/emshq/docs/Radiologic Medical Countermeasures 051403.pdf.

Dolnik, A. (2004). All God's poisons: re-evaluating the threat of religious terrorism with respect to non-conventional weapons. In Howard, R.D. and Sawyer, R.L. (eds.), *Terrorism and Counterterrorism: Understanding the New Security Environment* (Guilford, CT: McGraw-Hill).

Eden, L. (2004). *Whole World on Fire* (Ithaca, NY: Cornell University Press).
Einhorn, R. and Flourney, M. (2003). Protecting Against the Spread of Nuclear, Biological, and Chemical Weapons: An Action Agenda for the Global Partnership. Center for Strategic and International Studies.
Falkenrath, R.A. (1998). Confronting nuclear, biological and chemical terrorism. *Survival*, **40**, 3.
Falkenrath, R.A., Newman, R.D., and Thayer, B.A. (1998). *America's Achilles' Heel: Nuclear, Biological, and Chemical Terrorism and Covert Attack* (Cambridge, MA: MIT Press).
Ferguson, C.D. and Potter, W.C. (2005). *The Four Faces of Nuclear Terrorism* (New York: Routledge).
Forden, G. (3 May 2001). Reducing a common danger: improving Russia's early-warning system. *Policy Analysis*, 1–20.
Foxell, J.W. (1999). The debate on the potential for mass-casualty terrorism: the challenge to US security. *Terrorism and Political Violence*, **11**, 1.
Garwin, R.L. and Charpak G. (2001). *Megawatts and Megatons: A Turning Point in the Nuclear Age?* (New York: Alfred A. Knopf).
General Accounting Office. (2004). DOE Needs to take Action to Further Reduce the Use of Weapons-Usable Uranium in Civilian Research Reactors. GAO-04-807.
Gilinsky, V. (2004). Israel's bomb. Letter to the Editor, *The New York Review of Books*, 51/8. http://www.nybooks.com/articles/17104
Glasser, S. and Khan, K. (24 November 2001). Pakistan continues probe of nuclear scientists. *Washington Post*, p. A13.
Glasstone, S. and Dolan, P.J. (eds.) (1977). The Effects of Nuclear Weapons. U.S. Department Defense and Department of Energy and their own simulations.
Government Accountability Office. (2002). Impact of Terrorist Attacks on the World Trade Center. Report GAO-02-700R.
Gressang, D.S., IV (2001). Audience and message: assessing terrorist WMD potential. *Terrorism and Political Violence*, **13**(3), 83–106.
Gurr, N. and Cole, B. (2002). *The New Face of Terrorism: Threats from Weapons of Mass Destruction* (London: I. B. Tauris).
Helfand, I., Forrow, L., and Tiwari, J. (2002). Nuclear terrorism. *British Medicine Journal*, **324**, 357.
Herz, J. (1957). The rise and demise of the territorial state. *World Politics*, **9**, 473–493.
Hoffman, B. (1993a). *'Holy Terror': The Implications of Terrorism Motivated by a Religious Imperative* (Santa Monica: RAND).
Hoffman, B. (1993b). Terrorist targeting: tactics, trends, and potentialities. *Terrorism and Political Violence*, **5**, 12–29.
Hoffman, B. (1997). Terrorism and WMD: some preliminary hypotheses. *The Nonproliferation Review*, **4**(3), 45–50.
Hoffman, B. (1998). *Inside Terrorism* (New York: Columbia University).
Ingram, T.H. (December 1972). Nuclear hijacking: now within the grasp of any bright lunatic. *Washington Monthly*, pp. 20–28.

Jackson, B.A. (2001). Technology acquisition by terrorist groups: threat assessment informed by lessons from private sector technology adoption. *Studies in Conflict and Terrorism*, **24**, 3.

Jane's Intelligence Digest. (3 July 2003). Al-Qaeda and the Bomb.

Jenkins B. (1977). International terrorism: a new mode of conflict. In Carlton, D. and Schaerf, C. (eds.), *International Terrorism and World Security* (London: Croom Helm.

Jenkins, B. (1986). Defense against terrorism. *Political Science Quarterly*, **101**, 777.

Jones, F.D. (1995). Neuropsychiatric casualties of nuclear, biological, and chemical warfare. In *Textbook of Military Medicine: War Psychiatry*. Department of the Army, Office of The Surgeon General, Borden Institute.

Jones, S. (2006). Resolution 1540: universalizing export control standards? *Arms Control Today*. Available at http://www.armscontrol.org/act/2006_05/1540.asp

Kaplan, D.E. (2000). Aum Shinrikyo (1995). In Tucker, J. (ed.), *Toxic Terror: Assessing Terrorist Use of Chemical and Biological Weapons*, pp. 207–226 (Cambridge, MA: MIT Press).

Kristof, N.D. (11 August 2004). An American Hiroshima. *New York Times*.

Lapp, R.E. (4 February 1973). The ultimate blackmail. *The New York Times Magazine*.

Leader, S. (June 1999). Osama Bin Laden and the terrorist search for WMD. *Jane's Intelligence Review*.

Levanthal, P. and Alexander, Y. (1987). *Preventing Nuclear Terrorism* (Lexington, MA: Lexington Books).

Lifton, R.J. (1999). *Destroying the World to Save It: Aum Shinrikyo, Apocalyptic Violence, and the New Global Terrorism* (New York: Metropolitan Books).

Lugar, R.G. (2005). *The Lugar Survey on Proliferation Threats and Responses* (Washington, DC: U.S. Senate).

Macdonald, A. [pseudonym for Pierce] (1999). *The Turner Diaries: A Novel*. Hillsboro, W.V.: National Vanguard; originally published 1980.

Maerli, M.B. (Summer, 2000). Relearning the ABCs: terrorists and 'weapons of mass destruction'. *The Nonproliferation Review*.

Maerli, M.B. (2004). Crude Nukes on the Loose? Preventing Nuclear Terrorism by Means of Optimum Nuclear Husbandry, Transparency, and Non-Intrusive Fissile Material Verification. Dissertation, University of Oslo.

Mark, J.C., Taylor, T., Eyster, E., Maraman, W., and Wechsler, J. (1987). Can Terrorists Build Nuclear Weapons? in Leventhal and Alexander.

Marlo, F.H. (Autumn 1999). WMD terrorism and US intelligence collection. *Terrorism and Political Violence*, **11**, 3.

McCormick, G.H. (2003). Terrorist decision making. *Annual Reviews in Political Science*, **6**, 479–480.

McLoud, K. and Osborne, M. (2001). *WMD Terrorism and Usama bin Laden*. (Monterey, CA: Center for Nonproliferation Studies) Available at [http://cns.miis.edu/pubs/reports/binladen.htm.]

McPhee, J. (1974). *The Curve of Binding Energy* (New York: Farrar, Straus, and Giroux).

Mir, H. (10 November 2001). Osama Claims He Has Nukes: If US Uses N-Arms It Will Get Same Response. Dawn Internet Edition. Karachi, Pakistan.

National Research Council. (2002). *Committee on Science and Technology for Countering Terrorism. Making the Nation Safer: The Role of Science and Technology in Countering Terrorism* (Washington, DC: National Academy Press).

Negroponte, J. (2005). Annual Threat Assessment of the Director of National Intelligence. Statement to the Senate Armed Services Committee. Accessed at [http://armed-services.senate.gov/statemnt/2006/February/Negroponte%2002-28-06.pdf]

North, C. and Pfefferbaum, B. (2002). Research on the mental health effects of terrorism. *Journal of the American Medical Association*, **288** 633–636.

Pangi, R. (2002). After the attack: the psychological consequences of terrorism. *Perspectives on Preparedness*, **7**, 1–20.

Perkovich, G., Cirincione, J., Gottemoelle, R., Wolfsthal, J., and Mathews, J. (2004). *Universal Compliance* (Washington, DC: Carnegie Endowment).

Perry, W.J., Carter, A., and May, M. (12 June 2007). After the bomb. *New York Times*.

Petersen, J.L. (2000). *Out of the Blue* (Lanham, MD: Madison Books).

Post, J. (1987). Prospects for nuclear terrorism: psychological motivations and constraints. In Levanthal and Alexander.

Post, J. (2000). Psychological and motivational factors in terrorist decision-making: implications for CBW terrorism. In Tucker, J. (ed.), *Toxic Terror: Assessing Terrorist Use of Chemical and Biological Weapons* (Cambridge, MA: MIT Press).

Potter, W.C. (2001). A New Agenda for Disarmament and Non-Proliferation Education. Disarmament Forum, No. 3. pp. 5–12.

Potter, W.C. (2006). A Practical Approach to Combat Nuclear Terrorism: Phase Out HEU in the Civilian Nuclear Sector. Paper presented at the International Conference on the G8 Global Security Agenda: Challenges and Interests. Toward the St. Petersburg Summit. Moscow, April 20–22, 2006.

Potter, W.C. and Sokova, E. (Summer, 2002). Illicit nuclear trafficking in the NIS: what's new? what's true? *Nonproliferation Review*, 112–120.

Potter, W.C., Sokov, N., Mueller, H., and Schaper, A. (2000). *Tactical Nuclear Weapons: Options for Control* (Geneva: United Nations Institute for Disarmament Research).

Pry, P.V. (1999). *War Scare: Russia and America on the Nuclear Brink* (Westport, CT: Praeger).

Rapoport, D.C. (1999). Terrorism and weapons of the apocalypse. *Nonproliferation Review*, **6**(3), 49–67.

Rhodes, R. (1986). *The Making of the Atomic Bomb* (New York: Simon & Schuster).

Rosenbaum, D.M. (Winter 1977). Nuclear terror. *International Security*, pp. 140–161.

Sagan, S.D. (1993). *The Limits of Safety: Organizations, Accidents, and Nuclear Weapons* (Princeton, NJ: Princeton University Press).

Schelling, T. (1982). Thinking about nuclear terrorism. *International Security*, **6**(4), 61–77.

Scheuer, M. (2002). *Through Our Enemies' Eyes: Osama bin Laden, Radical Islam, and the Future of America* (Washington, DC: Potomac Books, Inc.).

Schlenger, W.E. (2002). Psychological reactions to terrorist attacks: findings from the national study of Americans' reactions to September 11. *Journal of the American Medical Association*, **288** 581–588.

Schollmeyer, J. (May/June 2005). Lights, camera, Armageddon. *Bulletin of the Atomic Scientists*, pp. 42–50.

Sokov, N. (2002). Suitcase Nukes: A Reassessment. Research Story of the Week, Center for Nonproliferation Studies, Monterey Institute of International Studies. Available at http://www.cns.miis.edu/pubs/week/020923.htm.

Sokov, N. (2004). 'Tactical Nuclear Weapons'. Available at the Nuclear Threat Initiative website http://www.nti.org/e_research/e3_10b.html

Spector, L.S. (1987). *Going Nuclear* (Cambridge, MA: Ballinger Publishing Co).

Sprinzak, E. (2000). On not overstating the problem In Roberts, B. (ed.), *Hype or Reality?: The 'New Terrorism' and Mass Casualty Attacks* (Alexandria, VA: Chemical and Biological Arms Control Institute).

Sprinzak, E. and Zertal, I. (2000). Avenging Israel's Blood (1946). In Tucker, J. (ed.), *Toxic Terror: Assessing Terrorist Use of Chemical and Biological Weapons*, pp. 17–42 (Cambridge, MA: MIT Press).

Stern, J.E. (1999). *The Ultimate Terrorists* (Cambridge, MA: Harvard University Press).

Stern, J.E. (2000). The Covenant, the Sword, and the Arm of the Lord (1985). In Tucker, J. (ed.), *Toxic Terror: Assessing Terrorist Use of Chemical and Biological Weapons*, pp. 139–157 (Cambridge, MA: MIT Press).

Sterngold, J. (18 April 2004). Assessing the risk on nuclear terrorism; experts differ on likelihood of 'dirty bomb' attack. *San Francisco Chronicle.*

Stober, D. (March/April 2003). No experience necessary. *Bulletin of the Atomic Scientists*, pp. 57–63.

Taleb, N.N. (2004). The Black Swan: Why Don't We Learn that We Don't Learn? in United States Department of Defense Highlands Forum papers.

The Frontier Post. (20 November 2001). Al Qaeda Network May Have Transported Nuclear, Biological, and Chemical Weapons to the United States, *The Frontier Post*, Peshawar.

The Middle East Media Research Institute. (12 June 2002). 'Why we fight America': Al-Qa'ida Spokesman Explains September 11 and Declares Intentions to Kill 4 Million Americans with Weapons of Mass Destruction. 2002. The Middle East Media Research Institute. Special Dispatch Series No. 388. Accessed at http://memri.org/bin/articles.cgi?Page=archives&Area=sd&ID=SP38802 on 20 August 2006.

Toki, M. and Potter, W. C. (March 2005). How we think about peace and security: The ABCs of initiatives for disarmament & non-proliferation education. *IAEA Bulletin*, 56–58.

U.S. Office of Technology Assessment. (1977). *Nuclear Proliferation and Safeguards. Vol.* 1 (New York: Praeger).

U.S. Office of Technology Assessment. (1979). *The Effects of Nuclear War.*

Union of Concerned Scientists. (2003). Scientists' Letter on Exporting Nuclear Material to W. J. 'Billy' Tauzin, 25 September 2003. Available at

von Hippel, F. (2001). Recommendations for Preventing Nuclear Terrorism. Federation of American Scientists Public Interest Report.

Wald, M.L. (23 January 2000). Suicidal nuclear threat is seen at weapons plants. *New York Times*, A9.

Walker, S. (2001). Regulating against nuclear terrorism: the domestic safeguards issue, 1970–1979. *Technology and Culture*, **42**, 107–132.

War Games. (1983). Film, Metro-Goldwyn-Mayer. Directed by John Badham.

Willrich, M. and Taylor, T.B. (1974). *Nuclear Theft: Risks and Safeguards* (Cambridge, MA: Ballinger Publishing Co.).
Zorza, V. (9 September 1972). World could be held to ransom. *The Washington Post*, p. A19.
Zorza, V. (2 May 1974). The basement nuclear bomb. *The Washington Post*.

·20·

Biotechnology and biosecurity

Ali Nouri and Christopher F. Chyba

20.1 Introduction

Biotechnological power is increasing exponentially, reminiscent of the increase in computing power since the invention of electronic computers. The co-founder of Intel Corporation, Gordon Moore, pointed out in 1965 that the number of transistors per computer chip – a measure of how much computation can be done in a given volume – has doubled roughly every 18 months (Moore, 1965). This exponential increase in computing power, now called 'Moore's Law', has continued to hold in the decades since then (Lundstrom, 2003) and is the reason that individuals now have more computing power available in their personal computers than that was available only to the most advanced nations only decades ago. Although biotechnology's exponential lift off began decades after that of computing, its rate of increase, as measured, for example, by the time needed to synthesize a given DNA sequence, is as fast or faster than that of Moore's Law (Carlson, 2003). Just as Moore's Law led to a world of personal computing and home appliance microprocessors, so biotechnological innovation is moving us into a world where the synthesis of DNA, as well as other biological manipulations, will be increasingly available to small groups of technically competent and even individual users.

There is already a list of well-known experiments – and many others that have received less public attention – that illustrates the potential dangers intrinsic to modern biological research and development. We review several examples of these in some detail below, including: genetic manipulations that have rendered certain viruses far more deadly to their animal hosts (Jackson et al., 2001); the synthesis of polio virus from readily purchased chemical supplies (Cello et al., 2002) – so that even if the World Health Organization (WHO) succeeds in its important task for eradicating polio worldwide, the virus can be reconstituted in laboratories around the world; the reduction in the time needed to synthesize a virus genome comparable in size to the polio virus from years to weeks; the laboratory re-synthesis of the 1918 human influenza virus that killed tens of millions of people worldwide (Tumpey et al., 2005); the discovery of 'RNA interference', which allows researchers to turn off certain genes in humans or other organisms (Sen et al., 2006); and the new field of 'synthetic biology', whose goal is to allow practitioners to fabricate small 'biological devices' and ultimately new types of microbes (Fu, 2006).

The increase in biological power illustrated by these experiments, and the global spread of their underlying technologies, is predicted to lead to breathtaking advances in medicine, food security, and other areas crucial to human health and economic development. For example, the manipulation of biological systems is a powerful tool that allows controlled analysis of the function – and therefore vulnerabilities of and potential defences against – disease organisms. However, this power also brings with it the potential for misuse (NRC, 2006). It remains unclear how civilization can ensure that it reaps the benefits of biotechnology while protecting itself from the worst misuse. Because of the rapid spread of technology this problem is an intrinsically international one. However, there are currently no good models from Cold War arms control or non-proliferation diplomacy that are suited to regulating this uniquely powerful and accessible technology (Chyba and Greninger, 2004). There are at least two severe challenges to any regulatory scheme (Chyba, 2006). The first is the mismatch between the rapid pace of biotechnological advances and the comparative sluggishness of multilateral negotiation and regime building. The second is the questionable utility of large-scale monitoring and inspections strategies to an increasingly widespread, small-scale technology.

However, this is not a counsel for despair. What is needed is a comprehensive strategy for the pursuit of biological security – which we assume here to be the protection of people, animals, agriculture and the environment against natural or intentional outbreaks of disease. Such a strategy is not yet in place, either nationally or globally, but its contours are clear. Importantly, this strategy must be attentive to different categories of risk, and pay attention to how responses within one category strengthen or weaken the response to another. These categories of risk include: naturally occurring diseases; illicit state biological weapons programmes; non-state actors and bio-hackers; and laboratory accidents or other inadvertent release of disease agents.

Just this listing alone emphasizes several important facts. The first is that while about 14 million people die annually from infectious diseases (WHO, 2004) (mostly in the developing world), only five people died in the 2001 anthrax attacks in the United States (Jernigan et al., 2002), and there have been very few other modern acts of biological terrorism. Any approach to the dual-use challenge of biotechnology, which substantially curtails the utility of biotechnology to treat and counter disease, runs the risk of sacrificing large numbers of lives to head off hypothetical risks. Yet it is already clear that humans can manipulate pathogens in ways that go beyond what evolution has so far wrought, so the hypothetical must nevertheless be taken seriously. A proper balance is needed, and an African meeting on these issues in October 2005 suggested one way to strike it. The Kampala Compact declared that while 'the potential devastation caused by biological weapons would be catastrophic for Africa', it is 'illegitimate' to address biological weapons threats without also addressing key public health issues such as infectious disease. The developed and developing world must find common ground.

A second important observation regarding biological terrorism is that there have, so far, been very few actual attacks by non-state groups. It is clearly important to understand why this has been the case, and to probe the extent to which it has been due to capabilities or motivations – and how whatever inhibitions may have been acting can be strengthened. Sceptical treatments of the biological terrorism threat, and of the dangers of apocalyptic dramatization more generally, can place important focus on these issues – though the examples of dual-use research already mentioned, plus recent US National Academy of Sciences studies and statements by the UN Secretary-General make it clear that the problem is real, not hype.[1]

While the focus of this chapter will be on biotechnological capabilities, and how those capabilities may be responsibly controlled, it should be remembered that a capabilities-based threat assessment only provides part of the comprehensive picture that is required. Indeed, it is striking to compare the focus on capabilities in many technology oriented threat assessments with the tenor of one of the most influential threat assessments in US history, George Kennan's 'X' article in *Foreign Affairs* in 1947. In this piece, Kennan crystallized the US policy of containment of the Soviet Union that prevailed for decades of the Cold War. On reading today, one is struck by how little of the X article addressed Soviet capabilities. Rather, nearly all of it concerned Soviet intentions and motives, with information based on Kennan's experience in Soviet society, his fluency in Russian, and his knowledge of Russian history and culture. To the extent that the biological security threat emanates from terrorist groups or irresponsible nations, a similar sophistication with respect to motives and behaviour must be brought to bear (see also Chapters 4 and 19 in this volume).

In this chapter, we first provide a survey of biological weapons in history and efforts to control their use by states via multilateral treaties. We then describe the biotechnological challenge in more detail. Finally, we survey a variety of approaches that have been considered to address these risks. As we will see, there are no easy answers.

[1] One important sceptical discussion of the bioterrorism threat is Milton Leitenberg, *Assessing the Biological Weapons and Bioterrorism Threat* (U.S. Army War College, 2005).

20.2 Biological weapons and risks

The evolutionary history of life on Earth has, in some instances, led to biological weapons in the form of harmful toxins (and their underlying genes) that are carried by simple organisms like bacteria and fungi, as well more complex ones like spiders and snakes. Humans learned that such natural phenomena could be used to their advantage; long before the identification of microbes and the chemical characterization of toxins, humans were engaged in rudimentary acts of biological warfare that included unleashing venomous snakes on adversaries, poisoning water wells with diseased animal flesh, and even catapulting plague-infested human bodies into enemy fortifications (Wheelis, 2002).

As scientists learned to optimize growth conditions for microbes, stockpiling and storing large quantities of infectious living organisms became feasible and dramatically increased the destructive potential of germ warfare. These advances and a better understanding of disease-causing microbes, together with the horror and carnage that was caused by non-conventional weapons during WWI, elevated fears of germ warfare and provided the impetus for the 1925 Geneva protocol, an international treaty that outlawed the use of chemical and bacteriological (biological) weapons in war. With the notable exception of Japan (Unit 731 Criminal Evidence Museum, 2005), warring states refrained from using biological weapons throughout WWII – but some continued to engage in offensive weapons programmes, which were not prohibited until the Bacteriological (Biological) and Toxins Weapons Convention (BWC) was opened for signature in 1972.

The BWC is the world's first international disarmament treaty outlawing one entire class of weapons – namely, the development, production, and stockpiling of biological agents and toxins for anything other than peaceful (i.e., prophylactic) purposes. Despite 155 ratifications out of the 171 states that are signatory to the convention, the BWC suffers from the lack of a monitoring and inspection mechanism to assess whether a country is engaged in illegal activities. This institutional weakness permitted sophisticated offensive programmes to continue long after the convention was signed, such as one in the former Soviet Union. Efforts to develop monitoring and verification protocols within the framework of the BWC began in 1991, but were suddenly terminated ten years later when the United States withdrew its support in July 2001, arguing that the additional measures would not help to verify compliance, would harm export control regimes, and place US national security and confidential business information at risk.[2]

[2] For a statement by Ambassador Donald Mahley, US Special Negotiator for Chemical and Biological Arms Control Issues, refer to http://www.state.gov/t/ac/rls/rm/2001/5497.htm

Similarly to what is found in the nuclear and chemical weapons realm, the BWC could also be strengthened by a rigorous verification process. However, the affordability and accessibility of biotechnologies, and the absence of any severe weapon-production bottlenecks analogous to that of the production of plutonium or high-enriched uranium in the nuclear case, render verification inherently more difficult in the biological realm. This is a distinguishing feature of biological weapons that is obscured by the tendency to include them with nuclear, chemical, and radiological weapons as a 'weapon of mass destruction' (Chyba, 2002).

20.3 Biological weapons are distinct from other so-called weapons of mass destruction

Producing a nuclear bomb is difficult; it requires expensive and technologically advanced infrastructure and involves uranium enrichment or plutonium production and reprocessing capacity that are difficult to hide. These features render traditional non-proliferation approaches feasible; despite being faced with many obstacles to non-proliferation, the International Atomic Energy Agency (IAEA) is able to conduct monitoring and verification inspections on a large number (over a thousand) of nuclear facilities throughout the world.

These traditional approaches are also reasonably effective in the chemical realm where the Organization for the Prohibition of Chemical Weapons (OPCW) can, inter alia, monitor and verify the destruction of declared chemical stockpiles.

But biological weapons proliferation is far more challenging for any future inspection regime – and it will only become more so as the underlying technologies continue to advance. In some respects, biological weapons proliferation poses challenges more similar to those presented by cyber attacks or cyber terrorism than to those due to nuclear or chemical weapons. An IAEA- or OPCW-like monitoring body against the proliferation of cyber attack capabilities would present a reductio ad absurdum for a verification and monitoring regime. Internet technology is so widely available that only a remarkably invasive inspection regime could possibly monitor it. Instead, society has decided to respond in other ways, including creating rapidly evolving defences like downloadable virus software and invoking law enforcement to pursue egregious violators.

Somewhat similar challenges are presented in the biological realm where the spread of life science research in areas like virology, microbiology, and molecular biology are contributing to a growing number of laboratories worldwide that engage in genetically based pathogen research. In addition, an expanding biotech industry and pharmaceutical sector is contributing to the spread of advanced and increasingly 'black box'[3] technologies that enable high consequence research to be carried out by a growing number of individuals; biotechnology is already commonplace in undergraduate institutions, it is beginning to enter high school classes, and is increasingly popular among amateur biotech enthusiasts. Moreover, an increasing number of countries are investing in biotechnology applications for health, agriculture, and more environment-friendly fuels. These trends are contributing to the increasing affordability of these technologies; the initial draft of the human genome cost an estimated $300 million (the final draft and all technologies that made it possible cost approximately $3 billion). Just six years later, one company hopes to finish an entire human genome for only $100,000 – a 3000-fold cost reduction. Researchers, spurred by government funding and award incentives from private foundations are now working towards a $1000 genome (Service, 2006).

These are exciting times for biologists; whereas the twentieth century saw great progress in physics, the early decades of the twenty-first century may well 'belong' to biology. These advances, however, provide unprecedented challenges for managing biotechnology's risks from misuse – challenges that are compounded by the ease with which biological materials can be hidden and the speed by which organisms can proliferate. Some bacteria can replicate in just twenty minutes, allowing microscopic amounts of organisms to be mass-produced in a brief period of time.

20.4 Benefits come with risks

Studies that uncovered DNA as life's genetic material and the discovery that genes encode for proteins that govern cellular characteristics and processes ushered in an era of modern molecular biology that saw rapid advances in our knowledge of living systems and our ability to manipulate them. At first, studying gene function involved introducing random mutations into the genomes of organisms and assessing physical and behavioural changes. Soon after, scientists learned to control gene function directly by introducing exogenous pieces of DNA into the organism of interest. Experiments like these began to shed light on the molecular mechanisms that underlie cellular processes and resulted in a better understanding of, and better tools to fight, human disease.

Modern molecular biology continues to develop medical solutions to global health issues such as newly occurring, re-emerging, and endemic infectious

[3] Meaning that the scientific or engineering details of what occurs 'inside' a particular component or technique need not be understood by the individual investigator in order to make use of it.

diseases. To address these threats, researchers are working to develop rational-design vaccines and antivirals. Microbiologists are exploring new avenues to counter antibiotic resistance in bacteria, while synthetic biologists are programming microorganisms to mass produce potent and otherwise rare anti-malarial drugs. Biotechnology's contribution to health is visible in medical genomics, where rapid improvements in DNA sequencing technology, combined with better characterization of genes, are beginning to unravel the genetic basis of disease. Other advances are apparent in the genetic modification of crops that render them resistant to disease and increase their yield. Biotechnology is even beneficial in industrial applications such as the development of new biological materials, potentially environment-friendly biological fuels, and bioremediation – the breakdown of pollutants by microorganisms.

But the same technologies and know-how that are driving the revolution in modern medicine are also capable of being misused to harm human health and agriculture (NRC, 2003a, 2003b). Traditional threats created by the misuse of biotechnology involve the acquisition, amplification, and release of harmful pathogens or toxins into the environment. One area of concern, for example, is a potential bioterrorist attack using pathogens or toxins on centralized food resources. Some toxins, like those produced by the bacterium *Clostridium botulinum*, are extremely damaging: small amounts are sufficient to inhibit communication between the nervous system and muscles, causing respiratory paralysis and death. In 2001, the United States found itself unprepared to cope with the intentional spread of anthrax, a bacterium[4] that can be obtained from the wild and amplified in laboratories. Anthrax can enter the body orally, or through cuts and skin lesions, after which it proliferates and releases illness-causing toxins. A more dangerous and deadly infection, however, can result if stable, dormant spores of the bacterium are 'weaponized', or chemically coated and milled into a fine powder consisting of small particles that can be suspended in air. These bacterial particles can travel long distances, be taken into the victim's respiratory airways and drawn into the lungs, where the spores germinate into active bacterium that divide and release toxic substances to surrounding cells. If left untreated, inhaled anthrax infects the lymph nodes, causing septic shock and death in the vast majority of its victims.

Whereas many bacterial pathogens, like anthrax, are free-living organisms that require proper conditions and nutrients for growth, other bioterrorism agents, like viruses, are parasitic and rely on their hosts' cellular machinery for replication and propagation. Viral propagation in laboratories involves propagating viruses in cells that are often maintained in incubators that

[4] Scientifically, one distinguishes between the microorganism, *Bacillus anthracis*, and the disease it causes, anthrax. Here we have adopted the more casual popular usage that conflates the organism itself with the name of the disease, at the risk of some loss of precision.

precisely mimic the host's physiological environment. A trained individual, with the proper know-how and the wrong intentions, could co-opt these life science tools to amplify, harvest, and release deadly pathogens into the environment. This threat is compounded by advances in microbiology, virology, and molecular biology, which enable directed changes in the genomes of organisms that can make them more stable, contagious and resistant to vaccines, antibiotics (in the case of bacteria), or antivirals.

Unless it is properly used, biotechnology's dual-use nature – the fact that beneficial advances can also be used to cause harm – poses a potential threat to human health and food security. But risk management measures aimed at minimizing these threats should not excessively impede biotechnology's benefits to health and food security and should also take care not to unnecessarily hinder scientific progress; inhibiting a developing country's access to health tools that are used in vaccines and pharmaceutical drug production would pose a serious ethical dilemma. Even if humanitarian arguments were set aside, solely from the perspective of self-interest in the developed world, restricting access to biotechnology could undermine desired security objectives by encouraging secrecy and impeding collaborative exchanges among different laboratories.

Biology's dual-use challenges extend beyond technology and include knowledge and know-how: better understanding of the molecular mechanisms that underlie cellular processes expose the human body's weaknesses and sensitivities, which can be exploited by those who intend to harm. Consider immunology research, which has characterized the interleukins, proteins that participate in the body's immune response to an infection. Foreign pathogens can disrupt the normal activity of these proteins and result in a defective immune response, serious illness, and death. A set of experiments that inadvertently illustrated some dual-use applications of this knowledge involved a group of Australian researchers who, in an attempt to sterilize rodents, added one of the interleukins, interleukin-4, to a mousepox virus (among other genetic modifications), hoping to elicit an autoimmune reaction that would destroy female eggs without eliminating the virus (Jackson et al., 2001). However, the virus unexpectedly exhibited more generalized effects; it inhibited the host's immune system and caused death – even in rodents that were naturally resistant to the virus or had previously been vaccinated. In a separate study, researchers showed that mice resistant to mousepox, if injected with neutralizing antibodies to the immune system regulator, IFN-γ, become susceptible to the virus and exhibit 100% lethalilty (Chaudhri et al., 2004).

Although there are genetic variations and different modes of infection between mousepox and its human counterpart, smallpox, these published experiments provide a possible road map for creating a more virulent and vaccine-resistant smallpox virus – a chilling notion given that the virus has been a major killer throughout human history (Tucker, 2001). Although the United

States has enough supplies to vaccinate its entire population (assuming the vaccine would be effective against a genetically modified virus), current world supplies can only cover 10% of the global population (Arita, 2005). (However, a 'ring vaccination' strategy, rather than a 'herd immunity' strategy, might be able to stop an outbreak well before complete vaccination was required.) Fortunately, smallpox has been eradicated from the natural world. The only known remaining stocks are located in US and Russian facilities. Although the WHO's decision-making body, the World Health Agency, had initially called for the destruction of these stocks by the end of 2002, it later suspended its decision, allowing smallpox research with the live virus to continue (Stone, 2002). Currently, researchers are using a variety of approaches to study the biology of smallpox, including research projects that involve infecting animals with the live virus (Rubins et al., 2004).

In addition to fears that smallpox could be accidentally or intentionally released or stolen from research facilities, there are also concerns that the smallpox virus could be regenerated from scratch. The latter requires piecing together the different fragments of the genome, which would be a difficult task, requiring knowledge of molecular biology, a standard molecular biology laboratory with appropriate reagents and equipment, and the skills and substantial time for trial and error. While synthesizing the smallpox virus from scratch in the laboratory is theoretically possible, it is fortunate that this challenge is both quantitatively and qualitatively much harder than for some other viruses, such as polio. Advances in the life sciences, however, are beginning to remove these hurdles.

20.5 Biotechnology risks go beyond traditional virology, micro- and molecular biology

To date, complete genomes from hundreds of bacteria, fungi, viruses, and a number of higher organisms have been sequenced and deposited in a public online database. While many of these genomes belong to inert microbes and laboratory research strains that cannot infect people or animals, others include those from some of the most pathogenic viruses known to humans, such as Ebola and Marburg, and even extinct ones like smallpox and the 1918 Spanish influenza virus. Alongside better DNA sequencing, biotechnology research has also seen the evolution of de novo DNA synthesis technologies; it is now possible to commercially order pieces of DNA as long as 40,000 bases[5] – longer than the genomes of many viruses; SARS for instance is roughly 30,000 bases, while the Ebola genome is less than 20,000 bases long. Moreover, the emerging rapid improvement of the technology over the next several years should enable

[5] A base, or a nucleotide, is the fundamental unit of a DNA molecule.

even the synthesis of bacterial genomes, many of which are around one million bases long. For example, just recently, scientists at the Venter institute transplanted an entire genome from one bacteria species into another, causing the host cell to effectively become the donor cell (Lartigue et al., 2007). The study demonstrates how a bacterial cell can be used as a platform to create new species for specialized functions (provided their genomes are available). The donor and host bacterial species that were chosen for the study are highly related to each other and contain relatively small genomes, features which facilitated the success of the transplantation experiment. Nevertheless, the study does point the way to more general applications, including transplantation of synthesized pathogen genomes, for the creation of otherwise difficult-to-obtain bacterial pathogens.

Automated DNA synthesis removes much of the time-consuming and technically difficult aspects of manipulating DNA; using commercial DNA synthesis, a researcher can copy a sequence of interest from an online public database and 'paste' it into the commercial DNA provider's website. Within days or weeks (depending on length of the sequence) the fragment, or even the entire genome of interest, is artificially synthesized and mail-delivered. For many viruses, a synthesized viral genome could then be introduced into a population of cells, which would treat the foreign DNA as if it were their own: 'reading' it, transcribing the genes into RNA molecules that are processed by the cell's internal machinery and translated into proteins. These proteins can then assemble themselves into infectious viral particles that are ejected from the cell, harvested and used in infection studies.[6]

In the wrong hands or in laboratories that lack proper safety precautions, this technology poses a serious security risk as it renders some of the traditional regulatory frameworks for the control of biological agents obsolete. A number of countries control the possession and movement of these substances. In the United States, these are referred to as 'select agents' and include a number of bacteria, viruses, fungi, and toxins that are harmful to humans, animals, or plants. Conducting research on these high-risk organisms and toxins require special licenses or security clearances. The ability to order genomes and create organisms de novo, however, necessitates revisiting these regulations. As we have seen, experiments have already been published in the highest profile international scientific journals that describe the re-creation of the poliovirus as well as the Spanish influenza virus, the agent that killed 50 million people in 1918. This virus, which was previously extinct, now exists and is used in research facilities in both the United States and Canada. The ability to

[6] Once they are inside their target cells, viruses hijack cellular proteins to convert their genomes into viral particles. However, viruses that contain negative strand RNA genomes, like Marburg and Ebola, cannot be turned into mature viruses with host proteins alone. Conversion of such genomes into virus particles also requires proteins that are normally packaged within the virus itself. Thus, the Ebola and Marburg genomes are not infectious on their own.

synthesize genomes and create organisms from them has spurred a US-based biosecurity advisory board to call for regulating the possession and movement of pathogen genomes, rather than pathogens themselves (Normile, 2006).

In addition to risks arising from the intentional misuse of these pathogens, there are serious laboratory safety considerations; many facilities worldwide lack the expensive safeguards needed for handling highly pathogenic organisms – even though they may have the technology to create these organisms. Moreover, the ease with which genomes can be synthesized raises the concern that highly pathogenic viruses and bacteria will become increasingly distributed in laboratories and among researchers interested in high consequence pathogen research. The accidental contamination of workers, and the subsequent escape of viruses from highly contained laboratories, have occurred a number of times. In one such case, a researcher at the National Defense University in Taipei was, unknowingly, infected with the SARS virus, after which he left Taiwan for a conference in Singapore. The event prompted quarantine of ninety individuals with whom the infected researcher had come into contact (Bhattacharjee, 2004). Although there were no known secondary infections in this particular case, the escape of pathogenic viruses or bacteria from contained laboratories could have serious consequences.

20.6 Addressing biotechnology risks

Dual-use risks posed by biotechnology may be addressed at a number of points. Efforts may be made to oversee, regulate, or prevent the most dangerous research altogether, or the publication of that research, or to restrict certain lines of research to particular individuals. One may also focus on recognizing disease outbreaks quickly whether natural or intentional when they occur, and responding to them effectively. This requires both improvements in surveillance and response capacity and infrastructure. Finally, one may encourage research, development, and production of appropriate vaccines, antibiotics, antivirals, and other approaches to mitigating an outbreak – along with the required infrastructure for meeting surges in both drug requirements and numbers of patients. Of course, none of these approaches is exclusive. Perhaps their one commonality is that each faces important drawbacks. We consider a variety of suggested approaches to each, and their challenges.

20.6.1 Oversight of research

The US National Research Council (NRC) has recommended a variety of monitoring mechanisms and guidelines for federally funded, high-risk research (NRC, 2003a). These 'experiments of concern' would be subjected to greater scrutiny at the funding stage, during the research phase, and at the publication stage.[7] They would include experiments that could make pathogens impervious to vaccines and antibiotics, allow pathogens to escape detection and diagnosis, increase the transmissibility or host range of a pathogen, and experiments that aim to 'weaponize' biological agents and toxins. But the NRC guidelines would only extend to laboratories that are funded by the National Institutes of Health, and they are therefore required to follow governmental guidelines. A comprehensive approach would take into account the commercial sector as well as increased funding from philanthropic and private foundations like the US-based Howard Hughes Medical Institute, or the Wellcome Trust of England, which annually distribute 500 million, and over one billion research dollars, respectively (Aschwanden, 2007). Comprehensive oversight mechanisms would also include governmental laboratories, including those involved in biodefence research. Finally, the biotechnology challenge is inherently global, so an effective research oversight regime would have to be international in scope.

To this end, John Steinbruner and his colleagues at the Center for International and Security Studies at Maryland (CISSM) have proposed a global system of internationally agreed rules for the oversight of potentially high-consequence pathogens research (Steinbruner et al., 2005). Although dedicated non-state groups would not be likely to be captured by such a system, they are unlikely to conduct forefront research. Instead, they might attempt to co-opt discoveries and techniques that are reported in the scientific literature. By overseeing certain high-risk research and its publication, society might therefore head off some of the worst misuse. A limited model for what oversight of the highest consequence biological research might look like is provided by the World Health Organization's international advisory committee that oversees smallpox research; it is important that this committee demonstrate that it is capable of real oversight.

The CISSM model oversight system calls for an International Pathogens Research Authority with administrative structures and legal foundations for participation of its states-parties. It is unlikely that such a system could be negotiated and ratified in the current climate, although plausibility of possible oversight mechanisms could change rapidly subsequent to a laboratory engineered pandemic; better that careful thinking be applied now before the urgency and fear that would become pervasive in the post-attack world. Other international approaches to provide some level of oversight have also been envisioned, including the creation of additional UN bodies, or the establishment of an 'International Biotechnology Agency' (IBTA) analogous to

[7] The US federal advisory group, NSABB, has called for self-regulation within the scientific community. Under the proposed plan, scientists themselves decide whether their research constitutes dual-use experiments of concern. For a discussion of NSABB's proposal, refer to Jocelyn Kaiser (2007). 'Biodefense: Proposed Biosecurity Review Plan Endorses Self-Regulation' *Science*, 316(5824), 529.

the IAEA. The IBTA could be established in a modular way, with initial modest goals of helping BWC states–parties meet their reporting requirements, and promoting best practices in laboratory safety. All these approaches require the creation of new international oversight bodies, a politically challenging requirement.

20.6.2 'Soft' oversight

At the other end of the spectrum from the CISSM oversight model are efforts at what might be called 'soft' oversight of high-risk research. Some of the most common among these are efforts to promote codes of ethics (or the more demanding, but rarer, codes of conduct or codes of practice) for scientists working in the relevant fields.[8] Many national and international groups have made efforts in this direction. If coupled with education about the possible misuse of scientific research, such codes would help provide the scientific community with tools to police itself. To this end, a US National Academy panel has recommended establishing a global internet-linked network of vigilant scientists to better protect against misuse within their community (NRC, 2006).

20.6.3 Multi-stakeholder partnerships for addressing biotechnology risks

The failure of the negotiations for a compliance protocol to the BWC shows some of the challenges now facing treaty negotiation and ratification. (This protocol, while it would have provided valuable transparency into certain high-end biological facilities, would not have – nor was it meant to – directly addressed the challenge of dual-use biotechnology.) One general result has been increasing interest in alternative policy models such as multi-stakeholder partnerships. Indeed, the international relations literature has seen an increasing volume of work devoted to the mismatch between important global problems and the absence of international mechanisms to address them in a timely and effective way.[9] Means of international governance without formal treaties are being sought.

In the biological security realm, efforts to forge multi-stakeholder partnerships are bringing together the academic science sector, commercial industry, the security community, and civil society, in order to raise

[8] For a discussion of codes of conduct in the case of biodefence research, see Roger Roffey, John Hart, and Frida Kuhlau, September 2006 'Crucial Guidance: A Code of Conduct for Biodefense Scientists', *Arms Control Today*. For a review and critical discussion of the broader need for codes applicable to all life scientists, see *Globalization, Biosecurity and the Future of the Life Sciences* (Washington, DC: National Academies Press, 2006), pp. 246–250.

[9] For key publications in this literature, see Wolfgang Reinicke, *Global Public Policy: Governing Without Government?* (Washington, DC: Brookings Institution, 1998); J.F. Rischard, *High Noon: Twenty Global Problems, Twenty Years to Solve Them* (New York: Basic Books, 2002); Anne-Marie Slaughter, *A New World Order* (Princeton: Princeton University Press, 2004).

awareness and facilitate feasible risk-management solutions to biology's dual-use problem. The former UN Secretary-General, Kofi Annan, recognized that the increasing distribution of biotechnology requires solutions that have an international dimension and called for a global forum to help extend the benefits of biotechnology and life-science research, while managing its security risks. The Secretary-General's unique convening power to bring together a diverse number of players from the appropriate sectors is instrumental for a successful bottom-up approach that aims to address biotechnology's challenges. This, combined with other efforts by the Royal Society, the InterAcademy Panel on International Issues, the International Council for the Life Sciences, The International Consortium for Infectious Diseases, and a number of others, are beginning to work towards an international framework in the absence of a formal, government-driven treaty process.

In addition to recognizing the urgency to address biotechnology's risks, some of these efforts have also highlighted the importance of risk-management strategies that do not hinder free flow of scientific communication and that do not impose excessively intrusive oversight mechanisms that would hurt scientific progress. An example of an effective risk-management strategy that manages risks without impacting potential benefits is a proposal that specifically addresses de novo DNA synthesis technology. The successful adoption of the proposal in the academic and commercial science sectors merits further attention, as the risk-management strategy might be applicable to some of biology's other dual-use areas.

20.6.4 A risk management framework for de novo DNA synthesis technologies

Currently, de novo DNA synthesis technologies capable of making complete pathogen genomes are concentrated in a relatively small number of companies. In 2004, the Harvard biologist and biotechnology developer, George Church, proposed a safeguards strategy to ensure that the technology is not used for the illegitimate synthesis of potentially harmful genomes (Church, 2005). This involves companies agreeing to install automated screening software that 'reads' the DNA sequence of incoming customer orders and compares them to genomes of a known list of pathogens (and, potentially, to a list of other potentially dangerous sequences, for example, those for particular genes). An exact match, or more likely a certain degree of sequence similarity, would elicit further inquiry and possibly result in the notification of proper authorities. Software in the synthesis machines could be installed, and updated, to make it impossible for the machines to synthesize certain sequences of particular concern.

This kind of DNA screening has already been adopted by a number of the DNA providers and has won endorsement within the synthetic biology community, which is a heavy user of DNA synthesis technologies (Declaration of the Second International Meeting on Synthetic Biology, 2006). Successful implementation of the protocol is in large part due to the proposal's non-intrusive nature; rather than requiring formal oversight structures, which many scientists oppose for fear that progress might be hindered, the screening tool allows the laboratory to go about business as usual as the computer software engages in the invisible detective work. The automated nature of DNA screening is also appealing to industry because it enables the protection of customer information. MIT synthetic biologist Drew Endy, together with George Church and other colleagues such as John Mulligan, CEO of a leading DNA synthesis company, are now working to extend the screening proposal to companies overseas. Indeed, lack of unity among the various DNA providers would risk jeopardizing the entire venture since it only takes a single non-compliant company to provide harmful materials to all interested customers. Possible strategies to address this deficiency include licensing all DNA providers or establishing a centralized international clearinghouse that receives and screens all DNA orders from the various providers (Bügl et al., 2006).

20.6.5 From voluntary codes of conduct to international regulations

While adopting safeguard strategies such as DNA screening exemplifies corporate responsibility, implementation of these measures is purely voluntary and without a legal framework. The UN Security Council Resolution 1540 provides the impetus to strengthen and globally extend such measures. Resolution 1540 requires UN member states to strengthen national legislation in order to address a number of issues, including biological terrorism. The legally binding implementation of the DNA screening protocol by countries that are users or providers of the technology could be cast in terms of a step in the implementation of resolution 1540. Alternatively or additionally, other international mechanisms such as the BWC could be adapted to carry out the operations of a centralized international clearinghouse for DNA synthesis screening.

20.6.6 Biotechnology risks go beyond creating novel pathogens

As biotechnological tools improve, the various methods that can be used to create novel organisms should be assessed further in order to make informed policy decisions regarding the risks. For example, a combination of virology and molecular biology could be used to create novel pathogens like hybrid viruses that are composed of inert laboratory strains loaded with additional toxic genes. Once inside its host, the hybrid virus would enter its target cell population, wherein viral genetic material would be converted into toxic proteins that disrupt normal cellular processes and cause disease (Block, 1999).

Similarly, viruses can be created that have the ability to shut down essential cellular genes (Block, 1999). Consider small interfering RNA (siRNA) technology, whose beneficial applications were recognized by the 2006 Nobel Prize for physiology or medicine. siRNA molecules turn off a gene by inactivating its RNA product (Sen and Blau, 2006). A highly contagious virus, supplemented with siRNA, or other 'gene knockdown' technologies, could shut down essential genes in particular cell populations of its host (Block, 1999). Theoretically, this could set off a novel epidemic for which no known vaccine or cure exists. Even more alarming is the ease with which commercial DNA sources can automate the synthesis of these novel pathogens, relieving even the novice from the laborious and methodical task of splicing genes into viral genomes (Tucker and Zilinskas, 2006). Thus, it is important that DNA screening safeguards encompass more than just naturally occurring pathogenic genomes and gene-encoding toxins.

20.6.7 Spread of biotechnology may enhance biological security

The spread of novel biotechnologies such as large-DNA synthesizers could, paradoxically, provide an opportunity to decrease the probability of misuse. If costs associated with commercial synthesis of large DNA fragments continue to decline, research laboratories will increasingly seek to outsource the laborious task of manipulating DNA sequences to more centralized, automated sources. Similar trends have been observed for DNA sequencing; laboratories that carried out sequencing operations in-house now outsource their needs to commercial sources that perform the task faster and at a fraction of the cost. A similar outcome for DNA synthesis could eventually replace a large number of diffuse and difficult-to-regulate DNA laboratories with more centralized DNA providers whose technologies are automated and more safeguard-friendly.

Not all dual-use issues will be addressed through technical solutions. But where possible, technologies that can be safeguarded should be promoted. This requires innovators and users of new biotechnologies to identify potential risks and develop appropriate safeguards. Biological security gatherings that bring together scientists and policy makers are useful for creating the right mechanism for this but they cannot replace hours of brainstorming of students searching for technical and feasible risk management solutions. Stronger communication links between the security community and biologists and more formal interdisciplinary education programmes should be fostered. Fortunately, some scientists at the forefront of fields such as synthetic biology have also been at the forefront of addressing the ethical and security implications of their research (Church, 2005; Endy, 2007).

20.7 Catastrophic biological attacks

It is difficult to forecast mortality figures resulting from potentially catastrophic bioterrorist incidents – however important such predictions may be for designing defensive public health measures. Unpredictability in human behaviour, for example, would impact morbidity and mortality figures, particularly if contagious agents are involved. For aerosolized pathogens, factors like wind speed and direction, as well as other environmental fluctuations, could result in very different attack outcomes. Limitations in our understanding of the biology of pathogens and their interaction with their hosts (i.e., precise mode of infection and, for transmissible pathogens, mode of spread) also render accurate predictions difficult. And as with any major disaster, it is difficult to know in advance the efficacy of emergency response plans and the competence with which they will be carried out (Clarke, 1999).

Moreover, modern society's experience with bioterrorism has, fortunately, so far been limited to a small number of events that were neither intended to, nor did result in high mortality figures, so they may not serve as good indicators for what a successful major attack would look like. The 2001 US Anthrax scare that caused five deaths, for instance, involved a non-contagious pathogen, and although milled into a fine powder, the bacterial spores were initially contained within envelopes that resulted in only local dissemination. By contrast, the Aum Shinrikyo cult, seeking to stage a mass-casualty attack in order to realize a prophecy, attempted to disperse *Bacillus anthracis*, from a building rooftop onto the dense urban population of Tokyo. The Aum, which later succeeded in dispersing Sarin nerve gas in Tokyo subways, was, fortunately, unsuccessful both in efforts to procure a pathogenic strain of *Bacillus anthracis*, and in its attempts to efficiently disseminate the bacterium. But a more rudimentary dispersal technique was successfully used by another group, the Rajneeshees, whose actions were motivated by a desire to keep a large block of individuals away from voting polls, in order to influence local elections. In 1984, members of the Oregon-based cult successfully spread the enteric bacterium, *Salmonella typhimurium*, onto salad bars, causing illness in over 750 Oregonians and sending many to hospitals. Had the Rajneeshees used a more virulent pathogen, or had the US Anthrax been more efficiently dispersed, major public health disasters may have ensued. In 1993, estimates from the US Congress' Office of Technology Assessment found that a single 100 kg load of anthrax spores, if delivered by aircraft over a crowded urban setting, depending on weather conditions, could result in fatalities ranging between 130,000 and 3 million individuals. However, these sort of dramatic results have been viewed as overly alarmist by those claiming that such high casualties would require optimal conditions and execution by the perpetrators, and that there would in fact be a very wide range of possible outcomes (Leitenberg, 2005).

Besides the Rajneeshees and the Aum Shinrikyo, another non-state group[10] that appears to have pursued biological weapons is Al Qaeda, apparently making use of one doctoral-level biologist and perhaps several others with undergraduate degrees. It is difficult from the open literature to determine either the level of sophistication or accomplishment of the programme, but what is available suggests that the programme was more aspirational than effective at the time that Al Qaeda was expelled from Afghanistan (National Commission on Terrorist Attacks Upon the United States, 2004; Commission on the Intelligence Capabilities of the United States Regarding Weapons of Mass Destruction, 2005).

While intentional biological attacks have yet to result in catastrophic scenarios, natural disease outbreaks can serve as proxies for what such events might look like. Consider smallpox, which infected 50 million individuals – annually – even as late as the early 1950s (WHO, 2007). Procuring (or creating), and releasing a vaccine-resistant or more lethal strain of this contagious virus in a dense urban environment might well be a cataclysmic event.

Whereas smallpox kills up to a third of its victims, certain strains of the hemorrhagic fever viruses, like Ebola-Zaire, can kill up to 90% of the infected – within several days after symptoms surface. Since 1976, when Ebola first appeared in Zaire, there have been intermittent outbreaks of the disease, often along Sub-Saharan African rainforests where the virus is transmitted from other primates to humans. The remoteness of these regions, and the rapid pace by which these viruses kill their human host, have thus far precluded a global pandemic. However, if these pathogens were procured, aerosolized, and released in busy urban centres or hubs, a catastrophic pandemic might ensue – particularly because attempts to generate vaccines to Ebola have, thus far, proven unsuccessful. In 1992, the Aum Shinrikyo sent a medical team to Zaire in what is believed to have been an attempt to procure Ebola virus (Kaplan, 2000). While the attempt was unsuccessful, the event provides an example of a terrorist group apparently intending to make use of a contagious virus.

In addition to the toll on human life, biological attacks can inflict serious psychological damage, hurt economies, cause political fallout, and disrupt

[10] We define 'sub-state' groups to be those that receive substantial assistance from a state or state entities; 'non-state' groups by contrast are those that do not. The Rajneeshees and Aum Shinrikyo were non-state groups. Because of its accommodation by the Taliban in Afghanistan, Al Qaeda arguably was, at least for a time, a sub-state group.

social order. A 1994 natural outbreak of pneumonic plague in Surat, India, provides a glimpse into what such a scenario might look like.[11] Pneumonic plague is an airborne variant, and the deadliest form, of the 'black death' – the disease caused by the bacterium, *Yersinia pestis*, that is believed to have wiped out a quarter of Europe's population in the fourteenth century (Anderson and May, 1991). The 1994 outbreak in Surat resulted in an estimated 300,000 individuals fleeing the city (Hazarika, 1995a), and even led to closure of schools, universities, and movie theatres in cities hundreds of miles away (*The New York Times*, September 20, 1994). Shock waves were felt globally as India faced increasing international isolation: its exports were banned; its tourism industry drastically declined (Hazarika, 1995b); and its citizens were subjected to scrutiny and surveillance at foreign airports (Altman, 1994).

But while much attention has been paid to human pathogens, threats to agriculture, livestock, and crops, which can cause major economic damage and loss of confidence in food security, should not be overlooked. In 1997, for example, an outbreak in Taiwan of the highly contagious foot-and-mouth disease, caused the slaughter of 8 million pigs and brought exports to a halt, with estimated costs of $20 billion (Gilmore, 2004). Crops can be particularly vulnerable to an attack; they inhabit large tracts of difficult-to-protect land, and suffer from low levels of disease surveillance, sometimes taking months, even years, before disease outbreaks are detected. In 2001, in an effort to contain a natural outbreak of *Xanthomonas axonopodis*, a bacterium that threatened Florida's citrus industry and for which there is no cure, 2 million trees were destroyed (Brown, 2001). There are well-defined steps that may be taken by countries (assuming the resources and capacity are available) to protect against such threats (NRC, 2003b).

In addition to actual attacks on human health and food security, biological 'hoaxes' can also exact an important societal toll. Between 1997 and 1998, as media attention to bioterrorism grew, the number of hoaxes in the United States increased from 1 to 150 (Chyba, 2001). In October and November of 2001, following the US Anthrax attacks, 750 hoax letters were sent worldwide, 550 of which went to US reproductive health clinics by a single group (Snyder and Pate, 2002). The high rate of these hoaxes requires defensive systems that can quickly distinguish a real attack from a fake one. This involves vigilance in disease detection and surveillance, as well as forensic discrimination. The remainder of this chapter explores such public health systems, as well as other strategies to defend against biological outbreaks.

[11] The implication of plague in Surat has been somewhat controversial. A number of studies, however, including Shivaji et al. (2000), have used DNA forensics to show that the causative agent of the disease outbreak was, in fact, *Yersinia pestis*.

20.8 Strengthening disease surveillance and response

The need to recognize and respond to disease outbreaks is the same regardless of whether the outbreak occurs naturally, by accident, or through an act of terrorism. Therefore, appropriate defence measures should include a strong public health sector that can react to the full spectrum of risks – whether they are relatively common infectious disease outbreaks or less familiar events like bioterrorist attacks.

Defence against biological attacks requires rapid detection of disease, efficient channels of communication, mechanisms for coordination, treating the infected, and protecting the uninfected. The World Health Organization's deliberate epidemics division provides specific guidelines in these areas.[12]

20.8.1 Surveillance and detection

Efficient response to a disease outbreak begins with disease surveillance. Early detection can greatly minimize the numbers of infected individuals – particularly when contagious pathogens that can cause secondary infections are involved. Clinicians and medical personnel are indispensable for diagnosing and detecting disease, but they can be complemented by improved surveillance of air, food, and water supplies. The United States employs BioWatch in 30 cities; BioWatch employs a device that concentrates outside air onto filters that are routinely tested for the presence of various bioterrorism agents. More advanced systems include the Autonomous Pathogen Detection System (APDS), an automated diagnostic device that conducts polymerase chain reaction (PCR), a method that amplifies DNA sequences as well as other forensic analysis within the device itself. In addition to rapidly recognizing disease agents, the effective use of these automated diagnostics also allows human capacity and laboratory resources to be utilized in other areas of need. Although they are partly effective for detecting aerosolized agents, automated diagnostics suffer from high rates of false positives, mistaking background and normal levels of pathogens for biological weapons (Brown, 2004). In addition to focusing future research on improving sensitivity and accuracy of detection devices, the range of pathogens that are surveyed should be broadened beyond the high probability bioterrorism agents – especially because novel technologies allow the synthesis of a growing number of organisms.

Affordability, availability, and proper implementation of better diagnostic tools represent some of biotechnology's biggest benefits for improving health (Daar et al., 2002). For example, effective diagnosis of acute lower respiratory infection, if followed by proper treatment, would save over 400,000 lives

[12] Much of the discussion here, regarding disease preparedness and response, has been based on WHO strategies that can be found here: http://www.who.int/csr/delibepidemics/biochemguide/en/index.html

each year (Lim et al., 2006). Equally optimistic predictions can be made for malaria, tuberculosis, and HIV (Girosi et al., 2006). Moreover, new generations of technologies in the pipeline, if they become more affordable, could revolutionize the future of disease detection. These include small nanotechnology based devices that can prepare and process biological samples and determine the nature of the pathogen in real time, effectively serving as an automated laboratory on a small chip (Yager et al., 2006).

Despite technological improvements, traditional methods such as surveillance of blood samples and diagnosis by medical professionals remain of utmost importance, and must not be overlooked in the face of 'high-tech' approaches of limited applicability. Diagnosis of anthrax in the 2001 bioterrorist attack, for example, was not made by sophisticated technologies, but by a vigilant clinician. Improvements in human diagnostic skills, however, is badly needed, especially for likely bioterrorist agents; a survey of 631 internal medicine residents in 2002–2003 demonstrated that only 47% were able to correctly diagnose simulated cases of smallpox, anthrax, botulism, and plague, better training for which increased the frequency to 79% (Bradbury, 2005; Cosgrove et al., 2005). Better diagnostic training, human capacity, and laboratory infrastructure are particularly important for parts of the developing world that suffer most from disease. Improving domestic and international disease surveillance capacity and infrastructure (in areas of human and animal health, as well as communication between the two communities) lacks the glamour of high-tech solutions, but remains one of the most important steps that needs to be taken (Chyba, 2001; Kahn, 2006).

20.8.2 Collaboration and communication are essential for managing outbreaks

While diagnosis of an unusual disease by a single astute physician can sound the alarm, detecting an epidemic for a disease that normally occurs at low frequencies requires the consolidation of regional surveillance data into a single database that is monitored for unusual trends. Once an outbreak is suspected or confirmed, it must be communicated to appropriate individuals and departments whose roles and responsibilities must be delineated in advance. Coordination, transparency, and timely sharing of information are of great importance – procedures which require oversight bodies and a clear chain of command.

The increasing volume in global trade and travel, and the rapid pace by which transmissible disease can spread, necessitate effective international communication and coordination. Electronic communication tools include the Program for Monitoring Emerging Diseases (ProMed) and the World Health Organization's Global Public Health Information Network (GPHIN). PROMED provides news, updates and discussion regarding global disease outbreaks, whereas the web-based early 'disease warning system', GPHIN, scans websites, blogs and media sources, gathers disease information, and reports unusual biological incidents. Coordination between countries is facilitated by the WHO's Global Outbreak and Response Network (GOARN), which links over 100 different health networks that provide support for disease detection and response. During the 2004 SARS outbreak, the WHO established laboratories to link efforts among different countries, which resulted in rapid identification of the disease agent as a coronavirus. The organization is in a position to provide similar support in the event of deliberate pandemics.

Coordination within and between countries is also necessary to facilitate sharing of disease samples, which are used for production of vaccines and treatments. Determining whether an outbreak is intentional or natural can be difficult, unless forensic analysis can be performed on the genome or protein composition of the organism. For example, rapid sequencing of a genome may uncover artificially added pieces of DNA that confer antibiotic resistance or enhanced stability to an organism. Such findings might impact the design of appropriate drugs and therapies, but require the prompt availability of disease data and samples; the US Center for Disease Control and Prevention, for instance, has been criticized for not sharing flu data with other scientists (Butler, 2005).

20.8.3 Mobilization of the public health sector

Once disease outbreaks are detected and communicated to the proper authorities, local agencies and individuals must assemble and respond to the public health crisis. Lack of preparation for responding to large-scale biological outbreaks can overwhelm the health care system, negatively impacting not just the disaster sector, but also the greater public health infrastructure. There have been speculations, for example, that Toronto's effective response in curbing SARS placed pressure on other critical health care areas that resulted in a number of preventable deaths (IOM, 2004, p. 34). Emergency relief procedures should be established and practiced in advance, particularly for procedures that deal with surge capacity – the ability to expand beyond normal operations in order to deal with emergencies and disasters. Surge capacity involves enlisting medical personnel from other sectors. The ability to co-opt existing networks of local health care workers to perform disaster relief is an important element of a successful surge-capacity strategy. While local health care providers can address general disaster relief functions, more specialized responders are also instrumental for proper isolation and handling of hazardous biological materials, for selection of appropriate decontamination reagents, and for assessing risks to health and to the environment (Fitch et al., 2003).

20.8.4 Containment of the disease outbreak

The rapid containment of a disease outbreak requires identification of the 'hot-zone' – the area that is contaminated. This is exceedingly difficult for contagious agents, particularly ones with long asymptomatic incubation periods during which disease can be transmitted to others and spread over large areas. Also difficult is containing novel agents whose mode of infection or transmissibility is not known. Both epidemiological tools and computer simulations may be used to help identify, isolate, and quarantine affected individuals, and to break the chain of infections.[13] This was successfully accomplished during the SARS outbreak with the WHO leadership issuing timely and aggressive guidelines concerning quarantine procedures, curfews, and travel advisories.

Halting disease spread also requires provisions to care for a large number of infected individuals, possibly in isolation from others in mobile hospitals or dedicated hospital wings, gymnasiums, or private homes. The public is most likely to respond well if there is effective dispersal of information through responsible media sources and credible internet sites such as the Center for Disease Control and Prevention and the World Health Organization. The use of telephone hotlines also proved to be an effective information dissemination tool during the SARS outbreak.

In addition to curbing social contacts, implementing curfews and quarantines, and halting public activities, other types of public health protection measures can be implemented in a disease outbreak. These include the decontamination of high-traffic areas like hospitals, schools, and mass-transit facilities and implementing personal precautionary measures as regards hand washing and protective clothing, gloves, and masks. Masks should be worn by both the infected and the uninfected; the N-95 masks, so-called for its ability to block particles greater than 0.3 microns in size 95% of the time, is particularly effective and provided protection against the SARS virus; even though that virus is smaller than 0.3 microns, SARS travels in clumps, resulting in larger-sized particles that become trapped (IOM, 2004, p. 18). Personal protection is particularly important for health care workers and first responders who are in the front lines and more at risk of becoming infected; during the early stages of the SARS pandemic, a single patient, the 'super-spreader', infected every one of 50 health workers who treated him. Fully contained suits and masks, depending on the nature of the pathogen, might be appropriate for health care workers. In addition, these individuals should also receive prophylaxis and immunizations, when available; the United States,

[13] There are a host of legal and ethical issues regarding implementation of quarantines. For a discussion, refer to Cécile M. Bensimon and Ross E.G. Upshur, 2007 'Evidence and effectiveness in decision making for quarantine'. *American Journal of Public Health* (Suppl. 1), pp. 44–48; Richard Scabs, 2003. SARS: prudence, not panic *Canadian Medical Association Journal*, 169(1), pp. 1432–1434.

for instance, encourages more than 40,000 medical and public health staff personnel to protect themselves against a potential smallpox outbreak by vaccination (Arita, 2005).

Beyond health workers, determining who should receive treatment can be difficult to assess, particularly when supplies are limited or when there are detrimental side-effects of receiving treatment. These difficult choices can be minimized and avoided through aggressive drug and vaccine research, development, and production strategies.

20.8.5 Research, vaccines, and drug development are essential components of an effective defence strategy

Defending against likely disease outbreaks involves the stockpiling of vaccines, antibiotics and antivirals in multiple repositories. But for outbreaks that are less probable, cost and shelf-life considerations may favour last-minute strategies to rapidly produce large quantities of drugs only when they are needed. Drug and therapy development strategies require coordination between public health experts, life scientists, and the commercial sectors. Equally important is investing in basic science research, which is the cornerstone to understanding disease; decades of basic science research on viruses, bacteria and other organisms has been instrumental for rapid identification and characterization of novel biological agents, and for developing appropriate treatments and cures. (It is, of course, also this research that may bring with it the danger of misuse.) Efforts are needed to generate and promote a stronger global research capacity. This requires better funding mechanisms worldwide so that local scientists can address local health needs that are neglected by pharmaceutical companies that focus on expensive drug markets in the industrialized world. Encouraging industry to address infectious diseases includes providing incentives such as advance market commitments. The United States uses the 'orphan drug legislation' to provide tax credits to companies to invest in rare diseases that are otherwise not deemed profitable.

Efforts to improve funding for addressing infectious and neglected diseases include those by The Bill and Melinda Gates Foundation, which provides grants with the precondition that any eventual product would be patent-free and publicly available. In the same spirit, the pharmaceutical giant Sanofi-Aventis and a non-profit drug development organization funded by 'Doctors Without Borders' have combined to create a cheap, patent-free Malaria pill (*New York Times*, March 5, 2007).

20.8.6 Biological security requires fostering collaborations

Due to the low availability of drugs and diagnostic tools, high population densities, and pre-existing health issues, developing countries may suffer the greatest consequences of a biological attack. In some places this is exacerbated by inadequate human resources and infrastructure, which contribute to less effective planning; more than 150 countries do not have national strategies to deal with a possible flu pandemic (Bonn, 2005). Based on current public health capabilities, it is estimated that were the 1918 Spanish Influenza to take place today, 95% of deaths would occur in the developing world (Murray et al., 2006). Moreover, global trends of increasing urbanization create high population densities that are breeding grounds for human pathogens and attractive targets for bioterrorists.

Improvements in disease surveillance, together with better communication and coordination, should be a global priority. In an increasingly interconnected world, small-scale outbreaks can rapidly turn into pandemics, affecting lives and resources worldwide. The SARS outbreak, a relatively small pandemic, is estimated to have cost $40 billion in 2003 alone (Murray et al., 2006). Even local or regionally confined outbreaks can result in a decrease in the trade of goods, travel, and tourism, the impact of which is felt globally. Strengthening global tools to fight disease outbreaks, therefore, is sound governmental and intergovernmental policy for humanitarian reasons as well as for national and international security.

20.9 Towards a biologically secure future

In addition to defensive measures like improved disease detection and response, a comprehensive biological security strategy must safeguard potentially dangerous biotechnologies. Despite some of the current difficulties in identifying and implementing safeguards, there are historical reasons for optimism regarding the response of the scientific community. In the past, when confronted with potentially hazardous research involving recombinant DNA technology,[14] biologists took precautions by adopting guidelines and self-regulatory measures on a particular class of experiment. One reason for this success was that from the beginning biologists enlisted support from prestigious scientific academies (Chyba, 1980). These continue to provide a powerful tool today.

A greater challenge for biotechnology nonproliferation will be the expansion of safeguards throughout the academic, commercial, and governmental scientific sectors, as well as the international implementation of these measures. Traditional nonproliferation conventions and arms control treaties predominantly address nation states and do not provide adequate models for

[14] Recombinant DNA technology facilitated the exchange of genetic material between vastly different organisms and opened new frontiers for molecular biology research, but it also brought with it a number of safety concerns regarding potential harm to laboratory workers and to the public.

dealing with the non-state aspects of the biotechnology dilemma (Chyba, 2006). But despite these shortcomings, novel and innovative efforts that safeguard biotechnology are beginning to take shape. Together with better disease detection and response, if accompanied by political will, these efforts may provide a multi-pronged approach of preventative and defensive measures that will help to ensure a more biologically secure future.

Suggestions for further reading

Christopher, F.C. (October 2006). Biotechnology and the Challenge to Arms Control. *Arms Control Today*. Available at http://www.armscontrol.org/act/2006_10/BioTech Feature.asp. This article deals with dual-use biotechnology, which, due to its increasingly accessible and affordable nature, provides unprecedented challenges to arms control. The article also reviews a number of strategies that are aimed at managing biotechnology risks, paying particular attention to proposals that have an international dimension.

Institute of Medicine of the National Academies. (2004). *Learning from SARS, Preparing for the next disease outbreak*. Washington, DC: National Academies Press. *Learning from SARS, Preparing for the next disease outbreak* explores the 2002–2003 outbreak of a novel virus, SARS. The rapidly spreading virus, against which there was no vaccine, posed a unique challenge to global health care systems. The report examines the economic and political repercussions of SARS, and the role of the scientific community, public health systems, and international institutions in the halting of its spread.

Lederberg, J. (1999). *Biological Weapons: Limiting the Threat* (Cambridge, MA: The MIT Press). This book is a collection of essays that examine the medical, scientific, and political aspects of the BW threat, and strategies aimed at mitigating these threats. These essays explore the history of the development and use of offensive biological weapons, and policies that might be pursued to contain them.

Leitenberg, M. (2005). *Assessing the Biological Weapons and Bioterrorism Threat*. U.S. Army War College. The spreading of Anthrax through the US postal system, and discoveries in Afghanistan that Al Qaeda was interested in procuring biological weapons, have contributed to shifting the context within which biological weapons are considered, to one that almost exclusively involves bioterrorism. This transformation in threat perception, together with a $30 billion, 4-year government spending package, arrived with inadequate threat assessments, which this book begins to provide.

National Research Council. (2006). Committee on Advances in Techology and the Prevention of their Application to Next Generation Biowarfare Threats, Globalization, Biosecurity, and the Future of the Life Sciences. Washington, DC: National Academies Press. *Globalization, Biosecurity, and the Future of Life Sciences* explores the current status and future projections of biomedical research in areas that can be applied to the production of biological weapons. The report explores and identifies strategies aimed at mitigating such threats.

National Research Council. (2003). *Biotechnology Research in an Age of Terrorism: Confronting the 'Dual Use' Dilemma*. Washington, DC: National Academies Press. The report addresses dual-use biotechnology and proposes a greater role for self-governance among scientists and journal editors. Other findings include identification of 'experiments of concern', which would be subjected to an approval process by appropriate committees. Proposals are put forward for the creation of an international forum aimed at mitigating biotechnology risks.

References

Altman, L.K. (15 November 1994). The doctor's world; was there or wasn't there a pneumonic plague epidemic? *The New York Times*.

Anderson, R.M. and May, R.M. (1991). *Infectious Diseases of Humans: Dynamics and Control* (Oxford: Oxford University Press).

Arita, I. (October 2005). Smallpox vaccine and its stockpile in 2005. *Lancet Infectious Disease*, **5**(10), 647–652.

Aschwanden, C. (9 February 2007). Freedom to fund. *Cell*, **128**(3), 421–423.

Bhattacharjee, Y. (2 January 2004). Infectious diseases: second lab accident fuels fears about SARS. *Science*, **303**(5654), 26–26.

Block, S.M. (1999). Living nightmares. In Drell, S.D., Sofaer, A.D., and Wilson, G.D. (eds.), *The New Terror: Facing the Threat of Biological and Chemical Weapons*. pp. 60–71 (Stanford, CA: Hoover Institution Press).

Bonn, D. (March 2005). Get ready now for the next flu pandemic. *Lancet Infectious Disease*, **5**(3), 139–139.

Bradbury, J. (November 2005). More bioterrorism education needed. *Lancet Infectious Disease*, **5**(11), 678–678.

Brown, K. (22 June 2001). Florida fights to stop citrus canker. *Science*, **292**(5525), 2275–2276.

Brown, K. (27 August 2004). Biosecurity: up in the air. *Science*, **305**(5688), 1228–1229.

Bügl, H., Danner, J.P., Molinari, R.J., Mulligan, J.T., Park, H.O., Reichert, B., Roth, D.A., Wagner, R., Budowle, B., Scripp, R.M., Smith, J.A., Steele, S.J., Church, G., and Endy, D. (4 December 2006). A Practical Perspective on DNA Synthesis and Biological Security. For the proposal's text, refer to http://pgen.us/PPDSS.htm

Butler, D. (22 September 2005). Flu researchers slam US agency for hoarding data. *Nature*, **437**(7058), 458–459.

Carlson, R. (2003). The pace and proliferation of biological technologies. *Biosecurity and Bioterrorism: Biodefense Strategy, Practice, and Science*, **1**(3), 203–214.

Cello, J.P., http://www.ncbi.nlm.nih.gov/sites/entrez?Db=pubmed&Cmd=Search&Term=A.V., Wimmer, E. (9 August 2002). Chemical synthesis of poliovirus cDNA: generation of infectious virus in the absence of natural template. *Science*, **297**(5583), 1016–1018.

Chaudhri, G., Panchanathan, V., Buller, R.M., van den Eertwegh, A.J., Claassen, E., Zhou, J., de Chazal, R., Laman, J.D., and Karupiah, G. (15 June 2004). Polarized

type 1 cytokine response and cell-mediated immunity determine genetic resistance to mousepox. *Proceedings of the National Academy of Sciences*, **101**(24), 9057–9062.
Church, G.M. (21 May 2005). A Synthetic Biohazard Nonproliferation Proposal. Found at http://arep.med.harvard.edu/SBP/Church_Biohazard04c.html.
Chyba, C.F. (1980). The recombinant DNA debate and the precedent of Leo Szilard. In Lakoff, S.A. (ed.), *Science and Ethical Responsibility*. pp. 251–264 (London: Addison-Wesley).
Chyba, C.F. (2001). Biological terrorism and public health. *Survival*, **43**(1), 93–106.
Chyba, C.F. (2002). Toward biological security. *Foreign Affairs*, **81**(3), 121–136.
Chyba, C.F. (October 2006). Biotechnology and the challenge to arms control. *Arms Control Today*, Available at http://www.armscontrol.org/act/2006_10/BioTech Feature.asp.
Chyba, C.F. and Greninger, A.L. (Summer 2004). Biotechnology and bioterrorism: an unprecedented world. *Survival*, **46**(2), 143–162.
Clarke, L. (1999). *Mission Improbable: Using Fantasy Documents to Tame Disaster* (Chicago: University of Chicago Press).
Commission on the Intelligence Capabilities of the United States Regarding Weapons of mass Destruction. (2005). *Report to the President of the United States* (Washington, DC: U.S. Government Printing Office).
Cosgrove, S.E., Perl, T.M., Song, X., and Sisson, S.D. (26 September 2005). Ability of physicians to diagnose and manage illness due to category A bioterrorism agents. *Archives of Internal Medicine*, **165**(17), 2002–2006.
Daar, A.S., Thorsteinsdóttir, H., Martin, D.K., Smith, A.C., Nast, S., and Singer, P.A. (October 2002). Top ten biotechnologies for improving health in developing countries. *Nature Genetics*, **32**(2), 229–232.
Declaration of the Second International Meeting on Synthetic Biology (29 May 2006). Berkeley, CA. Available at https://dspace.mit.edu/handle/1721.1/18185
Endy, D. Presentation available at http://openwetware.org/images/d/de/Enabling.pdf http://openwetware.org/images/d/de/Enabling.pdf Accessed June 2007.
Fitch, J.P., Raber, E., and Imbro, D.R. (21 November 2003). Technology challenges in responding to biological or chemical attacks in the civilian sector. *Science*, **302**(5649), 1350–1354.
Fu, P. (June 2006). A perspective of synthetic biology: assembling building blocks for novel functions. *Biotechnology Journal*, **1**(6), 690–699.
Gilmore R. (December 2004). US food safety under siege? *Nature Biotechnology*, **22**(12), 1503–1505.
Girosi, F., Olmsted, S.S., Keeler, E., Hay Burgess, D.C., Lim, Y.W., Aledort, J.E., Rafael, M.E., Ricci, K. A., Boer, R., Hilborne, L., Derose, K.P., Shea, M.V., Beighley, C.M., Dahl, C.A., and Wasserman, J. (23 November 2006). Developing and interpreting models to improve diagnostics in developing countries. *Nature*, **444**(Suppl.), 3–8.
Hazarika, S. (14 March 1995a). Plague's origins a mystery. *The New York Times*.
Hazarika S. (5 January 1995b). Bypassed by plague, and tours. *The New York Times*.
Institute of Medicine. (2004). *Learning from SARS, Preparing for the Next Disease Outbreak* (Washington, DC: National Academies Press).
Jackson, R.J., Ramsay, A.J., Christensen, C.D., Beaton, S., Hall, D.F., and Ramshaw, I.A. (February 2001). Expression of mouse interleukin-4 by a recombinant ectromelia

virus suppresses cytolytic lymphocyte responses and overcomes genetic resistance to mousepox. *Journal of Virology*, **75**(3), 1205–1210.

Jernigan, D.B., Raghunathan, P.L., Bell, B.P., Brechner, R., Bresnitz, E.A., Butler, J.C., Cetron, M., Cohen, M., Doyle, T., Fischer, M., Greene, C., Griffith, K.S., Guarner, J., Hadler, J.L., Hayslett, J.A., Meyer, R., Petersen, L.R., Phillips, M., Pinner, R., Popovic, T., Quinn, C.P., Reefhuis, J., Reissman, D., Rosenstein, N., Schuchat, A., Shieh, W.J., Siegal, L., Swerdlow, D.L., Tenover, F.C., Traeger, M., Ward, J.W., Weisfuse, I., Wiersma, S., Yeskey, K., Zaki, S., Ashford, D.A., Perkins, B.A., Ostroff, S., Hughes, J., Fleming, D., Koplan, J.P., Gerberding, J.L., and National Anthrax Epidemiologic Investigation Team (October 2002). Investigation of bioterrorism-related anthrax, United States, 2001: epidemiologic findings. *Emerging Infectious Diseases*, **8**(10), 1019–1028.

Kahn, L.H. (April 2006). Confronting zoonoses, linking human and veterinary medicine. *Emerging Infectious Diseases*, **12**(4), 556–561.

Kaiser, J. (27 April 2007). Biodefense: proposed biosecurity review plan endorses self-regulation. *Science*, **316**(5824), 529–529.

Kampala Compact: The Global Bargain for Biosecurity and Bioscience. (1 October 2005). Available at http://www.icsu-africa.org/resourcecentre.htm

Kaplan, D.E. and Marshall, A. (1996). The Cult at the End of the World: The Terrifying Story of the Aum Doomsday Cult, from the Subways of Tokyo to the Nuclear Arsenals of Russia. Random House. New York.

Lartigue, C., Glass, J.I., Alperovich, N., Pieper, R., Parmar, P.P., Hutchison, C.A. III, Smith, H.O., and Venter, J.C. (28 June 2007). Genome transplantation in bacteria: changing one species to another. *Science* (electronic publication ahead of print).

Leitenberg, M. (2005). Assessing the Biological Weapons and Bioterrorism Threat. U.S. Army War College.

Lim, Y.W., Steinhoff, M., Girosi, F., Holtzman, D., Campbell, H., Boer, R., Black, R., and Mulholland, K. (23 November 2006). Reducing the global burden of acute lower respiratory infections in children: the contribution of new diagnostics. *Nature*, **444** (Suppl. 1), 9–18.

Lundstrom, M. (January 2003). Enhanced: Moore's law forever? *Science*, **299**(5604), 210–211.

Moore, G. (19 April 1965). Cramming more components onto integrated circuits. *Electronics Magazine*, 114–117.

Murray, C.J., Lopez, A.D., Chin, B., Feehan, D., and Hill, K.H. (23 December 2006). Estimation of potential global pandemic influenza mortality on the basis of vital registry data from the 1918-20 pandemic: a quantitative analysis. *The Lancet*, **368**(9554), 2211–2218.

National Commission on Terrorist Attacks Upon the United States. (2004). *The 9/11 Commission Report* (New York: W.W. Norton).

National Research Council. (2003a). *Biotechnology Research in an Age of Terrorism: Confronting the 'Dual Use' Dilemma.* (Washington, DC: National Academies Press).

National Research Council. (2003b). *Countering Agricultural Bioterrorism* (Washington, DC: National Academies Press).

National Research Council. (2006) Committee on Advances in Techology and the Prevention of their Application to Next Generation Biowarfare Threats, Globalization, Biosecurity, and the Future of the Life Sciences.

Normile, D. (3 November 2006). Bioterrrorism agents: panel wants security rules applied to genomes, not pathogens. *Science*, **314**(5800), 743a.

Reinicke, W. (1998). *Global Public Policy: Governing Without Government?* (Washington, DC: Brookings Institution).

Rischard, J.F. (2002). *High Noon: Twenty Global Problems, Twenty Years to Solve Them* (New York: Basic Books).

Roffey, R., Hart, J., and Kuhlau, F. (September 2006). Crucial guidance: a code of conduct for biodefense scientists. *Arms Control Today*.

Rubins, K.H., Hensley, L.E., Jahrling, P.B., Whitney, A.R., Geisbert, T.W., Huggins, J.W., Owen, A., Leduc, J.W., Brown, P.O., and Relman, D.A. (19 October 2004). The host response to smallpox: analysis of the gene expression program in peripheral blood cells in a nonhuman primate model. *Proceedings of the National Academy of Sciences*, **101**(42), 15190–15195.

Sen, G.L. and Blau, H.M. (July 2006). A brief history of RNAi: the silence of the genes. *FASEB Journal*, **20**(9), 1293–1299.

Service, R.F. (17 March 2006). Gene sequencing: the race for the $1000 genome. *Science*, **311**(5767), 1544–1546.

Shivaji, S., Bhanu, N.V., and Aggarwal, R.K. (15 August 2000). Identification of *Yersinia pestis* as the causative organism of plague in India as determined by 16S rDNA sequencing and RAPD-based genomic fingerprinting. *FEMS Microbiology Letters*, **189**(2), 247–252.

Slaughter, A.M. (2004). *A New World Order* (Princeton: Princeton University Press).

Snyder, L. and Pate, J. (2002). Tracking Anthrax Hoaxes and Attacks. Available at http://cns.miis.edu/pubs/week/020520.htm

Steinbruner, J., Harris, E.D., Gallagher, N., and Okutani, S.M. (2005). Controlling Dangerous Pathogens: A Prototype Protective Oversight System. Available at http://www.cissm.umd.edu/papers/files/pathogens_project_monograph.pdf

Stone, R. (24 May 2002). World health body fires starting gun. *Science*, **296**(5572), 1383.

Tucker, J.B. (2001). *Scourge: The Once and Future Threat of Smallpox* (New York: Grove Press).

Tucker, J.B. and Zilinskas, R.A. (Spring 2006). The promise and perils of synthetic biology. *The New Atlantis*, **12**, 25–45.

Tumpey, T.M., Basler, C.F., Aguilar, P.V., Zeng, H., Solórzano, A., Swayne, D.E., Cox, N.J., Katz, J.M., Taubenberger, J.K., Palese, P., and García-Sastre, A. (7 October 2005). Characterization of the reconstructed 1918 Spanish influenza pandemic virus. *Science*, **310**(5745), 77–80.

Unit 731 Criminal Evidence Museum. (2005). *Unit 731: Japanese Germ Warfare Unit in China*. China Intercontinental Press.

Wheelis, M. (September 2002). Biological warfare at the 1346 Siege of Caffa. *Emerging Infectious Diseases*, **8**(9), 971–975.

Word Health Organization (WHO). (2004). World Health Report 2004, Statistical Annex. pp. 120–121.

WHO smallpox fact sheet. http://www.who.int/mediacentre/factsheets/smallpox/en/ http://www.who.int/mediacentre/factsheets/smallpox/en/ Accessed on 7 July 2007.

Yager, P., Edwards, T., Fu, E., Helton, K., Nelson, K., Tam, M.R., and Weigl, B.H. (27 July 2006). Microfluidic diagnostic technologies for global public health. *Nature*, **442**(7101), 412–418.

·21·
Nanotechnology as global catastrophic risk

Chris Phoenix and Mike Treder

The word 'nanotechnology' covers a broad range of scientific and technical disciplines. Fortunately, it will not be necessary to consider each separately in order to discuss the global catastrophic risks of nanotechnology, because most of them, which we will refer to collectively as *nanoscale technologies*, do not appear to pose significant global catastrophic risks. One discipline, however, which we will refer to as *molecular manufacturing*, may pose several risks of global scope and high probability.

The 'nano' in nanotechnology refers to the numeric prefix, one-billionth, as applied to length: most structures produced by nanotechnology are conveniently measured in nanometres. Because numerous research groups, corporations, and governmental initiatives have adopted the word to describe a wide range of efforts, there is no single definition; nanotechnology fits loosely between miniaturization and chemistry. In modern usage, any method of making or studying sufficiently small structures can claim, with equal justice, to be considered nanotechnology. Although nanoscale structures and nanoscale technologies have a wide variety of interesting properties, most such technologies do not pose risks of a novel class or scope.

Interest in nanotechnology comes from several sources. One is that objects smaller than a few hundred nanometres cannot be seen by conventional microscopy, because the wavelength of visible light is too large. This has made such structures difficult to study until recently. Another source of interest is that sufficiently small structures frequently exhibit different properties, such as colour or chemical reactivity, than their larger counterparts. A third source of interest, and the one that motivates molecular manufacturing, is that a nanometre is only a few atoms wide: it is conceptually (and often practically) possible to specify and build nanoscale structures at the atomic level.

Most nanoscale technologies involve the use of large machines to make tiny and relatively simple substances and components. These products are usually developed to be integral components of larger products. As a result, the damage that can be done by most nanoscale technologies is thus limited by the means of production and by the other more familiar technologies with which it will be integrated; most nanoscale technologies do not, in and of themselves, appear to pose catastrophic risks, though the new features and augmented power of nano-enabled products could exacerbate a variety of other risks.

Molecular manufacturing aims to exploit the atomic granularity and precision of nanoscale structures, not only to build new products, but also to build them by means of intricate nanoscale machines, themselves the products of molecular manufacturing techniques. In other words, precisely specified nanoscale machines would build more nanoscale machines by guiding molecular (chemical) processes. This implies that once a certain level of functionality is reached (specifically, the level at which a general-purpose machine can build its physical duplicate and variants thereof), the nanoscale can become far more accessible, products can become far more intricate, and development of further capabilities could be rapid. The molecular manufacturing approach to nanotechnology may unleash the full power of the nanoscale on the problems of manufacturing – creating products of extreme power in unprecedented abundance. As we will see, the success of this approach could present several global catastrophic risks.

It should not be overlooked that molecular manufacturing could create many positive products as well. Stronger materials and more efficient mechanisms could reduce resource usage, and non-scarce manufacturing capacity could lead to rapid replacement of inefficient infrastructures. It is also plausible that distributed (even portable), general-purpose, lower-cost manufacturing could allow less-developed and impoverished populations to bootstrap themselves out of deprivation. Reductions in poverty and resource constraints could work to reduce some of the risks described in this chapter. Although the focus here is on the most severe dangers, we support responsible development of molecular manufacturing technology for its benefits, and we promote study and understanding of the technology as an antidote to the threats it poses.

21.1 Nanoscale technologies

21.1.1 Necessary simplicity of products

A common characteristic of nanoscale technologies is that they are not suitable for manufacturing finished products. In general, the method of production for nanoscale components is simple and special-purpose, and can only build a material or component that subsequently must be included in a larger product through other manufacturing methods. For example, a carbon nanotube may be deposited or grown by an innovative process on a computer chip, but the chip then must be packaged and installed in the computer by traditional means. Ultraviolet-blocking nanoparticles may be manufactured by an innovative process, but then must be mixed into sunscreen with mostly ordinary ingredients. A related point is that nanoscale technologies make use of machines far larger in scale than their output.

Access to the nanoscale is typically indirect, painstaking, or both, and most manufacturing processes are unable to convey detailed information to the nanometre realm. The resulting nanoscale products, therefore, cannot be highly structured and information-rich, even by comparison with today's manufactured products; every feature must be either miniscule, highly repetitive, or random. Large quantities of identical or nearly identical particles and molecular constructs can be built, but they will be either amorphous, highly ordered to the point of simplicity (like a crystal), organized by a pre-built template, or have a partially random structure influenced by only a few parameters. Computer chips, which now contain millions of nanoscale features, are an exception to this, but they can only be built with the aid of 'masks' that are immensely expensive, slow, and difficult to produce. This is another reason why nanoscale technologies generally are not suitable for manufacturing finished products.

Because the outputs of nanoscale technologies can typically be used only as inputs to traditional manufacturing or material-use steps, their scope and impact necessarily will be limited by the non-nanotech components of the products and the non-nanotech manufacturing steps. For the most part, nanoscale technologies can be viewed as an addition to the existing array of industrial technologies. In some cases, nanoscale structures including nanoparticles may be released into the environment. This could happen deliberately as part of their use, due to accidents including unintended product destruction, or as part of the product lifecycle (including manufacturing processes and product disposal).

21.1.2 Risks associated with nanoscale technologies

The risks of nanoscale technologies appear in two broad classes, which are analogous to existing industrial risks. The first class is risks resulting from the availability and use of the eventual products. In this way, the nanoscale technology contributes only indirectly, by making products more powerful, more dangerous, more widely used, and so on. The second class is risks resulting from new materials that may cause inadvertent harm.

Depending on the application, risks may be increased by making improved products. Improved medical technology frequently must confront ethical issues. Some applications of nanoscale technology (such as using nanoparticles to treat cancer) will have strong demand and little downside beyond the standard medical issues of creating good treatment protocols. The use of nanoscale technologies for military medicine (ISN, 2005) or advanced goals such as anti-aging may be more controversial. But in any case, these risks and issues are only incremental over existing risks; they are not new classes of risk, and there do not appear to be any global catastrophic risks in nanoscale health care technology.

Nanoscale technologies may contribute indirectly to the development of weaponized pathogens, which arguably could lead to an existential risk. Smaller and more powerful research tools could be used to fine-tune or accelerate the development process. However, the same tools also would be useful to counter a biological attack. The SARS virus was sequenced in only six days in 2003 (Bailey, 2003). The development of sufficiently fast, sensitive, and inexpensive tests could greatly reduce the threat of almost any infectious disease. Thus, it is not yet clear whether nanoscale technologies will increase or reduce the risk of globally catastrophic pandemics.

Improved computers may produce new classes of risk, some of which arguably might be existential. Surveillance technology is improving rapidly, and will continue to improve as data mining, data processing, and networking become cheaper. A state that knows every movement and action of its populace would have power unequalled by history's most repressive regimes. Misuse of that power on a global scale could pose a catastrophic risk. However, this risk cannot be blamed entirely on nanoscale technologies, since it seems likely that sufficiently powerful computers could be developed anyway. Another possible risk stemming from improved computers is artificial intelligence (see Chapter 15, this volume).

Environmental or health risks from inadvertent releases of nanoscale materials will not be existential. That is not to say that there is no risk. Just like any industrial material, new nanoscale materials should be evaluated for toxicology and environmental impacts. Some nanoparticles may have high stability and environmental persistence, may migrate through soils or membranes, and may be chemically active; these are all reasons to study them closely. Information about the health and environmental impacts of nanoparticles is still far from complete, but at this writing their risks seem to fall into the familiar range of industrial chemicals. Some nanoparticles will be mostly harmless, while others may be quite toxic. Although it is possible to imagine scenarios in which the manufacture of sufficient quantities of sufficiently toxic particles would threaten the world in the event of accidental release, such scenarios seem unlikely in practice.

21.2 Molecular manufacturing

As nanotechnology develops, it is becoming possible to plot a continuum between near-term nanoscale technologies and molecular manufacturing. It begins with a convergence of increasingly complicated molecules, improved facility in mechanical (e.g., scanning probe) chemistry, and more powerful and reliable simulations of chemistry. It is now possible for mainstream nanotechnology researchers to imagine building machine-like constructions with atomic precision. Once this is achieved, the next step would be to harness these machines to carry out an increasing fraction of manufacturing operations. Further advances in machine design and construction could lead to integrated large-scale manufacturing systems building integrated large-scale products, with both the manufacturing systems and the products taking advantage of the (expected) high performance of atomically precise nanoscale machinery and materials.

The molecular manufacturing approach is expected by several researchers, including Robert Freitas, Ralph Merkle, and Eric Drexler, to lead to 'nanofactories' weighing a kilogram or more, capable of producing their own weight in product from simple molecules in a matter of hours. Because of the immense number of operations required, the factory is expected to be entirely computer-controlled; this is thought to be possible without advanced error-correcting software because of the high precision inherent in molecular construction (Phoenix, 2006). And because the manufacturing process would be general-purpose and programmable, it is expected to be able to produce a range of products including nanofactories and their support structure.

Any step along the pathway to molecular manufacturing would represent an advance in nanotechnology. But the point at which nanofactories become able to build more nanofactories seems particularly noteworthy, because it is at this point that high-tech manufacturing systems could become, for the first time in history, non-scarce. Rather than requiring cutting-edge laboratory or industrial equipment to produce small amounts of nanoscale products, a nanofactory would be able to build another nanofactory as easily as any other product, requiring only blueprints, energy, and feedstock. Thus, the cost of nanofactory-built products would owe very little to either labour or (physical) capital.

The supply of nanofactory-built products, as well as their cost, would depend on which resource – *information*, *feedstock*, or *energy* – was most difficult to obtain at point of use (the nanofactory itself).

- *Information* costs very little to store and copy; although it might be limited by regulations such as intellectual property laws, this would not represent a natural limit to the use of nanofactory-based manufacturing.
- *Feedstock* would be a small molecule, used in bulk. The molecule has not been specified yet, so its cost and availability cannot be determined, but the relatively low cost and high production of manufactured gases such as acetylene and ammonia indicates that nanofactory feedstock may contribute only a few dollars per kilogram to the price of the product.
- The largest cost of product fabrication may be *energy*. A preliminary analysis of a primitive nanofactory architecture was carried out recently (Phoenix, 2005), which calculated that building a product might require approximately 200 kWh/kg.

> Although this is significant, it is comparable with the energy cost of refining aluminum, and the material fabricated by the nanofactory would probably have significantly greater strength than aluminum (comparable to carbon nanotubes), resulting in lower weight products.

As general-purpose manufacturing systems, it is quite possible that nanofactories would be able to build feedstock processing plants and solar collectors. In that case, the quantity of products built would seem to be potentially unlimited by any aspect of the present-day industrial infrastructure, but rather by the resources available from the environment: sunlight and light elements such as carbon, none of which is in short supply. The time required for a nanofactory to build another nanofactory might be measured in days or perhaps even hours (Phoenix, 2005). With strong, lightweight materials, the time required to build another nanofactory and all its supporting infrastructure, thus exponentially doubling the available manufacturing capacity, might be as little as a few days (though without detailed designs this can be only speculation).

21.2.1 Products of molecular manufacturing

Nanofactory-built products (including nanofactories) would potentially enjoy a number of advantages over today's products. Certain atomically precise surfaces have been observed to have extremely low friction and wear (Dienwiebel et al., 2004), and it is hoped that these attributes could be constructed into nanomachines. Smaller machines work better in several respects, including greater power density, greater operating frequency, far greater functional density, and greater strength with respect to gravity (Drexler, 1992). Being built at the scale of atoms and molecules, the machines would be able to perform a number of medically significant procedures (Freitas, 1999). Atomically precise materials in appropriate structures could be far stronger than today's building materials (Drexler, 1992).

Building on the basic set of capabilities analysed by Drexler (1992), a number of researchers have worked out designs for products in varying degrees of detail. Freitas has analysed a number of basic capabilities relevant to medical devices, and has described several medical nanorobots in detail. Freitas has proposed devices to collect and neutralize harmful microbes (Freitas, 2005), to supplement the gas-transport function of blood (Freitas, 1998), and to replace damaged chromosomes (Freitas, 2007), among others. For everyday use, masses of relatively simple robots a few microns wide could cooperate to reconfigure themselves in order to simulate a wide range of shapes and conditions (Hall, 1996). Hall has also undertaken preliminary investigation of small aircraft with several innovative features.

Detailed designs for larger products have been scarce to date. An exception is the architectural and scaling study of nanofactory design mentioned earlier

(Phoenix, 2005); this 85-page paper considered numerous factors including energy use and cooling, physical layout, construction methods, and reliability, and concluded that a kilogram-scale, desktop-size, monolithic nanofactory could be built and could build duplicates in a few hours. Larger nanofactories building larger products, up to ton-scale and beyond, appear feasible. But in the absence of detailed plans for building and assembling nanomachines, it has not generally been feasible to design other products at high levels of detail. Nevertheless, analyses of extreme high performance at the nanoscale, as well as expectations of being able to integrate nearly unlimited numbers of functional components, have led to a variety of proposed applications and products, including the rapid construction of lightweight aerospace hardware (Drexler, 1986).

General-purpose manufacturing would allow the rapid development of new designs. A manufacturing system that was fully automated and capable of making complete products from simple feedstock could begin making a new product as soon as the design was ready, with no need for retraining, retooling, or obtaining components. A product designer developing a new design could see it produced directly from blueprints in a matter of hours. The cost of manufacture would be no higher for a prototype than for full-scale manufacturing. This would benefit designers in several ways. They would be able to see and evaluate their products at several stages in the development process. They would not have to spend as much effort on getting the design right before each prototype was produced. They would not have to make an additional design for full-scale manufacture. With these constraints removed, they could be considerably more aggressive in the technologies and techniques they chose to include in a product; if an attempted design failed, they would lose relatively little time and money.

Once a design was developed, tested, and approved, its blueprint could be distributed at minimal cost (as computer files on the Internet), ready for construction as and where needed. This means that the cost to deliver a new product also would be minimal. Rather than having to manufacture, ship, and warehouse many copies of a design whose success is not assured, it could be built only when purchased. This would further reduce the risk associated with developing innovative products. If a design was successful, it could be manufactured immediately in as many copies as desired.

21.2.2 Nano-built weaponry

Because weapons figure in several global catastrophic risks, it is necessary to discuss briefly the kinds of weapons that might be built with molecular manufacturing. Increased material strength could increase the performance of almost all types of weapons. More compact computers and actuators could make weapons increasingly autonomous and add new capabilities. Weapons could be built on a variety of scales and in large quantities. It is possible, indeed easy, to imagine combining such capabilities: for example, one could imagine an uncrewed airplane in which 95% of the dry weight is cargo, the said cargo consisting of thousands of sub-kilogram or even sub-gram airplanes that could, upon release, disperse and cooperatively seek targets via optical identification, and then deploy additional weapons capabilities likewise limited mainly by imagination.

The size of the gap between such speculation and actual development is open to debate. Smart weapons presumably would be more effective in general than uncontrolled weapons. However, it will be a lot easier to cut-and-paste a motor in a computer-aided design programme than to control that motor as part of a real-world robot. It seems likely, in fact, that software will require the lion's share of the development effort for 'smart' weapons that respond to their environment. Thus, the development of novel weapon functionality may be limited by the speed of software development.

To date, there does not appear to have been a detailed study of molecular manufacturing-built weapons published, but it seems plausible that a single briefcase full of weaponry could kill a large percentage of a stadium full of unprotected people (to take one scenario among many that could be proposed). Small robots could implement some of the worst properties of land mines (delayed autonomous action), cluster bombs (dispersal into small lethal units), and poison gas (mobile and requiring inconvenient degrees of personal protection). A wide variety of other weapons may also be possible, but this will suffice to put a lower bound on the apparent potential destructive power of molecular manufacturing-built products.

An idea that has caused significant concern (Joy, 2000) since it was introduced two decades ago (Drexler, 1986) is the possibility that small, self-contained, mobile, self-copying manufacturing systems might be able to gain sufficient resources from the ecosphere to replicate beyond human control. Drexler's original concern of accidental release was based on a now-obsolete model of manufacturing systems (Phoenix and Drexler, 2004). However, there is at least the theoretical possibility that someone will design and release such a thing deliberately, as a weapon (though for most purposes it would be more cumbersome and less effective than non-replicating weapons) or simply as a hobby. Depending on how small such a device could be made, it might be quite difficult to clean up completely; furthermore, if made of substances not susceptible to biological digestion, it might not have to be very efficient in order to perpetuate itself successfully.

21.2.3 Global catastrophic risks

Molecular manufacturing, if it reaches its expected potential, may produce three kinds of risk: (1) as with nanoscale technologies, molecular manufacturing may augment other technologies and thus contribute to the risks they present; (2) molecular manufacturing may be used to build new products that may introduce new risk scenarios depending on how they are used by people; (3) molecular manufacturing may lead to self-perpetuating processes with destructive side effects. At the same time, however, molecular manufacturing may help to alleviate several other catastrophic risks.

Rapid prototyping and rapid production of high-performance nanoscale and larger products could speed up innovation and R&D in a wide variety of technologies. Medicine is an obvious candidate. Aerospace is another: constructing new airframes and spacecraft tends to be extremely labour-intensive and expensive, and the ability to rapid-prototype finished test hardware at relatively low cost may allow significantly more aggressive experimentation. If molecular manufacturing is developed within the next 20 years, then it would be capable of building computers far more advanced than Moore's Law would predict (Drexler, 1992). Each of these technologies, along with several others, is associated with catastrophic risks; conversely, medicine and aerospace may help to avert risks of plague, asteroid impact, and perhaps climate change.

If molecular manufacturing fulfils its promise, the products of molecular manufacturing will be inexpensive and plentiful, as well as unprecedentedly powerful. New applications could be created, such as fleets of high-altitude uncrewed aircraft acting as solar collectors and sunshades, or dense sensor networks on a planetary scale. General-purpose manufacturing capacity could be stockpiled and then used to build large volumes of products quite rapidly. Robotics could be advanced by the greatly increased functional density and decreased cost per feature associated with computer-controlled nanoscale manufacturing. In the extreme case, it may even make sense to speak of planet-scale engineering, and of modest resources sufficing to build weapons of globally catastrophic power.

21.2.3.1 Global war

If molecular manufacturing works at all, it surely will be used to build weapons. A single manufacturing system that combines rapid prototyping, mass manufacturing, and powerful products could provide a major advantage to any side that possessed it. If more than one side had access to the technology, a fast-moving arms race could ensue. Unfortunately, such a situation is likely to be unstable at several different points (Altmann, 2006). A number of players would want to enter the race. Uncertainty over the future, combined with a temporary perceived advantage, could lead to preemptive strikes. And even if no one deliberately launched a strike, interpenetrating forces with the necessary autonomy and fast reaction times could produce accidental escalation.

During the Cold War period, the world's military power was largely concentrated in two camps, one dominated by the United States and the other by the Soviet Union. As both sides continued to develop and stockpile massive amounts of nuclear weapons, the doctrine of Mutually Assured Destruction (MAD) emerged. Full-scale war could have resulted in the annihilation of both powers, and so neither one made the first move.

Unfortunately, many of the factors that allowed MAD to work in deterring World War III may not be present in an arms race involving nano-built weapons:

1. The Cold War involved only two primary players; once a rough parity was achieved (or perceived), the resulting standoff was comparatively stable. Unless the ability to make nano-weapons is somehow rigorously restricted, a large number of nations can be expected to join the nanotech arms race, and this could create a situation of extreme instability.
2. Acquiring the capability to use nuclear weapons is an expensive, slow, and difficult process. It is therefore relatively easy to track nations that are seeking to gain or expand a nuclear fighting potential. By contrast, the capability to make weapons with molecular manufacturing will be very inexpensive, easy to hide (in the absence of near-total surveillance), and can be expanded rapidly. A 'starter' nanofactory could be smuggled from place to place more easily than a stick of gum, then used to build more and larger nanofactories.
3. Rapidly shifting balances of military power may create an atmosphere of distrust. Greater uncertainty of the capabilities of adversaries could foster caution – but it also could increase the temptation for preemptive strikes to prevent proliferation.

Following the collapse of the Soviet Union, another factor emerged to keep the likelihood of full-scale war at a low level. This was the growing economic interdependence of nations: the global economy. It is said that democracies rarely attack one another, and that's also true for trading partners (Lake, 1996; Orr, 2003).

But the proliferation of civilian molecular manufacturing capability could reduce global trade (McCarthy, 2005), at least in physical goods. Any nation with nanofactories would be able to provide virtually all their own material needs, using inexpensive, readily available raw materials. As economic interdependence disappears, a major motivation for partnership and trust also may be substantially reduced (Treder, 2005). Trade in information may represent an important economic exchange between countries that have compatible intellectual property systems but cannot be counted on to stabilize relations between any given pair of nations; indeed, the ease of copying information may lead to increased tensions due to 'theft' of potential earnings.

Today, the destruction wrought by war is an incentive not to engage in it; a nation may very well make more profit by trading with an intact neighbour than by owning a shattered neighbour. But if molecular manufacturing enables rapid inexpensive manufacture, it might be possible to rebuild quickly enough that war's destruction would represent less of an economic penalty for the victor.

Better sensors, effectors, communications, and computing systems – made at very low cost – may enable deployment of teleoperated robot 'soldiers' able to occupy territory and carry out missions without risk to human soldiers. There could be a wide variety of such 'soldiers', including ground-, water-, and air-based robots in a wide range of sizes from a few grams to many tons. Each robot might be directly teleoperated or partially autonomous (e.g., able to navigate to a programmed location). It is important to realize that, while stronger materials and greater power efficiency would increase the performance of the robot somewhat, computers many orders of magnitude, more compact and powerful than today's, would enable algorithms that are barely in research today to be deployed in even the smallest robots. (Robots smaller than a gram might have difficulty in locomotion, and small robots would also be limited in the payloads they could carry.)

A consequence of removing humans from the battlefield is to make war potentially less costly to the aggressor, and therefore more likely. Removal of humans from the battlefield (at least on one side), and the likely advances in 'less lethal' weapons aided by advanced nanotechnology, could reduce the moral sanctions against wars of aggression. It is tempting to think that removal of humans from the battlefield would make wars less destructive, but history shows that this argument is at best overly simple. Automated or remote-controlled weapons, rather than removing humans from the field of battle, instead may make it easier to take the battlefield to the humans. Although these new weapons might shift the focus of conflict away from conventional battlefields, new battlefields would likely develop, and many of them could overlap and overwhelm civilian populations.

Finally, big wars often arise from small wars: with trust factors rapidly declining, the potential for escalation of internecine or regional wars into larger conflagrations will be substantially higher. A significant (and sustainable) imbalance of power, wherein some government (whether national or international) has access to far more force than any of the combatants, could prevent small wars and thus their growth into big wars. However, recent history shows that it is quite difficult for even a powerful government to prevent local conflict or civil war, and it is far from certain that even the technological power of molecular manufacturing, as used by fallible political institutions, would be able to prevent small wars.

The desire to avoid unexpected destructive conflict would in theory provide a counterbalance to these destabilizing factors. But recognition of the danger probably will not be sufficient to avoid it, particularly when failing to develop and deploy advanced weapon systems may be tantamount to unilateral disarmament. An alternative, which might be especially attractive to an actor perceiving that it has a definite lead in nano-based military technology, could be consolidation of that lead by force. Attempts to do so might succeed, but if they did not, the outcome may well be exactly the destructive escalating conflict that the pre-emptive actor wanted to avoid.

21.2.3.2 Economic and social disruption

It is unclear at this point how rapidly molecular manufacturing might displace other kinds of manufacturing, and how rapidly its products might displace established infrastructures and sources of employment. A sufficiently general manufacturing technology, combined with advanced inexpensive robotics (even without major Artificial Intelligence advances), could in theory displace almost all manufacturing, extraction, and transportation jobs, and many service jobs as well. If this happened slowly, there might be time to find new sources of employment and new social systems. But if it happened quickly, large numbers of people might find themselves economically superfluous. This would tend to facilitate their oppression as well as reducing their motivation to produce and innovate. In an extreme scenario, the resulting loss of human potential might be considered catastrophic.

Another potential problem raised by distributed general-purpose manufacturing is a variety of new forms of crime. Even something as simple as a portable diamond saw capable of quickly cutting through concrete could facilitate breaking and entering. Medical devices might be used for illegal psychoactive purposes. Sensors could be used to violate privacy or gather passwords or other information. New kinds of weapons might enable new forms of terrorism. These are not new classes of problems, but they might be exacerbated if nanofactory technology were available to the general public or to organized crime. Although it seems unlikely that crime by itself could constitute a catastrophic risk, sufficient levels of crime could lead to social breakdown and/or oppressive governance, which might result in significant risk scenarios.

21.2.3.3 Destructive global governance

Any structure of governance is limited in scope by the technology available to it. As technology becomes increasingly powerful, it raises the possibility of effective governance on a global scale. There are a number of possible reasons why such a thing might be tried, a number of forms it could take, and likewise a number of pathways leading up to it and a number of potential bad effects. Molecular manufacturing itself, in addition to supplying new tools of governance, may supply several new incentives which might tend to promote attempts at global governance.

As discussed above, molecular manufacturing may enable the creation of new forms of weapons and/or newly powerful weapons on an exceptionally large scale. Along with other kinds of nanotechnology and miniaturization, it also may produce small and capable sensor networks. It should be noted that a single unrestricted nanofactory, being fully automated, would be able to build any weapon or other device in its repertoire with no skill on the part of the user (use of the weapon might or might not require special skills). Building large weapons would simply require building a bigger nanofactory first. Most governments would have a strong incentive to keep such capabilities out of the hands of their citizens; in addition, most governments would not want potentially hostile foreign populations to have access to them.

In addition to keeping excessively destructive capabilities out of private hands, both at home and abroad, many governments will be concerned with avoiding the possibility of a high-tech attack by other governments. If an arms race is seen as unstable, the alternative would seem to be the disarming of rival governments. This in turn would imply a global power structure capable of imposing policy on nations.

The analysis to this point suggests a winner-take-all situation, in which the first nation (or other group) to get the upper hand has every incentive to consolidate its power by removing molecular manufacturing from everyone else. Gubrud (1997) argues that development of molecular manufacturing will not be so uneven and that no one will have a clear advantage in the race, but at this point that is unclear. Given the potential for rapid exponential growth of manufacturing capacity, and given foresighted preparation for rapid design, it seems plausible that even a few months' advantage could be decisive. Unless some foolproof safeguard can be devised, a policy of ruthless pre-emption may be adopted by one or more parties. If this calculation were generally accepted, then the first to develop the capability would know that they were first, because otherwise they would have been pre-empted; this temporary certainty would encourage an immediate strike. The ensuing struggle, if it did not result in a terminal catastrophic war, could quite possibly produce a global dictator.

One pathway to global governance, then, is a pre-emptive strike to consolidate a temporary advantage. Another pathway, of course, is consolidation of advantage after winning a more symmetrical (and presumably more destructive) war. But there are also pathways that do not involve war. One is evolutionary: as the world becomes more tightly interconnected, international infrastructures may expand to the point that they influence or even supersede national policy. This does not seem catastrophically risky in and of itself; indeed, in some respects it is going on today. Deliberate molecular manufacturing-related policy also may be implemented, including arms control regimes; the closest present-day analogy is perhaps the International Atomic Energy Agency (IAEA).

Global governance appears to present two major risks of opposite character. The first is that it may create stagnation or oppression causing massive loss of human potential. For example, an arms-control regime might choose not to allow space access except under such tight restrictions as to prevent the colonization of space. The second risk is that, in attempting to create a safe and predictable (static) situation, policy inadequately planned or implemented may instead lead to instability and backlash. The effects of the backlash might be exacerbated by the loss of distributed experience in dealing with hostile uses of the technology that would have been a result of restrictive policies.

We cannot know in advance whether a dictatorial regime, in order to ensure security and stability, would choose to engage in genocide, but if the desire existed, the means would be there. Even if it were simply a matter of permanent 'humane' subjugation, the prospect of a world hopelessly enslaved can be regarded as a terminal catastrophe. (Chapter 22 from this volume has more on the threat of totalitarianism.)

21.2.3.4 Radically enhanced intelligences

Nanotechnology, together with bioengineering, may give us the ability to radically enhance our bodies and brains. Direct human–computer connections almost certainly will be employed, and it is possible to imagine a situation in which one or more enhanced humans continue the process of augmentation to an extreme conclusion. Considering the power that merely human demagogues, tycoons, and authorities can wield, it is worth considering the potential results of enhanced post-humans who retain human competitive drives. Fully artificial intelligences – a likely consequence of nano-built computers – may pose similar issues, and Eliezer Yudkowsky (Chapter 15, this volume) argues that they may also cause existential risks due to unwisely specified goals.

21.2.3.5 Environmental degradation

Extensive use of molecular manufacturing could produce environmental impacts in any of several ways. Some types of environmental impact might represent disastrous, even existential, risks. Deliberate destruction of the environment on a globally significant scale would probably happen only as a side effect of a war that would do greater damage in other ways. It is conceivable that unwise development could do enough damage to threaten planetary ecology, but this is already a risk today.

The ease and speed with which planet-scale engineering could be carried out may tempt huge well-intended projects that backfire in ecological catastrophe. If the earth's climate exists at a delicate balance between a runaway greenhouse effect and a sudden ice age, as some experts think (FEASTA, 2004), then large-scale releases of waste heat from overuse of billions of nanofactories and other nanomachinescould inadvertently tip the scale. Or, a seemingly benign effort such as the creation of huge areas of land or sea covered with solar collecting material might change our planet's albedo enough to drastically alter the climate. (Climate change risk is covered in more depth in Chapter 13, this volume.)

Another possibility is ecosystem collapse as a secondary effect of massive infrastructure change. Such change could be rapid, perhaps too rapid to notice its effects until they are well underway. Computer models (Sinclair and Arcese, 1995) have shown that declining biodiversity, for example, can reach a tipping point and slide rapidly into widespread devastation, affecting whole food chains, including perhaps our own.

21.2.3.6 Ecophagy

When nanotechnology-based manufacturing was first proposed (Drexler, 1986), a concern arose that tiny manufacturing systems might run amok and 'eat' the biosphere, reducing it to copies of themselves. However, current research by Drexler (Burch, 2005) and others (Phoenix, 2003) makes it clear that small self-contained 'replicating assemblers' will not be needed for manufacturing – nanofactories will be far more efficient at building products, and a nanofactory contains almost none of the functionality of a free-range replicating robot. Development and use of molecular manufacturing appears to pose no risk of creating free-range replicators by accident at any point.

Deliberately designing a functional self-replicating free-range nanobot would be no small task. In addition to making copies of itself, the robot also would have to survive in the environment, move around (either actively or by drifting – if it were small enough), find usable raw materials, and convert what it finds into feedstock and power, which entails sophisticated chemistry. The robot also would require a relatively large computer to store and process the full blueprint of such a complex device. A nanobot or nanomachine missing any part of this functionality could not function as a free-range replicator (Phoenix and Drexler, 2004). Despite this, there is no known reason why such a thing would be theoretically impossible.

Although free-range replicators have no apparent commercial value, no significant military value, and only limited terrorist value, they might someday be produced by irresponsible hobbyists or by sects with apocalyptic creeds (compare Chapter 4, this volume), or might even be used as a tool for large-scale blackmail. Cleaning up an outbreak would likely be impossible with today's technology, and would be difficult or perhaps impossible even with molecular manufacturing technology. At the least, it probably would require severe physical disruption in the area of the outbreak (airborne and waterborne devices deserve special concern for this reason).

Possible ways to cope with outbreaks include irradiating or heating the infested area, encapsulating the area, spreading some chemical that would gum up the replicator's material intakes (if such a weakness can be discovered), or using a large population of robots for cleanup. Although cleanup robots would require advanced designs, they might be the least disruptive alternative, especially for infestations that are widespread by the time corrective measures are deployed. An exponentially replicating free-range replicator population would *not* require an exponentially replicating population of robots to combat it. Since the time to nullify a robot is dominated by the time to find it in the first place, which is inversely proportional to the concentration in the environment, a constant population of cleanup robots would exert a constant pressure on the replicator population regardless of replicator concentration (see Freitas, 2000 for further discussion of these issues).

In theory, a replicator that was built of materials indigestible to biology (such as diamond), and was not combated by a technological cleanup method, might be able to sequester or destroy enough biomass to destroy the planetary ecology. A hypothetical airborne replicator might block significant amounts of sunlight, and a widespread replicator operating at full speed could produce destructive amounts of heat even before it consumed most of the biomass (Freitas, 2000). Although much more research will be needed to understand the possibilities and countermeasures, at this point the future possibility of a global catastrophe cannot be ruled out.

21.3 Mitigation of molecular manufacturing risks

In the preceding discussion of nanotech-related global catastrophic risks, the greatest near-term threats appear to be war and dictatorship. As we have seen, the two are closely related, and attempts to prevent one may lead to the other. Although the focus of this chapter is on understanding risk, we will take a brief look at some potential mitigating solutions.

One of the biggest unanswered questions is the balance between offence and defence in a nano-enabled war. If it turns out that defence is relatively easy compared to offence, then multiple coexisting agents could be stable even if not all of them are reliable. Indeed, preparing to defend against unreliable or hostile states might create an infrastructure that could mitigate the impacts of private misuse of molecular manufacturing. Conversely, if it turns out that resisting offensive weaponry requires a lot more resources than deploying the weaponry, there will be a strong temptation to maintain lopsided concentrations of power, comparable to or even more extreme than today's imbalance between states and citizens, or between nuclear superpowers and non-nuclear nations. Such power imbalances might increase the risk of oppression.

One possible solution to maintaining a stable defence in a situation where defence is difficult is to design new institutions that can maintain and administer centralized global power without becoming oppressive, and to plan and implement a safe evolution to those institutions. Another possibility would be to devise safe and predictable versions of today's untrustworthy institutions, and transition to them. Each of these solutions has serious practical problems. Some kind of limits will have to be set in order to prevent runaway processes, including human economic activity, from doing excessive amounts of damage. Self-regulation may not be sufficient in all cases, implying at least some need for central coordination or administration.

A research organization that focused on developing defensive technologies and disseminated its work freely to everyone might help to improve stability. It is admittedly unusual in today's military or commercial systems to have an organization that is completely transparent; however, transparency could lead to trust, which would be a key factor in the success of this approach. If defence could be made easier relative to offence, then stability may be increased. Groups that wanted to defend themselves would be able to do so without engaging in intensive military research. Some researchers who would have worked on offensive weaponry might be attracted to the defence organization instead. Ideally, most nations and other organizations should support the goals of such an organization, especially if the alternative is known to be instability leading to war, and this level of support might be sufficient to protect against efforts to subvert or destroy the organization.

For private or civilian use of molecular manufacturing, it might be a good idea to make more convenient and less powerful capabilities than nanofactories available. For example, nanofactories could be used to make micron-scale blocks containing advanced functionality that could be rapidly assembled into products. The block-assembly operation would likely be far faster and less energy-intensive than the manufacture of the blocks. Product design would also be simpler, since products would contain far fewer blocks than atoms, and the designers would not have to think about nanoscale physics. Meanwhile, especially dangerous capabilities could be omitted from the blocks.

Another unanswered question is how quickly the technological components of molecular manufacturing will be developed and adopted. Competition from other technologies and/or a widespread failure to recognize molecular manufacturing's full potential may blunt development efforts. In the absence of targeted and energetic development, it seems likely that preparation of a supporting infrastructure for molecular manufacturing – from CAD software to feedstock supply – would be uneven, and this could limit the consequences of rapid development and deployment of new products. On the other hand, slower and more limited development would retard the ability of molecular manufacturing technology to mitigate risks, including counteracting ecophagic devices.

These suggestions are admittedly preliminary. The primary need at this point is for further studies to understand the technical capabilities that molecular manufacturing will enable, how rapidly those capabilities may be achieved, and what social and political systems can be stable with those capabilities in the world.

21.4 Discussion and conclusion

Vast parallel arrays of precise molecular tools using inexpensive, readily available feedstock and operating under computer control inside a nanofactory could be capable of manufacturing advanced, high-performance products of all sizes and in large quantity. It is this type of nanotechnology – molecular manufacturing – that offers the greatest potential benefits and also poses the worst dangers. None of the many other nanoscale technologies currently in development or use appear likely to present global catastrophic risks.

A nanofactory, as currently conceived, would be able to produce another nanofactory on command, leading to rapid exponential growth of the means of production. In a large project requiring scale-up to large amounts of fabrication capacity, the nanofactories could still weigh a fraction of the final product, thanks to their high throughput. This implies that the resources required to produce the product – not the nanofactories – would be the limiting factor. With nanofactory-aided rapid prototyping, the development and production of revolutionary, transformative products could appear very quickly. If solar cells and chemical processing plants are within the range of nanofactory products, then energy and feedstock could be gathered from abundant sources; with no shortage of manufacturing capital, projects of almost any scale could be undertaken.

As a general-purpose fully automated manufacturing technology, molecular manufacturing could become extremely pervasive and significant. Among the risks examined in this chapter, two stand out well above the others: global war fought with nano-weapons, and domination by nano-enabled governments. The probability of one or the other of these (or both) occurring before the end of the twenty-first century appears to be high, since they seem to follow directly from the military potential of molecular manufacturing. (The argument for instability of a nano-driven arms race in particular deserves careful scrutiny, since if it is incorrect but accepted as plausible, it may lead to unnecessary and destructive attempts at pre-emption.)

In the absence of some type of preventive or protective force, the power of molecular manufacturing products could allow a large number of actors of varying types – including individuals, groups, corporations, and nations – to obtain sufficient capability to destroy all unprotected humans. The likelihood of at least one powerful actor being insane is not small. The likelihood that devastating weapons will be built and released accidentally (possibly through overly sensitive automated systems) is also considerable. Finally, the likelihood of a conflict between two MAD-enabled powers escalating until one feels compelled to exercise a doomsday option is also non-zero. This indicates that unless adequate defences can be prepared against weapons intended to be ultimately destructive – a point that urgently needs research – the number of actors trying to possess such weapons must be minimized.

In assessing the above risks, a major variable is how long it will take to develop nanotechnology to the point of exponential molecular manufacturing (nanofactories building nanofactories). Opinions vary widely, but nanotechnologists who have studied molecular manufacturing most closely tend to have the shortest estimates. It appears technically plausible to us that molecular manufacturing might be developed prior to 2020, assuming that increasing recognition of its near-term feasibility and value leads to a substantial ($100–1000 million) programme starting around 2010.

If molecular manufacturing were to be delayed until 2040 or 2050, many of its strengths would be achieved by other technologies, and it would not have as much impact. However, this much delay does not seem plausible. Well before 2030, rapid advances in nanoscale technologies and computer modelling will likely make molecular manufacturing development relatively straightforward and inexpensive. It should be noted that once the first nanofactory is developed, the cost of further development may drop sharply while the payback may increase dramatically, thanks to the convergence of high performance, atomic precision, exponential manufacturing, and full automation. As the cost of nanofactory development falls, the number of potential investors increases very rapidly. This implies that the development of capabilities and products based on molecular manufacturing may come quite soon. Given the present lack of interest in related policy research, it seems likely that risks which might be mitigated by wise policy could in fact be faced unprepared.

Suggestions for further reading

Elementary level

Drexler, K.E. (1986). *Engines of Creation* (New York, NY: Anchor Books). While outdated in some respects, this book remains a classic introduction to the potential of nanotechnology. The manufacturing mechanism described, a collection of semi-autonomous cooperating 'assemblers', has been superseded by a model far more similar to conventional factories; this development obsoletes the book's most famous warning, about runaway self-replicators or 'grey goo'. Although *Engines* was written in the political context of the US–Soviet conflict, its broader warnings about military implications and the need for wise policy still have great value. Many of the author's predictions and recommendations are relevant and forward-looking two decades after publication (entire book online at http://www.e-drexler.com/d/06/00/EOC/EOC_Table_of_Contents.html).

Hall, J.S. (2005). *Nanofuture: What's Next for Technology?* (New York: Prometheus Books). This combines a layman's introduction to molecular manufacturing with a description of some of its possible applications. The discussion of implications is scanty and rather optimistic, but the reader will get a good sense of how and why molecular manufactured products could transform – and often revolutionize – many of the technologies we use in daily life.

Mulhall, D. (2002). *Our Molecular Future: How Nanotechnology, Robotics, Genetics, and Artificial Intelligence Will Transform our World* (New York: Prometheus Books). This book explores some of the implications and applications of molecular manufacturing, including its applicability to mitigating a variety of natural disasters. The proposal to deal with threatening asteroids by breaking them apart with self-replicating robots appears both unworkable and unwise, but aside from this, the book provides interesting food for thought as to how molecular manufacturing might be used.

Burch, J. and Drexler, K.E. (2005). *Productive Nanosystems: From Molecules to Superproducts* (available at http://lizardfire.com/html_nano/nano.html). This remarkable animation shows clearly how a nanofactory might work to combine vast numbers of atoms into a single integrated product. The molecular manipulation steps shown have been simulated with advanced quantum chemistry methods (online at http://tinyurl.com/9xgs4, http://ourmolecularfuture.com).

Phoenix, C. and Treder, M. (2003). Safe utilization of advanced nanotechnology (available at http://crnano.org/safe.htm). Early descriptions of molecular manufacturing raised fears that the manufacturing systems might somehow get loose and start converting valuable biomass into pointless copies of themselves. This paper describes a manufacturing system made of components that would all be entirely inert if separated from the factory. In addition, the paper describes some ways to make the factory system less prone to deliberate misuse or abuse.

Intermediate level

Freitas, R.A., Jr and Merkle, R.C. (2004). *Kinematic Self-Replicating Machines* (Georgetown, TX: Landes Bioscience). In its broadest sense, 'replication' means simply the building of a copy; the replicator need not be autonomous or even physical (it could be simulated). KSRM is worth reading for several reasons. First, it contains a wide-ranging and near-encyclopedic survey of replicators of all types. Second, it contains a taxonomy of 12 design dimensions containing 137 design properties, most of which can be varied independently to describe new classes of replicators. Third, it includes a discussion on the design and function of several molecular manufacturing-style replicators (again, in the broad sense; these are not the dangerous autonomous kind). The book puts molecular manufacturing in a broader context and is a rich source of insights (entire book online at http://molecularassembler.com/KSRM.htm).

Phoenix, C. (2003). Design of a primitive nanofactory. *J. Evol. Technol.*, 13 (available at http://jetpress.org/volume13/Nanofactory.htm). This lengthy paper explores many of the design issues that will have to be solved in order to build a nanofactory. Design concepts have improved somewhat since the paper was written, but this only strengthens its conclusion that nanofactories will be able to be pressed into service rapidly in a wide variety of applications (online at http://jetpress.org/volume13/Nanofactory.htm).

Phoenix, C. and Treder, M. (2003). Three systems of action: A proposed application for effective administration of molecular nanotechnology (available at http://crnano.org/studies.htm). Different types of issues need to be addressed in different ways. As explored in this paper, the approaches required for security issues, economic activity, and easily copied information are so dissimilar that an organization would have a very hard time encompassing them all without ethical conflicts. Molecular manufacturing will create issues of all three types, which will require a delicate and carefully planned interplay of the three approaches (online at http://crnano.org/systems.htm).

Phoenix, C. (2003). Thirty essential studies (available at http://crnano.org/studies.htm). Molecular manufacturing needs to be studied from many different angles, including technological capability, strategies for governance, and interactions with existing world systems. These studies raise a broad spectrum of necessary questions, and provide preliminary answers – which are not especially reassuring (online at http://crnano.org/studies.htm).

Advanced level

Freitas, R. A. Jr (1999). *Nanomedicine, Volume I: Basic Capabilities* (Austin, TX: Landes Bioscience). This book was written to lay a foundation for subsequent volumes applying molecular manufacturing to medicine. As a fortunate side-effect, it also lays the foundation for many other applications of the technology. Although some of the chapters are heavily medical, much of the book deals with technical capabilities such as sensing, power transmission, and molecular sorting. Scattered throughout are many useful physics formulas applied to real-world nanoscale, microscale, and macroscale problems. As such, it is useful on its own or as a companion to Nanosystems (entire book online at http://nanomedicine.com/NMI.htm).

Drexler, K. E. (1992). *Nanosystems: Molecular Machinery, Manufacturing and Computation* (New York: John Wiley & Sons). This book is the foundational textbook and reference book of the field. Although not accessible to non-technical readers, it makes a clear technical case, grounded in well-established physics and chemistry, that molecular manufacturing can work as claimed. Several of its key theoretical extrapolations, such as superlubricity, have since been demonstrated by experiment. Although others remain to be confirmed, no substantial error has yet been found (partially available online at www.edrexler.com/d/06/00/Nanosystems/toc.html).

References

Altmann, J. (2006). *Military Nanotechnology: Potential Applications and Preventive Arms Control* (Hampshire: Andover).

Bailey, R. (2003). SARS Wars. *ReasonOnline*, Apr. 30. http://www.reason.com/news/show/34803.html

Burch, J. (Lizard Fire Studios) and Drexler, E. (Nanorex) (2005). *Productive Nanosystems: From Molecules to Superproducts* (animated film). http://lizardfire.com/html_nano/nano.html

Dienwiebel et al. (2004). Superlubricity of graphite. *Physical Review Letters*, **92**, 126101.

Drexler, K.E. (1986). *Engines of Creation* (New York, NY: Anchor Books).

Drexler, K.E. (1992). *Nanosystems: Molecular Machinery, Manufacturing, and Computation* (New York: John Wiley & Sons).

FEASTA (Foundation for the Economics of Sustainability) (2004). World climate liable to sudden, rapid change. http://www.feasta.org/documents/feastareview/climatechangepanel.htm

Freitas, R.A (1998). Exploratory design in medical nanotechnology: a mechanical artificial red cell. *Artificial Cells, Blood Substitutes, and Immobilization Biotechnology*, **26**, 411–430. http://www.foresight. org/Nanomedicine/Respirocytes.html

Freitas, R.A. (1999). *Nanomedicine, Volume I: Basic Capabilities* (Austin, TX: Landes Bioscience).

Freitas, R.A (2000). Some limits to global ecophagy by biovorous nanoreplicators, with public policy recommendations. http://www.rfreitas.com/Nano/Ecophagy.htm

Freitas, R.A., Jr and Merkle, R.C. (2004). *Kinematic Self-Replicating Machines*. (Georgetown, TX: Landes Bioscience).

Freitas, R.A. (2005). Microbivores: artificial mechanical phagocytes using digest and discharge protocol. *J. Evol. Technol.*, **14**. http://www.jetpress.org/volume14/freitas.html

Freitas, R.A. (2007). The ideal gene delivery vector: chromallocytes, cell repair nanorobots for chromosome replacement therapy. *J.Evol. Technol.*, **16**(1), 1–97. http://jetpress.org/volume16/freitas.html

Gubrud, M. (1997). Nanotechnology and international security. http://www.foresight.org/conference/MNT05/Papers/Gubrud/

Hall, J.S. (1996). Utility fog: the stuff that dreams are made of. In Crandall, B.C. (ed.), *Nanotechnology: Molecular Speculations on Global Abundance*, pp. 161–184 (Cambridge, MA: MIT Press).

Hall, J.S. (2005) *Nanofuture: What's Next for Nanotechnology*. (New York: Prometheus Books).

ISN (Institute for Soldier Nanotechnologies, Massachusetts Institute of Technology) (2005). Biomaterials and nanodevices for soldier medical technology. http://web.mit.edu/isn/research/team04/index.html

Joy, B. (2000). Why the future doesn't need us. *Wired Magazine*, April. http://www.wired.com/wired/archive/8.04/joy.html

Lake, A. (1996). Defining Missions, Setting Deadlines. *Defense Issues*, **11**(14), 1–4.

McCarthy, T. (2005). Molecular nanotechnology and the world system. http://www.mccarthy.cx/WorldSystem/

Mulhall, D. (2002). *Our Molecular Future: How Nanotechnology, Robotics, Genetics and Artificial Intelligence Will Transform Our World.* (New York: Prometheus).

Orr, S.D. (2003). Globalization and democracy. *A World Connected.* http://aworld connected.org/article.php/527.html

Phoenix, C. (2003). Design of a primitive nanofactory. *J. Evol. Technol.*, 13. http://www.jetpress.org/volume13/Nanofactory.htm

Phoenix, C. and Treder, M. (2003). Three Systems of Action: A Proposed Application for the Effective Administration of Molecular Nanotechnology. Center for Responsible Nanotechnology.

Phoenix, C. and Treder, T. (2003). Safe Utilization of Advanced Nanotechnology. Center for Responsible Nanotechnology.

Phoenix, C. (2005). Molecular manufacturing: what, why and how. http://wise-nano.org/w/Doing_MM

Phoenix, C. (2006). Preventing errors in molecular manufacturing. http://www.crnano.org/essays06.htm#11,Nov

Phoenix, C. and Drexler, E. (2004). Safe exponential manufacturing. *Nanotechnology*, **15**, 869–872.

Sinclair, A.R.E. and Arcese P. (eds.) (1995). *Serengeti II: Dynamics, Management and Conservation of an Ecosystem* (Chicago, IL: Chicago University Press).

Treder, M. (2005). War, interdependence, and nanotechnology. http://www.futurebrief.com/miketrederwar002.asp

·22·
The totalitarian threat

Bryan Caplan

If you want a picture of the future, imagine a boot stamping on a human face – forever.
–*George Orwell,* 1984 (1983, *p.* 220)

22.1 Totalitarianism: what happened and why it (mostly) ended

During the twentieth century, many nations – including Russia, Germany, and China – lived under extraordinarily brutal and oppressive governments. Over 100 million civilians died at the hands of these governments, but only a small fraction of their brutality and oppression was necessary to retain power. The main function of the brutality and oppression, rather, was to radically change human behaviour, to transform normal human beings with their selfish concerns into willing servants of their rulers. The goals and methods of these governments were so extreme that they were often described – by friend and foe alike – as 'total' or 'totalitarian' (Gregor, 2000).

The connection between totalitarian goals and totalitarian methods is straightforward. People do not *want* to radically change their behaviour. To make them change requires credible threats of harsh punishment – and the main way to make such threats credible is to carry them out on a massive scale. Furthermore, even if people believe your threats, some will resist anyway or seem likely to foment resistance later on. Indeed, some are simply *unable* to change. An aristocrat cannot choose to have proletarian origins, or a Jew to be an Aryan. To handle these recalcitrant problems requires special prisons to isolate dangerous elements, or mass murder to eliminate them.

Totalitarian regimes have many structural characteristics in common. Richard Pipes gives a standard inventory: '[A]n official all-embracing ideology; a single party of the elect headed by a "leader" and dominating the state; police terror; the ruling party's control of the means of communication and the armed forces; central command of the economy' (1994, p. 245). All of these naturally flow from the goal of remaking human nature. The official ideology is the rationale for radical change. It must be 'all-embracing' – that is, suppress competing ideologies and values – to prevent people from being side-tracked by conflicting goals. The leader is necessary to create and interpret the official ideology, and control of the means of communication to disseminate it. The party is comprised of the 'early-adopters' – the people who claim to have 'seen the light' and want to make it a reality. Police terror and control of the armed forces are necessary to enforce obedience to the party's orders. Finally, control of the economy is crucial for a whole list of reasons: to give the party the resources it needs to move forward; to suppress rival power centres; to ensure that economic actors do not make plans that conflict with the party's; and to make citizens dependent on the state for their livelihood.

This description admittedly glosses over the hypocrisy of totalitarian regimes. In reality, many people join the party because of the economic benefits of membership, not because they sincerely share the party's goals. While it is usually hard to doubt the ideological sincerity of the founding members of totalitarian movements, over time the leadership shifts its focus from remaking human nature to keeping control. Furthermore, while totalitarian movements often describe their brutality and oppression as transitional measures to be abandoned once they purify the hearts and minds of the people, their methods usually severely alienate the subject population. The 'transition' soon becomes a way of life.

The Soviet Union and Nazi Germany are the two most-studied totalitarian regimes. By modern calculations, the Soviets killed approximately 20 million civilians, the Nazis 25 million (Courtois et al., 1999, pp. 4–5, 14–15; Payne, 1995). However, these numbers are biased by the relative difficulty of data collection. Scholars could freely investigate most Nazi atrocities beginning in 1945, but had to wait until the 1990s to document those of the Soviets. In all likelihood, the death toll of the Soviets actually exceeded those of the Nazis.

One of the main differences between the Soviet Union and Nazi Germany was that the former became totalitarian very rapidly. Lenin embarked upon radical social change as soon as he had power (Malia, 1994). In contrast, totalitarianism developed gradually in Nazi Germany; only in the last years of World War II did the state try to control virtually every area of life (Arendt, 1973). The other main difference is that most of the atrocities of the Soviet Union were directed inwards at its own citizens, whereas most Nazi atrocities were directed outwards at the citizens of occupied countries (Noakes and Pridham, 2001; Friedlander, 1995).

But despite historians' focus on Russia and Germany, Maoist China was actually responsible for more civilian killings than the Soviet Union and Nazi Germany put together. Modern estimates put its death toll at 65 million (Margolin, 1999a). The West is primarily familiar with the cruelties inflicted on Chinese intellectuals and Party members during the Cultural Revolution, but its death toll was probably under one million. The greatest of Mao's atrocities was the Great Leap Forward, which claimed 30 million lives through man-made starvation (Becker, 1996).

Besides mass murder, totalitarian regimes typically engage in a long list of other offences. Slave labour was an important part of both the Soviet and Nazi economies. Communist regimes typically placed heavy restrictions on migration – most notably making it difficult for peasants to move to cities and for anyone to travel abroad. Freedom of expression and religion were heavily restricted. Despite propaganda emphasizing rapid economic growth, living standards of non-party members frequently fell to the starvation level. Totalitarian regimes focus on military production and internal security, not consumer well-being.

Another notable problem with totalitarian regimes was their failure to anticipate and counteract events that even their leaders saw as catastrophic. Stalin infamously ignored overwhelming evidence that Hitler was planning to invade the Soviet Union. Hitler ensured his own defeat by declaring war on the United States. Part of the reason for these lapses of judgement was concentration of power, which allowed leaders' idiosyncrasies to decide the fates of millions. But this was amplified by the fact that people in totalitarian regimes are afraid to share negative information. To call attention to looming disasters verges on dissent, and dissent is dangerously close to disloyalty.

From the viewpoint of the ruling party, this may be a fair trade: more and worse disasters are the price of social control. From the viewpoint of anyone concerned about global catastrophic risks, however, this means that totalitarianism is worse than it first appears. To the direct cost of totalitarianism we must add the indirect cost of amplifying other risks. It is important not to push this argument too far, however. For goals that can be achieved by brute force or mobilizing resources, totalitarian methods have proven highly effective. For example, Stalin was able to develop nuclear weapons with amazing speed simply by making this the overarching priority of the Soviet economy (Holloway, 1994). Indeed, for goals that can only be achieved by radically changing human behaviour, *nothing but* totalitarian methods have proven highly effective. Overall, totalitarian regimes are less likely to foresee disasters, but are in some ways better-equipped to deal with disasters that they take seriously.

22.2 Stable totalitarianism

There are only four ways in which a ruling group can fall from power. Either it is conquered from without, or it governs so inefficiently that the masses are stirred to revolt, or it allows a strong and discontented Middle Group to come into being, or it loses its own self-confidence and willingness to govern ... A ruling class which could guard against all of them would remain in power permanently. Ultimately the determining factor is the mental attitude of the ruling class itself.

–George Orwell, 1984 *(1983, p. 170)*

The best thing one can say about totalitarian regimes is that the main ones did not last very long.[1] The Soviet Union greatly reduced its level of internal killing after the death of Stalin, and the Communist Party fell from power in 1991. After Mao Zedong's death, Deng Xiaoping allowed the Chinese to resume relatively normal lives, and began moving in the direction of a market economy. Hitler's Thousand-Year Reich lasted less than thirteen years, before it ended with military defeat in World War II.

The deep question, however, is whether this short duration was *inherent* or *accidental*. If the short lifespan of totalitarianism is inherent, it probably does not count as a 'global catastrophic risk' at all. On the other hand, if the rapid demise of totalitarianism was a lucky accident, if future totalitarians could learn from history to indefinitely prolong their rule, then totalitarianism is one of the most important global catastrophic risks to stop before it starts.

The main obstacle to answering this question is the small number of observations. Indeed, the collapse of the Soviet bloc was so inter-connected that it basically counts as only one data point. However, most of the historical evidence supports the view that totalitarianism could have been much more durable than it was.

This is clearest in the case of Nazi Germany. Only crushing military defeat forced the Nazis from power. Once Hitler became dictator, there was no serious internal opposition to his rule. If he had simply pursued a less aggressive foreign policy, there is every reason to think he would have remained dictator for life. One might argue that grassroots pressure forced Hitler to bite off more than he could militarily chew, but in fact the pressure went the other way. His generals in particular favoured a *less* aggressive posture (Bullock, 1993, pp. 393–394, 568–574, 582).

The history of the Soviet Union and Maoist China confirms this analysis. They were far less expansionist than Nazi Germany, and their most tyrannical leaders – Stalin and Mao – ruled until their deaths. But at the same time, the demise of Stalin and Mao reveals the stumbling block that the Nazis would have eventually faced too: succession. How can a totalitarian regime ensure that each generation of leaders remains stridently totalitarian? Both Stalin and Mao fumbled here, and perhaps Hitler would have done the same.

A number of leading Communists wrestled for Stalin's position, and the eventual winner was Nikita Khrushchev. Khrushchev kept the basic structure of Stalinist Russia intact, but killed far fewer people and released most of the

[1] In contrast, authoritarianism has historically been quite durable, and even today arguably remains the most common form of government.

slave labourers. He even readmitted many of Stalin's victims back into the Party, and allowed anti-Stalinists like Solzhenitsyn to publish some of their writings. The result was a demoralization of the Party faithful, both inside the Soviet Union and abroad: Stalin was a tyrant not merely according to the West, but to the new Party line as well (Werth, 1999, pp. 250–260).

Khrushchev was eventually peacefully removed from power by other leading Communists who might be described as 'anti-anti-Stalinists'. While they did not restore mass murder of Soviet citizens or large-scale slave labour, they squelched public discussion of the Party's 'mistakes'. As the Party leadership aged, however, it became increasingly difficult to find a reliable veteran of the Stalin years to take the helm. The 54-year-old Mikhail Gorbachev was finally appointed General Secretary in 1985. While it is still unclear what his full intentions were, Gorbachev's moderate liberalization measures snowballed. The Soviet satellite states in Eastern Europe collapsed in 1989, and the Soviet Union itself disintegrated in 1991.

The end of totalitarianism in Maoist China happened even more quickly. After Mao's death in 1976, a brief power struggle led to the ascent of the pragmatist Deng Xiaoping. Deng heavily reduced the importance of Maoist ideology in daily life, de facto privatized agriculture in this still largely agricultural economy, and gradually moved towards more free-market policies, under the guise of 'socialism with Chinese characteristics'. China remained a dictatorship, but had clearly evolved from totalitarian to authoritarian (Salisbury, 1992).

It is tempting for Westerners to argue that the Soviet Union and Maoist China changed course because their systems proved unworkable, but this is fundamentally incorrect. These systems were *most* stable when their performance was *worst*. Communist rule was very secure when Stalin and Mao were starving millions to death. Conditions were comparatively good when reforms began. Totalitarianism ended not because totalitarian policies were unaffordable, but because new leaders were unwilling to keep paying the price in lives and wealth.

Perhaps there was no reliable way for totalitarian regimes to solve the problem of succession, but they could have tried a lot harder. If they had read George Orwell, they would have known that the key danger to the system is 'the growth of liberalism and scepticism in their own ranks' (1983, p. 171). Khrushchev's apostasy from Stalinism was perhaps unforeseeable, but the collapse of the Soviet Union under Gorbachev could have been avoided if the Politburo considered only hard-line candidates. The Soviet Union collapsed largely because a reformist took the helm, but a reformist was able to take the helm only because his peers failed to make holding power their top priority. Mao, similarly, could have made continuity more likely by sending suspected 'capitalist-roaders' like Deng to their deaths without exception.

Probably the most important reason why a change in leaders often led totalitarian regimes to moderate their policies is that they existed side by side with non-totalitarian regimes. It was obvious by comparison that people in the non-totalitarian world were richer and happier. Totalitarian regimes limited contact with foreigners, but news of the disparities inevitably leaked in. Even more corrosively, party elites were *especially* likely to see the outside world first-hand. As a result, officials at the highest levels lost faith in their own system.

This problem could have been largely solved by cutting off contact with the non-totalitarian world, becoming 'hermit kingdoms' like North Korea or Communist Albania. But the hermit strategy has a major drawback. Totalitarian regimes have trouble growing and learning as it is; if they cannot borrow ideas from the rest of the world, progress slows to a crawl. But if other societies are growing and learning and yours is not, you will lose the race for political, economic, and military dominance. You may even fall so far behind that foreign nations gain the ability to remove you from power at little risk to themselves.

Thus, a totalitarian regime that tried to preserve itself by turning inwards could probably increase its life expectancy. For a few generations the pool of potential successors would be less corrupted by alien ideas. But in the long run the non-totalitarian neighbours of a hermit kingdom would overwhelm it.

The totalitarian dilemma, then, is that succession is the key to longevity. But as long as totalitarian states co-exist with non-totalitarian ones, they have to expose potential successors to demoralizing outside influences to avoid falling dangerously behind their rivals.[2]

To understand this dilemma, however, is also to understand its solution: totalitarianism would be much more stable if there *were* no non-totalitarian world. The worse-case scenario for human freedom would be a global totalitarian state. Without an outside world for comparison, totalitarian elites would have no direct evidence that any better way of life was on the menu. It would no longer be possible to borrow new ideas from the non-totalitarian world, but it would also no longer be necessary. The global government could economically and scientifically stagnate without falling behind. Indeed, stagnation could easily increase stability. The rule of thumb 'Avoid all change' is easier to correctly apply than the rule, 'Avoid all change that makes the regime less likely to stay in power.'

It is fashionable to paint concerns about world government as xenophobic or even childish. Robert Wright maintains that the costs would be trivial and

[2] Given their durability, it is not surprising that authoritarian regimes face only a highly attenuated version of this dilemma. Many authoritarian regimes provide reasonable levels of prosperity and happiness for their citizens, so exposure to the outside world is no more than mildly demoralizing. Authoritarian regimes can therefore safely allow their people enough contact with other countries to economically and scientifically keep up.

the benefits large:

> Do you cherish the freedom to live without fear of dying in a biological weapons attack? Or do you prefer the freedom to live without fear of having your freezer searched for anthrax by an international inspectorate in the unlikely event that evidence casts suspicion in your direction? ... Which sort of sovereignty would you rather lose? Sovereignty over your freezer, or sovereignty over your life? (2000, p. 227)

But this is best-case thinking. In reality, a world government might impose major costs for minor gains. If mankind is particularly unlucky, world government will decay into totalitarianism, without a non-totalitarian world to provide a safety valve.

Totalitarianism would also be more stable than it was in the twentieth century if the world were divided between a small number of totalitarian states. This is Orwell's scenario in 1984: Oceania, Eurasia, and Eastasia control the earth's surface and wage perpetual war against one another. In a world like this, elites would not be disillusioned by international comparisons, because every country would feature the poverty and misery typical of totalitarian societies. Deviating from the hermit strategy would no longer taint the pool of successors. Orwell adds the interesting argument that the existence of external enemies helps totalitarian regimes maintain ideological fervour: 'So long as they remain in conflict they prop one another up, like three sheaves of corn' (1983, p. 162).

For this scenario to endure, totalitarian regimes would still have to make sure they did not fall far behind their rivals. But they would only have to keep pace with other relatively stagnant societies. Indeed, if one state fell so far behind its rivals that it could be conquered at little risk, the world would simply be one step closer to a unified totalitarian government. The greater danger would be an increase in the number of states via secession. The more independent countries exist, the greater the risk that one will liberalize and begin making the rest of the world look bad.

22.3 Risk factors for stable totalitarianism

On balance, totalitarianism could have been a lot more stable than it was, but also bumped into some fundamental difficulties. However, it is quite conceivable that technological and political changes will defuse these difficulties, greatly extending the lifespan of totalitarian regimes. Technologically, the great danger is anything that helps solve the problem of succession. Politically, the great danger is movement in the direction of world government.

22.3.1 Technology

Orwell's 1984 described how new technologies would advance the cause of totalitarianism. The most vivid was the 'telescreen', a two-way television set. Anyone watching the screen was automatically subject to observation by the Thought Police. Protagonist Winston Smith was only able to keep his diary of thought crimes because his telescreen was in an unusual position which allowed him to write without being spied upon.

Improved surveillance technology like the telescreen would clearly make it easier to root out dissent, but is unlikely to make totalitarianism last longer. Even *without* telescreens, totalitarian regimes were extremely stable as long as their leaders remained committed totalitarians. Indeed, one of the main lessons of the post-Stalin era was that a nation can be kept in fear by jailing a few thousand dissidents per year.

Better surveillance would do little to expose the real threat to totalitarian regimes: closet sceptics within the party. However, other technological advances might solve this problem. In Orwell's 1984, one of the few scientific questions still being researched is 'how to discover, against his will, what another human being is thinking' (1983, p. 159). Advances in brain research and related fields have the potential to do just this. Brain scans, for example, might one day be used to screen closet sceptics out of the party. Alternately, the new and improved psychiatric drugs of the future might increase docility without noticeably reducing productivity.

Behavioural genetics could yield similar fruit. Instead of searching for sceptical thoughts, a totalitarian regime might use genetic testing to defend itself. Political orientation is already known to have a significant genetic component (Pinker, 2002, pp. 283–305). A 'moderate' totalitarian regime could exclude citizens with a genetic predisposition for critical thinking and individualism from the party. A more ambitious solution – and totalitarian regimes are nothing if not ambitious – would be genetic engineering. The most primitive version would be sterilization and murder of carriers of 'anti-party' genes, but you could get the same effect from selective abortion. A technologically advanced totalitarian regime could take over the whole process of reproduction, breeding loyal citizens of the future in test tubes and raising them in state-run 'orphanages'. This would not have to go on for long before the odds of closet sceptics rising to the top of their system and taking over would be extremely small.

A very different route to totalitarian stability is extending the lifespan of the leader so that the problem of succession rarely if ever comes up. Both Stalin and Mao ruled for decades until their deaths, facing no serious internal threats to their power. If life extension technology had been advanced enough to keep them in peak condition forever, it is reasonable to believe that they would still be in power today.

Some of Orwell's modern critics argue that he was simply wrong about the effect of technology (Huber, 1994). In practice, technology – from fax machines to photocopiers – undermined totalitarianism by helping people behind the Iron Curtain learn about the outside world and organize resistance. Whatever the effect of past technology was, however, the effect of future technology is hard to predict. Perhaps genetic screening will be used not to prevent the births of a future Alexandr Solzhenitsyn or Andrei Sakharov, but a future Stalin. Nevertheless, future technology is likely to eventually provide the *ingredients* that totalitarianism needs to be stable.

At the same time, it should be acknowledged that some of these technologies might lead totalitarianism to be less violent than it was historically. Suppose psychiatric drugs or genetic engineering created a docile, homogeneous population. Totalitarian ambitions could then be realized without extreme brutality, because people would want to do what their government asked – a possibility explored at length in the dystopian novel *Brave New World* (Huxley, 1932).

22.3.2 Politics

To repeat, one of the main checks on totalitarian regimes has been the existence of non-totalitarian regimes. It is hard to maintain commitment to totalitarian ideologies when relatively free societies patently deliver higher levels of wealth and happiness with lower levels of brutality and oppression. The best way for totalitarian societies to maintain morale is the hermit strategy: cut off contact with the non-totalitarian world. But this leads to stagnation, which is only viable in the long run if the rest of the world is stagnant as well.

From this perspective, the most dangerous political development to avoid is world government.[3] A world totalitarian government could permanently ignore the trade-off between stability and openness.

How likely is the emergence of world government? Recent trends towards secession make it improbable in the immediate future (Alesina and Spolaore, 2003). At the same time, however, nominally independent countries have

[3] In correspondence, Nick Bostrom raises the possibility that the creation of a democratic world government might provide better protection against the emergence of a totalitarian world government than the status quo does. In my view, however, major moves in the direction of world government will happen either democratically or not at all. Any world government is going to begin democratically. The problem is that once a world democratic government exists, there is at least a modest probability that it becomes totalitarian, and if it does, the existence of a non-totalitarian world will no longer exist to provide a safety valve. Bostrom (2006) suggests that a technological breakthrough in something like nanotechnology or artificial intelligence might be a non-democratic route to world government. The discovering nation or bloc of nations could leverage the breakthrough to subjugate the rest of the world. In my view, though, modern communications make it highly implausible that the first mover could retain a monopoly on such a breakthrough for long enough to realign world politics.

begun to surrender surprising amounts of sovereignty (Barrett, 2003; Wright, 2000). The most striking example is the European Union. Without military conquest, the long-quarrelling peoples of Western and Central Europe have moved perhaps half-way to regional government. Membership is attractive enough that many neighbouring countries want to join as well. It is quite conceivable that within a century the continent of Europe will be as unified as the United States is today.

European unification also increases the probability of copycat unions in other parts of the world. For example, if the European Union adopts free trade internally and protectionism externally, other nations will be tempted to create trading blocs of their own. Once economic unions take root, moreover, they are likely to gradually expand into political unions – just as the European Economic Community became the European Union. If it seems fanciful to imagine North and South America becoming a single country in one century, consider how improbable the rise of the European Union would have seemed in 1945. Once the idea of gradual peaceful unification takes root, moreover, it is easy to see how rival super-nations might eventually merge all the way to world government.

The growing belief that the world faces problems that are too big for any one country to handle also makes the emergence of world government more likely. Robert Wright (2000, p. 217) observes that an extraterrestrial invasion would make world government a respectable idea overnight, and argues that other, less fanciful, dangers will gradually do the same:

> [T]he end of the second millennium has brought the rough equivalent of hostile extraterrestrials – not a single impending assault, but lots of new threats that, together, add up to a big planetary security problem. They range from terrorists (with their menu of increasingly spooky weapons) to a new breed of transnational criminals (many of whom will commit their crimes in that inherently transnational realm, cyberspace) to environmental problems (global warming, ozone depletion, and lots of merely regional but supranational issues) to health problems (epidemics that exploit modern thoroughfares).

So far, environmental concerns have probably gotten the most international attention. Environmentalists have argued strongly on behalf of global environmental agreements like the Kyoto Treaty, which has been ratified by much of the world (Barrett, 2003, p. 373–374). But environmentalists often add that even if it were ratified, it would only address one of the environment's numerous problems. A natural inference is that it is hopeless to attack environmental problems one at a time with unanimous treaties. It seems more workable to have an omnibus treaty that binds signatories to a common environmental policy. The next step would be to use economic and other pressures to force hold-outs to join (Barrett, 2003). Once a supranational 'Global Environmental Protection Agency' was in place, it could eventually turn into a full-fledged political union. Indeed, since environmental policy has major implications for the economy, a world environmental agency with the power to accomplish its goals would wind up running a great deal of the traditionally domestic policy of member nations. World government would not inevitably follow, but it would become substantially more likely.

Another political risk factor for totalitarianism is the rise of radical ideologies. At least until they gain power, a tenet common to all totalitarians is that the status quo is terribly wrong and must be changed by any means necessary. For the Communists, the evil to abolish was private ownership of the means of production; for the Nazis, it was the decay and eventual extinction of the Aryan race; for totalitarian religious movements, the great evils are secularization and pluralism. Whenever large movements accept the idea that the world faces a grave danger that can only be solved with great sacrifices, the risk of totalitarianism goes up.

One disturbing implication is that there may be a trade-off between preventing totalitarianism and preventing the other 'global catastrophic risks' discussed in this volume. Yes, in practice, totalitarian regimes failed to spot some major disasters until it was too late. But this fact is unlikely to deter those who purport to *know* that radical steps are necessary to save humanity from a particular danger. Extreme pessimism about the environment, for example, could become the rationale for a Green totalitarianism.

Of course, if humanity is really doomed without decisive action, a small probability of totalitarianism is the lesser evil. But one of the main lessons of the history of totalitarianism is that moderation and inaction are underrated. Few problems turned out to be as 'intolerable' as they seemed to people at the time, and many 'problems' were better than the alternative. Countries that 'did nothing' about poverty during the twentieth century frequently became rich through gradual economic growth. Countries that waged 'total war' on poverty frequently not only choked off economic growth, but starved.

Along these lines, one particularly scary scenario for the future is that overblown doomsday worries become the rationale for world government, paving the way for an unanticipated global catastrophe: totalitarianism. Those who call for the countries of the world to unite against threats to humanity should consider the possibility that unification itself is the greater threat.

22.4 Totalitarian risk management

22.4.1 Technology

Dwelling on technology-driven dystopias can make almost anyone feel like a Luddite. But in the decentralized modern world, it is extremely difficult to prevent the development of any new technology for which there is market demand. Research on the brain, genetics, and life extension all fit that description. Furthermore, all of these new technologies have enormous direct benefits. If people lived forever, stable totalitarianism would be a little more likely to emerge, but it would be madness to force everyone to die of old age in order to avert a small risk of being murdered by the secret police in 1000 years.

In my judgement, the safest approach to the new technologies I have discussed is freedom for individuals combined with heavy scrutiny for government. In the hands of individuals, new technology helps people pursue their diverse ends more effectively. In the hands of government, however, new technology risks putting us on the slippery slope to totalitarianism.

Take genetic engineering. Allowing parents to genetically engineer their children would lead to healthier, smarter, and better-looking kids. But the demand for other traits would be about as diverse as those of the parents themselves. On the other hand, genetic engineering in the hands of government is much more likely to be used to root out individuality and dissent. 'Reproductive freedom' is a valuable slogan, capturing both parents' right to use new technology if they want to, and government's duty not to interfere with parents' decisions.

Critics of genetic engineering often argue that *both* private and government use lie on a slippery slope. In one sense, they are correct. Once we allow parents to screen for genetic defects, some will want to go further and screen for high IQ, and before you know it, parents are ordering 'designer babies'. Similarly, once we allow government to genetically screen out violent temperaments, it will be tempting to go further and screen for conformity. The difference between these slippery slopes, however, is where the slope ends. If *parents* had complete control over their babies' genetic makeup, the end result would be a healthier, smarter, better-looking version of the diverse world of today. If *governments* had complete control over babies' genetic makeup, the end result could easily be a population docile and conformist enough to make totalitarianism stable.

22.4.2 Politics

World government hardly seems like a realistic threat today. Over the course of a few centuries, however, it seems likely to gradually emerge. Nationalism prevents it from happening today, but the emerging global culture has already begun to dilute national identities (Wright, 2000). Cultural protectionism is unlikely to slow this process very much, and in any case the direct benefits of cultural competition are large (Caplan and Cowen, 2004).

In the long run, it is better is to try to influence global culture instead of trying to fragment it. Xenophobia may be the main *barrier* to world government for the time being, but it is a weak *argument*. In fact, there would be good reasons to avoid political centralization even if the world were culturally homogeneous.

Two of the most *visible* reasons to preserve competition between governments are to (1) pressure governments to treat their citizens well in order to retain population and capital; and (2) make it easier to figure out which policies work best by allowing different approaches to exist side by side (Inman and Rubinfeld, 1997). Yes, there are large benefits of international economic integration, but economic integration does not require political integration. Economists have been pointing out for decades that unilateral free trade is a viable alternative to cumbersome free trade agreements (Irwin, 1996).

Reducing the risk of totalitarianism is admittedly one of the less visible benefits of preserving inter-governmental competition. But it is an *important* benefit, and there are ways to make it more salient. One is to publicize the facts about totalitarian regimes. The extent of Hitler's crimes is well known to the general public, but Lenin's, Stalin's, and Mao's are not. Another is to explain how these terrible events came to pass. Even though the risk of genocide is very small in most of the world, historians of the Holocaust have helped the world to see the connection between racial hatred and genocide. Historians could perform a similar service by highlighting the connection between political centralization and totalitarianism. Perhaps such lectures on the lessons of history would fall on deaf ears, but it is worth a try.

22.5 'What's your *p*?'

I am an economist, and economists like to make people assign quantitative probabilities to risks. 'What's your *p* of *X*?' we often ask, meaning 'What probability do you assign to *X* happening?' The point is not that anyone has the definitive numbers. The point, rather, is that explicit probabilities clarify debate, and impose discipline on how beliefs should change as new evidence emerges (Tetlock, 2005). If a person says that two risks are both 'serious', it is unclear which one he sees as the greater threat; but we can stop guessing once he assigns a probability of 2% to one and 0.1% to another. Similarly, if a person says that the probability of an event is 2%, and relevant new information arrives, consistency requires him to revise his probability.

How seriously do I take the possibility that a world totalitarian government will emerge during the next 1000 years and last for a 1000 years or more? Despite the complexity and guesswork inherent in answering this question, I will hazard a response. My unconditional probability – that is, the probability I assign given all the information I now have – is 5%. I am also willing to offer conditional probabilities. For example, if genetic screening for personality traits becomes cheap and accurate, but the principle of reproductive freedom prevails, my probability falls to 3%. Given the same technology with extensive government regulation, my probability rises to 10%. Similarly, if the number of independent countries on Earth does not decrease during the next 1,000 years, my probability falls to 0.1%, but if the number of countries falls to one, my probability rises to 25%.

It is obviously harder to refine my numbers than it is to refine estimates of the probability of an extinction-level asteroid impact. The regularities of social science are neither as exact nor as enduring as the regularities of physical science. But this is a poor argument for taking social disasters like totalitarianism less seriously than physical disasters like asteroids. We compare accurately measured to inaccurately measured things all the time. Which is worse for a scientist to lose: one point of IQ, or his 'creative spark'? Even though IQ is measured with high accuracy, and creativity is not, loss of creativity is probably more important.

Finally, it is tempting to minimize the harm of a social disaster like totalitarianism, because it would probably not lead to human extinction. Even in Cambodia, the totalitarian regime with the highest death rate per-capita, 75% of the population remained alive after 3 years of rule by the Khmer Rouge (Margolin, 1999b). But perhaps an eternity of totalitarianism would be worse than extinction. It is hard to read Orwell and not to wonder:

> Do you begin to see, then, what kind of world we are creating? It is the exact opposite of the stupid hedonistic Utopias that the old reformers imagined. A world of fear and treachery and torment, a world of trampling and being trampled upon, a world which will grow not less but *more* merciless as it refines itself. Progress in our world will be progress towards more pain. The old civilizations claimed that they were founded on love or justice. Ours is founded upon hatred. In our world there will be no emotions except fear, rage, triumph and self-abasement. Everything else we shall destroy – everything ... There will be no loyalty, except loyalty towards the Party. There will be no love, except the love of Big Brother. There will be no laughter, except for the laugh of triumph over a defeated enemy. There will be no art, no literature, no science. When we are omnipotent we shall have no more need of science. There will be no distinction between beauty and ugliness. There will be no curiosity, no enjoyment of the process of life. All competing pleasures will be destroyed. (1983, p. 220)

Acknowledgements

For discussion and useful suggestions I would like to thank Tyler Cowen, Robin Hanson, Alex Tabarrok, Dan Houser, Don Boudreaux and Ilia Rainer. Geoffrey Lea provided excellent research assistance.

Suggestions for further reading

Becker, J. (1996). *Hungry Ghosts: Mao's Secret Famine* (New York: The Free Press). An eye-opening history of Chinese Communism's responsibility for the greatest famine in human history.

Courtois, S., Werth, N., Panné, J.-L., Paczkowski, A., Bartošek, K., and Margolin, J.-L. *The Black Book of Communism: Crimes, Terror, Repression*. The most comprehensive and up-to-date survey of the history of Communist regimes around the world.

Gregor, A.J. *The Faces of Janus: Marxism and Fascism in the Twentieth Century*. An excellent survey of the parallels between 'left-wing' and 'right-wing' totalitarianism, with an emphasis on intellectual history.

Noakes, J., and Pridham, G. (2001). *Nazism* 1919-1945, *volumes* 1-4 (Exeter, UK: University of Exeter Press). A comprehensive study of Nazism, with insightful commentary interspersed between critical historical documents.

Orwell, G. 1984. The greatest and most insightful of the dystopian novels.

Payne, S. *A History of Fascism* 1914-1945. A wide-ranging comparative study of Fascist Italy, Nazi Germany, and their numerous imitators in Europe and around the globe.

References

Alesina, A. and Enrico, S. (2003). *The Size of Nations* (Cambridge: MIT Press).

Arendt, H. (1973). *The Origins of Totalitarianism* (New York: Harcourt, Brace & World Press).

Barrett, S. (2003). *Environment and Statecraft: The Strategy of Environmental Treaty-Making* (Oxford: Oxford University Press).

Becker, J. (1996). *Hungry Ghosts: Mao's Secret Famine* (New York: The Free Press).

Bostrom, N. (2006). What is a Singleton? *Linguistic and Philosophical Investigations*. **5**(2), 48–54.

Bullock, A. (1993). *Hitler and Stalin: Parallel Lives* (New York: Vintage Books).

Caplan, B. and Cowen, T. (2004). Do We Underestimate the Benefits of Cultural Competition? *Am Econ Rev*, **94**(2), 402–407.

Courtois, S., Werth, N., Panné, J.-L., Paczkowski, A., Bartošek, K., and Margolin, J.-L. (1999). *The Black Book of Communism: Crimes, Terror, Repression* (Cambridge: Harvard University Press).

Friedlander, H. (1995). *The Origins of Nazi Genocide: From Euthanasia to the Final Solution* (Chapel Hill, NC: University of North Carolina Press).

Gregor, A.J. (2000). *The Faces of Janus: Marxism and Fascism in the Twentieth Century* (New Haven, CT: Yale University Press).

Holloway, D. (1994). *Stalin and the Bomb: The Soviet Union and Atomic Energy*, 1939-1956 (New Haven, CT: Yale University Press).

Huber, P. (1994). *Orwell's Revenge: The* 1984 *Palimpsest* (New York: The Free Press).

Huxley, A. (1996 [1932]). *Brave New World* (New York: Chelsea House Publishers).

Inman, R. and Daniel, R. (1997). Rethinking federalism. *J Econ Perspect*, **11**(4), 43–64.

Irwin, D. (1996). *Against the Tide: An Intellectual History of Free Trade* (Princeton, NJ: Princeton University Press).

Malia, M. (1994). *The Soviet Tragedy: A History of Socialism in Russia*, 1917-1991 (NY: The Free Press).

Margolin, J.-L. (1999a). China: a long march into night. In Courtois, S., Werth, N., Panné, J.-L., Paczkowski, A., Bartošek, K., and Margolin, J.-L. *The Black Book of Communism: Crimes, Terror, Repression*, pp. 463–546 (Cambridge: Harvard University Press).

Margolin, J.-L. (1999b). Cambodia: the country of disconcerting crimes. In Courtois, S., Werth, N., Panné, J.-L., Paczkowski, A., Bartošek, K., and Margolin, J.-L. *The Black Book of Communism: Crimes, Terror, Repression*, pp. 577–635 (Cambridge: Harvard University Press).

Noakes, J. and Pridham, G. (2001). *Nazism* 1919-1945, *vol. 3: Foreign Policy, War and Racial Extermination* (Exeter, UK: University of Exeter Press).

Orwell, G. (1983). 1984(New York: Signet Classic).

Payne, S. (1995). *A History of Fascism* 1914-1945 (Madison, WI: University of Wisconsin Press).

Pinker, S. (2002). *The Blank Slate: The Modern Denial of Human Nature* (New York: Viking).

Pipes, R. (1994). *Russia Under the Bolshevik Regime* (New York: Vintage Books).

Salisbury, H. (1992). *The New Emperors: China in the Era of Mao and Deng* (New York: Avon Books).

Tetlock, P. (2005). *Expert Political Judgment: How Good is It? How Can We Know?* (Princeton, NJ: Princeton University Press).

Werth, N. (1999). A state against its people: violence, repression, and terror in the Soviet Union. In Courtois, S., Werth, N., Panné, J.-L., Paczkowski, A., Bartošek, K., and Margolin, J.-L. *The Black Book of Communism: Crimes, Terror, Repression*, pp.33–628 (Cambridge: Harvard University Press).

Wright, R. (2000). *Nonzero: The Logic of Human Destiny* (New York: Pantheon Books).

Authors' biographies

Gary Ackerman is research director of the National Consortium for the Study of Terrorism and Responses to Terrorism (START), a US Department of Homeland Security Center of Excellence. Ackerman concurrently also holds the post of director of the Center for Terrorism and Intelligence Studies, a private research and analysis institute. Before taking up his current positions, Ackerman was director of the Weapons of Mass Destruction Terrorism Research Program at the Center for Nonproliferation Studies in Monterey, California, and earlier he served as the chief of operations of the South Africa-based African-Asian Society. He received his M.A. in International Relations (Strategic Studies – Terrorism) from Yale University and his Bachelors (Law, Mathematics, International Relations) and Honors (International Relations) degrees from the University of the Witwatersrand in Johannesburg, South Africa. Originally hailing from South Africa, Ackerman possesses an eclectic academic background, including past studies in the fields of mathematics, history, law, and international relations, and has won numerous academic awards. His research encompasses various areas relating to terrorism and counterterrorism, including terrorist threat assessment, terrorist technologies and motivations, terrorism involving chemical, biological, radiological, and nuclear (CBRN) weapons, terrorist financing, environmental extremism, and the modelling and simulation of terrorist behaviour.

Fred Adams was born in Redwood City and received his undergraduate training in mathematics and physics from Iowa State University in 1983 and his Ph.D. in Physics from the University of California, Berkeley, in 1988. For his Ph.D. dissertation research, he received the Robert J. Trumpler Award from the Astronomical Society of the Pacific. After a post-doctoral fellowship at the Harvard–Smithsonian Center for Astrophysics, he joined the faculty in the Physics Department at the University of Michigan in 1991. Adams is the recipient of the Helen B. Warner Prize from the American Astronomical Society and the National Science Foundation Young Investigator Award. He has also been awarded both the Excellence in Education Award and the Excellence in Research Award from the College of Literature, Arts, and Sciences (at Michigan). In 2002, he was given The Faculty Recognition

Award from the University of Michigan. Adams works in the general area of theoretical astrophysics with a focus on star formation and cosmology. He is internationally recognized for his work on the radiative signature of the star formation process, the dynamics of circumstellar disks, and the physics of molecular clouds. His recent work in star formation includes the development of a theory for the initial mass function for forming stars and studies of extra-solar planetary systems. In cosmology, he has studied many aspects of the inflationary universe, cosmological phase transitions, magnetic monopoles, cosmic rays, anti-matter, and the nature of cosmic background radiation fields. His work in cosmology includes a treatise on the long-term fate and evolution of the universe and its constituent astrophysical objects.

Myles R. Allen is joint head of the Climate Dynamics Group, Atmospheric, Oceanic, and Planetary Physics, Department of Physics, University of Oxford. His research focuses on the attribution of causes of recent climate change, particularly changing risks of extreme weather, and assessing what these changes mean for the future. He has worked at the Energy Unit of the United Nations Environment Programme, the Rutherford Appleton Laboratory, and the Massachusetts Institute of Technology. He has contributed to the Intergovernmental Panel on Climate Change as lead author on detection of change and attribution of causes for the 2001 Assessment, and as review editor on global climate projections for the 2007 Assessment. He is principal investigator of the climate*prediction*.net project, also known as the BBC Climate Change Experiment, using public resource distributed computing for climate change modelling. He is married to Professor Irene Tracey and has three children.

Nick Bostrom is director of the Future of Humanity Institute at Oxford University. He previously taught in the Faculty of Philosophy and in the Institute for Social and Policy Studies at Yale University. He has a background in physics and computational neuroscience as well as philosophy. Bostrom's research covers the foundations of probability theory, scientific methodology, and risk analysis, and he is one of the world's leading experts on ethical issues related to human enhancement and emerging technologies such as artificial intelligence and nanotechnology. He has published some 100 papers and articles, including papers in *Nature, Mind, Journal of Philosophy, Bioethics, Journal of Medical Ethics, Astrophysics & Space Science,* one monograph, *Anthropic Bias* (Routledge, New York, 2002), and two edited volumes with Oxford University Press. One of his papers, written in 2001, introduced the concept of an existential risk. His writings have been translated into more than fourteen languages. Bostrom has worked briefly as an expert consultant for the European Commission in Brussels and for the Central Intelligence Agency in Washington, DC. He is also frequently consulted as a commentator by the media. Preprints of many of his papers can be found on his website, http://www.nickbostrom.com.

Bryan Caplan received his Ph.D. in Economics in 1997 from Princeton University, and is now an associate professor of Economics at George Mason University. Most of his work questions the prevailing academic assumption of voter rationality; contrary to many economists and political scientists, mistaken voter beliefs do not harmlessly balance each other out. Caplan's research has appeared in the *American Economic Review*, the *Economic Journal*, the *Journal of Law and Economics*, *Social Science Quarterly*, and numerous other outlets. He has recently completed *The Logic of Collective Belief*, a book on voter irrationality. Caplan is a regular blogger at Econlog, http://www.econlog.econlib.org; his website is http://www.bcaplan.com.

Christopher F. Chyba is professor of astrophysical sciences and international affairs at Princeton University, where he also directs the Program on Science and Global Security at the Woodrow Wilson School of Public and International Affairs. Previously, he was co-director of the Center for International Security and Cooperation at Stanford University. He was a member of the White House staff from 1993 to 1995, entering as a White House Fellow, serving on the National Security Council staff and then in the Office of Science and Technology Policy's National Security Division. He is a member of the Committee on International Security and Arms Control of the US National Academy of Sciences. His degrees are in physics, mathematics, history and philosophy of science, and astronomy and space sciences. His work addresses international security (with a focus on nuclear and biological weapons proliferation and policy) as well as solar system physics and astrobiology. He has published in *Science, Nature, Icarus, Foreign Affairs, Survival, International Security*, and elsewhere. In October 2001 he was named a MacArthur Fellow for his work in both planetary sciences and international security. Along with Ambassador George Bunn, he is co-editor of the recent volume *US Nuclear Weapons Policy: Confronting Today's Threats.*

Joseph Cirincione is a senior fellow and director for nuclear policy at the Center for American Progress and the author of *Bomb Scare: The History and Future of Nuclear Weapons* (Columbia University Press, Spring 2007). Before joining the Center in May 2006, he served as director for non-proliferation at the Carnegie Endowment for International Peace for eight years. He teaches a graduate seminar at the Georgetown University School of Foreign Service, has written over 200 articles on defence issues, produced two DVDs on proliferation, appears frequently in the media, and has given over 100 lectures around the world in the past two years. Cirincione worked for nine years in the US House of Representatives on the professional staff of the Committee on Armed Services and the Committee on Government Operations and served as staff director of the Military Reform Caucus. He is the co-author of *Contain and Engage: A New Strategy for Resolving the Nuclear Crisis with Iran* (March 2007), two editions of *Deadly Arsenals: Nuclear, Biological and Chemical Threats* (2005 and 2002), *Universal Compliance: A Strategy for Nuclear Security* (March 2005) and *WMD in Iraq*, (January 2004). He was featured in the 2006 award-winning documentary *Why We Fight*. He is a member of the Council on Foreign Relations and the International Institute for Strategic Studies.

Milan M. Ćirković is a research associate of the Astronomical Observatory of Belgrade, (Serbia) and a professor of cosmology at the Department of Physics, University of Novi Sad (Serbia). He received his Ph. D. in Physics from the State University of New York at Stony Brook (USA), M.S. in Earth and Space Sciences from the same university, and his B.S. in Theoretical Physics from the University of Belgrade. His primary research interests are in the fields of astrophysical cosmology (baryonic dark matter, star formation, future of the universe), astrobiology (anthropic principles, SETI studies, catastrophic episodes in the history of life), as well as philosophy of science (risk analysis, foundational issues in quantum mechanics and cosmology). A unifying theme in these fields is the nature of physical time, the relationship of time and complexity, and various aspects of entropy-increasing processes taking place throughout the universe. He wrote one monograph (*QSO Absorption Spectroscopy and Baryonic Dark Matter*; Belgrade, 2005) and translated several books, including titles by Richard P. Feynman and Roger Penrose. In recent years, his research has been published in *Monthly Notices of the Royal Astronomical Society, Physics Letters A, Astrobiology, New Astronomy, Foundations of Physics, Philosophical Quarterly* and other major journals.

Arnon Dar is a professor of physics at the Department of Physics and the Asher Space Research Institute of the Technion, Israel Institute of Technology, Haifa and is the incumbent of the Naite and Beatrice Sherman Chair in physics. He received his Ph.D. in 1964 from the Weizmann Institute of Science in Rehovot for inventing the Diffraction Model of direct nuclear reactions. After its generalization to high energy particle reactions, he worked on the quark model of elementary particles at the Weizmann Institute and MIT. In the late 1960s and early 1970s, he applied the quark model to the interaction of high energy elementary particles, nuclei and cosmic rays with atomic nuclei, while working at the Technion, MIT, the University of Paris at Orsay, and Imperial College, London. In the late 1970s, he became interested in neutrino physics and neutrino astronomy. Since the early 1980s his main research interest has been particle astrophysics and cosmology, particularly astrophysical and cosmological tests of the standard particle-physics model, the standard Big Bang model, and general relativity. These included studies of cosmic puzzles such as the solar neutrino puzzle, the origin of cosmic rays and gamma-ray bursts, dark matter and dark energy. The research was done at the Technion, the University of Pennsylvania, the Institute of Advanced Study in Princeton, NASA's Goddard Space Flight Center, the Institute of Astronomy, Cambridge, UK, and the European research centre, CERN in Geneva. In collaboration with various authors, he suggested the existence of cosmic backgrounds of energetic neutrinos from stellar evolution, supernova explosions, and cosmic ray interactions in external galaxies, the day–night Effect in solar neutrinos, tests of neutrino oscillations with atmospheric neutrinos, gravitational lensing tests of general relativity at very large distances and the supernova–gamma ray burst–cosmic rays–mass extinction connection. His most recent work has been on a unified theory of cosmic accelerators, gamma-ray bursts, and cosmic rays. He published more than 150 scientific papers in these various fields in professional journals and gave more than 100 invited talks, published in the proceedings of international conferences. He won several scientific awards and served the Technion, Israel Defense Ministry, Israel Atomic Energy Committee, and numerous national and international scientific committees and advisory boards.

David Frame holds a Ph.D. in Physics and a bachelors degree in philosophy and physics from the University of Canterbury, in New Zealand. He spent two years working in the Policy Coordination and Development section of the New Zealand Treasury as an economic and social policy analyst, followed by a stint in the Department of Meteorology at the University of Reading, working on the PREDICATE project. In 2002 he moved to the Climate Dynamics group in the Atmospheric, Oceanic and Planetary Physics, sub-department of Physics at the University of Oxford, where he managed the climateprediction.net experiment.

Yacov Y. Haimes is the Lawrence R. Quarles Professor of Systems and Information Engineering, and founding director (1987) of the Center for Risk Management of Engineering Systems at the University of Virginia. He received his M.S. and Ph.D. (with Distinction) degrees in Systems Engineering from UCLA. On the faculty of Case Western Reserve University for 17 years (1970–1987), he served as chair of the Systems Engineering Department. As AAAS-AGU Congressional Science Fellow (1977–1978), Haimes served in the Office of Science and Technology Policy, Executive Office of the President, and on the Science and Technology Committee, US House of Representatives. He is a fellow of seven societies, including the IEEE, INCOSE, and the Society for Risk Analysis (where he is a past President). The second edition of his most recent book, *Risk Modeling, Assessment, and Management,* was published by Wiley & Sons in 2004 (the first edition was published in 1998). Haimes is the recipient of the 2001 Norbert Weiner Award, the highest award presented by the Institute of Electrical and Electronics Engineers and Systems, Man, and Cybernetics Society; the 2000 Distinguished Achievement Award, the highest award presented by the Society for Risk Analysis; the 1997 Warren A. Hall Medal, the highest award presented by Universities Council on Water Resources; the 1995 Georg Cantor Award, presented by the International Society on Multiple Criteria Decision Making; and the 1994 Outstanding Contribution Award presented by the Institute of Electrical and Electronics Engineers and Systems, Man, and Cybernetics Society, among others. He is the Engineering Area Editor of *Risk Analysis: An International Journal*, member of the Editorial Board of *Journal of Homeland Security and Emergency Management*, and Associate Editor of *Reliability Engineering and Systems Safety*. He has served on and chaired numerous national boards and committees, and as a consultant to public and private organizations. He has authored (and co-authored) six books and over 250 editorials and technical publications, edited twenty volumes, and has served as dissertation/thesis advisor to over thirty Ph.D. and seventy M.S. students. Under Haimes' direction, the Center for Risk Management of Engineering Systems has focused most of its research during the last decade on risks to infrastructures and safety–critical systems.

Robin Hanson is an assistant professor of Economics, and received his Ph.D. in 1997 in social sciences from Caltech. He joined George Mason's economics faculty in 1999 after completing a two year postdoc at University of California, Berkeley. His major fields of interest include health policy, regulation, and formal political theory. He is known as an expert on idea futures markets and was involved in the creation of the Foresight Exchange and DARPA's FutureMAP project. He is also known for inventing Market Scoring Rules such as LMSR (Logarithmic Market Scoring Rule) used by prediction markets such as Inkling Markets and Washington Stock Exchange, and has conducted research on signalling.

James J. Hughes is a bioethicist and sociologist at Trinity College in Hartford, Connecticut, where he teaches health policy. He holds a doctorate in sociology from the University of Chicago, where he also taught bioethics and health policy at the MacLean Center for Clinical Medical Ethics. Hughes serves as the Executive Director of the Institute for Ethics and Emerging Technologies and its affiliated World Transhumanist Association. He is the author of *Citizen Cyborg: Why Democratic Societies Must Respond to the Redesigned Human of the Future* (Westview Press, 2004) and produces a syndicated weekly programme, Changesurfer Radio. In the 1980s, while working in Sri Lanka, Hughes was briefly ordained as a Buddhist monk, and he is working on a second book 'Cyborg Buddha: Spirituality and the Neurosciences'. He is a fellow of the World Academy of Arts and Sciences, and a member of the American Society of Bioethics and Humanities and the Working Group on Ethics and Technology at Yale University. Hughes lives in rural eastern Connecticut with his wife, the artist Monica Bock, and their two children.

Edwin Dennis Kilbourne has spent his professional lifetime in the study of infectious diseases, with particular reference to virus infections. His early studies of coxsackieviruses and herpes simplex preceded intensive study of influenza in all of its manifestations. His primary contributions have been to the understanding of influenza virus structure and genetics and the practical application of these studies to the development of influenza vaccines and to the understanding of the molecular epidemiology and pathogenesis of influenza. His studies of influenza virus genetics resulted in the first genetically engineered vaccine of any kind for the prevention of human disease. The approach was not patented, and recombinant viruses from his laboratory have been used by all influenza vaccine manufacturers since 1971. A novel strategy for infection permissive influenza immunization has received two US Patents. Following his graduation from Cornell University Medical College in 1944, and an internship and residency in medicine at the New York Hospital, he served two years in the Army of the United States. After three years at the former Rockefeller Institute, he served successively as associate professor of Medicine at Tulane University, as Professor of Public Health at Cornell University Medical College, and as founding Chairman of the Department of Microbiology at Mount Sinai School of Medicine at which he was awarded the rank of Distinguished Service Professor. His most recent academic positions were as Research Professor and then as Emeritus Professor at New York Medical College. He is a member of the Association of American Physicians and the National Academy of Sciences and was elected to membership in the American Philosophical Society in 1994. He is the recipient of the Borden Award of the Association of American Medical Colleges for Outstanding Research in Medical Sciences, and an honorary degree from Rockefeller University in addition to other honors and lectureships. As an avocation, Kilbourne has published light verse and essays and articles for the general public on various aspects of biological science – some recently collected in a book on *Strategies of Sex*.

William Napier is an astronomer whose research interests are mostly to do with the interaction of comets and asteroids with the Earth. He co-authored the first paper (Napier & Clube, 1979) to point out that the impact rates then being found were high enough to be relevant on timescales from the evolutionary to the historical, and has co-authored three books on the subject and written about 100 papers. His career covers twenty-five years at the Royal Observatory, Edinburgh, two at Oxford and nine at Armagh Observatory, from which he retired in March 2005. He now writes novels with a scientific background

(*Nemesis* was referred to in a House of Lords debate on the impact hazard). This allows him to pursue a peripatetic research career, working with colleagues in La Jolla, Armagh and Cardiff. He is an honorary professor at Cardiff University.

Ali Nouri is a post-doctoral fellow at Princeton's Program on Science and Global Security where he works on issues related to biological security. His interests include designing non-proliferation schemes to curb the potential misuse of biotechnology. Before joining the program, Nouri was engaged with biotechnology-related activities at the United Nations office of the Secretary General. Nouri holds a Ph.D. in molecular biology from Princeton University, where he studied the role of various tumour suppressor genes during animal development. Before Princeton, he was a research assistant at the Oregon Health Sciences University and studied the molecular basis of retroviral entry into cells. He holds a B.A. in Biology from Reed College.

Chris Phoenix, director of research at the Center for Responsible Nanotechnology (CRN), has studied nanotechnology for more than fifteen years. He obtained his B.S. in Symbolic Systems and M.S. in Computer Science from Stanford University in 1991. From 1991 to 1997, Phoenix worked as an embedded software engineer at Electronics for Imaging. In 1997, he left the software field to concentrate on dyslexia correction and research. Since 2000, he has focused exclusively on studying and writing about molecular manufacturing. Phoenix, a published author in nanotechnology and nanomedical research, serves on the scientific advisory board for Nanorex, Inc., and maintains close contacts with many leading researchers in the field. He lives in Miami, Florida.

Richard A. Posner is a judge of the U.S. Court of Appeals for the Seventh Circuit, in Chicago, and a senior lecturer at the University of Chicago Law School. He is the author of numerous books and articles, mainly dealing with the application of economics to law and public policy. His recent books include the sixth edition of *Economic Analysis of Law* (Aspen Publishers, 2003); *Catastrophe: Risk and Response* (Oxford University Press, 2004); and *Preventing Surprise Attacks: Intelligence Reform in the Wake of* 9/11 (Hoover Institution and Rowman; Littlefield, 2005). He is a former president of the American Law and Economics Association, a former editor of the *American Law and Economics Review*, and a Corresponding Fellow of the British Academy. He is the recipient of a number of honorary degrees and prizes, including the Thomas C. Schelling Award for scholarly contributions that have had an impact on public policy from the John F. Kennedy School of Government at Harvard University.

William Potter is institute professor and director of the Center for Nonproliferation Studies at the Monterey Institute of International Studies

(MIIS). He is the author or editor of 14 books, the most recent of which is *The Four Faces of Nuclear Terrorism* (2005). Potter has been a member of numerous committees of the National Academy of Sciences and currently serves on the Non-proliferation Panel of the Academy's Committee on International Security and Arms Control. He is a member of the Council on Foreign Relations, the Pacific Council on International Policy, and the International Institute for Strategic Studies, and has served for five years on the UN Secretary-General's Advisory Board on Disarmament Matters and the Board of Trustees of the UN Institute for Disarmament Research. He currently serves on the International Advisory Board of the Center for Policy Studies in Russia (Moscow). He was an advisor to the delegation of Kyrgyzstan to the 1995 NPT Review and Extension Conference and to the 1997, 1998, 1999, 2002, 2003 and 2004 sessions of the NPT Preparatory Committee, as well as to the 2000 and 2005 NPT Review Conferences.

Michael R. Rampino is an associate professor of Biology working with the Environmental Studies Program at New York University (New York City, USA), and a research associate at the NASA, Goddard Institute for Space Studies in New York City. He has visiting appointments at the Universities of Florence and Urbino (Italy), the University of Vienna (Austria), and Yamaguchi University (Japan). He received his Ph.D. in Geological Sciences from Columbia University (New York City, USA), and completed a post-doctoral appointment at the Goddard Institute. His research interests are in the fields of the geological influences on climatic change (such as explosive volcanism), astrobiology (evolution of the Universe, planetary science, catastrophic episodes in the history of life) as well as the history and philosophy of science. He has edited or co-edited several books on climate (*Climate: History, Periodicity and Predictability*; Van Nostrand Reinhold, 1987); catastrophic events in Earth history (*Large Ecosystem Perturbations*; The Geological Society of America, 2007; and *K-T Boundary Events (Special Issue)*; Springer, 2007), and co-authored a text in astrobiology with astrophysicist Robert Jastrow (*Stars, Planets and Life: Evolution of the Universe*; Cambridge University Press, 2007). He is also the co-editor of the *Encyclopedia of Earth Sciences Series* for Springer. His research has been published in *Nature, Science, Proceedings of the National Academy of Sciences (USA), Geology* and other major journals.

Sir Martin J. Rees is professor of cosmology and astrophysics and Master of Trinity College at the University of Cambridge. He holds the honorary title of Astronomer Royal and is also visiting professor at Imperial College London and at Leicester University. He has been director of the Institute of Astronomy and a research professor at Cambridge. He is a foreign associate of the National Academy of Sciences, the American Philosophical Society, and the American Academy of Arts and Sciences, and is an honorary member of the Russian Academy of Sciences, the Pontifical Academy, and several other foreign academies. His awards include the Balzan International Prize, the Bower Award for Science of the Franklin Institute, the Cosmology Prize of the Peter Gruber Foundation, the Einstein Award of the World Cultural Council and the Crafoord Prize (Royal Swedish Academy). He has been president of the British Association for the Advancement of Science (1994–1995) and the Royal Astronomical Society (1992–1994). In 2005 he was appointed to the House of Lords and elected President of the Royal Society. His professional research interests are in high energy astrophysics and cosmology. He is the author or co-author of more than 500 research papers, and numerous magazine and newspaper articles on scientific and general subjects. He has also written eight books – including *Our Final Century?*, which highlighted threats posed by technological advances.

Peter Taylor is a research associate at the Future of Humanity Institute in Oxford, with a background in science, mathematics, insurance, and risk analysis. Following a B.A. in Chemistry and a D.Phil. in Physical Science, Taylor spent twenty-five years in the City of London first as a management consultant and then as a director of insurance broking, underwriting, and market organizations in the London insurance market. During this time, he was responsible for IT, analysis, and loss modelling departments and led and participated in many projects. Taylor is also Deputy Chairman of the Lighthill Risk Network (www.lighthillrisknetwork.org), created in 2006 to link the business and science communities for their mutual benefit.

Mike Treder, executive director of the Center for Responsible Nanotechnology (CRN), is a professional writer, speaker, and activist with a background in technology and communications company management. In addition to his work with CRN, Treder is a consultant to the Millennium Project of the American Council for the United Nations University, serves on the Scientific Advisory Board for the Lifeboat Foundation, is a research fellow with the Institute for Ethics and Emerging Technologies, and is a consultant to the Future Technologies Advisory Group. As an accomplished presenter on the societal implications of emerging technologies, he has addressed conferences and groups in the United States, Canada, Great Britain, Germany, Italy, and Brazil. Treder lives in New York City.

Frank Wilczek is considered one of the world's most eminent theoretical physicists. He is known, among other things, for the discovery of asymptotic freedom, the development of quantum chromodynamics, the invention of axions, and the discovery and exploitation of new forms of quantum statistics (anyons). When he was only twenty-one years old and a graduate student at Princeton University, working with David Gross he defined the properties of colour gluons, which hold atomic nuclei together. He received his B.S. degree from the University of Chicago and his Ph.D. from Princeton University. He taught at Princeton from 1974 to 1981. During the period 1981–1988, he was the Chancellor Robert Huttenback Professor of Physics at the University of California at Santa Barbara, and the first permanent member of the National Science Foundation's Institute for Theoretical Physics. In the fall of 2000, he moved from the Institute for Advanced Study, where he was the J.R. Oppenheimer Professor, to the Massachusetts Institute of Technology, where he is the Herman Feshbach Professor of Physics. Since 2002, he has been an Adjunct Professor in the Centro de Estudios Científicos of Valdivia, Chile. In 2004 he received the Nobel Prize in Physics and in 2005 the King Faisal Prize. He is a member of the National Academy of Sciences, the Netherlands Academy of Sciences, and the American Academy of Arts and Sciences, and is a Trustee of the University of Chicago. He contributes regularly to *Physics Today* and to *Nature*, explaining topics at the frontiers of physics to wider scientific audiences. He received the Lilienfeld Prize of the American Physical Society for these activities.

Christopher Wills is a professor of biological sciences at the University of California, San Diego. Trained as a population geneticist and evolutionary biologist, he has published more than 150 papers in *Nature, Science*, the *Proceedings of the U.S. National Academy of Sciences* and elsewhere, on a wide diversity of subjects. These include the artificial selection of enzymes with new catalytic capabilities, the distribution and function of microsatellite repeat DNA regions, and most recently the maintenance of diversity in tropical forest ecosystems. He has also written many popular articles for magazines such as *Discover* and *Scientific American* and has published six popular books on subjects ranging from the past and future evolution of our species to (with Jeff Bada) the origin of life itself. In 1999 he received the Award for Public Understanding of Science from the American Association for the Advancement of Science, and in 2000 his book *Children of Prometheus* was a finalist for the Aventis Prize. He lives in La Jolla, California, and his hobbies include travel, scuba diving, and photography.

Eliezer Yudkowsky is a research fellow of the Singularity Institute for Artificial Intelligence, a non-profit organization devoted to supporting full-time research on very-long-term challenges posed by Artificial Intelligence. Yudkowsky's current focus as of 2006 is developing a reflective decision theory, a foundation for describing fully recursive self-modifying agents that retain stable preferences while rewriting their source code. He is the author of the papers, 'Levels of organization in general intelligence' and 'Creating friendly AI', and assorted informal essays on human rationality.

Index

Note: page numbers in *italics* refer to Figures and Tables.